Theory and Practice of Piano Tuning

a manual on the art, techniques and theory

Brian Capleton PhD

Brian Capleton PhD MMusRCM DipTCL LTCL MIMIT lectured for many yeas in Piano Technology at the Royal National College, where piano tuning was taught for more than a century. He is an alumnus of Wolfson College Oxford, the Royal College of Music, Trinity College of Music London, and Dartington College of Arts.

Other books by the author include

Piano Action Regulating

a reference source for students and professionals

Theory and Practice of Piano Tuning

a manual on the art, techniques and theory

Brian Capleton PhD

Amarilli Books
www.amarilli-books.co.uk

Published by Amarilli Books

Malvern, UK

amarilli-books.co.uk

Copyright © 2007, Brian Capleton

2nd Edition

The right of Brian Capleton to be identified
as the author of this work
has been asserted in accordance with the Copyright, Designs
and Patents Act, 1988.

ISBN 9780992814106

A CIP catalogue record for this book is available from the
British Library.

All rights reserved. No part of this publication may be reproduced,
transmitted, or stored in a retrieval system, in any form or by any means,
without permission in writing from
Brian Capleton.

Acknowledgments

The following people are especially acknowledged for their influence and help in leading to the creation of this book:

Jacqueline Capleton, Jake Capleton, Gerald Gifford, Lewis Jones, Philip Kennedy, Jeremy Montagu, Terrance Pamplin, John Spice, Michael Wigmore, Gabriel Weinreich, Jenny Zarek, and the students and staff of the Royal National College.

Paul Read is gratefully acknowledged for his help in making studio digital recordings that were used in the acoustical research.

Contents

Acknowledgments	5
Piano tuning and this book	17
PART 1 - BACKGROUND THEORY	**21**
Chapter 1 - The invisible art and science	**23**
The tempered clavier	26
Talent	29
The art and science – qualities and properties	31
The "traditional" theory	32
Technology - ETDs	33
The need for piano tuning	36
Time to learn	38
The importance of theory and technique	39
Chapter 2 - The essential ideas	**41**
Note numbering	41
Note naming	43
Construction terminology	43
Upright piano	44
Grand piano	46
Top stringing, backstringing and speaking length	47
Piano scaling	47
The breaks	50
Essential terminology – tools	50
Piano tuning – from the top down	51
Beat rates	61
For students of tuning	63
Training in tuning	63
Equal temperament	64
Harmonics, partials, inharmonicity, falseness	65
Use of the tuning lever	65
Pitch raising	67
Chapter 3 – Sound	**69**
Why study the science of sound?	69
Sound waves	70
Pure tones	74
The *Principle of Superposition*	74
Simple Harmonic Motion	76
Vibrations are just bounded motion	78
Superposition of vibrations – normal modes	78

Periodic and aperiodic vibrations	78
Frequency and cycle	79
The fundamental	80
Wavelength, velocity, and amplitude	81
Decay	82
Frequency and pitch	86
"Perfect pitch"	87
Phase	88
Mixing sinusoids to produce beats	92
Mixed pure tones, beats and difference tones	93
Pure tones as essential ingredients of the sound recipe	94
Pure tone frequency and perceived pitch	96
Audible *Partials*	98
Chapter 4 - Temperament Theory	**100**
What is a musical temperament?	100
Temperament theory versus tuning theory	101
Some practical basics	102
Tempering is not the art of piano tuning	105
The approach we need	105
The psychological idea of musical interval "size"	106
The idea of wide and narrow	108
Tuning historical temperaments	109
Harmonic ratios	110
Historical context	113
Pythagoras and *the music of the spheres*	115
The harmonic series	117
Normal modes	119
Harmonic tuning theory	120
Demonstrating "harmonics" on the piano	122
The chord of nature	123
"Coinciding harmonics"	124
Why "coinciding harmonics" are important	128
Just Intonation	131
Cents	132
The Pythagorean or ditonic comma	133
Equal Temperament	138
The syntonic comma (*aka* the comma of Didymus)	139
The Temperament Circle	141
Pythagorean tuning	145
Combining the Pythagorean and syntonic commas	145
Interpreting the Temperament Circle	146
Equal Temperament	147
The complete Temperament Circle	149

Temperament classification	150
The mean tone temperaments	151
The "well temperaments"	155
Valotti temperament	156
Tuning equal temperament	157
Temperament generalisation	158

Chapter 5 - "Traditional" piano tuning theory and elementary practice **160**

The idea versus the reality	161
The "traditional" acoustical model	161
Interval ratios and beat rates	164
The equally tempered semitone	166
The rule for finding coinciding harmonics	168
Calculating the beat rate	170
Revision	170
Worked example	170
General beat rate computation for harmonic notes	173
Beat rate relationships	175
Beat Rate Table	176
Major thirds – major sixths	178
Fourth – fifth - octave	179
Minor third – major sixth	179
Compound major thirds	180
Beat pitch relationships	180
Elementary practical tuning procedure for the scale	181
RNC scale sequence	183
The real scale tuning situation	185
Tuning the octaves	187
The real octave tuning situation	188
The physical situation in the real piano	188
Practical tuning summary	190

Chapter 6 - The soundscape, spectrum and tone **192**

The transient	196
Changing tone - *movement*	196
Piano tone partials – revision of essential facts	197
Harmonics	198
Partials	199
The musical pitch arrangement of piano tone partials	200
Hearing partials and beats	203
The 3-D spectrum of partials	204
Intervals	209
Adjustable partials	209

Finding adjustable partials	211
Proportions of the spectrum that are adjustable	214
The partials of the perfect fifth	217

Chapter 7 - Partial decay patterns — 220

Piano tuning beats are not simple heterodyne beats	223
More than just beat rates	224
Two ways of representing partials	224
Dual decays	228
Regular beats	230
The single null	233
Progression of beat patterns in fine tuning	235
Summary of beat patterns	238
Unstable beat rate	240
The significance of decay patterns	240
The adjustable partials	240
Falseness	241
Mistuning and decay patterns	241
The coupling region	243
Behaviour within the coupling region	244

PART 2 – ADVANCED TUNING PRACTICE — 245

Chapter 8 - Unison Tuning — 247

Listening to specific partials	250
Relationship of unison beat rates	251
First steps – summing up some practical points	251
The seascape analogy	252
Beating in fine tuning	253
Unisons and beat inheritance	255
Two beat species - *false beats*	255
Visual conceptualisation of falseness in the soundscape	260
Falseness and context	261
Tonal beauty need not be analysed	262
Judgement	263
Some scientific studies of unison tuning	264
Summary of the difficulties in the scientific study	266
Mistuning - the technical meaning	266
Movement as tuning changes are made	267
How partials are mapped from normal modes	268
Inheritance of false beats	271
Sequence of tuning unison strings where one is false	276
The skill – general beat attenuation	276
False beat attenuation – an overview	277

Eliminating the beat: rate reduction and attenuation	278
A little theory – Weinreich behaviour	281
The coupling region	286
Listening to the whole	288
Unison tone and the *vowel* or *cat* in the unison	289
Magnified partial envelope shapes	292
Examples of false beat attenuation	292
Example of the coupling region – effect on beat rate	297
Practical effects of bridge coupling	299
A Steinway trichord	300
The singing quality – *single nulls* and *aftersound*	304
"Hammer irregularities" and tone – the Weinreich model	305
The symmetric and antisymmetric modes	306
Movement in the spectrum – some important points highlighted	308
Adjusting tone through decay rates	309
Unisons in the high treble	312
Falseness in the high treble	313
Unisons in the bass	314
Three string unison groups (trichords)	316
Trichord string tuning sequence	316
Inharmonicity, stringing errors and unisons	320
Utilising unison tuning to fine tune other intervals	320
Tuning Blüthner unisons with aliquot strings	321
Chapter 9 - Tuning the Scale	**322**
The learning curve	322
The approach	323
The basic principles	325
Beat behaviour	329
False beats	330
Aural scale tuning - background	332
How to "ear train" and find partials	335
The scale tuning sequence	339
Fast and slow intervals	341
Beating partials in the tuning intervals	342
Beat rate relationships	346
Tuning tempered perfect fifths: harmonic ratio 3:2	349
Beat rates for the fifths	350
Beat patterns in the fifths	351
The actual behaviour of partials in fine tuning the fifth	358
How to tell a wide fifth	362
Plasticity in tuning fifths	363
Practical approach to tuning fifths	364
Inharmonicity mistaken for falseness	367

Tuning tempered perfect fourths: harmonic ratio 4:3	368
Tuning tempered thirds and sixths	369
Judging beat rates for thirds and sixths	378
The generic beat form	379
The dual or phantom beat rate.	380
Tuning the scale – the construction principles	382
Beat rate *effect*	385
Interval quality – coarseness as distinct from beat rate	385
Scale *logic*	385
Scale plasticity; three note "closed box" relationships	386
Scale *plasticity* and interval *sensitivity*	387
Improving a scale – the advanced approach	390
The two "error" types	390
The psychology of beat rates – judging progression	391
The importance of interval quality	392
Very small pianos	393
Scale sequence – limitations	395
Scale tuning sequence example number 1	396
Major thirds progression on a very small instrument:	398
Scale tuning sequence example number 2	399
Generalised scale building principles	400
A complete scale covers an octave and a major sixth.	401
Relationship between the major thirds and the fifths.	401
The position of the scale area	405
The non-"traditional" features have consequences	405
"Error" diagnosis	407
Chapter 10 - Octave tuning	**408**
The 2:1 fallacy	408
Initial issues	408
Arguments about stretching	411
The essential idea of octave stretching	412
Tailored stretching	413
The need to stretch octaves in fine tuning	414
Musical expectations and inharmonicity	415
Octave stretching in rough tuning	415
Octave tuning and the so-called "checks"	416
The octave's partials and false beating	417
Which of the octave's partials are adjustable?	418
The octave's dependence on unison tuning	419
Tone and intonation	419
Pitch flattening	420
The top octave mistake	421
Pitch versus beating	422

The fine tuning region	423
Tuning octaves by intonation	425
Tone versus intonation of an octave	426
Inharmonicity - relation of single and multiple octaves	427
The octave as a building block	430
Exercise : Finding the fine tuning region	431
Exercise : The naturally beating octave	432
Example analysis	435
The case of no falseness – larger stretching	440
The case of no falseness - small stretching	444
Beat amplitudes and rates – tone quality	445
How to address pitch flattening – inharmonicity	446
The wide beating "pitch correction" mistake	449
Inharmonicity and instrument size	450
Beat rate prominence – maj 21st and dominant sevenths	451
Procedure for controlled stretching - beginning	452
The minor third / major sixth check	452
The function of the minor third / major sixth relation	453
Continuing down to the break	454
Continuing down to the bass	455
Tuning upwards from the scale	455
Caution for the first few octaves	457
Continuation of stretching	458
From lower to higher treble – a summary	461
Tuning major tenths, seventeenths, and triple octaves	462
Bass octaves – first principles	463
Falseness in the bass	464
Pitch and tone in the bass	464
Intonation versus tone in the bass	465
Pitch ambiguity	466
Bass octave reinforcement – the simple difference tone	467
A diversion	468
Tuning the bass first – especially on mini pianos	469
Unusual trebles – high inharmonicity inequality	470
The top treble octaves – additional considerations	471
Moving the head - position of listening	472
The musical picture	473
Tuning two or more pianos together	475
Summary of common "checks" for octaves in the treble	476
Summary of common "checks" for octaves in the bass	476
Chapter 11 - Setting the Pin	**477**
A practical overview	479
Theory in detail	486

Basic practical observations	487
Understanding friction – the main characteristics	490
Elasticity – the main characteristics	491
Mechanical hysteresis in the string-bridge system – technical description	491
Summary of hysteresis	496
Assessing hysteresis	497
Recoil	498
Summary of essential points so far	502
The effect of the hammer strike	505
Large hysteresis compared to the elastic range	506
Small hysteresis compared to the elastic range	506
From theory to practice: setting the pin	507
Scenario 1: *Spring* smaller than the *travel*	510
Scenario 2: *Spring* greater than the *travel*	510
Scenario 3: *Spring* equal to the *travel*	511
Tuning slowly and smoothly	511
Coaxing	512
The audible soundscape	515
Pin position - avoiding recoil	516
Striking and tuning	517
Catching the movement	518
Bridge coupling and hysteresis	518
Differences between bass, mid compass, and treble.	519
Multiple friction points	520
Backstringing and "re-setting" the string tension	520
Latency	521
Latency in "settle"	522
Practical guidelines summary	524

Chapter 12 - Setting the pitch — 525

Chapter 13 - Small piano syndrome — 529
General approach to tuning very small pianos	534
Seeing inharmonicity changes	535
The scale puzzle and the trick of solving it	536
Tuning the scale – an example	536
Tuning pianos together	540

Chapter 14 - Hearing — 541

Chapter 15 - The Kirk Experiment — 546

PART 3 - ADVANCED THEORY — 551

Chapter 16 - The single piano string in one plane — 552
- The ideal string - general proof of harmonic partials — 554
- The harmonic spectrum — 557
- The Fourier transform of the ideal string spectrum — 558
- The partial spectrum of the real string in one plane — 560

Chapter 17 - The Weinreich Model — 565
- Two normal modes in one plane – Weinreich's model — 567
- Effects of an 'imperfect' hammer strike – partial phase difference — 568
- Effects of an "imperfect" hammer strike – partial amplitudes — 570
- Frequency difference between the partials — 571
- Purely resistive idealised bridge — 577
- Purely reactive idealised bridge — 579
- Realistic bridge – both resistive and reactive — 580
- Decay curves — 586
- Decay curves for an idealised, purely resistive bridge — 589
- Decay curves for a realistic bridge, both resistive and reactive — 590
- Generic decay patterns for two string partial unisons — 593
- Tuning in practice — 595

Chapter 18 - Two strings, two planes — 598
- Degrees of falseness over the compass — 603
- One string false – "hiding" falseness — 604
- Two strings false – the perception of the unison — 610
- Falseness generalised — 612

Chapter 19 - The Trichord — 614
- Chaotic trichords? — 621

Chapter 20 - Further comments on false partials — 623
- Frequency variation in false partials — 623
- Falseness and boundary conditions — 627

Chapter 21 - Inharmonicity — 630
- General formula for inharmonic beat rates — 632
- Tuning strategy — 651
- Octaves and inharmonicity — 652

Scale tuning sequences — 655

Glossary of some key concepts — 661

Select bibliography — 677

Piano tuning and this book

Piano tuning is an art easily misapprehended, one whose true nature and techniques are readily obscured by popular myths surrounding it. The art of fine tuning is often assumed by the layperson to be just a straightforward musical skill of tensioning strings "to the right pitch". Alternatively, it is frequently confused by those with perhaps a little technical knowledge of it, with its elementary mathematical model that was developed in the 19th century.

In the UK and probably elsewhere, there is a long standing, tacit, and largely unchallenged (outside the piano industry itself) "tradition" of spurious lay beliefs about its nature, what skills it requires, what sort of people do it, what sort of people *should* do it, its place in relation to the piano and its music, and its place in relation to academia.

This book contributes towards eradicating some of the myths and redressing this position, by elucidating the art and its techniques in ways that will be accessible and useful both to students and professionals of piano tuning, and to anyone else who holds a serious academic interest. Many of the descriptions are supported by illustrating the actual acoustical phenomena involved using digital analysis, and the discussion puts the musical purpose of the art into the context of modern acoustics.

The belief that fine piano tuning can be achieved by simply applying some "worked out" numbers from the elementary mathematical model, has been alive and well since the birth of the model over a hundred years ago. It is a belief that manages to survive in an environment where there is a wide diversity of musical standards and expectations, especially regarding the appreciation of musical tone and intonation. The reality is that piano tuning, despite using a mathematical model on which to base its initial techniques, is very much a highly skilled *art* that takes the best part of a working lifetime to master.

This book does not purport that tuners who are already masters of their art should do anything differently. Rather, its main role is as a comprehensive text book on the practical art and its associated theory, exploring some of the practice that gifted tuners develop through many years of experience, practice that extends well beyond the limitations of the traditional theoretical models for tuning. Much has already been said on the divide between the "traditional" piano tuning theory that was established in the 19th century, and actual tuning practice. In particular, it is now well known that a phenomenon called *inharmonicity* can have a substantial effect on tuning, rendering many of the edicts of "traditional" theory inaccurate, to a degree that can be highly significant in practice. However, inharmonicity is by no means the only significant aspect of piano tone that is ignored by the "traditional" theory.

Other important factors include decay mechanisms, coupling mechanisms, psycho-acoustic effects, and not least, something that has "traditionally", and perhaps misleadingly been known as "falseness". The latter has long been recognised in piano tuning in its more severe occurrences, and has been implicitly accepted as a kind of "add on" problem or "fault", that sometimes occurs. In actuality, what has been called "falseness" in this way, is just an acute instance of a wider, normal, and ubiquitous part of piano tone behaviour, that in its more subtle forms can affect every part of fine tuning.

When one is in the process of both training the sense of hearing, and attempting to fit what is heard with a theoretical model, one can be at the mercy of any faults in the model. Suggestion can be a powerful thing, and it is easy to be "suggested" into hearing something that is not actually "there", or into mis-hearing what *is* there, because one is trying to make sense of what is heard according to the model one has learnt, or according to one's own expectation.

In teaching supporting theory to piano tuning at the Royal National College, I have attempted to avoid this kind of misdirection by using digital technology to remove the initial difficulty in hearing, so that what is actually "there", in the sound itself, can be recognised without doubt or confusion. This trains the ear without misdirection. Questions inevitably then arise concerning the causes of what one is hearing, since the "traditional" theory often does not explain it. This book goes some way towards providing a better understanding of these causes.

In the teaching of the subject there has been a pressing need of a single text book that puts the "traditional" theory in its proper context, and contains in one volume, up-to-date information on relevant theory and practice, suitable for study to an advanced level, by students of piano technology in today's education system. In particular, the majority of available texts on the subject have approached tuning theoretically "from the bottom up", where the bottom is mainly reiterated 19th century theory. This kind of theoretical "bottom up" approach had a proper place in the works of great 19th century theorists like Helmholtz, but applied to an art that has been maturing for more than a century and a half, and presented in a world where science itself has progressed over that time, it would now be misplaced.

In this book, the approach taken is "top down". Approaching tuning "bottom up" from simplified theory has one, and one only, real advantage for practical tuning. It provides a rough but at least clear "map" for tuning a chromatic scale. But it also has considerable disadvantages in the way it has created certain rigid ideas that have become spurious mental paradigms for tuning. Science is a wonderful tool for exploring and illustrating an art like piano tuning, and it strongly features in this book, but it is not a method for proving how an art *is*, or how it *should be*. Not, that is, unless you subscribe to scientism. What it can do well, of course, is to *disprove* long-standing notions or false beliefs based on bad science. Piano tuning is as much a question for Arts and Humanities as it is for the Sciences, and to this end I would like to think this book encourages high standards of artistic expectation in the art, as well as rigorous scientific assessment of any ideas that purport to put the art itself on a scientific footing.

Although this book takes the subject to an advanced level, relatively little prior knowledge of the subject is assumed, other than familiarity with the piano as a musical instrument, and some knowledge of music theory. The reader is taken through relevant areas that would be encountered by students studying on a full time course for a recognised qualification in piano technology. It also includes more advanced information both for tuning students who are interested in this, and for those whose main interest is of a more general, or specifically academic nature. The book includes a good deal of practical instruction on the art of tuning from elementary to advanced levels, and includes much of practical value that will probably not be readily found elsewhere.

No text book, however, can be considered a complete answer to the desire to learn the art of piano tuning. Those who wish to learn the art are best taught by those who are adept and in touch with the art and its theory, and have access to the necessary resources. It is impossible to learn effectively just from a book, or by practising on just one or two instruments, because pianos differ in characteristics as much as persons.

Trainees who have had, say, six months full time experience at tuning, will probably have already discovered some of the "traditional" theory's limitations, and should be ready to discover much of what can be found here. For others, much of the book may serve as an advanced manual, fitting theory to practice that is already in place, and perhaps introducing some new ideas.

Above all, this book emphasises a fact often forgotten during the intensity of training in the use of piano tuning's underlying *techniques*. That is, that the art of tuning is not merely the efficient application of techniques or theory, or technical procedures. The master tuner is an *audiophile of piano tone*, working in a *musical context*, whose techniques, as in any art, are merely the means to an end that is definitely well beyond those techniques.

One cannot simply produce a stage-by-stage algorithm for the art itself. The various topics fit together more like the pieces of a jigsaw puzzle, and make most sense when the whole is viewed. I have included a good deal of deliberate repetition of key ideas in varied forms, as one would in teaching, because this is often helpful in the learning process.

Piano tuning is a large and complex subject when explicated. The information in this book is therefore provided in relatively small textual sections suitable for individual study as necessary, combined with use of the Glossary. Nevertheless, the book is ordered in such a way that there is a natural progression of information from beginning to end, and this would be the best way to read the book.

Part 1 - Background Theory

"Here's the piano. Presumably you have perfect pitch..."

Chapter 1 - The invisible art and science

Piano tuning is a practical art, but it probably could never have existed as practiced today, without the initial support of science and theoretical models, which helped develop the underlying techniques for the precision tuning of high tension instruments. This most definitely does not mean fine tuning a piano is wholly an applied science, but it does indicate there is something more to the art than its appearance suggests.

Piano tuning looks superficially like a process of simply adjusting musical pitches, and many people do assume pitch adjustment to be its essence. In fact – and as conscientious students of tuning soon discover in training - with a little experience, tuning is soon found to be an art that mushrooms out into an extensive world of sonic complexity and puzzle solving, in which musical pitch itself, plays only one part. In the musical world pitch is by no means the only important feature of musical notes and intervals, as musicians know very well. Just as important, and often *more* important, is *tone*, and it is in fact *tone* that occupies the major part of the art of fine tuning.

We tend to think of musical *pitch* as a single, simple property of a musical note. In fact, properly speaking, the pitch of a piano note is not even a property of the note itself, but is a complex psycho-acoustic *human response* to a usually large and *complex set* of sound ingredients. When we talk of "pitch", we are using the same kind of colloquial "license" that we use when we speak of "coloured" light. Different coloured light, is of course actually light of different *wavelengths*, the experiential quality we call "colour" being provided by human consciousness, or the human eye-brain system, in response to those wavelengths. Similarly, what we perceive as "pitch" is our sensory response to what in piano tone is a complex mixture of acoustical ingredients. In contrast, a common misconception is that each musical note is defined by *a frequency*, and that its pitch is something that maps simply and directly from that frequency.

The ingredients that affect pitch also affect tone, but in quite different ways. Fine tuning requires that the tuner makes micro-adjustments to tone and intonation by listening critically and analytically to the soundscapes of notes and intervals (throughout this book, *interval* refers to an interval with both notes sounding together, unless otherwise stated – see the *Glossary*). In training, tuners are introduced to this through the concept of *beats*, which are a particular kind of fluctuation pattern in the audible soundscape.

Training in the perception of beats is essential to the development of acute tone perception, but it can have an unfortunate side effect. Training in the techniques of adjusting beats is often so intense that their proper context can become forgotten, and many tuners end up "hearing" piano tone and intonation as though they were just a collection of beats, or relative absence of beats. The ability to hear what are called "beats" is indeed the initial door into the inner world of piano tone and intonation, through which one must pass. But that world, of course, consists of much more than just "beats", as it indeed consists of much more than any single one of its individual properties.

It is the behaviour of beats in the tuning process that is described *to some extent*, through early scientific theory that models the acoustics of piano tone, and provides an initial "map" for the piano tuner to follow. Nevertheless, in musical practice, the qualities of a piano important in music, that are directly affected by tuning, are *tone* and *intonation*, as *musical* qualities. Ultimately, piano tuning is about beauty of *tone* and *intonation*, and not about any theory that may attempt to "model" this in a simplified way. The gap between the science and the art is in many ways similar to the gap between science in general, and art in general. It lies in the *mapping* between the scientific model or theory, and the actuality of the experience produced by the art.

Professional tuners themselves are often content to confine any technical discussion about the inner workings of the art, or of the science on which it is based, to the professional circle itself. There are good reasons for this. The auditory world of piano tuning would appear to the layperson to be exactly the same world of sound to which the musician listens, but this appearance is deceiving. The art of tuning requires an ability to listen and hear in a way that is literally extraordinary, and actually beyond the experience of most musicians. Divorced from this practical experience and intimacy with the inner acoustical world of piano tone, theoretical discussions tend to be idealistic or merely academic. The purpose of the art, as I have emphasised, is not simply to produce notes that sound at the desired musical pitches, but even more importantly, to produce *good tone*, and good *intonation*. Much of this part of the art in any case lies beyond the formal techniques that can be described by the associated theory.

The tone and intonation of a piano is naturally dependent on the physical quality and condition of the instrument, but it also depends very much on the quality of the fine tuning. Not surprisingly, given that tone is to some extent psycho-acoustic in nature, different tuners, like different artists, can produce different results. Nevertheless, as in fine art, there is in the work of the piano tuner a more objective, "visible", technical dimension, obvious to all those who are truly expert, that reflects ability in underlying technique, and knowledge of theory. Professional artists, in music or the visual arts, are quick to see strengths and weaknesses in the *technical aspects* of the work of other artists. A similar situation arises in piano tuning.

Technical merit in a piano tuning, or the lack of it, is to professional tuners very obvious, even though it may not be recognised by musicians in general. Even so, as in much fine art, any notion of technical "perfection" would be misleading. In the case of piano tuning, the concept of the technically "perfect" tuning for an instrument whose acoustics is as complex as the piano's, is, *even as a theoretical entity*, what philosophers would call *chimerical*; it is a mythical beast that arises out of limited understanding of the acoustics and psycho-acoustics. It comes from the use of crude or idealised theoretical models to describe a complex psycho-acoustic actuality.

Despite the use of technical procedures, the master piano tuner approaching a tuning is much less like a scientist applying a theory, than like a sculptor approaching a variable medium that must be worked in a particular way, because of its complex and varying physical properties. Theory, through providing an understanding of the general physical nature of the medium, can provide help on certain techniques for working it. But even with a remit to sculpt a particular object, the final result will still depend to a significant extent on the *individual* qualities of the given medium. Although there may be a firm idea of what the final object will look like, some of the finest details may be largely determined by the medium itself. However, the more skilled sculptor will produce the better result, through a better understanding of the medium, and a higher level of skill in working it.

In piano tuning, tone properties are themselves part of the medium with which the tuner works. In learning to tune, a great deal of attention must be paid to the tone property called "beating", which is controlled and adjusted as part of the tuning process. Learning to control tone properties, first just with proper technique applied to beating, and then more artistically, takes years of commitment. In the beginning it requires following "traditional" directions based on simplified rules, drawn from the idealised, theoretical model. This is rather like "painting with numbers". Eventually, an empirical knowledge of how tone properties like beating actually behave in the tuning process, and of what can or cannot actually be done with these properties, develops through experience. This takes considerable time, and results in knowledge extending well beyond the idealised theory.

What takes a good deal longer, so much that it may never be realised by some, is the natural reassembling of that analytical acuity of hearing, resulting from intense training and practise in listening to beating and other tone properties, into a proper, holistic appreciation of *tone quality* and *intonation* themselves. At this stage true *tone production* can begin in a positive way. This is not something achieved in just a couple years, but it is the true state of the art, as every expert tuner who loves the art eventually discovers.

The tempered clavier

Piano tuners listen to many details of acoustical behaviour and tone, rather than just pitch alone. The awareness of this, combined with the way the "traditional" theoretical model for tuning is typically presented, has given rise to some long standing "informed" urban myths about why piano tuners do this, and what it is they are actually doing. It may appear that the tuner must use such techniques in order to improve the "accuracy of pitch". It is also sometimes stated that the tuner uses these analytical methods in order to achieve a particular set of string frequencies. In this book, we shall inevitably see beyond these myths.

So what kind of assumptions *can* we safely make about piano tuning before going deeper into an understanding of it?

Firstly, it could be useful to at least suspend any of the "usual" preconceptions. For example, JS Bach's Forty Eight Preludes and Fugues also known as the *Well Tempered Clavier*, is a work whose title many musicians quite rightly associate with keyboard tuning. We might therefore already have an idea that piano tuning is like Bach's harpsichord tuning, and that, as Bach's title suggests for harpsichord tuning, it is all about *temperament*, micro-intervals called *commas*, or a semi-esoteric science called *temperament theory*.[1] Of course this *does* have something to do with it, but it should soon become clear that this is certainly not what piano tuning is *all about*.

It is also especially important to appreciate that the theoretical study of musical temperaments, or *temperament theory*, which is a very old branch of mathematics, is not a modern acoustical study that can be used to accurately predict piano string or piano tone behaviour, or its aspects that master tuners have to address.

Temperament theory, with which JS Bach may well have been familiar, was in the past an important area of study for music theorists, and continues to this day be a developed area of what is essentially pure mathematics. Nevertheless, it is based on simplified and idealised initial assumptions about the acoustics of musical intervals, that in the modern scientific arena can only be considered as crudely modelling the much more complex, actual acoustical and psycho-acoustical reality of musical tone.

Some of the assertions of temperament theory are still utilised in piano tuning theory, but the former can also be *very misleading* when applied to piano tuning, or to the complex acoustical behaviour of piano strings and tone. There are also many other musical instruments (for example metallophones) for which temperament theory says nothing very meaningful, or to which it simply does not apply.

Connected with temperament theory is another, rather "fuzzy" idea often encountered, which is that piano tuners deliberately tune certain intervals "out of tune". The idea of what constitutes "in tune" or "out of tune", of course needs clarification here.

The usual idea that piano intervals are deliberately made "out of tune" rests on the notion that musical intervals only properly qualify as "in tune", if they are tuned in a special way, that we shall later meet as *acoustical consonance*. In piano tuning, intervals are indeed deliberately tuned *away* from acoustical consonance. "Acoustical consonance" is itself a technical, acoustical term, referring to a particular acoustical condition of tuning (one that piano tuners call *beatless, pure* or *just*). "Acoustical consonance" itself can be well defined in a way that does not involve

---

[1] *Temperament theory* in this book refers to theory concerning musical intervals, and is not to be confused with the much more recent "temperament theory" concerning human psychology.

aesthetic or value judgements. In contrast, the term "in tune" as used in common parlance, implies not only "correctly tuned", but also "tuned in the nicest sounding way". Similarly "out of tune" implies "not correctly tuned" and "not sounding nice". The old assertion that piano tuners deliberately tune intervals "out of tune", therefore mixes up impartial acoustical facts with the very knotty philosophical problem of aesthetic preferences and value judgements. It attempts to *link* aesthetic value judgement to acoustical fact. It unashamedly makes a huge inductive leap from initial scientific or acoustical fact to aesthetic preference, or *vice versa*, without establishing any real scientific continuity from one to the other.

The whole idea contains a subtle slant in thinking about musical intervals, and involves *philosophically interpreting* their acoustical structure. It is an idea traceable back to the ancient (Pythagorean) philosophy that there is a divinely ordained "perfect tuning" that each musical interval really *ought* to have, described by a single, neat, "mystical", Pythagorean whole number ratio, which is its hallmark. The attachment to such notions, and the confusing of value-judgement phrases like "in tune" with such *rough* acoustical observations, has its essential origins much more in ancient philosophy, "Pythagoreanism", and in what used to be known as *natural philosophy*, and in cultural influences, than in hard scientific fact.

I am not suggesting that acoustically consonant intervals are anything other than beautiful. Nor am I suggesting that equally tempered intervals as tuned on the piano are not beautiful. I am not insisting we should make aesthetic judgements about the tuning of musical intervals one way or the other. I may like one way best, whilst you may prefer another. Alternatively, there may be a consensus. That is the nature of aesthetics. It is the tacit assertion that roughly observed scientific facts can somehow give authority to an aesthetic preference, that is problematic. The facts of natural phenomena are invariably much more complex than they appear to be on rough observation. Perhaps not surprisingly, the acoustical assertions made in many of these cherished ideas turn out to be uncomfortably ambiguous in the arena of modern acoustical science. As we shall be seeing, some of the premisses for these ideas, that are frequently repeated even today, arguably border on the realms of a kind of acoustical alchemy or belief system, rather than being consistent with the acoustical facts. The assertion that properly tuned (equally tempered) piano intervals are "out of tune", *and that this is a scientific or acoustical fact*, has no scientific validity that has yet been established. It is a seductive claim initially, but on closer scrutiny it collapses.

Hard science, of course, does not in any case provide all the useful answers related to music, or the art of piano tuning, and it would be nothing less than scientism to suggest this. We should always be aware that ideas purporting to be either *scientific descriptions* or descriptions of *fact*, rather than purely philosophical or cultural ideas, do deserve to come under the full scrutiny of modern scientific rigour, and to have any weaknesses or false assertions exposed, not least from empirical evidence. In this context, *temperament theory*, for example, remains very firmly within the confines of *idealised theory* and pure mathematics. Its apparent precisions are sometimes entirely lost under the weight of other important acoustical factors, especially when it is applied to actual piano tones.

Talent

Just as some individuals are "natural" musicians, can we assume that some piano tuners are "natural" tuners? Any experienced teacher of tuning will know that the answer to this is most definitely "yes", but nevertheless the ability to tune to the highest standards is still not something that can happen without years of effort.

One of the commonest myths about piano tuning is that piano tuners are people who can tune pianos because they possess an ability called "perfect pitch". This is the ability found in many musicians, to "abstractly" recognise and name musical notes and pitches without any given reference pitch. It is in fact properly known as "absolute pitch", and only somewhat more ambiguously called "perfect pitch". It is something that can initially help students acquire some of the practical skills necessary for piano tuning, not least if they are visually impaired. It is not, however, the requisite ground for fine tuning a piano, not least because piano tuning is much more than merely adjusting pitches. In fact, there are a number of reasons, that we shall be looking at in detail later, why any very fine pitch adjustments that *are* made in piano tuning, must be adjusted as *relative pitch*.[2] There is no evidence that individuals who have "absolute pitch"

---

[2] See the Glossary for the meaning of *relative pitch*.

sense have any better sense of *relative pitch* than those who do not. In fact, the reverse may well be true.[3]

Fine piano tuning involves not just changes to perceived pitch, but very much requires adjustments to *tone*. Initially this happens by listening to, and adjusting certain tone properties that are perceivable with training, like vowel sounds and "beat patterns". Eventually (at a much higher level), in "natural" tuners, tone itself, as a *musical quality*, becomes the focus of attention. This, however, is a stage that usually matures only after many years of disciplined listening to more "abstract" tone properties, and after this has distilled into something more naturally musical - a truly musical and audiophile appreciation of piano tone itself.

Any pitch sense (relative or absolute) is simply not sufficient, in itself, to substitute for the need to address tone properties and qualities, directly (the main reasons for this are outlined in Chapter 2). Even with the best pitch sense in the world, the intimate and analytical knowledge of tone necessary for excellent piano tuning still requires training, and years of experience. Whilst "absolute pitch" can certainly help with initial tuning and with the musical "geography" of tuning whilst learning, those with "absolute" pitch sense often find that in fine tuning it can in some instances be a hindrance, rather than a help.

So what is innate talent in piano tuning? It begins with fine abilities in listening and hearing, and necessarily includes very good ear to hand and arm muscle co-ordination. Real talent in piano tuning then very much lies in the ability to understand and control acoustical features like *beating* not merely as an abstract acoustical feature, but as a feature of piano *tone* and *intonation*. To begin with, a purely analytical approach to *beating* as an abstract feature of tone may be an unavoidable necessity for most inexperienced and learner tuners. The phenomenon we call beating, however, is much more varied and complex in its form and behaviour than the way it is portrayed for elementary tuning instruction. Talented tuners soon realise this, and naturally enter into an exploration of the real acoustical world with which they are dealing. This increases their ability to manipulate it as an art in the musical context, rather than approaching tuning as a mere technical task.

---

[3] From unpublished research by the author.

The art and science – qualities and properties

We do need to distinguish between tone *properties* and tone *qualities*. Throughout this book I will use these words consistently. On the one hand, tone *properties* are physically measurable properties of a musical sound that can affect our perception of its sonic quality. In other words, when we speak of tone *properties*, we will be referring to the acoustical properties behind our perception of tone, that are (at the very least *in principle*) scientifically measurable. "Tone *quality*", on the other hand, is a name for how we subjectively *perceive* the sonic quality of a musical tone. Tone *properties* are objective, acoustical features that can be directly measured in a hard, scientific way, even without the need for anyone to *hear* the sound. Tone *quality*, in contrast, is in part determined by psycho-acoustic or subjective factors, and is often best described using similes, or non-scientific adjectives, just as musicians describe it. Tone quality is something that is *perceived*, rather than measured, and it requires a human ear-brain system, and human consciousness, for its perception.

There will always be some listeners who regard as "good", tone qualities that piano tuners and most other musicians would readily agree are poor. Similarly, a few piano owners will actually prefer the tone of their piano in a neglected tuning condition, to its tone when it has just been tuned by an expert tuner. Some piano owners cannot tell very reliably, the tonal differences between an inexpensive piano in poor condition, and a top quality instrument in excellent condition. The truth is that judgements of "good", "poor", etc., when applied to tone quality, are contextual. They depend on things like cultural expectation, experience, conditioning, education, discernment, and acuity of hearing. There is, however, usually a "general consensus" in these matters. It is necessary for the piano tuner to work *within the general consensus*, not outside it. But it is also necessary for a piano tuner to be the custodian of an intimate knowledge of piano tone beyond that which has been necessary to form the consensus.

Fine tuning is not just about manipulating *properties* of the piano tone. It is largely about perceived *qualities*. Like any art, it is at least partly subjective. It has often been said, quite misleadingly, that the difference between tuning theory and tuning actuality lies in piano tuners necessarily *compromising* in their work, that is, that their tuning is necessarily expected to "fall short" of some "theoretical ideal". In reality, the work of a master tuner need not fall short of anything, except perhaps the tuner's best effort. It is what it is, and that is the drawing out of an instrument's best possible tone and intonation, even if the instrument is less than good. There is no compromise here, because no absolute "target" exists in the first instance, except as a hypothetical notion of partly informed theorists, a notion that is essentially fictitious. From the master tuner's point of view, the only unavoidable compromise is in some instruments not allowing the kind of result that might have been possible on a higher quality piano.

In general, different individuals may have a different idea of what could be called "in tune" or "out of tune", but in piano tuning this is not *merely* a matter of subjective preference. It is as much a matter of tone as it is of pitch, and though both are in part subjective, in piano tuning neither are free to be adjusted *only* by subjective preferences. Attempting to do the latter is a trap into which the tuner without proper training readily falls. In contrast, the expert tuner understands the complex *network of relationships* between notes, in terms of tone properties, and knows that this network must be handled by combining the artistic control of sound, with control of sound *properties*, using firm logical decisions based on good *theoretical knowledge of the network*. It is, put simply, a question of *art*, founded on proper *technique* and knowledge of theory.

The "traditional" theory

The basic model for "traditional" piano tuning theory, relies on a description, dating back to the nineteenth century, for the behaviour of a hypothetical vibrating string. Science has of course progressed considerably since that time, and it is now known that real piano strings *in situ* behave in a way that is much more complicated than the early model is able to describe.

To some extent, "improvements" to the early idealised theory can be made by "adding on" modifications or "improvements", the best known being the phenomenon associated with real strings, called *inharmonicity*. We do not, for the time being, need to elaborate further on inharmonicity, except to say that in piano tuning and technology it is widely recognised, and its effects are well understood by theorists. Even when *inharmonicity* is taken into account, the resulting "improved" theory is nevertheless still a long way from being able to describe the actual, scientifically observable behaviour of piano string vibrations *in situ*, and is just as far from being able to describe the audible (and scientifically observable) behaviour of piano tone in the fine tuning process. As this book illustrates, this shortfall in the theory is far from being merely academic. On the contrary, it sometimes turns out to be very important because it leaves the elementary theory unable to describe behaviour that is both aurally discernable to the master tuner, and controllable as part of the master tuner's art.

That the basic theoretical paradigm inherited from the 19th century is flawed, is widely appreciated by expert tuners today. It is, however, often believed that there is this single shortcoming in the theory, encompassed by this now well known complication of *inharmonicity*. What is not so widely acknowledged is that the "traditional" model is in the first instance built from a whole set of simplifying assumptions and idealisations. This results is a theory that, compared to the actual art of fine tuning, and compared to the actual, observable nature of piano string and piano tone behaviour, is no more than an approximate explanation. It describes what in many cases master tuners would consider as being only rough tuning.

The "traditional" theory, even though oversimplified, has been in existence long enough to have taken root in the "piano tuning" psyche, and as a result sometimes becomes an actual impediment to better understanding and the pursuit of higher practical tuning standards. All good tuners become good tuners through great experience, through taking their empirical knowledge of the acoustical system well beyond the "traditional" theoretical model.

It is quite possible to believe we are hearing something that is *not actually there*, or to be unable to hear something that *is* there, purely from the persuasive force of what we think we already "know" *should* be there, based on what we have been taught. It is not easy to hear the finest nuances of piano tone necessary to master the control of it. It is generally easier if there is some guidance or "map", but only if the guidance or map does not deceive or mislead us. Fortunately, now, digital analysis and contemporary theory can illustrate in detail some aspects of the true nature of piano tone with which the master tuner must work. It can show precisely what we are listening to. We can provide a better "map" for those who wish to take the art to a higher level, perhaps a little sooner than experience alone would allow.

Technology - ETDs

Electronic Tuning Devices (ETDs) can be used for tuning. I have taught both students of aural tuning, and those who use ETDs. Hard working ETD users can "get ahead" of the aural tuners, initially. However, this initial impressiveness of ETDs in producing better results more quickly needs to be kept in context. The first thing is to understand the nature of the ETD.

The function of an ETD is much more sophisticated than might be supposed by the layperson. There is no one "set of string frequencies" that we can simply apply to all pianos, and produce good tuning, good tone, and good intonation. The ideal "frequency set" notion is a myth, a theoretical paradigm that simply does not work in practice, and good ETDs do not work by applying this principle. Master tuners do not of course measure string frequencies, in any case, as a means of tuning. They listen to tone and intonation, and in the course of applying the *techniques*, to "beat" phenomena. The best ETDs actually seek to emulate the quality of aural tuning by the master tuner.

Having passed right through the mastery of tuning techniques, so that they are no longer an objective in their own right, the master aural tuner naturally and deliberately produces beauty of tone and intonation, quite directly. This is not just a side effect of robotically applying techniques. Indeed, merely applying the *techniques* will not even produce tone and intonation in a positive, consciously controlled way. The natural production of tone and intonation in the tuning process comes about by the tuner possessing a direct, *audiophile* knowledge of musical tone and intonation specifically related to the piano. It comes from a deep knowledge of their behaviour in the tuning process, how they can be manipulated, and how the best of both tone and intonation can be brought out in any individual instrument. The most accomplished tuning is not tuning by techniques, but tuning by tone and intonation, *using* techniques, even though one cannot necessarily arrive at this before mastering and passing through the process of tuning by precision technique alone, which typically takes years.

Conversely, the ETD, in order to produce beauty of tone and intonation, must rely on a *theoretical mapping* from specific, measured acoustical *properties*, typically frequencies, to beauty of tone and intonation. Less sophisticated ETDs (many of which have been produced) are not very successful in this attempted mapping. The best ETDs today produce a much better mapping mainly by taking into account *inharmonicity*, not just by modelling it, but by actually *measuring it* on the piano before tuning commences. In the hands of users who have considerable patience in learning, and who have built up experience and technique in using the ETD, the very best of these ETDs can often create their own brand of appreciable beauty of tone and intonation over most of the compass. Nevertheless, this is then typically supplemented with some aural tuning.

The result may be ephemeral, however. This is because a very important part of the art of tuning lies in making each note *stabilise* in exactly the condition one requires, in a process (rather misleadingly) called *setting the pin*. The ETD itself does not, of course, physically do the tuning. It cannot manipulate the different tensions in the various portions of the string between the friction points, which is necessary in the *setting the pin* process. It can itself do nothing to *set the pin*, as the process is called, this being a task of considerable complexity in itself. Consequently, whilst the ETD substitutes for an audiophile knowledge of piano tone and intonation, its user still cannot escape from the onus of setting the pin, an art in itself, that takes aural tuners years to master.

Some pianos are "good mannered", and naturally very stable, giving the false impression that setting the pin is something really quite easy to achieve. But many are not. We need to be very clear about what stability through setting the pin means. When pins are set, subsequent forceful striking of tuned notes, even some time after tuning, will not cause rapid falling out of tune to the struck note. How hard should the strike be to test this? Well, on a modern action in proper condition you can strike very hard with a couple of fingers instead of one, on a single note, without fear of damaging the action. A tuning that a powerful pianist can just "knock out of tune" is no use, so the tuning needs to withstand blows harder than those delivered by such a pianist. Of course, any tuning can be "knocked out", eventually, but where there is no "setting the pin", the change, when it takes place, is rapid – several quick *fortissimo* strikes and the note is noticeably or even badly out of tune. This can happen immediately after tuning, or a few days after. Often, the "falling out of tune" is a *tone quality deterioration* in a note, rather than a "pitch fall".

ETDs do not therefore simply remove the need of skill in tuning, but rather, they substitute for the ears only, and demand a different approach to pin setting. This still takes a long time to learn. Part of the art of pin setting in aural tuning is enabled by the fact that the sheer acoustical complexity of piano sound can quite naturally be perceived by the ear *holistically* as just one thing - *tone*. The acoustical complexity then appears in the fact that *tone* can have a myriad of different, changing qualities. It is nuances of *tonal change* that indicate which pin movements are necessary in order to stabilise the tone.

This natural, acoustical complexity of piano tone, is not of course perceived by the ETD holistically as *tone*. The ETD cannot provide the *psycho-acoustic* part of the whole business of tuning, necessary for *tone* perception, since it is not conscious and does not have a human ear and brain. Rather, the complexity of piano sound is translated by the ETD into a simple, moving visual display, which represents a single acoustical property, or a select set of properties, that the ETD is measuring. Typically, the ETD display moves one way or another, simply indicating the need to increase or decrease the string tension. Also typically, it often does this without ever reaching a truly stationary display state for a given string, for *any* tuning condition. There is in general, often a "margin of ambiguity" or "instability" around the indicated tuning, which reflects not so much a limitation in the ETD technology, as the naturally complex and constantly changing nature of piano tone itself, even of a single string.

The ETD then, translates the complexity of piano tone to a simple, visual display, and it is not surprising that something can get lost in translation. Dealing with this "ambiguous" feature of ETD tuning, itself demands considerable judgement and experience. As a result of such differences, good ETD tuning generally takes much longer to complete to a given standard of technical merit, than expert aural tuning.

ETD tuning is not the same thing as aural tuning, but it has a place. *Skilled* ETD tuning at the *high end* of the technology can even produce tuning better than that of the inexperienced or less talented aural tuner, given enough time spent on the tuning. But it firmly remains something that aspires to emulate the work of the master aural tuner, and not *vice versa*.

The need for piano tuning

Piano tuners who are masters of their art are custodians of an esoteric but necessary knowledge and art within the wider field of music. Their work is an essential adjunct to one of the most important musical instruments in Western music, and it is the very nature of this instrument that continues to create the need of good tuners. The modern piano is indeed remarkable – it is an instrument using complex mechanical, wood, felt, and cloth technology, and yet it still has a valued place in a contemporary world of microchips, electronics, photonics and high technology. It is unusual in that its status is quite unlike that of many musical instruments that in the past fell into disuse, or were replaced by different instruments using more contemporary technology of the time. This special status, already established from the piano's sheer versatility, and the breadth of its repertoire, is now enhanced by today's awareness of the need for informed *performance practice* in "classical" music. This makes it less likely than ever that the piano will simply become considered "obsolete" and replaced by another instrument, or some evolution from it, using technology that is more "advanced".

The consideration of "authentic" performance practice and its effect on music, began of course in "early music" revival movement of the 20th century, and now, both early music performance and *performance practice* awareness generally, are established and thriving features of the classical music industry worldwide. As far as what is now considered the "early period" is concerned, harpsichords, lutes, viols, baroque violins, and all the other renaissance and baroque instruments, are now no longer "obsolete", but have a proper and renewed place in the "classical" music world.

Performance practice awareness now applies to all periods, and embraces the piano itself, even though it has never had the "disadvantage" of having been considered obsolete. There is always some room for acceptable performance practice compromise, and there is always a range of instrument technologies that can be regarded as suitably "authentic" for a given period. Nevertheless, it certainly does not look as though the piano is in danger of being replaced, say, by a supposed "electronic equivalent", however "good" the emulating technology.

The higher up the hierarchy one looks, from occasional domestic usage of the piano, to professional and concert use, the greater becomes the demand for, and musical status of, the "authentic" instrument. Today, any attempt to "evolve" the piano, will face scrutiny of a kind often overlooked by those who propose such changes, and are perhaps not aware as they could be, of the influence of performance practice standards.

The art of piano tuning will therefore continue to be needed while we continue to live in a world where there are music conservatoires recognising the piano as an important instrument; where there is the pursuit of scholarly performance practice, and where there are regular professional and amateur performances from the piano's vast repertoire of music. Piano tuning is needed simply because of the presence of pianos in concert halls, institutions and private homes all over the world.

In such a diversity of situations and pianos there is of course a corresponding diversity in the approach to tuning. There is a great difference between tuning in its crudest form roughly applied to poor or mediocre instruments, and the work of the artist tuner on an artist's instrument, in a highly critical environment. The two should not be confused - the difference is as great as the difference between the poorest pianos and the best concert instruments. Excellence in piano building naturally deserves a corresponding standard of excellence in tuning. Excellence in tuning will not only enhance the most expensive pianos, it will improve the mediocre ones too. There are always two major factors determining the result – the quality of the piano and the quality of the tuning.

If we demand excellence in piano building, in musical performance, and in the appreciation of music and its instruments, then we need the art of piano tuning at its finest to complement these. We should indeed be concerned that it does not become a lost art. We should also be concerned that it does not become obscured by an attitude of scientism, that knowing very little about music practice, would have us believe there is such a thing as a "theoretically correct" tuning. The musical world itself should not really be as content as it has been, for awareness of the true nature of this art – the true *state of the art* - to remain locked up in the experiential custody of artist tuners.

Time to learn

Learning to tune to a reliable and professionally acceptable standard takes a long time. One could perhaps learn to paint, play the piano, dance or do mathematics, probably in the course of a year or so, but the end result after merely a year could hardly be expected to represent the true state of these disciplines in the world. All these activities take many, many years to learn, and many more years to mature. Even after maturation they exist at many different levels from mediocre to masterful.

Precisely the same is true of the art of piano tuning. Nevertheless, the situation at this time (and it seems unlikely to change) is that apparently anyone can *instantly* become a piano tuner simply by announcing that they are, just as anyone can become a "sculptor" or an "artist", simply by advertising themselves as such. If you wish, you can even obtain a "do it yourself" tuning manual relatively easily, and sell your presumed skills to the public. The reality of course is that these notions are misleading, and to the artist tuner, they are absurd.

The average student learning to tune, with constant guidance and tuition from experienced professionals, and with full training facilities, takes at least two or three years full-time to achieve sufficient *basic skills*, which then, as with any art, act only as a foundation for continued improvement and development which can only be gained with sufficient experience in the field. The need for further improvement after training in such arts, is recognised in practice, which is why an organisation such as the Institute of Musical Instrument Technology in the UK requires seven years professional experience for full membership. It is not unrealistic to say it might take ten, twenty or even thirty years for a piano tuner to approach the "state of the art".

The importance of theory and technique

The *art* of tuning is ultimately its essence. But like all art, piano tuning absolutely demands *technique*. The main part of technique is not about how to sit, stand, or hold the tuning lever, any more than for the fine artist it is about how to hold the brush. Of course these things are important, but they are elementary, primary conditions, and technique is much more than this, as any accomplished painter or musician will know. Technique comes from understanding the nature of the medium with which one is working, and the technical nature of the methods used for dealing with it.

In piano tuning, this requires a good grasp of associated theory, and a great deal of experience in physically handling the medium, which is the tensioned strings and the audible soundscape adjusted in the tuning process. Technique, however, being closely allied to theory, can often be explained in an objective way, rather than using the language of art, and concentration on it in training can lead to the misunderstanding that piano tuning is *exclusively* a matter of technique.

The truth is that all practice of tuning by technique alone, is only a set of steps on the ladder towards the *art*. Once the art is attained, which takes a very long time, we can throw away the ladder[4] and, remarkably, the tuner then does indeed tune simply according to what "sounds right", but *never* by abandoning technique and science. Only by having first climbed the ladder does the tuner know *immediately* what "sounds right", without falling into wholly subjective (and unreliable) judgement. Thus the layperson's perception that the tuner is simply making adjustments until the piano "sounds right" is *literally* correct, but highly misleading, because the layperson's notion of what "sounds right" *means*, will be nowhere near the truth.

---

[4] As both Schopenhauer and Wittgenstein suggested for the attainment of wisdom or knowledge.

Professional artists and classical musicians will already know the importance of technique. There are many musicians and painters who are talented, but who will always be limited by their lack of training, technique, and theoretical knowledge, which constitutes the *science* necessary for their art. The best musicians, at the very least in classical music, will always have a thorough knowledge of such things as scales, harmony, performance practice, and music theory.

Truly accomplished painters will not lack a solid knowledge of colour theory, perspective, and sciagraphy, and will probably employ a scientific knowledge of geometry. These sciences are essential for the *technique*, without which the highest level of the art would be impossible. None of the world's great artists lacked technique and knowledge of theory appropriate to their time and place. Precisely the same necessity for technique and grasp of theory applies to tuning the modern piano, even though piano tuning is not thought of as a *fine art*. The accomplished tuner's ability to understand the instrument being tuned, to understand the nature of the sound, and to control and produce tone, is only achieved by supporting the art with its due foundation of theory and technique.

Chapter 2 - The essential ideas

Note numbering

Much of the academic literature on the science of music and related subjects adopts *Helmholtz notation* for identifying the pitches of notes. Whilst Helmholtz notation has a long history as an academic standard, it is not used in this book as there are now two other standards that in the context of piano technology, are today more appropriate and widely used. We should, however, be familiar with Helmholtz notation, so a brief summary is included here.

In Helmholtz notation a diatonic scale rising from middle C has accented lower case letters *c'*, *d'*, *e'*, *f'*, *g'*, *a'*, *b'*, *c"*. The number of "accents" (the comma indexes like apostrophes) signifies the octave in which the note falls. Notes with one accent start on middle C (written c'), and rise to the B above (b'), the C an octave above middle C then being "double accented" (c"). The C two octaves above middle C is then *c'''*, and so on. Notes in the octave below middle C are lower case with no accent, and in the octave below this are given upper case letters. Notes an octave below this are then notated CC, DD, EE, etc.

Scientific notation notates the pitch of the bottom A on the full piano compass as "A0" (A-zero). The adjacent A# above is then A# (or B-flat) zero, and the B is B zero. The lowest C is then C1, the adjacent C# is C#1, the D is D1, and so on, for an octave, until the next C an octave above C1 is reached, which is then notated C2. The number in the scientific system designates the octave position of a note. Middle C would then be C4 and top C would be C8, the B immediately below it being B7.

In the piano manufacturing industry the notes of the piano are traditionally numbered consecutively from the lowest A to the to the highest note. For example, middle C is C40, middle C# is C#41, bottom A is A1, and top C is C88. In the UK professional tuner-technicians and those training for the profession are expected to be familiar with the this numbering system, and to be able to use it in practice. The ability to recognise notes numbered in this way has traditionally been viewed in the UK as a sign of the professional piano tuner-technician's competence and familiarity with the instrument. In this book this system will be used primarily, but together with the scientific notation, which appears in brackets afterwards.

If you want an *easy* method for converting from the piano construction notation to scientific notation, then you just need to know your 12 times table. The method works for all notes from C4 (*scientific* C1) upwards.

1. Take the piano construction number and subtract 3.
2. Then divide by 12 and ignore any remainder.
3. If there was no remainder, you will be on a B, and the answer is the scientific notation number. If there was a remainder ignore it, but add one, to get the scientific notation number.

Here is an example. Take the note A61. Subtract 3 from 61, which gives 58. Divide by 12 and ignore the remainder, which gives 4. Add one and the answer is 5. So the note A61 is A5 in scientific notation.

Here is another example, which illustrates the method, but really the answer is obvious to begin with. Take C4 in piano construction notation, which C1 in scientific notation. Starting with the piano construction number, 4, we subtract 3. 4 minus 3 equals 1. One divided by 2 is zero plus some remainder ("1 into 12 doesn't go"). Zero plus 1 is 1, so the note in scientific is C1.

Now consider the note B39, next to middle C. Subtracting 3 gives 36, which divides by 12 exactly as 3. So the scientific notation number for this note is B3.

Some readers who are not piano tuners or technicians may find this unfamiliar, but if one is to become intimately acquainted with the piano and its tuning as a musical instrument, from the technician's point of view, there are advantages to using and thinking with this system. Numbering in this way is useful because it relates to the physical numbering of specific keys, string groups, and action parts. Once an action is out of the piano and separated from the keyboard, there is no convenient visual octave pattern as there is on the keyboard, to make scientific notation useful. In fact, scientific notation then becomes inappropriate. The same is true when one is locating strings without the use of the keyboard. Without the visible keyboard pattern, notes are most usefully identified by simply numbering upwards from the lowest note.

Use of the manufacturing system also encourages a subtle but important slant to the way one thinks about musical notes and tones in relation to the piano. Rather than thinking of a musical pitch as a generic entity or property in its own right, identified, say, by a scientific notation, we should very much associate the perceived tone with the physical note and action parts on the instrument itself. As tuners, we are especially interested in the properties of the tone, as connected to the physical properties of the action and stringing producing that note, rather than just the generic musical pitch used to identify it.

Note naming

When identifying physical notes on the piano in the context of its construction, those with accidentals (the sharps and flats with raised keys) are "traditionally" named using only the sharp enharmonic names.

The name ascribed to a physical piano note in music theory, however, depends on the musical context. Thus, for example, the B-flat may sometimes be called an A-sharp, or at other times a C double–flat. For largely historical reasons the default names for the raised notes used by musicians, when no other musical context is present, are F#, Ab, Bb, C# and Eb. In terms of "traditional" piano construction identification they are F#, G#, A#, C# and D#, often inscribed on the piano itself. On many instruments they can even be found inscribed as Fx, Gx, Ax, Cx and Dx, which though done for purely pragmatic reasons, can appear to the musician as a series of double sharps! The musical convention may therefore conflict with the piano construction convention in these instances.

Construction terminology

The subject of this book is specifically piano tuning, which is a large subject in its own right, so it is not intended to include much discussion on the construction and design of the piano itself, beyond that which very directly impinges on tuning practice. However, in order to discuss tuning, it is necessary to understand certain basic construction and design terminology. Everyone who wishes to understand the piano better should examine as many different pianos as possible and make a complementary study of piano design and construction as a subject in its own right.

A string has its tension altered by turning the *tuning pin* (*wrest pin*). The convention for all pianos is for a clockwise turn to increase the tension, and an anticlockwise turn to decrease it.

The tuning pins (wrest pins) are inserted in the *wrest plank* (the *pin block*), which in the modern piano is constructed from various special multilaminated hardwood materials, depending on the manufacturer. Different manufacturers use different lamination patterns, with the common purpose of creating a wooden material that is highly resistant to splitting, and very stable.

The visible end of the tuning pin is square in cross section (some earlier, obsolete instruments use oblong section pins), but this part of the pin is also tapered. The part of the pin inserted into the wood (which may be behind the iron plate) is "threaded" with very shallow, multiple helix grooving, so that extended anticlockwise turning of the pin would withdraw it from the plank (pin block). Tuning pins are available in different diameters or gauges.

The arrangement of the tuning pins (wrest pins) in the wrest plank (pin block) reflects the fact that the strings are grouped per note, as *trichords* (three strings to one note), *bichords* (two strings to one note), and *monochords* or *singles* (one string to each note). From treble to bass, the notes fall into a section of trichords (the largest section), a section of bichords, and a section of monochords for the lowest bass (the smallest section). The pattern of pin arrangement in the plank (pin block) is designed to keep maximum possible distance between the pins, within the bounds of other restrictions, to maximise the strength of the plank (pin block).

Upright piano

On the upright piano, the tuning pins of a trichord are arranged so that the top pin is for the left-hand string, the middle pin for the middle string, and the bottom pin for the right-hand string. Usually, in the bass section where there are bichords, the top pin of each note is for the left hand string, but there are exceptions to this. Corresponding pins of adjacent notes are not generally horizontal, in order to increase the spacing between pins without increasing the horizontal distance between them, and improve the structural security of the wrest plank (pin block).

A treble string on a modern upright piano can be traced downwards from where it coils around the tuning pin (wrest pin). The bend in the wire where the string enters the hole in the tuning pin (wrest pin) is the *knuckle*. From the knuckle, the string is coiled around the tuning pin (wrest pin), and leaving the tuning pin (wrest pin) closer to the wrest plank (pin block) than the knuckle, it extends downwards towards the *pressure bar* or *agraffe*. This will be the next metal component with which the string comes into contact.

The pressure bar is a bar whose length rests at a small angle to the horizontal, and is held in place with many *pressure bar screws*. Where there is a pressure bar, the string passes behind it, and then almost immediately over the *top bridge* which is an integral part of the iron frame. Alternatively, each individual note has a brass *agraffe*, which looks like an individual circular "stud", one for each note, which replaces the pressure bar and top bridge. The string passes through a hole in the agraffe and then, on the upright piano agraffe, over a small bridge that is part of the agraffe.

The purpose of the agraffe or the top bridge is to provide a rigid point of no acoustical movement for the string. The string is unable to move at this point, and so this point is an acoustical *node*. This point is the top end of the *speaking length* of the string, the part of string responsible for generating the sound. Tracing the string still downwards, the other end of the speaking length is at the *long bridge*, attached to the soundboard.

On its way downwards, the string first encounters the top edge of the bridge surface, and a metal *side draft pin* (bridge pin) protruding no more than a few millimetres from the bridge surface. Once in contact with the bridge, the string is drafted sideways by this first pin to another side draft pin (bridge pin) on the lower edge of the bridge, where it is drafted back again so that it becomes parallel, or approximately parallel to its original path, but displaced slightly to one side. From top to bottom, the side draft is from left to right.

The purpose of the side drafting is ensure a good fixing of the string on the bridge. At the top bridge, the string passed under the pressure bar and over the bridge, the pressure bar clamping the string down over the bridge at an angle up to a maximum of around 15 degrees. The tension on the string and this angle, known as the top bridge *downbearing* angle, is sufficient to ensure good contact of the string with the bridge, and a proper termination to the speaking length. The design of the agraffe also ensures a rigid termination to the speaking length. At the long bridge, however, the downbearing angle is necessarily much smaller, because the acoustical function of this termination to the speaking length is not the same as at the top bridge or agraffe. Side drafting is therefore necessary.

Whilst the upper edge of the long bridge terminates the lower end of the string's speaking length, this end of the speaking length is not designed to be a *node*. Unlike the point at which the string crosses the top bridge or agraffe, this point is allowed to move, in order for energy and movement to be transmitted from the vibrating string to the *soundboard*. This movement is too small to be seen by eye.

From the lower side draft pin the string continues downwards to the *hitch pin*. The hitch pin, inserted into the *hitch plate* area of the iron frame, is the lowest point reached by the string. A *return string* will pass around the hitch pin and back up to form the next adjacent string. Alternatively, it may pass around more than one hitch pin, and then return as a string of another note. In between the string and its return, will then be other hitch pins and strings.

The precise arrangement varies. The most usual arrangement is for a return string to pass around just one hitch pin, and then to return as an adjacent string. Return strings under full tension do not slip around the hitch pin. There is too much friction, and the wire, which is very stiff, is unavoidably kinked at this point. Some strings do not return, but are fixed to the hitch pin with an eye. Some instruments (notably Blüthners) are strung entirely with such *eye strings*.

The copper covered bass strings are individual eye strings. The normal arrangement for upright piano bass strings is essentially the same, except that there is a side draft pin rather than a pressure bar at the bass top bridge, where the string passes through a large side draft angle. There is a separate soundboard bridge for the bass strings, called the *bass bridge*, or *floating bridge*. The latter refers to a design feature that allows maximum speaking length whilst maintaining a position of contact of the bridge on the soundboard, not too close to the edge of the soundboard.

Grand piano

In the grand piano the string runs approximately horizontally from the tuning pin (wrest pin) to the hitch pin, but passes through angles in both the vertical and horizontal planes. Return strings are also used on grand pianos, but some instruments, notably Blüthner and Bösendorfer pianos, are strung entirely with eye strings. The terms top stringing, backstringing, long bridge, bass bridge (or floating bridge) and overstrung apply in the corresponding way, but there are differences in the arrangement at the top stringing end of the speaking length.

The string passes from the tuning pin (wrest pin) over the *bearing plate*, separated from it by cloth or felt, before reaching the agraffe, or the *capo d'astro* bar, at this top end of the speaking length. The *capo d'astro* is an inverted bridge on the underside of a bar at the treble of the frame, that runs approximately parallel to the keyboard. It is cast as part of the iron frame. The grand agraffe differs from the upright one, because the bearing angle of the string through the agraffe is larger than on the upright, and in the opposite direction relative to the soundboard surface. The grand agraffe does not have a bridging surface included in its design, other than the hole through which the string passes.

The backstringing of a large part of the compass above the bass, on many grand pianos, includes a *duplex scaling* arrangement that consists of an additional metal bridge affixed to the iron frame on the hitch plate, just in front of the hitch pin. This creates a secondary length of string in addition to the main speaking length, that is allowed by design to be set into motion by the bridge. Duplex scaling is designed to enhance the tone properties of the instrument. The length of this duplex stringing is a specific proportion of the main speaking length, which causes wave energy from the main speaking length to be transmitted along it, and reflected back at the duplex bridge.

Many Blüthner pianos incorporate another tone enhancement feature called *aliquot stringing*, which is the inclusion of an additional (fourth), unstruck, sympathetic string, to each note over most of the part of the compass using trichords.

Top stringing, backstringing and speaking length

In piano tuning discussion, it is important to differentiate between *top stringing*, *backstringing*, and the *speaking length*. The length of stringing from the tuning pin (wrest pin) to the top bridge is known as *top stringing*. The length from the lower side draft pin to the hitch pin is known as the *backstringing*. The length of string between the soundboard bridge and the top bridge or agraffe, the length that the hammer strikes, and that produces the main vibration, is the *speaking length*.

Piano scaling

The term *scaling* refers to the design aspects of the piano that determine string lengths, dimensions, types, and positions. The scaling is itself affected by considerations of piano size, but is very much open to variations in design even among pianos of similar size.

The modern piano has a normal maximum compass of 8 ¼ octaves, A1 – C88, but an 8 octave compass A1 – A85 is also standard for instruments not in concert use. Over the greater part of the compass there are three strings per note, each group of three being a *trichord*. Much of the bass section of the compass has two strings per note (*bichords*), and only the lower bass employs just one string per note (*monochords* or *singles*). The strings over most of the compass above the bass are made from high tensile steel of different gauges. The bass notes have steel strings that are "covered" with one or two, and sometimes three, layers of (usually) copper windings, in order to add mass to the string.

Most notes on the modern instrument are therefore called "plain steel" (actually high-tensile steel) trichords. These will be found from around the central part of the compass or lower, up to the top of the treble. Most bichords are covered strings. Some instruments may include a very small number of plain steel bichords (two strings per note) in the part of the compass below the plain steel trichords and before beginning of the covered bichord section. Some pianos are designed with covered trichords in the upper part of the covered section. There is no fixed rule applicable to all pianos, determining precisely where changes in the stringing take place in the scaling of the instrument. These features are part of the specific instrument design.

The scaling design of some instruments will be found to vary from the standard modern principles. Some older or smaller instruments utilise bichords only, most of which are plain steel, for all of the treble, and some employ trichords only for the very top section of the treble.

The modern instrument is *overstrung*, which means the bass strings are fixed on a plane further from the soundboard surface, and cross over the treble strings, which are in arranged a plane closer to the soundboard. This arrangement allows longer bass strings to be used, within the size and shape constraints of the instrument. Longer bass strings produce a better bass tone. Where an instrument is not overstrung, it may be *straight strung*, or *oblique strung* if some attempt has been made in the design to increase string lengths by angling the strings.

Modern pianos are *full iron frame* construction. This means that the instrument incorporates a cast iron frame as an integral tension bearing structure, which is necessary because the overall collective tension of the strings can be in the order of 20 tonnes. Cast iron is still the most appropriate material from which to construct the frame. Cast iron is adequately rigid and strong in compression, and is suitable for the construction and casting details necessary in a piano frame. The high density of cast iron gives the frame great weight, the frame itself being the component mostly responsible for the characteristic heaviness of pianos.

Far from being an unfortunate price to pay for the compressional strength of cast iron, this high mass and hence inertia of the frame, together with its rigidity, is a positive design feature as far as the acoustical function of the frame is concerned. The frame not only serves to bear the tension of the strings, but also acts as a rigid and inertial acoustical "ground" relative to which the soundboard and string vibrations can take place. For acoustical reasons one end of the speaking length of each string must be rigidly fixed, and the frame provides this rigidity. Similarly, the perimeter of the soundboard must be rigidly fixed, and the frame contributes to the construction in this respect also.

Where the iron frame is not genuinely a single, integral component in tension bearing, extending from above the tuning pin (wrest pin) (on a grand piano the side of the tuning pin (wrest pin) closest to the keyboard) to below the hitch pins, it is not a full iron frame in the modern sense. Some earlier pianos are described on the instrument itself as "full iron frame", whilst they do not fulfil this criterion. Some upright pianos, now obsolete, have only a "three-quarter iron frame", extending to beneath the wrest plank, or a "half iron frame", extending only approximately half way up the instrument. Obsolete "wooden frame" instruments, constructed without an iron frame, also exist. There are a number of earlier grands still in use in private households, without an integral full iron frame, notably, in the UK, Broadwoods.

The obsolete designs without the integral, full iron frame, are generally lower tension instruments. The tuning characteristics of all these pianos of different types, can vary considerably. In particular, only the high tension, high quality instrument, has the tuning characteristics associated with its design. Designs of lesser quality, or of lower tension, are not capable of supporting the same tonal characteristics and tuning results, or tuning stability, that can be achieved on the high tension, high quality instrument. The approach to tuning taken in this book assumes, in the first instance, the high quality, high-tension, full iron frame, overstrung instrument, but we shall also be considering certain characteristic more likely to be found on lesser instruments.

The breaks

Part of the scaling design of a piano determines where physically different sections of stringing begin and end. The position where one section ends and another begins is known as a *break*. The *treble break* and *tenor break* (if these are present) are breaks in the regularity of the trichord spacing due to the position of a structural bar in the iron frame. A small number of instruments do not include these breaks. The *bass break* occurs between the strings on the long bridge and those on the bass bridge. On the overstrung piano, the crossover of treble strings and bass strings takes place here.

The breaks in the scaling are of particular consequence in the tuning process. This is (a) because strings close to the break often have different tuning characteristics due to the effect of the break on the bridging, string lengths, and position of the string on the bridge, and (b) because around the area approaching the bass break from the treble, rapid changes of string type, diameter, and speaking length may take place, which affect the tuning behaviour.

Essential terminology – tools

Tuning itself, as distinct from the repair and action regulation work that often accompanies tuning, requires only three essential tool, even though most professional tuners will possess many more. The first is the *tuning fork*, which is used to set the tuning of the first note. The best tuning forks are made of an alloy of iron and nickel (*invar*), which has a very small coefficient of thermal expansion. This ensures constancy of the fork's frequency, with changes of temperature. The professional tuner may carry a number of different tuning forks.

The second tool is the *tuning lever*, which is used to turn the tuning pin (wrest pin). There are a number of different kinds of lever available. The tuning levers used in the profession typically have interchangeable *tuning heads*. The tuning head is the metal component that fits over the tuning pin (wrest pin). The standard modern head contains a star-shaped hole that allows the lever to be fitted over the square cross-section of the tuning pin (wrest pin) in eight different positions. There are at least three standard "sizes" of head, large, medium and small, the "size" referring to the size of the hole.

Variation of tuning head size is necessary because of variations in the sizes (the diameters) of tuning pin (wrest pin). Some tolerance is allowed owing to the fact that the tuning pin (wrest pin) and the tuning head hole, are tapered. Some auxiliary sizes of tuning head are also available, together with special variations of the length of the head, and its diameter. Tuning heads with square holes, and with oblong holes, are also available for special circumstances, such as the tuning of an early instrument, or in cases where the lever may be used for stringing.

Last but not least is the *wedge*. For tuning grands, a rubber or felt wedge (a *grand wedge*) is used, which may or may not have a handle. The purpose of the wedge is to be inserted between the strings in order to mute selected strings. On the upright piano, a *Papp's wedge* can be used. This is a sprung device, effectively two connected wedges, that can be quickly passed through the action and inserted between strings. The modern Papp's wedge is usually made from plastic or nylon. When the wedge is "closed" the ends fit between two adjacent strings of a trichord. When it is sprung open, it fits between two trichords or bichords, or between the two strings of a bichord.

Up to a certain stage in training, student tuners may use *muting felt* or *listing felt* in tuning the upright piano. This is a strip of felt inserted between the trichords of a central section of the compass, so that only the central string of each trichord sounds. This enables the tuning of intervals without the added complication of trichord tuning. On the grand piano, the equivalent tool is the *gang mute*, which is in effect a row of connected rubber wedges. Expert tuners do not use either, because the tuning qualities of intervals will often differ depending on whether just the one or all three strings of the unison are sounding, so unisons *must* be tuned as part of the process of tuning intervals, and not treated as a separate process. This, however, is usually reserved for later stages of training, as it requires a considerably higher level of skill, and the success of the interval tuning is then *dependent* on reliable and excellent unison tuning.

Piano tuning – from the top down

Descriptions of piano tuning "from the bottom up", where the bottom is "traditional" 19th century tuning theory, or even just *temperament theory*, can be misleading, and can leave a large gap between the understanding of an enquirer, and that of a master piano tuner. Here we take a "top down" view that keeps the elementary theory in context.

Intonation and tone

Piano tuning should never be thought of as merely "tuning each note to the right pitch". The intuitive idea that there is a set of musical "pitches" that just sort of "exists" somewhere in a "musical ether", and that all we need to do is to tune our notes to these pitches, is in fact, wildly wrong.

The most *critical things* in the tuning of the piano are not the individual or isolated perceived *pitches* of the notes, but (1) the *tones* of the notes, (2) the *intonations* of the intervals between the notes, (3) the *tone qualities* of these intervals when both notes are played together (see *interval* in the Glossary), and not least, (4) the compass-wide intonation of the piano and its tone as a whole.

Scale tuning

The business of how to tune a musical *scale*, is often mistakenly identified as the sole task of piano tuning. "Traditional" piano tuning theory itself goes little further than addressing this task. In practice, the four points just mentioned above, concerning intonation and tone, require much more than the business of tuning a scale, for their success. Nevertheless, we *are* now going to look at the question of tuning a musical scale on the piano, because it is nonetheless a natural starting point.

Good intonation and tone in the intervals

The first thing to realise is that when a piano interval sounds obviously "out of tune" compared to the usual expectation, it does so, *in musical terms*, because the relative intonation of its two notes is altered away from that expected, *and because the tone of the interval is unpleasant*. Let's illustrate this further. Imagine we are tuning two piano notes, each with just a single string.

If we take a good sounding interval[5] and deliberately mistune it, its intonation is altered, *but its tone is also altered*. The tone becomes more turbulent, amongst other things. Piano tuners, of course, would say it has too much *beating*. If we were to carry on increasing the mistuning, the intonation would become less and less like what we would expect for that

---

[5] "Interval" generally refers to both notes sounding together. See *interval* in the glossary.

interval, *and the tone becomes more and more turbulent*. But only up to a point.

If we carry on increasing the mistuning by altering one note in the same direction, flat or sharp, the tuning will eventually pass through the "worst" point, at which it sounds not really like any commonly recognised musical interval, as it starts to approach another, different musical interval.[6]

Thus we can change from, say, a major third to a perfect fourth. There will be a small range of tunings in which the major third sounds good, and a small range in which the perfect fourth sounds good. But in between the two, is a range that sounds less aesthetically pleasing (to the consensus of musical taste), or even sounds "horrible". It is both the unfamiliar intonation, *and the tone of the interval* that produces the effect.

Achieving good intonation and tone in an interval sounds as though it ought to be straightforward, if we have a good musical ear. Certainly most talented musicians, especially string players, can tune a good perfect fifth and a good octave on instruments whose tuning processes are familiar. However, consider this:

Every note of the scale has a relationship with, or makes an *interval* with, every other note of the scale. Fig. (2.1) shows all the intervals middle C makes with all the notes to G above, and G below.

Fig. 2.1

G Ab A Bb B **C** C# D Eb E F F# G

The figure represents (rather crudely) not only the physical distances of the notes on the keyboard from the middle C, but also the different "musical distances" or pitch separations between the various notes and middle C, or in other words the "sizes" of the *musical intervals* between the various notes and middle C. To meet modern musical expectations, each and every one of these intervals must have *good intonation* and *good tone*. This would, on the face of it, be straightforward enough to

---

[6] If you are thinking we might break the string, in fact, we would not expect to, on piano in good condition. The strings in the mid-compass are at least around a major third below their breaking tension.

arrange, just by tuning each note with the middle C, so that it made the right interval with middle C, with good intonation, and good tone.

One of the first problems in tuning is that whilst this is all very well, *other* intervals between the tuned notes, notes other than ones including middle C, are also formed, and as it happens, they will not necessarily just obligingly turn out to also have good intonation and tone, just as a result our tuning nice intervals with the middle C. In fact, usually, what happens is that some of these other intervals rather too easily turn out to have *grossly* unpleasant intonations and tones. This may seem strange, but there is in fact no scientific reason why aurally arranging 13 musical tones to make a scale in this way, *should* just naturally and easily "work out" so that *every* interval sounds fine.

What happens, for example, if we consider also the relationship of say, the E, with every other note in the scale? We now need to consider all the intervals with middle C in them, plus all the intervals that include the E, and we would want them all to be tuned with good intonation and good tone. This is shown in Fig. (2.2).

Fig. 2.2

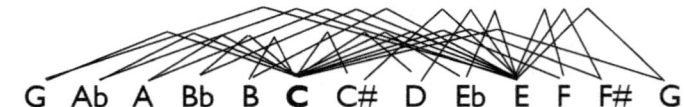

G Ab A Bb B **C** C# D Eb E F F# G

We can see that things are beginning to look a little more complicated now, even though we have so far considered only two starting notes! Imagine the complexity if we considered *all* the intervals for *all* the notes. There are 78 intervals in all. Remember that tuning one set by intonation so that all the intervals sound very good, very easily (in fact, invariably) causes some intervals in another set to be very poor. The fact is, that arranging the tuning of the 13 notes so that all 78 intervals they form, have equally good intonation and good tone, is not as easy as it might have seemed at first. I should point out here that in professional practice on the piano we have an additional burden. We are not merely aiming at a rough, "reasonable" result, which we might get if we tweaked and re-tweaked the tuning for long enough. We are aiming at a tuning in which the intonations of every interval of any given species, are very precisely the same, which means our requirement is *very* exacting.

"Surely", you might say, "all we have to do tune each note to the *correct* musical pitch, and it must then follow that *all* the intervals will be good?" This is an apparently "common sense" observation to make. But if this is the case, then we must be more specific about what these supposedly "correct" musical pitches *are*, and more to the point, for practical tuning purposes, we would have to say *how we get to them*, or *how we tune them*.

I should also point out that "traditional" theory's idea that there is one fixed set of *frequencies* for the notes of any scale, that is the solution, is quite mistaken. "Traditional" theory's *frequency set* applied to a piano, invariably produces only a very *rough* solution, at best not much better than we get if we just carry on tweaking the scale by ear for long enough, until we arrived at a "reasonable" but somewhat "random" result.

Could we get someone with "perfect pitch" to do the job?

More or less every year I have taught tuning, I have had some students with "absolute pitch", a facility often ambiguously called "perfect pitch". This is the ability to instantly name any musical note on hearing it, without reference to any other source of pitch information. There is no mistaking the "full blown" ability when it is properly tested. Being able to name just the odd note correctly, without a reference pitch, does not in itself indicate full "perfect pitch" sense. The ability, if it is genuine, means that sequences of unseen notes, atonally and "randomly" related, can be played on the piano, and then named faultlessly, more or less instantly, every time. The "atonal" arrangement of notes is necessary because a competent musician without "perfect pitch" but with good *relative* pitch, using nothing more than standard musicianship skills, can name notes successfully, provided one reference pitch is already given or memorised, and provided the notes are related in a *memorable melody*.

Being able to write down a heard melody was a required musicianship skill when I was a music student. The difference between those with *relative pitch* and those with "perfect pitch", was immediately obvious in that those who possessed "perfect pitch" could accomplish the task immediately, *even when the melody was atonal*, and *no starting note was named*. Those with *relative* pitch, in contrast, even when told the starting note, would have to "work out" what the intervals were, first, which is difficult when there is no easy, memorable melody. For someone with "perfect pitch", in contrast, every note is immediately and simply identifiable in its own right, so recognising a "random" string of notes is not difficult.[7]

---

[7] The task may of course involve more than one type of memory.

Usually, at some stage during teaching, I set those with reliable "perfect pitch" the apparently obvious task of tuning a complete chromatic scale, one note at a time, without listening to any intervals. I don't do this as part of some covert plan to "catch them out" or to "test" them, and always the reasons for the exercise are explained. I have not yet found anyone who was not happy to try it.

I have also not yet found anyone with "perfect pitch" capable of producing by this method a scale that approaches the standard required after the first year of training in tuning – not because the absolute pitches of the notes appears poor, but because of the *relative intonation* and *tone properties and qualities* in the resultant intervals. The result, every time, is simply not good enough. Often, the result contains one or two intervals that are so poor on the first attempt, that re-tuning is imperative. To clarify this again, it is not that the absolute pitches of the notes sound wrong. The problem lies in the *relative intonation*, and the *tones* of the intervals, when both notes are sounding together.

The result (with some re-tuning and tweaking allowed, if necessary) might sound quite acceptable on a harpsichord, and arguably might well have been appropriate on harpsichords in late 18th century England. The problem is that the recognised standard of tuning *equal temperament* on modern pianos, which applies right up to that required for the concert platform or recording studio, is *much more* exacting than this. It is more exacting mainly in respect of the *tone properties* and *qualities* of the intervals (when both notes are sounding together), which in professional piano tuning are *directly adjusted* in the tuning process. If we tune by absolute pitch alone, we do not adequately control the tone properties and qualities of the intervals formed, but if we tune the latter directly, the pitches fall into place, and *both* are sufficiently controlled.

Now this might seem to be implying that "perfect pitch" is "not accurate". In fact, what we are dealing with here has very little to do with "pitch accuracy", which is a misleading concept in itself. The whole idea of "pitch accuracy" and "perfect pitch" is a cauldron of confusion. So-called "perfect pitch" ability is something many people confuse with *pitch acuity*, which is the ability to hear, distinguish, and judge small differences of pitch. The confusion arises in the following way.

Janet and John are both excellent musicians each with a sensitive "musical ear". Janet, who has "perfect pitch" can say exactly which note is being played on the piano, just by hearing it in isolation. John, who has "relative pitch" can only say roughly where on the keyboard the note was played, say, just for argument's sake, within an octave (most talented musicians with relative pitch would probably be more accurate than this). This superficially *makes it look as though* Janet's pitch acuity or sensitivity is much better than John's, because it seems "she can hear pitch to an accuracy of less than a semitone". But it is of course not the case that John cannot tell pitch differences smaller than an octave! He would have little appreciation of music if this were the case.

More likely, equally with Janet, he can easily detect pitch differences of a mere few hundredths of a semitone. When a note is heard *in the context of a piece of music*, rather than in isolation, John will be just as demanding and discerning as Janet about the precise pitch of the note. The difference between Janet and John here, is not one of pitch sensitivity. It is one of *reference*. Janet has an apparently "built in" *absolute* system of reference for pitch, whilst John has a *relative* frame of reference, for pitches relative to other pitches, or to what musicians call a *tonal centre*.

"Absolute" here is a technical term meaning the opposite of "relative". It does not mean "absolutely correct". It is not the case that a *relative* reference frame is necessarily any less *accurate* than an *absolute* system, *except* in its ability to give absolute (rather than relative) co-ordinates. The minority who possess "perfect pitch" will also possess *pitch acuity*, but the fact is that *most people* possess pitch acuity in any case, whether or not they have "perfect pitch". It is not the same thing.

Many people do make the inductive leap, after seeing someone with "perfect pitch" recognising note pitches without the help of a given reference pitch, to the assumption that this "perfect pitch" sense must be a form of hyper pitch-acuity. This is, indeed, an entirely unfounded conclusion to which many people jump, and it is helped along when the person with "perfect pitch" is then also keen to give their opinion as to whether the pitch is a little too high or a little too low. Somehow, the experience of seeing the first ability, which we find we can't understand, is accepted as providing the authority for the second proclamation about whether the pitch is a little too high or low. This, in turn, then feeds the belief that the "correct" pitch is something we can absolutely pin down, somewhere, somehow, objectively, maybe perhaps in the acoustical properties of the note.

When someone with "perfect pitch" decides to fine tune an isolated note "up a bit" or "down a bit", what are they doing? They are (ironically) using pitch acuity, to judge the pitch, as it were, relative to their "pre-wired" expectation of what the pitch should be. But does the "correct pitch" of the note *exist* as something in their head, so to speak, so that this personally held *mental pitch* is what we should all subscribe to, and unquestioningly accept as "correct"? Or does it "exist" as something in a kind of "invisible musical ether" somewhere, that they somehow "hear" or "know", whilst others do not? Or does it exist as an acoustical property of the played note itself, that they simply *recognise* when they hear it?

As it turns out, *all* of these ideas are misleading, and not really very useful. Pitch *may* arguably exist as a *qualia* in someone's head or psyche when there is no physical sound wave entering the ear, and this may even be associated with certain brain activity. Absolute pitch may well be associated with neural configuration. It indeed *does* (normally) exist as a subjective *sensation* or *sense perception* when the sound wave is present. But we are a long way from being able to quantify "mental pitches" or of course invisible ethers for that matter.[8]

Most importantly, pitch certainly does *not exist* as a scientifically measurable *property* of sound waves themselves. It is not a property of the played note itself. Neither pitch nor tone quality are *properties* of sound waves, but rather, they are a human sensory *response*; they are the ear-brain system's response to the complex sound wave's properties. When we get to the real nitty gritty of the idea of an absolute, "exact pitch" of an isolated note, we are not even talking about an acoustical property of the note, but about a psychological response.[9] It is true that over much of the hearing range there is a relationship between the frequencies present in a sound, and perceived pitch, but this is far from having the reliability and precision of a single law of physics.

The role of subjectivity in pitch perception, both "absolute" and "relative", is in general underestimated. We do not fully understand how the brain processes sound information to produce pitch or tone sensation, but it is possible that absolute pitch recognition is a different neural process to relative pitch perception. The acoustical properties on which "absolute pitch" processing in the brain is based, are not necessarily the

---

[8] Certain neurons in the brain are "tuned" to certain frequencies (Bear, MF; Connors, BW; Paradiso, MA, *Neuroscience*, Baltimore; Philadelphia, 2001, p. 381), but the connection between this and what is *experienced* as the quality we call *pitch*, is as difficult as the question of the relationship between neural matter and consciousness.

[9] At the time of writing the OED itself unfortunately perpetuates misconception that is widespread outside the disciplines of acoustics and psychoacoustics themselves, by defining "absolute pitch" as 'pitch according to a fixed standard defined by the frequency of the sound vibration'.

same as those that most critically affect tone properties (like beating) or the brain's interpretation of tone and relative intonation in musical intervals. Simply tuning by absolute pitch sense should not therefore even be *expected* to necessarily produce full control of the acoustical properties that affect the tone properties and qualities, and relative intonation in the intervals.

The solution used in piano tuning

The technique used in piano tuning to achieve the scale tuning, is to tune the intervals by listening to specific acoustical *properties* of tone. We can also listen to tone qualities and to intonation, and even absolute pitch, if we wish, but listening to, and adjusting the special tone *properties*, is essential.

Fig. (2.2) showed the beginning of a complex *network* of intervals between the notes of the scale. If we wanted to show the whole network, with all the connections, it would be clearer and more elegant to represent it as a circle, as in Fig. (2.3):

Fig. 2.3

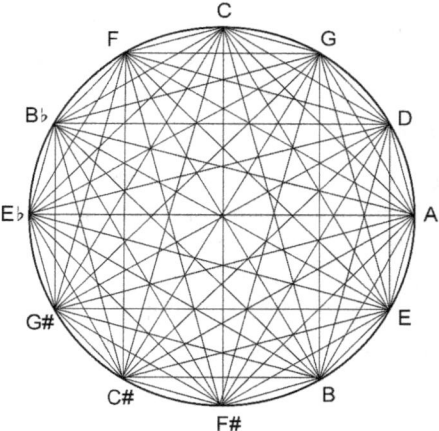

What happens in piano tuning technique is that we tune just a critical selection from all these intervals, concentrating on the adjustable tone properties. These are perfect fifths, perfect fourths, thirds and sixths, and if we get these right, the rest falls sufficiently into place. The complex network of interval relationships that we actually tune, could then be more simply, but still elegantly represented as in Fig. (2.4)

Fig. 2.4

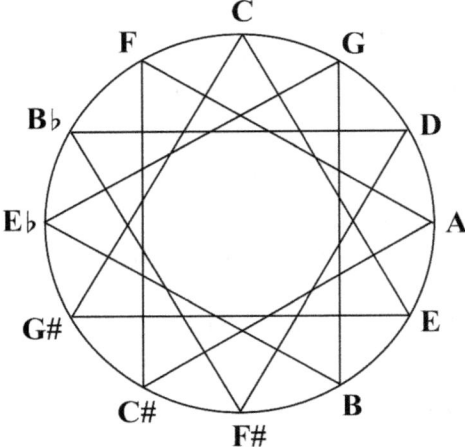

The business of how to arrange the tuning of the notes in order to provide nice sounding intervals precisely where you want them in the scale, whilst leaving any less pleasant sounding intervals where you are less likely to use them, is historically what *temperament theory* has been largely about. As there is a chapter dealing with this in detail, we will avoid getting into these issues now, and instead, just point out that the accepted way of tuning modern pianos is by a unique chosen system called *equal temperament*.

In *equal temperament* tuning, a "solution" to the "problem" of how to arrange the tuning of one's notes in the scale, is provided by the idea (which will probably have already occurred to you) of making the semitones, *i.e.* the musical "distance" or interval "size" between any two adjacent notes in the chromatic scale, all the same. In other words, the octave is divided into 12 equally sized semitones, and as a result, the intonations and tones of the intervals are those that are today heard on a well tuned piano.

Now you might think that this is surely not only the "obvious" solution, but also the perfectly "natural" one. This is not quite what it might seem, however, for two intertwined reasons. The first reason concerns *culture*, and the second is about acoustics. This *equal temperament* system of arranging the tuning of the notes is ubiquitously adopted as the standard in Western music today.[10] So for most listeners in Western culture, this is

---

[10] It is ubiquitous, but not universal.

all they have ever heard. It is, in fact, now a cultural expectation, but this was not always the case. We know that it was a system in use (to quite good approximation) on some fretted instruments in the West, from the very beginning of the rise of written instrumental music,[11] but it was by no means a ubiquitously adopted standard for music then, as it is today. It was not in (precision) use on keyboard instruments at the time, and in fact did not come into general precision use on keyboard instruments until the nineteenth century.

Apart from the fact that there are technical difficulties in achieving *equal temperament* with precision on keyboard instruments, difficulties that were not really overcome until the nineteenth century, it is not the case that the intonation and tone qualities produced by equal temperament, have always been considered "natural" or the most pleasing. As it happens, quite the reverse is true. For a long time, Western culture held beliefs about how musical intervals should be tuned, both in terms of philosophical ideals and aesthetic preferences, which tuning in *equal temperament* would not in any case satisfy. In particular, the resultant intonation and tone of the major thirds that are produced by *equal temperament* is far from that which was considered pleasant and acceptable in certain earlier periods, particularly in the Middle Ages and the renaissance.

The second reason *equal temperament* may not be as much the obvious "natural" choice as it might seem, is because the basic acoustical laws of nature governing the tones of musical strings (and indeed pipes also), tend to point towards certain interval tunings as being mathematically or statistically more significant than others, and these tunings are not those found in equal temperament. One could of course argue, however, that mathematically, equal temperament is an *especially elegant* solution to the problem.

Beat rates

The *theory of piano tuning* works out for us, as much as possible, how to adjust the necessary acoustical properties of the intervals. The *theory of temperament* works out, in its own way, a *description* of the chromatic scale we will produce, but it cannot tell us how to precision tune it in practice, because it does not deal with *tone properties*. It is possible for

---

[11] See Lindley, Mark, *Lutes, viols & temperaments*, Cambridge, 1984.

piano tuning theory, on the other hand, to work out what our tone properties need to be, without actually needing temperament theory.

Remember that tone properties are acoustical features that can relatively easily be scientifically measured, whilst tone *qualities* are what we *perceive* in the tone. None of what I have so far said means that in tuning we do not also listen to the subjective tone *qualities* of the intervals, and sometimes make adjustments based on this. *Satisfactory* tuning of the scale can be done just by listening to and adjusting the tone *properties*, but *musically excellent* tuning may well require consideration also of the tone *quality*, and the ability to alter it through the tone *properties*. There must, however, be a thorough knowledge of what effect on the network making such a change will have. In other words, we can leave behind the directions given to us by *tuning theory*, for the sake of *tone quality*, provided *we know what we are doing*.

The technique for tuning equal temperament on the piano relies on listening to properties called *beat rates*. Beat rates are the rates (in beats per second) of "beats" or audible fluctuations that occur in the soundscapes of the tuned intervals. The beat rate (properly speaking there is more than one beat rate in each interval) in an interval varies with the tuning of the interval. There is one special tuning for each interval at which its beat rate will be, roughly speaking, at a "minimum". This is called the *harmonic*, *beatless*, or *pure* tuning for that interval. In an equal temperament scale (a chromatic scale with equal sized semitones), *none* of the intervals formed between its notes (except the octave itself) will be beatless.

"Traditional" piano tuning theory calculates which beat rates must be present in each interval, when the semitones in the scale are equal in size (the chapter on "traditional" theory goes into this in detail). The tuning task, then, according to "traditional" theory, involves tuning particular beat rates in particular intervals. Octaves are then tuned outwards from the scale, into the bass and treble, both the beat rates for unison string groups, and the beat rates for the octaves, being specified, by "traditional" theory, as zero. This last specification, however, is, as we shall be seeing, rather misleading.

The musical intonation and tone of the piano as a whole, is then largely dependent on *precisely* how the unisons and the octaves, and other large compound intervals, are actually tuned.

For students of tuning

A beat rate is a number. It is a *rate*, a *number* of so-called "beats" per second. Much of the acoustical phenomena in piano tone, to which this notion of a *beat rate* is applied, is actually much more complicated in its behaviour than such a simple notion can fully describe. In reality, even the phenomenon of so called "beats" itself, is not always reducible to a straightforward single number that could properly, with any accuracy, be called a single *rate*.

One should also be aware that during training, it is often the case that one becomes so immersed in so-called *beats* and *beat rates*, that one can completely loose sight of *tone* and *intonation*. At least in part, this may be unavoidable to begin with. For many, the utilitarian attempt to quantify *beat rates* simply displaces concern with the beauty of piano tone and intonation. The whole approach to the sounds of the piano then becomes reduced simply to a crude, numerical question of beats and beat rates.

It should be remembered that "beat rates" were brought into the awareness of keyboard tuners effectively in order to solve the *network problem* in the intervals, in accordance with the idea of *equal temperament*. Beat rates themselves are just *numbers* used in this process, generated by an idealised theory. Listening just to beat rates is not to listen to tone or intonation. There is much more to what we are hearing, *and adjusting*, than the simple idea of a *beat rate*, with a fixed number, would suggest.

Training in tuning

Training in tuning "traditionally" begins with *chipping up,* which is a rough, plucked tuning of the strings on a *strung back*, *i.e.* the iron frame and soundboard without the rest of the piano. This was a stage in the "traditional" piano manufacturing process, still used, but now assisted by electronics. The first learning stage is usually to tune unisons. This is then followed by learning octave tuning, and finally *scale tuning*. The same sequence of learning then may be applied in tuning the piano itself, which requires higher standards of precision.

The required tuning process on a piano is not the same as the learning sequence. It begins with *tuning the scale*, otherwise called *setting the scale*, or in older sources *laying the bearings*. This is the tuning of an initial minimum of thirteen adjacent notes forming a chromatic scale in the central part of the compass. The first note is typically set *to the fork*, using the tuning fork as the initial reference. Tuning usually commences from middle C or A, but other notes may be used. The tuning of *the scale* must then proceed by tuning *intervals*. The rest of the compass is then tuned outwards from the scale, in a sequence of octaves, multiple octaves, and other intervals.

Equal temperament

Equal temperament is now the adopted standard tuning system for modern pianos. Equal temperament is defined in different ways in different contexts. In modern *temperament theory* the "equally tempered semitone" is defined by a number, the twelfth root of 2, that is:

$$\left(\frac{2}{1}\right)^{1/12}$$

$$\approx 1.059463$$

This figure does not in practice properly *define* the acoustical condition of an equally tempered semitone on the piano, or for that matter on most other instruments, because a single number or ratio cannot in general contain all the information necessary to define the acoustical properties of any musical interval. The practice of *defining* musical intervals with single ratios is inherited from very early acoustics and temperament theory. It is not sufficient in modern acoustics when the aim is to show in detail how an interval's tone and acoustical properties are affected by its tuning.

The figure *can* be used as a coefficient to obtain approximate frequencies for string vibrations, in the process of designing an instrument. The figure *cannot* be used to obtain precise frequencies for strings of an instrument tuned in an excellent state of "equal temperament" tuning. The latter is determined by *how intervals sound* on a given instrument, and the actual musical sound produced is the purpose of the piano in the first instance. The single ratio cannot be used to define the acoustical properties that determine how an interval will sound, simply because it leaves out too much information.

Temperament theory expresses "equal temperament" in terms of a micro-interval called the *Pythagorean comma*. This is dealt with in detail in the chapter of temperament. Equal temperament has in the past often been misquoted as the only tuning system in which playing in all keys is possible. There are in fact many other temperaments that in previous times were in common use, that allow successful playing in all keys. The distinguishing characteristic of equal temperament is the equality of intonation for any species of interval, in all keys, and on the piano, the distinguishing tone qualities it produces in the intervals, and over the compass as a whole.

Harmonics, partials, inharmonicity, falseness

Chapter 3, on sound, and chapters 6 and 7, deal with the audible ingredients of tone recipes to which piano tuners listen. The early "traditional" piano tuning theory calls these ingredients of piano tone *harmonics* or *overtones*, and assumes that the strings are perfectly flexible. Piano wire is of course far from flexible. Later theory therefore uses the term *partials* rather than *harmonics*, to distinguish the sound ingredients produced by stiff strings. The technical term for the effects of string stiffness on audible piano tone ingredients is known as *inharmonicity*.

The term *partials* in modern acoustics generally, however, has much broader meaning than this. Correspondingly, real piano string partials are characterised not only by the fact that they are subject to *inharmonicity*, but also, equally importantly, because they may contain a phenomenon called *false beating*, sometimes just called *falseness*.

Use of the tuning lever

Professional tuning levers have interchangeable *tuning heads*, that fit onto the tuning pin. This allows use of the same lever on different gauge pins, and also allows different length or outer diameter heads to be used in certain situations, where required. The chosen head must be the best fit on the pin, and extra long heads are to be avoided except for occasional pins. Thin-walled heads may be useful where there is little clearance around a pin, but should only be used in these situations.

Much emphasis has been "traditionally" placed on issues such as how to hold the tuning lever, how to sit, whether to sit or stand, and in which hand the lever should be held. There are different "schools of thought" on some of these issues, and those that I believe are deserving of particular scrutiny I have dealt with in detail in the appropriate places, in the text of the chapters. It will suffice here just to outline that certain "traditional" edicts on physical practice are questionable because they are based on flawed or incomplete theory, and this is not difficult to show. In the teaching I have been involved with, we have not, for example, insisted on the use of a particular hand for tuning ("traditional" edict, for example, once held that upright pianos should only be tuned left-handed).

Nowadays in education we are all obliged to be *health and safety* conscious, and in keeping with this, it is perhaps worth saying something here about lever technique. It is generally recognised that holding the lever as far as possible in line with the string has advantages, and I deal with these particularly in the chapter on setting the pin. However, this is not always possible, because of physical space constraints, for example in the upper treble of many grand pianos. One should therefore also develop techniques for equally effective tuning using other lever positions. In short, one should be adept in using the lever in *any* position, and should know how to deal with differing effects of different positions.

The lever position on the pin, and its position in the compass, will in turn have an impact on the hand position. My current view is that didactic instruction on a fixed "lever grip", given for the sake of reiterating "traditional" ideas that have not been fully scrutinised, is not especially helpful. Much more important is the question of the physical ergonomics of *individual technique*, just as it would be important in sports and the playing of musical instruments.

Pianos are high tension instruments and many require considerable physical effort to tune. Nevertheless, we are not tightening wheel nuts on a lorry or truck, and we do not need the kind of grip or technique that might be appropriate there. We *are* using the lever to make repetitive, very small adjustments, on 220 or so pins in every tuning. The risk of Repetitive Strain Injury (RSI) in such a scenario is arguably much more of an issue than questionable theories about pin control.

I would suggest the most important thing about lever technique, therefore, is that there should be *no tension* in the arm or hand other than what is purely necessary to pull or push the lever with complete control. The hand should be *essentially relaxed at all times*. Resting the arm on the casework is part of the technique, sometimes, but one should certainly not have to rely on resting on the casework in order to effect *relaxed* and fully controlled lever movements.

Most importantly, tension in the *grip* of the lever itself is entirely unnecessary and can be damaging to the wrist. The lever does not *need* to be muscularly *gripped*. It will invariably stay on the pin by itself, even on an upright piano. *Gripping* the lever does not enhance control. One should not fight the pin and the tension. If the hand and arm are not essentially relaxed, then you cannot be sure of problem-free tuning, as far as physical health is concerned, in the long term. If you tune with a tensioned grip, or a tensioned arm, then you may be at risk of inviting muscular or joint problems.

Tuning requires great concentration, which can lead to the ignoring of pain. The back, neck, arm, hand, or any other part should never ache as a consequence of tuning. Any aches or pains should have their cause investigated immediately, and corrective action taken. Assuming good health otherwise, then discomfort is a sign of incorrect technique.

Lastly, one must always remain aware of the basic mechanics involved. More detailed discussion on this can be found in the chapter on setting the pin, but a few comments here may be helpful. The further the hand from the lever, the less force will be required from the hand, for a given required turning force on the pin. A longer lever can therefore make a tight pin easier to turn, but will not necessarily give finer control in tuning. Using a normal length lever and applying force furthest from the head is usually best.

Any turning force applied to the lever will produce a "bending" force on the pin, parallel to the wrest plank (pin block). Generally, we want to keep this to a minimum. The bending force is an *unavoidable* consequence of the mechanics of applying torque to the lever handle which is *above* (on the grand piano) or *in front of* (on the upright piano) the tuning pin.

Whilst lever position will affect the *direction* of the bending force, *no lever technique will circumvent it*. A bending moment will be produced that is proportional to the product of the turning force applied to the lever, and the distance of the hand from the open end of the tuning head, measured perpendicular to the plank (pin block). The bending moment will therefore be made greater for a longer tuning head, and also will be made greater, the greater the angle between the tuning head and the lever arm. The worst scenario, therefore, is to use a very long tuning head, on a lever that has an angle considerably greater than a right angle, between the head and the lever arm.

Pitch raising

When tuning involves raising all the string tensions by an amount corresponding to a frequency change of more than a few cycles per second, this is regarded as a "pitch change". A "pitch raise" of around a semitone may be necessary sometimes, if an instrument has been neglected. A re-strung instrument will of course need to have the string tensions raised from practically zero, in a process called a "strain".

The principle of pitch raising is to increase the tension as evenly as possible over the frame. For semitone changes, the "traditional" method would be to *chip up*, by plucking the strings, in an upright piano with the action removed. This does not, of course, increase the tension evenly over the compass, but for a tension change restricted to a semitone this is accepted.

For a *strain*, the process should be more evenly arranged over the compass, by, for example, raising all the Cs, then all the Gs, then all the Es, *etc*. When the pitch is raised a long way, the tension "falls back" rapidly, so this is countered by raising above the target tuning by roughly one extra cycle per second for every three that the starting tuning was "below pitch". When the pitch of a string is dropped a long way, it similarly rises again.

Many neglected pianos will typically be much flatter in the treble than in the bass, but the reverse can occur. The approach on any particular instrument is therefore largely a matter of assessment and experience. Pitch raising and fine tuning are mutually exclusive, and would normally be carried out as two (or more) separate processes. Pitch raising is a skill that requires training and experience, and whilst it is a necessary part of the skills repertoire of the professional tuner, this book is devoted mainly to issues of *fine tuning*.

Chapter 3 — Sound

Why study the science of sound?

At the core of the art of piano tuning is sound itself. It is therefore fitting that someone aiming to become an expert in tuning should know something about the nature of sound, the medium in which the work takes place. The science of sound contained in this chapter is detailed, but limited to what is most related to piano tuning itself. In the field of piano tuning, and indeed in order to communicate effectively about it, it is often necessary to use terms such as *frequency, pitch, amplitude, beats, decay* and *partials*. Without proper understanding, these terms can often be misused, which can lead to confusion and misunderstanding about piano tuning itself. This chapter explains exactly what these terms mean, and how they relate to each other.

A general approach

The approach to piano tuning in "traditional" theory begins by considering the supposed motion of the piano string itself, and how this motion produces audible "harmonics", "overtones" or "partials", which we supposedly hear. Briefly, the model used is roughly as follows:

As a result of wave motion travelling along the string, a type of vibration called *standing wave motion* is set up on the string. The standing waves are said to naturally divide the string's speaking length into equal-length fractional segments, each segment of the string vibrating with a specific frequency, and producing one audible partial. (The model in detail can be found in the last section of this book).

We are not going to pursue this approach for two reasons. The standard model for vibrating piano strings is too oversimplified and idealised to be of use for understanding the finer details of piano tone behaviour with which piano tuners work in practice. To be an expert tuner is to be conversant and familiar with aspects of piano tone behaviour that the standard model of vibrating strings is simply unable to describe or even suggest. It might be possible to start with a "better" model of vibrating strings, more in keeping with what is now known about actual piano string behaviour *in situ*, but such a model, if it is to explain the more advanced aspects of the tuning skill, would have to be one more fitting for study by postgraduates in acoustics. Fortunately, this is not necessary, because there is a simpler, but more powerful approach available, but one that is still in line with the analytical techniques of contemporary acoustics.

We are certainly not saying that waves do not travel along the string, or that standing waves are not involved. In fact, waves do turn out to be important for our understanding of sound. Nevertheless, the practical process of tuning involves listening to sounds, rather than observing the motions of the strings. Those audible sounds are not produced *directly* and immediately by the strings, but by a more complicated vibrational system comprised of the strings plus the bridges and soundboard, and other components, whose collective behaviour is different to that of single strings.

When we hear a piano tone, we are primarily hearing sound radiated into the surrounding air by the soundboard, rather than sound radiated directly by the string. The approach here, therefore, is to begin with a more generalised "top down" description of the *sound* that is actually heard, drawn from contemporary acoustics. There are advantages in gleaning a picture from this position, rather than first describing an oversimplified model of a vibrating string, and then having to add on more and more modifications, to explain what is aurally or scientifically observable. We need to gain an insight in a relatively short time, into the actual acoustical behaviour of the *tone* to which piano tuners listen, rather than concentrating on a model that will ultimately never be able to represent what is heard.

Sound waves

Putting first things first, we should initially deal with the essential nature of audible sound. The sound that we hear when tuning the piano comes from *sound waves* travelling through the air. When a physical body such as a piano soundboard or a loudspeaker cone vibrates, the surface area of the body, being in contact with the surrounding air, exerts oscillating forces on the air, which results in waves of motion in the air travelling outwards through the air from the vibrating body. A similar phenomenon is seen to happen when a stick is moved rapidly back and forth in water. Waves are sent outwards through the fluid from the motion of the stick. On a small lake, for example, waves setting off in any direction, continue to move in that direction, even if the stick stops moving, or changes motion in some way.

On water, the waves moving across or through the water, make the water itself move up and down. Only where the waves are breaking, as on a shore, or if there happens to be current in the water, does the water itself very noticeably move in a direction other than up and down,[12] but at the same time waves can clearly be seen travelling through or across the water. A floating ball, buoy or small boat on water, for example, will bob up and down as the waves pass it, but will not generally be carried along with the waves motion across the water, unless it is also blown by the wind or drawn by a current. The main motion is up and down, rather than along the surface of the water. The crest of each wave can be seen moving across the water, because it is not a stationary area of the water's surface, but is an effect of the way in which different parts of the water are moving up and down at different times. It is sometimes useful to think of waves themselves as invisible waves of energy that pass through the medium - in this case water - that only become visible in the way they make the medium move.

Water waves are called *transverse* waves, because the main motion of the water is at right angles to, or *transverse* to the direction in which the waves are actually travelling through the water. Sound waves in air are somewhat different in nature. These are called *longitudinal* waves, because the main motion of the air carrying sound waves is back and forth along the same line of direction in which the waves themselves travel. The waves can be travelling in all three spatial dimensions at one, that is, they are generally *three-dimensional* waves.

A simplified analogy might make this clearer. Imagine a line of soldiers side by side standing to attention. The soldier on one end has a soldier on his left, whom he pushes by thrusting his left hand sharply away from himself, and then back into position by his side. The soldier to the left is unbalanced by this, and leaning briefly leftwards attempts to stop himself toppling by doing the same thing as the first soldier – thrusting his left hand out to steady himself, and then returning to a disciplined attention position. The third soldier, unsteadied by the push from the second soldier, then does the same thing, and so on. A "wave" of motion will pass

---

[12] The situation regarding water wave motion is in practice a little more complex, the actual locus of motion being in general a non-circular ellipse, but the vertical motion nevertheless serves as a good example for understanding transverse waves.

along the line of soldiers, while each soldier remains standing on the same spot. Each soldier, however, in sequence, moves briefly leftwards and then corrects himself by moving back to the right. The "wave" of motion through the line of soldiers is a longitudinal wave because the "medium" - in this case the soldiers - moves back and forth on the same line of direction through which the wave moves. Overall, however, the medium, which here is the line of soldiers, stays where it is.

Another example could be seen in a line of railway train carriages. If one end of the train is shunted, a longitudinal wave of motion passes quickly from the shunted end to the other end of the train. The carriage furthest from the shunted end does not move instantaneously as the shunt takes place. There is a short delay between the shunt and the last carriage being jolted, as the longitudinal wave travels along the length of the train. If the carriages have good buffers at each end, these absorb energy as the wave passes, so the effect may be diminished further down the train.

As sound waves pass through air, the air as a whole, stays where it is, but small regions in the air move back and forth, like the soldiers, in the same line of direction through which the sound wave is moving. The wave itself, unlike water waves, does not consist of vertical crests and troughs, but of small regions of slightly "squashed" and slightly "stretched" air, just as the "wave" of motion in the line of soldiers, would consist of soldiers briefly falling too close together, and then separating themselves apart again.[13] If we suddenly "stretch" a small portion of air, the air pressure in that portion briefly drops.

Similarly, if we suddenly "squash" a portion, the pressure briefly rises. The sound wave in air consists of a moving "train" or series of regions of pressure variation, of higher pressure and lower pressure air. Unlike waves in water, whose path of travel tends to be in two dimensions across the water surface, sound waves in air are generally three-dimensional, moving in three different directions. When these waves enter the ear, they cause the eardrum to move, and a sound is heard.

We can represent the simplest kind of sound wave, two-dimensionally, by drawing a graph of the pressure at some fixed point in the air through

---

[13] The true picture is slightly more complicated. To represent the air motion, each soldier would have to move to the right of the upright position, and then to the left of the upright position, and so on. Similarly, the air molecules oscillate both sides of an equilibrium position.

which the wave passes, versus time. The graph then looks like a transverse wave, and in some ways, becomes easier to study. An example, Fig. (3.1), shows a theoretical pressure variation for the first 1/50th of a second (or 20 milliseconds, 1 mS = 1/1000th of a second) in a simple sound wave.

Fig. 3.1

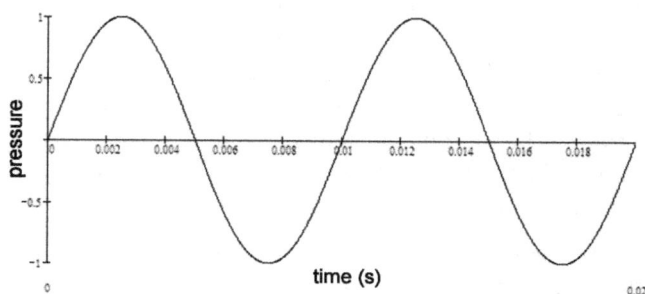

Fig. 3.1. A sine wave variation of pressure versus time.

The pressure on the vertical axis has both positive and negative values, corresponding to high and low pressure regions respectively, with pressure values set for convenience to range from plus 1 to minus 1, representing the maximum pressure variation range from the normal air pressure. The zero line is the normal pressure in the air as it would be when no sound wave is present. In this example, it can be seen that the pressure varies in a very smooth and regular way. This kind of wave might result from the vibration of an object in the air, whose motion is similarly smooth and regular, for example, the prongs of a tuning fork.

The simple and regular looking wave in Fig. (3.1) is of a special kind, and is associated with vibratory motion of a special kind. The wave is one particular example of the simplest kind of wave, known as a *sine* wave. The reason it is called a sine wave, is that this kind of wave can be obtained as a graph plotting $y = A \sin x$, with y on the vertical axis and x on the horizontal, where A is a constant. Certain other trigonometric functions will also produce similar kinds of wave graphs called *sinusoidal* waves, or just *sinusoids*. The graph itself is a *transverse* wave, because the pressure values on the vertical or y axis vary at right angles to the time values on the horizontal axis. In the *longitudinal* sound wave itself, the local air pressure varies along the length of the wave, as the sine of the position along the wave's length. When we hear the sine wave as a sound, it is called a *pure tone*.

Pure tones

We are going to be considering the sound we hear from the piano, as a *recipe* made from *ingredients*, and part of piano tuning involves listening to individual ingredients, or groups of ingredients, or indeed to the psycho-acoustic effects of certain ingredients. Any tone we hear from the piano, is a recipe that can, in principle, be "split" into its component ingredients. It is not just piano tone, or sound generated from vibrating strings that can be "spit" into these ingredients. The principle that sound consists of the sum of a set of simple ingredients, applies to all sound, and has its roots in the mathematics of waves. It is certainly not exclusively a consequence of how piano strings vibrate. The simplest ingredients of any sound are pure tones.

No matter how complicated the vibration of piano strings *in situ*, and no matter what the contributing effects of the soundboard that radiates the sound, the sound we hear is still a wave form. This wave form, however complicated, can be shown to be a summation,[14] or mixture of sine waves of the kind shown in Fig. (3.1). This relationship was discovered by the mathematician Fourier in the first quarter of the nineteenth century. In the study of acoustics, mixing together simple sine waves to make a more complicated wave form, is known as *superpositioning*.

The *Principle of Superposition*

This principle that a complicated wave is equivalent to a mixture of simple sine wave ingredients, is known as the *principle of superposition*. It applies to all waves, whether waves drawn on a graph, or actual, physical waves. *Any* complicated wave is equal to a summation of simple sinusoids,[15] but starting with the complicated wave and *finding* the particular set of sine waves that add up to it, requires special techniques. For complicated waves in general, this process is known as *spectrum analysis*, and must be carried out through a mathematical process known as Fourier transformation, or by using technology to carry out the electronic equivalent. Today, the commonest electronic method is known as the *fast Fourier transform*, or simply, *FFT*.

We are saying that the simplest ingredient of a sound, that cannot be "split" into any more simple ingredients, is called a *pure tone*. A sound of

---

[14] In the case of piano tone the summation, mathematically, will be an integral.

[15] See the note below, on pathological functions.

the kind that can be represented by a sine wave (Fig. 3.1) is a *pure tone*. Pure tones are mathematically the *simplest* kinds of sounds. Their audible tonal quality is also simple, rather like that of a tuning fork, but may be much higher or lower in pitch than the tones that tuning forks can produce, and pure tones can also occur outside the range of human hearing.

All musical sounds can be shown to be equivalent to a mixture or superposition of pure tones, but only in specific kinds of sounds will this mixture be *constant* or *unchanging*. Also, in some sounds it may be difficult or impossible to identify individual pure tones separately, if the mixture consists of a very large number of pure tones, with very small differences between them. Pure tones can nevertheless still be regarded as elemental ingredients of the sound.[16] Most musical sounds are complicated sounds that can be considered as being made from a *complex recipe* of pure tones.

Recipes of sound can have pure tone ingredients that are "well ordered" or "chaotic", or anywhere in between, depending on the sound. However, the recipe of any *piano tone* changes over time. When the recipe for a sound is not constant, as is the case with piano tone, one of the things that characterises the sound, and contributes to its perceived tone quality, is the *way* the recipe changes in time.

For a complicated sound recipe, the variation in localised air pressure may happen in a complicated way, leading to a complicated waveform. Fig. (3.2) shows a section from a wave produced by a piano tone, as displayed on analysis equipment. This kind of wave might result from an object whose vibratory motion is itself of a more complicated nature, which is the case with piano string and soundboard vibrations. Notice in Fig. (3.2) an interesting feature of the wave, which is that the shape of the wave contains parts that *almost* repeat their shape, but they never repeat exactly. This is typical of the kind of sound waves we encounter in piano tone. The sound was recorded using a microphone rather than a device designed to measure air pressure, so the vertical axis can now be considered in a more general way, as a measure of what is called the sound's *amplitude*, rather than being specifically a measure of air pressure. We will have more to say about this, later.

---

[16] It may be possible to create "artificial" exceptions to this principle - mathematically, it may be possible to create *pathological functions* that will cause the principle to fail.

Fig. 3.2

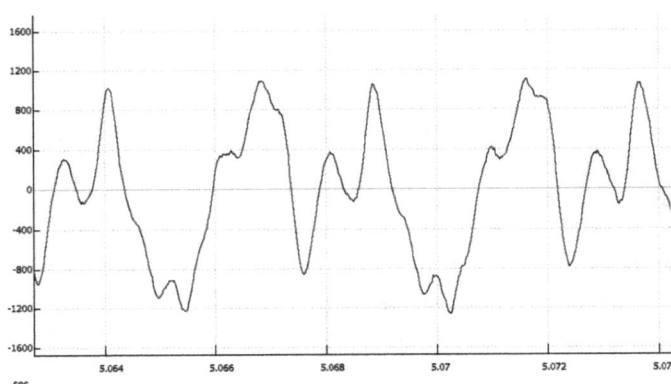

Fig. 3.2. A more complicated wave. The shape approximately repeats itself regularly, but not quite perfectly. This type of almost repetitive waveform is typical in piano tone. The sound that this wave represents would be a recipe with many pure tone ingredients. Mathematically, this wave is a "recipe" made by adding together many different "ingredient" sine waves (like that in Fig. 3.1).

Simple Harmonic Motion

The sine wave [the simple kind of wave as in Fig. (3.1)] is associated not only with the *pure tone*, but also with the simplest kind of vibratory motion, known as *simple harmonic motion*. Strictly speaking, a wave like this, if it is to be properly called a *sine* wave, must on a graph begin with the vertical value equal to zero, when the value along the horizontal, is zero. If it does not, then we ought to just call it a *sinusoidal* wave, or *sinusoid*. All sine waves are sinusoids, but not all sinusoids are sine waves. Nevertheless, the term *sine wave* is often used to refer generally to a sinusoidal wave.

If a body or a point on a body is vibrating with *simple harmonic motion*, and if we plot a graph of its position (on the vertical axis) against time (on the horizontal axis), the graph will be a *sinusoidal* wave. Also, if a sound wave were to be emitted through the air from this vibrating body, it would, if we plotted it as in Fig. (3.1), be a sinusoidal wave. Such a sound, if audible, would have a "rounded", steady, and simple quality rather like the tone of a tuning fork, but such tones may also appear at pitches that tuning forks are incapable of producing. Not all *simple harmonic motion* necessarily produces audible sound. Some examples of objects in *simple harmonic motion* would be small motions of a swinging pendulum, or a weight bobbing up and down on the end of a spring, or the prongs of a tuning fork.

A body (or a point on a body) vibrating with *simple harmonic motion*, will be vibrating back and forth about a *point of equilibrium*. The point of equilibrium is the point where it will remain when it is at rest and not vibrating (or swinging, as in the case of the pendulum). When it is vibrating, the further the body gets away from the point of equilibrium, the greater will be the force acting on it to bring it back to the point of equilibrium. Take the example of a spiral spring. The more it is extended, the greater will be the force attempting to return the spring to its normal equilibrium position, or unextended state. In the case of the pendulum, gravity provides the force always attempting to return the bob to its equilibrium position. Gravity acts downwards at a constant force, whilst the sideways component caused by the angle of the pendulum string, for relatively small motions, increases as the pendulum moves away from its position of equilibrium.

In the case of the pendulum, when the bob is in free motion, it can only go so far in one direction, before the "returning" forces overcome its motion, and pull it back towards the point of equilibrium. At the furthest point, where its outward motion stops and it begins motion back towards the point of equilibrium, it momentarily stops, and its velocity is zero. However, by the time it reaches the point of equilibrium – the point where it would normally come to rest – its velocity and momentum is at a maximum, and it "overshoots" the equilibrium position, continuing on, until "returning" forces overcome its motion again.

This is the nature of *simple harmonic motion*, and the reason that such vibrations come to stop eventually is because energy is lost or drained away from the system. This might be, for example, through friction or air damping. In the case of a vibrating piano string energy also drains away from the string through the soundboard and into the surrounding air. The piano is deliberately designed this way. Energy is taken from the string by allowing the soundboard bridge to move a little at one end of the string's speaking length. The moving soundboard, having a larger surface area than the string, then acts as an amplifier, creating a louder sound from the string's motion, than the string, with its small surface area.

Vibrations are just bounded motion

The distinguishing feature of *vibrations*, as distinct from any kind of motion in general, is that *vibrations* are *bounded* motion in which the direction of motion frequently changes. In *simple harmonic motion*, for example, because the vibrating object can only move so far in any one direction, the motion is said to be *bounded*. Bounded motion itself, however, although it may be vibratory, does not necessarily have to be *simple harmonic motion*. The vibratory motion of an object can be much more complicated than *simple harmonic motion*, but still be bounded, as long as the object can only move a limited distance in any one direction. An example is the piano string or the piano soundboard.

Superposition of vibrations – normal modes

Remember that *simple harmonic motion* can lead to simple sounds – *pure tones*. Complicated motion may lead to a "complicated" sound wave, that if plotted as a graph, would appear complicated in form, like for example, the one in Fig. (3.2). However, since complicated sound waves are a *superposition* or mixture of simple pure tones, and pure tones are produced by *simple harmonic motion*, can we say that complicated vibratory motion is a mixture of Simple Harmonic vibratory motions? The answer is yes. The principle of superposition applies to sounds, wave forms, and vibratory motion. Most sounds we hear in the everyday world, including piano tones, are complicated sound waves, produced by complicated vibrations and waves in objects. All can nevertheless be "unravelled" or "split up" into ingredient pure tones, *simple harmonic motion*s, or simple, sinusoidal wave forms. The individual *simple harmonic motion* ingredients in a recipe for a complicated vibratory motion, are known as the motion's *normal modes*. In other words, the complicated motion of a piano string or soundboard, is a *superposition* of the string's or soundboard's *normal modes*.

Periodic and aperiodic vibrations

Simple harmonic motion, by its very nature, repeats itself in time, and so it is one example of what is called *periodic* bounded motion (*i.e.* vibration). *Periodic* motion is any bounded motion that repeats itself exactly, in time. Periodic motion, however, does not necessarily have to be *simple harmonic motion* - there can also be complex, periodic motion.

Even very complicated bounded motion, if it repeats itself exactly in time, will be *periodic*. The corresponding wave form will be one that repeats its shape exactly at regular intervals. Fig. (3.1) is a wave form that is periodic and simple – a simple sine wave. *Aperiodic* motion is any motion that does *not* repeat itself exactly, in time. Fig. (3.2) is an aperiodic wave form, and would be associated with a vibration that does not repeat itself exactly in time. A wave form that is both complicated, and periodic, is shown in Fig. (3.3). Many motions or waves are *almost* periodic, but not perfectly so, as in the example wave form Fig. (3.2). Piano tone is one example.

Fig. 3.3

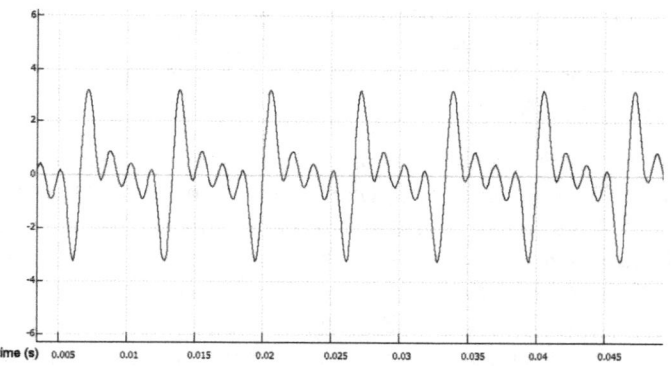

Fig. 3.3. A periodic, but complicated wave. The waveform repeats itself exactly, in time.

Frequency and cycle

Most people are familiar with the term *frequency* from radio and television. In talking about radio and television, *frequency* refers to electromagnetic (radio) waves, but here we are concerned with sound waves. We need to be clear about the precise meaning of *frequency*, and it is the same in both cases.

Because periodic motion or a periodic wave form keeps repeating itself in time (or along the horizontal axis, in the case of a graph), we can say that it has a *frequency*, which is the number of complete motions or waves or vibrations that occur in a given length of time, or within some other unit along the horizontal axis. For example, if we pull a pendulum to one side, and let go, we could count the number of swings that happen in one minute. A complete "cycle", as it is called, or a complete swing of the pendulum, has to be from any point, all the way through the motion, and back again to the same point again. So in the case of the pendulum, swinging completely from one side to the other, is only half a cycle.

A complete cycle, or swing of the pendulum, is all the way from one side across to the other side, and back again. If, say, we counted 10 ½ such swings in one minute, we could say the frequency of the pendulum is 10 ½ swings per minute. More generally, because periodic motion happens in repeating *cycles*, for example cycles of pressure variation (as in the case of sound waves in air), we usually speak simply of *cycles* per second, or cycles per some other unit of time. Look again at Fig. (3.1). Starting from zero, a complete wave shape has already been formed by 10 mS (milliseconds) on the horizontal axis. After this, the same wave shape forms all over again. We would refer to this as 1 *cycle* in 10 mS, and since there are 1000 milliseconds in one second, this frequency is equivalent to 100 *cycles per second*.

For brevity, it is usual to shorten "100 *cycles per second*'" to "100 Hz", where 'Hz' is short for 'Hertz', which is the name of a 19th century German physicist. Thus, we would just say "one hundred Hertz". Tuning forks usually have a frequency printed or engraved on them. This is the number of times per second that the prongs vibrate towards and away from each other. The term "A 440" used to denote so-called "concert pitch", means that A49 above middle C40 is tuned so that it has a vibrational frequency of 440 Hz.

Notice that it is more difficult to try to specify a frequency in the case of the wave in Fig. (3.2), as there are no points where the wave starts to exactly repeat itself again – it is an *aperiodic* wave. In the case of piano tones, their waves *are* aperiodic, but we can get around this, and still specify certain frequencies, by splitting the sound up into its component sinusoids, or ingredients, which are periodic. We can then specify frequencies for these.

The fundamental

Because a complicated sound, wave or vibration, can be a superposition of many simpler ingredients, it is possible to *number* the ingredients. The ingredient with the *lowest* frequency also has a special name – it is called the *fundamental*, and its frequency is called the *fundamental frequency*. Numbering ingredients is a matter of convention, because numbering can start from zero or from one. The convention we will use starts numbering from one. Thus, the fundamental is ingredient number 1. Depending on the circumstances, this might later be called *harmonic* number 1, or *partial* number 1.

Wavelength, velocity, and amplitude

As it happens, for given air temperature and pressure conditions, sound waves will always move at the same velocity through air. This velocity is often called the *speed of sound*,[17] and is about 330 m/s or roughly 740 mph at sea level in moderate temperature conditions.

Remember that one *cycle*, is from one point in a vibration or wave, or air pressure variation, to another point, where the motion or pressure change then begins to repeat itself exactly, starting over again from the same starting point. The distance a moving sound wave moves in the time it takes to measure one cycle of air pressure, will therefore depend on how fast the wave is moving and how much time it takes to measure one cycle.

The time taken for one cycle to complete, is called the *period*. If the period is long, the wave will move a long way in one period. If the period is very short, the wave will move only a short distance in one period, since the waves always move with the same velocity. The distance the wave moves in one period, is called the wavelength – this is also the length of the wave from one point on the wave, to another at which it starts to repeat itself. As the wave form passing a fixed point in the air repeats itself, its effects on the air at that point *cycle* through the high and low pressures.

Frequency, wavelength, and the velocity of the wave (sometimes confusingly called the *phase velocity*), are thus all related. In most circumstances the velocity of sound in air is fixed, so if the frequency is increasing, the wavelength must be getting correspondingly shorter, and the period decreasing also. If the frequency decreases, the wavelength must get correspondingly longer and the period increases. Mathematically we can write this relationship as the simple equation

$$f = \frac{c}{\lambda}$$

---

[17] Velocity is a vector quantity that specifies a direction. The "speed" of sound assumes the direction in which the speed is measured is the longitudinal direction of the waves.

where c is the velocity of sound in the air, f is the frequency, and λ (Greek letter lambda) is the wavelength. The frequency in Hz is the number of waves that pass a given point in the medium (the air) in one second. Each wave causes a cycle of change in the air pressure, so the frequency is the number of cycles (of air pressure) per second. The velocity c can be measured in meters per second, and the wavelength λ, in meters. Audible sound wavelengths in air range from about 1 cm to about 16 meters. The longest wavelength associated with piano tones would be about 12 meters.

The *amplitude* of a waveform in a graphical representation is vertical the measurement from the horizontal or time axis, to the highest or lowest point on the waveform. If the graph represents a sound wave, the greater the amplitude, the louder would be the sound. Amplitude and perceived loudness are not proportional, but they do correlate, in other words, doubling the amplitude will not double the loudness, but it will generally increase it. Fig. (3.4) shows the same wave of Fig. (3.3), indicating amplitude and wavelength measurements.

Fig. 3.4

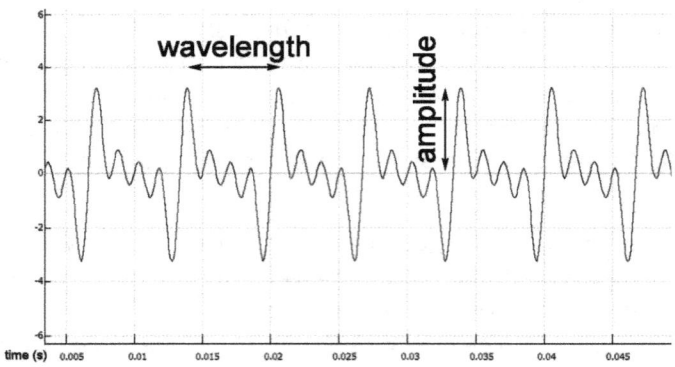

Fig. 3.4. Wavelength is the length of a complete part of the waveform, before it starts to repeat itself. Amplitude is the vertical distance from the horizontal axis to the maximum height of the wave.

Decay

The amplitude and wavelength of a wave does not necessarily have to remain constant. Usually, we would be looking at waveforms for sounds with hundreds or thousands, or tens of thousands of cycles per second.

The only way we can see individual waveforms like those in Figs. (3.2) or (3.3) or (3.4) for frequencies like this, is by looking at a very small time segment of the wave, e.g. a section where the change in time along the length of the horizontal axis, is very small. If we represent the wave over a longer time period, say, several seconds, then the wave of Fig. (3.4) would appear as in Fig. (3.5), where there are in effect now too many waves per millimetre of length along the horizontal, to distinguish them separately. We then get a continuous black area.

Fig. 3.5

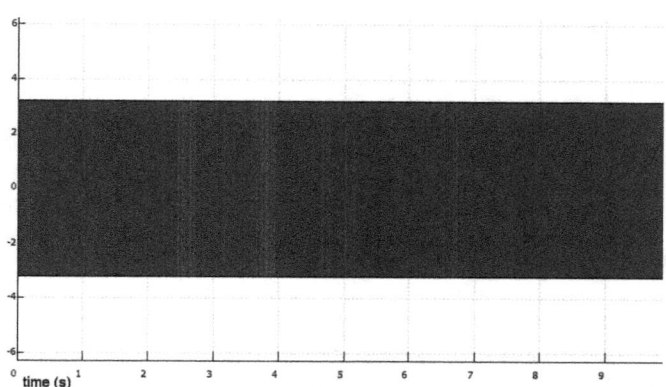

Fig. 3.5. When the horizontal axis covers a sufficient time, the graphical representation of a sound appears as a solid shape, because the individual curves of the waveform are too close together to appear separately.

A piano tone is not continuous and without change. It dies away or *decays*. In this case, the amplitude does not remain constant, but decreases as can be seen in Fig. (3.6), a display of an actual piano tone.

Fig. 3.6

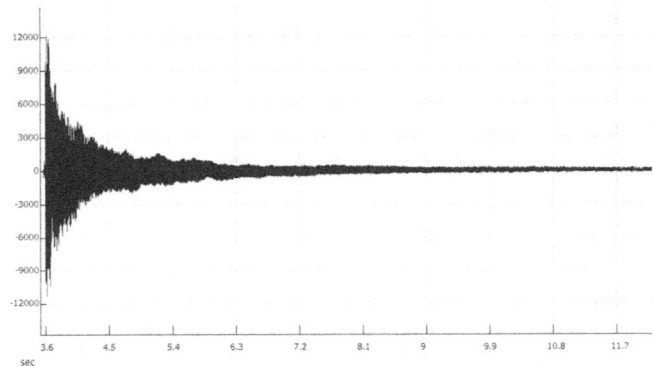

Fig. 3.6. Graphical display of an actual piano tone, showing the overall *envelope* shape of the decay, which is exponential, and has ragged edges indicating non periodicity.

What such displays show is the overall shape – known as the *envelope* - of the wave over a time period that is long relative to the wavelengths in the waveform itself. Note that the outline of the envelope is ragged, indicating that the waveform contained in it is irregular, or non periodic. Also, note that the envelope amplitude decreases by an *exponential curve*, rather than a straight line. This is always recognisable by the fact that it starts steep, and gets progressively less steep. The rate of change of its steepness decreases along the horizontal axis. This is normal, and is known as *exponential decay*.

The envelope shows decay, because the ingredients are each decaying in their own right. In this case, a sinusoid ingredient may look like Fig. (3.7).

Fig. 3.7

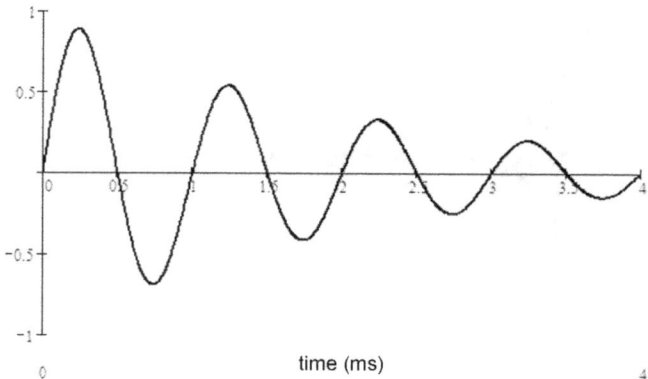

Fig. 3.7. A decaying, sinusoidal waveform. The wavelength (and hence frequency) remains constant, but the amplitude decreases exponentially with time.

Fig. (3.7) show a sine-like waveform but with an additional feature. Now the amplitude of the wave depends on the value on the horizontal axis. As the amplitude decreases in time along the horizontal axis, the wave *decays*. It is in effect a sine wave whose amplitude decays exponentially. Careful observation of where the wave crosses the horizontal axis will show that the wavelength of the wave nevertheless remains constant, as the amplitude decreases. This means it can still have a constant, definite frequency. The wave form is *not* periodic, because its amplitude changes, but it *does* have a constant frequency. The audible tone that this wave could represent is still, for our purposes, considered to be a pure tone, but a decaying pure tone – one that "dies away" or gets progressively quieter in time. Any pure tones that we encounter in piano tones, will always be decaying pure tones.

Any change to the shape of the envelope is sometimes called *amplitude modulation*, or the envelope may be said to be *amplitude modulated*. Fig. (3.8) shows an example of an amplitude modulated envelope, in this case showing a *beat pattern* of fairly regular bumps, or rising and falling of the amplitude. Notice that in addition to the bumps of the beat pattern, the overall envelope shape still looks like an exponential decay, just like in Fig. (3.6). This is typical of the kind of *decay pattern* or beat pattern encountered in piano tuning.

Fig. 3.8

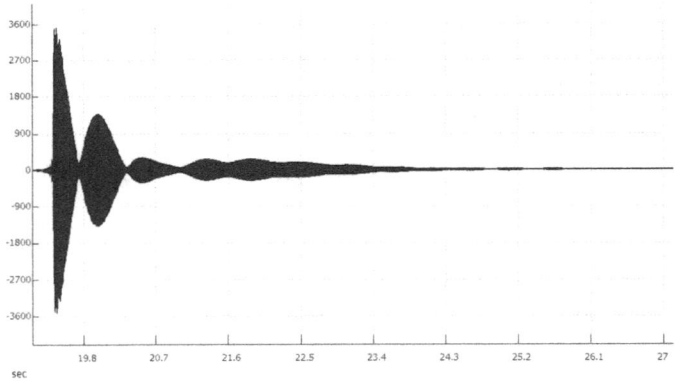

Fig. 3.8. An amplitude modulated, decaying waveform. The amplitude modulation is a *beat pattern* of the kind found in adjustable partials in piano tuning. The illustration is of an actual piano partial.

Frequency and pitch

Frequency and perceived "musical" pitch are related, but not with a fixed, objective, mathematical law. Human hearing responds to frequencies from about 20 Hz – 30 Hz at the lowest frequency end of the range, to about 20 KHz – 30 KHz at the highest end (1 KHz = 1000 Hz, or "one kilohertz"). Low frequencies are "low pitched" sounds, whilst high frequencies are "high pitched" sounds. However, the perception of pitch is to a significant degree subjective, or *psycho-acoustic*, being dependant on the individual, and even the individual ear. There is a good consensus of pitch perception between different listeners, but pitch itself, is *not* a property of the sound or tone being heard. What we sense as *pitch* is internal, it is due to the *response* by the listener to the *properties* of the sound heard.

In the middle of the piano compass, if we were to measure the lowest frequency (called the *fundamental*) produced by each note, we would find that two notes whose (fundamental) frequencies were in the ratio 2:1, sounded an octave apart. Mathematically, we could say that if two frequencies are f and $2f$, the latter frequency will generally sound an octave above the first, at least within a certain central range of frequencies.

There are simple ratios associated with all the musical intervals. Simple ratios associated with musical intervals, not only occur in relation to *frequencies*, but also in relation to musical string lengths and musical pipe lengths. For example, if two musical strings at the same tension are identical, except that the sounding length of one string is twice as long as the other, then the longer string will sound an octave lower than the shorter string. If the lengths were in the ratio 3:2, then the shorter string would sound a perfect fifth above the longer string. Similar ratios would be found between lengths of organ pipes.

This relationship is not always exact, and musical intervals depend in any case on perceived, relative musical pitches, which are in part, subjective. Our sense of pitch is usually very sensitive, but it must be remembered that most musical sounds are complex, and contain many ingredient pure tones. Sensitivity to pitch is a result of the effects of all, or many of the ingredients. It is not generally possible to quote a single ratio for a single perceived pitch interval. Any two notes forming a musical interval on the piano, will contain many different ingredients and many different ratios within the recipe. Two pure tones, however, will have a single frequency ratio between them. Whilst particular frequency ratios between pure tones often produce the sensation of a specific musical interval, the relationship is not entirely reliable. It can vary with frequency range, individuals, and even each ear of the same person.

"Perfect pitch"

Some musicians have the ability to recognise musical notes and name them, without any other reference. This ability is commonly called "perfect pitch" or "absolute pitch". A short random selection of notes played on a piano, would be easily named by someone with "perfect pitch", even if no other clue had been provided, and even if the notes formed no recognisable musical tune in the context of tonal music. Musicians with "relative pitch" may be able to name notes in music, provided they initially know what key the music is in, or what one of the notes is, and are adept in musicianship skills. "Perfect pitch" itself does not require any other musicianship skills or knowledge of intervals, etc., even though these may be also possessed by someone with "perfect pitch". Those with "perfect pitch" instantly recognise notes, or indeed any sounds to which a pitch might be attributed, without effort, and they usually have no idea how they do this.

"Perfect pitch" involves naming pitches in the context of a "pitch standard" on which there is a consensus. A piano tone today, with a lowest frequency ingredient (fundamental) of 440 Hz, would be named as an "A", rather than say, a "G" or and A-flat. This is a matter of "Man-made" (international) convention, rather than something that has its basis entirely in natural laws. At other times, in the past, different places had different pitch conventions, so that what was labelled an "A" would be recognised at, say, 415 Hz, or another frequency, rather than 440 Hz.

At some other times and places, 440 Hz may itself even have been regarded as being somewhere between two recognised pitches, for example, between G# and A. At 440 Hz a note might have been be considered "out of tune". The association of certain perceived pitches, or certain frequencies, with particular note names, always takes place in the context of cultural influences, and is not exclusively due to acoustical causes. There is no absolute "natural law" of acoustics or physics determining that 440 Hz, for example, rather than say, 444 Hz or 431 Hz, or some other value, should be "musically significant" or a "correct" frequency for a given musical note name.

Like our other senses, the sense of pitch, even "perfect pitch" sense, is affected by physiological and psychological influences. Students of piano tuning who have "perfect pitch", whom I have taught, have often experienced a "shift" of pitch recognition by up to a semitone, when physically unwell, or under psychological stress.

Phase

The last basic property of waves or vibratory motion to consider, is *phase*. Phase is closely related to the *cycle*. If we pulled two identical pendulums to the left and released the weights at the same time, the two pendulums would be said to be swinging *in phase* with each other. If we pulled one to the left and one to the right and then released them at the same time, they would be said to be swinging *out of phase* with each other. Fig. (3.9) shows two sine waves of different amplitudes that are *in phase* with each other – they both rise and fall at the same time. Fig. (3.10) shows two sine waves of different amplitudes that are *out of phase* with each other – they rise and fall at opposite times.

Fig. 3.9

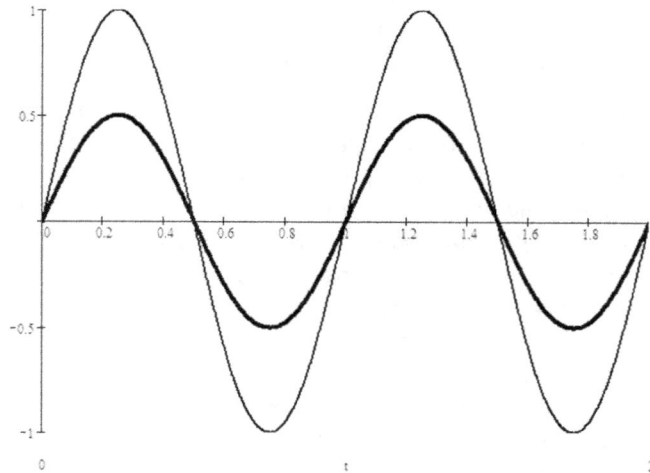

Fig. 3.9. Two waves that are *in phase*. Crests coincide with crests and troughs coincide with troughs.

Fig. 3.10

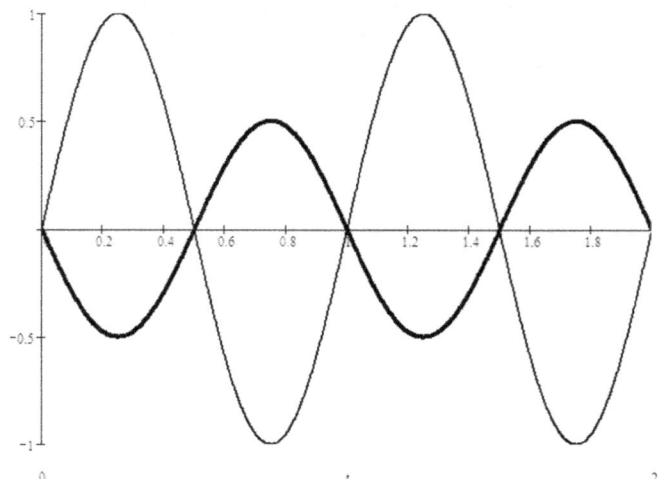

Fig. 3.10. Two waves that are *out of phase*. Crests of one wave coincide with troughs of the other, and *vice versa*.

Phase describes the relationship between two waves or vibrations, in terms of where each begins and ends a given *cycle* of change. In Fig. (3.9) the two waves begin and end their cycles in the same position along the horizontal time axis. If we wanted to choose to measure a cycle from one uppermost peak to the next, in a given waveform, then in Fig. (3.10), we can see that the positions of these peaks in the smaller, bold wave, is exactly half way (along the horizontal axis) between the peaks of the larger wave. These two waves are *out of phase*.

The *phase* of a periodic wave is actually expressed as an angle, in degrees or radians. The reason for this is because the relative positions of the waves are cyclic, that is, there is a limit to how far two waves can be out of position relative to each other, before they will be aligned again. This happens because the shape of a periodic wave is a repeating pattern. Angular measurement is suitable to describe this kind of relation, because angular measurement is also cyclic. Consider measuring the angle between two radii of a circle. As the angle increases, once we reach 360 degrees, we are back where we started with the positions of the two radii. Similarly, two periodic waves of the same frequency and wavelength, are in the same position relative to each other, whether they are at zero degrees phase angle, or at 360 degrees phase angle.

Two waves that have a relative phase angle of 180 degrees, however, are said to be exactly *out of phase* with each other. Fig. (9) illustrates this situation. The terms *exactly out of phase* and *180 degrees out of phase* and in *180 degree phase relation*, all mean the same thing. Two waves at 45 degrees phase angle to each other are said to be *in quadrature*. Fig. (3.11) illustrates this.

Fig. 3.11

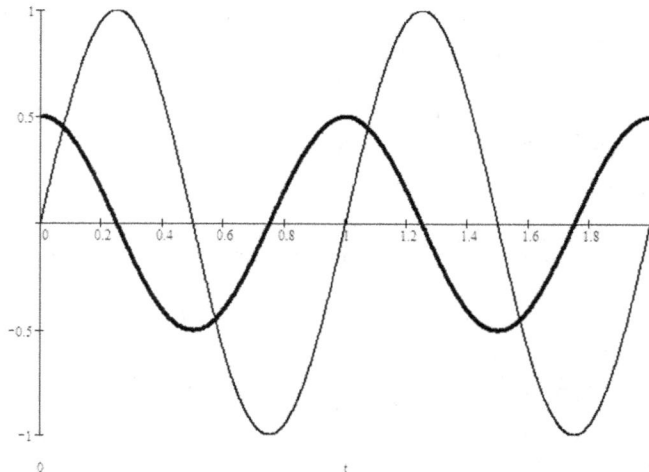

Fig. 3.11. Two waves that are *in quadrature*, or 45 degree phase relation.

A phase angle can still be quoted for a non periodic wave, or between two (or more) non periodic waves, if the wavelength of the wave remains the same throughout the period of measurement, and if the non periodicity is due only to changes in the wave's amplitude.

A sine wave graph crosses the origin of the graph axes, in other words, $y = 0$ when $x = 0$, as illustrated by the larger amplitude wave in Fig. (3.11). If we take a sine wave and give it a *phase shift* by moving it along the horizontal axis, as illustrated by the smaller amplitude wave in Fig. (3.11), then it still retains its essential sinusoidal shape, but no longer crosses the vertical axis with a value of zero. It is this essential sinusoidal shape, rather than its phase position, that determines whether or not it could represent a *pure tone*. If a sound wave plots as a sinusoid, it is a *pure tone*.

In general, the human ear is not directly sensitive to phase information in sound waves, but we can nevertheless hear the effects of phase in some circumstances. In particular, two waves that are in phase may "reinforce" one another, producing a stronger sound, whilst two waves that are out of phase may tend to cancel each other, producing a weaker or quieter sound. Phase effects of this kind are often audible when a sound can be heard as varying in loudness, according to the position of one's ears or head.

Mixing sinusoids to produce beats

Fig. (3.12) shows two sine waves with slightly different frequencies and wavelengths, superposed, with time in milliseconds (mS) along the horizontal axis. The waves are also of different amplitudes, to make them easier to distinguish. In practice, sine waves that become mixed in piano tuning, are seldom of equal amplitude.

Fig. 3.12

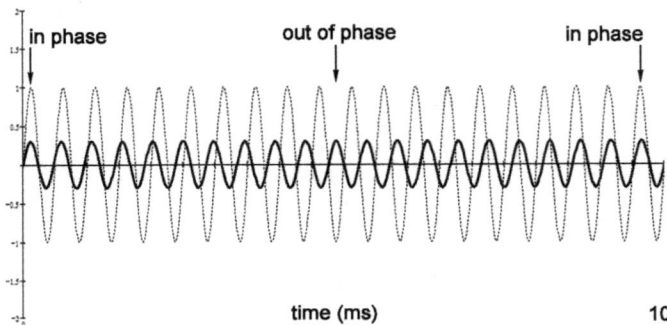

Fig. 3.12. Two waves that are of slightly different wavelengths (and hence frequencies). The phase relation alters cyclically.

The largest amplitude wave completes 21 cycles in 10 milliseconds, whilst the smaller amplitude wave completes only 20 cycles in the same time. As a result, whilst the two waves begin in phase, by 5 mS (half way) along the horizontal axis, they are exactly out pf phase, and then at 10 mS they are in phase again. Over a longer time they would constantly cycle in and out of phase.

If we add these two together, by adding their values on the vertical axis at each point on the time axis, we would obtain a new wave, representing the result of mixing the two waves. We will find that where the two original waves are *in phase*, the new wave has a larger amplitude, whilst where the original waves are *out of phase*, the new wave has a smaller amplitude. The new wave will be *amplitude modulated*. It will exhibit a *beat pattern*, as shown in Fig. (3.13).

Fig. 3.13

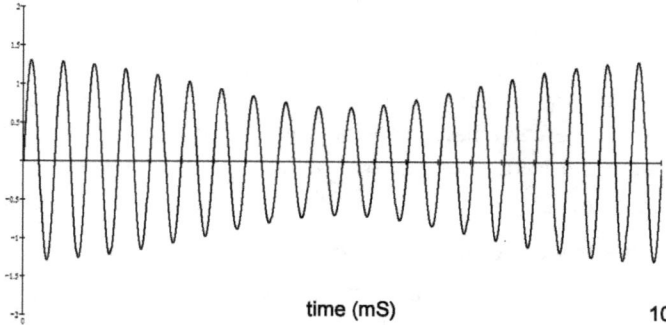

Fig. 3.13. The resulting waveform when the two waves in Fig. (3.12) are added. A new wave at the average of the two wavelengths (or frequencies) is formed, whose amplitude is modulated in a beat pattern. The beat frequency (number of beats per second) is equal to the difference between the two frequencies added.

Fig. (3.13) shows one complete cycle of a beat, from large amplitude, to small, and back to large again. In other words, there is one *beat* in 10 mS. Over a longer time, this would be 100 beats in one second, or in other words, there would be a *beat frequency* of 100 cycles per second.

The two original waves would have had 2100 and 2000 cycles, respectively, in one second. The difference between them is 100 cycles per second. The beat frequency is given by the difference between the two frequencies added together, make the new, beating wave.

The fact that mixed sinusoids can produce beat patterns is most important, and is used as an acoustical tool in piano tuning. In the actual sounds to which we would listen in piano tuning, we would not, however, be concerned with such large frequency differences, or such large beat frequencies. We listen to beat frequencies typically of between zero and a maximum of around, say, 25 cycles per second, or 25 Hz.

Mixed pure tones, beats and difference tones

Two close-frequency pure tones will mix to create a sinusoid whose *loudness* will fluctuate regularly up and down, in the *beat* pattern. As we have said, the frequency of the beats, or the number of beats per second, is equal to the frequency difference between the two pure tones. This is also called the beat frequency, or in piano tuning, the *beat rate*.

As the two pure tones become further apart in frequency, the beat rate increases, until eventually, when it has a frequency of around 20 – 30 Hz, it becomes an audible frequency in its own right, and a perceptible tone in its own right, called a *simple difference tone*. Difference tones are not so important in piano tuning. A simple difference tone is in any case not always perceivable, especially in piano tones. A more readily perceivable difference tone known as the *cubic difference tone*,[18] occurs at a frequency $2f_L - f_U$, where f_L is the frequency of the lower note and f_U the frequency of the upper. The cubic difference tone is generally much easier to hear, and is important in some areas of music, for example in recorder playing. Its pitch is a fourth below the lowest of two pure tones a major third apart, and because the tone of the recorder is close to being a pure tone, professional recorder players often use difference tones to adjust the intonation of major thirds between recorders playing in ensemble. Both simple and cubic difference tones are collectively known as *combination tones*.

We are not concerned with combination tones here, but are concerned with the effects of superposing or adding together pure tones much closer in frequency. At sufficiently small frequency differences, the pulsation or beating between two pure tones becomes slow enough to count. At these small differences of frequency, the two pure tones can no longer be distinguished as two simultaneous but separate pitches, but rather, appear as one tone at one pitch that is beating. Finally, at zero frequency difference, the two pure tones combine to become *one pure tone*, and beating vanishes.

We have said that pure tone ingredients of piano tones always decay. We have also said that mixing two close-frequency pure tones produces one new beating tone. Mixing two close-frequency, decaying pure tones, also produces a beating tone, but ones in which the beats themselves, as distinct from the tone in which they occur, may grow or decay.

Pure tones as essential ingredients of the sound recipe

The most important thing about pure tones is that they are elemental ingredients of sound. Just as we do not normally see all the individual colours of the rainbow in white light, even though these are essential

---

[18] The term 'cubic' is an historical misnomer – there are no cubes involved in the production of the cubic difference tone.

ingredients of white light, we do not normally hear the pure tones that are the ingredients of musical tones and other sounds, unless we are trained to do so. The seven colours of the rainbow are present in white light, and can be seen if the light is passed through a refracting medium like a prism or a raindrop, that splits up the light so that the inherent colours become obvious. In the case of musical sound, with considerable training the ear-brain system itself can be used like a "prism" to "split" sound into some of its component pure tones, which can then be heard at their respective pitches.

The situation for sound, however, is considerably more complicated than the example of light, which is in some ways only a poor analogy. To begin with, there are only seven colours of the rainbow, whilst there are an infinite number of possible pure tones that could be ingredients in a sound, and there are an infinite number of possible sounds, all with different *recipes* of ingredients. Furthermore, the pure tone ingredients of sound can change, and even appear and disappear, during the course of the sound. Some sounds are thus easier to "split" by ear into some of their component pure tones, than others, depending on the constancy and "orderliness" of the mixture.

The tone of a tuning fork, correctly struck, is almost exactly a pure tone, because the motion of the fork prongs vibrating in and out, is almost exactly *simple harmonic motion*. The sound wave emitted from the fork, if plotted it as a graph, would appear as a sinusoidal wave, or would be very close in appearance to a sine wave. In effect then, all sounds are made up of tones rather similar to the tones of tuning forks, except that the pitches may not be those encountered with tuning forks.

This may seem far-fetched on first thinking about it, and the reason is that there are also some other factors that we have not yet mentioned, which are also very important, in determining how we perceive a sound. Also, the possibilities for mixing pure tones, and the results this can produce, are not altogether what one might expect from experience of the tuning fork tone, alone. So what are these other factors?

Most sounds are not fixed and unchanging. Many musical tones have a beginning, a middle and an end, all of which may be different, and may be changing in themselves. In sound in general, pure tone ingredients can abruptly appear and disappear in time, and even in "musical" sounds where the pure tone mixture may be relatively steady, the pure tone ingredients often decay, or die away, or they can grow in time. When the mixture of pure tones is less orderly, or even chaotically or randomly changing, the overall sound tends towards what we might describe as a "noise" rather than a musical note or tone.

A single piano tone, for example, begins with a noise, called the *transient*, as the hammer hits the string, whose pure tones would be impossible to distinguish by ear. This is followed by a *decay tone* as the strings vibrate, whose pure tone ingredients are altogether more constant and orderly than those of the transient, and can often be individually heard by the trained ear. This is then usually followed by another "cut off" noise, as the damper comes back into contact with the strings and they are prevented from continuing their vibration.

For piano tuning purposes, it is the decay tone that is of most interest, and this is the part of the piano tone in which it is easiest to hear pure tones or partials. It is very difficult, or even impossible, to hear individual pure tone ingredients in many everyday noises like tearing paper or a boiling kettle, but relatively easy to hear them in other everyday noises like a copper pipe or spanner dropped on a hard floor, or the touching of wine glasses in a dinner party toast.

Pure tones are relatively easy to hear in bell tones, organ pipe tones, and with sufficient practise, in piano tones. However, in practice, if you hear what sounds like a pure tone in any of these sounds, it is also very likely that you are in fact listening to a close-frequency pair or group of two or more pure tones. If you are listening to something that sounds like the pure tone of a tuning fork, but it is beating or fluctuating in some other way, then it will actually be a mixture of two or more pure tones combined. Below, we will see that this distinct, audible component, is in piano tuning terminology, properly called a *partial*.

Pure tone frequency and perceived pitch

Musical *pitch* and *frequency*, are not the same thing, and nor is there any *absolute* relationship between them. Different individuals may hear pure tones at different pitches. What makes listeners able to agree reliably on the pitches of *musical* tones, is the fact that these tones are usually a complex recipe of many ingredients. They contain many different pure tone frequencies that combine to create the overall information the ear-brain system perceives as *pitch*.

Musical intervals are often said to be associated with *harmonic ratios*, which are simple, whole number ratios. For example, two frequencies in the ratio 2:1 are often supposed to be heard an octave apart. Similarly, the frequency ratio 3:2 is supposed to produce a perfect fifth. There is a whole set of simple whole number ratios, known as harmonic ratios, each associated with a musical interval. The ratios can be found (to various degrees of accuracy) in tensioned string lengths or (organ) pipe lengths tuned to musical intervals, and in the frequencies of electronically generated notes rich in pure tones. But applied to pure tones, harmonic frequency ratios can often produce fairly unpredictable results.

Suppose we hear a pure tone whose frequency is 261.5 Hz, which is normally perceived at about the pitch of middle C. Another pure tone twice this frequency, 523 Hz, might typically be perceived at an octave above the first tone. Similarly, a third tone, doubling the frequency again, would be at a frequency of 1046 Hz. Would this third tone be heard at an octave above the second tone, and hence two octaves above the first? Not necessarily. Some individuals, for example, will hear the 1st, 2nd and 3rd tones as each separated by an octave, but will hear the separation between the 1st and 3rd tones as considerably *less* than 2 octaves. It may also be that the 2nd tone is heard an octave above the 1st, but that the 3rd tone is heard less than an octave above the 2nd. The results may depend on the time of day, as hearing perception may be different in the morning to late at night.

Some individuals may even hear the same pure tone at a different pitch in each ear (known as binaural diplacusis). It turns out that pitch perception, even in one ear of a single individual, may vary with frequency, previous exposure to sound, and other factors, physiological and psychological, affecting the individual. There is, in other words, a *correlation* between the frequency and perceived pitch of pure tones, but this correlation cannot be specified as one fixed mathematical *function*. *Frequency* is a scientific measurement of *cycles per second*, but pitch is a psycho-acoustic descriptor, and not an objective scientific measurement. There is no scientific "unit" of pitch. One often sees "pitch" changes cited in *cents*, but this is just a "colloquial" convention, or an "indicator". The use of the "cent" as a measurement is dealt with later.

Many musicians unquestioningly think that musical intervals between pitches are "musical distances" or "spaces" between notes or tones, that are "out there" and are not affected by the listener. Scientifically speaking, this is a psychological misconception, as experiments with electronically generated pure tones can easily demonstrate. However, most musical tones from acoustic musical instruments are not pure tones. Rather, they consist of a large number of pure tones mixed together. The ear's response to musical tones that are rich in pure tones, is more involved than its response to pure tones, and there is usually very good agreement between musicians on the perceived sizes, or "intonation" of musical intervals.

In effect, musical sounds rich in pure tones provide more "information" to the ear and brain than a single pure tone. The general agreement amongst musicians on the pitch of a note, reinforces the notion that intervals between pitches are "objective" and not affected by the listener. The fact remains, however, that what exists "objectively" and can be precisely measured, is not pitch but *frequency*, or the relationship between frequencies, whilst "pitch" itself remains a psycho-acoustic response. The perception of *pitch* and *musical interval*, is not simple, but is complex. It is a psycho-acoustic response that is affected not just by a single *frequency*, or even a set of frequencies, but by other things also, including physiological, and psychological factors.

Audible *Partials*

In a cooking recipe, we could split up the recipe into ingredients including, say egg and milk. Egg and milk, however, are not really single substances. Egg contains albumen plus yolk, whilst milk can also contain cream. When sound is "split up" into its discrete audible ingredients, these ingredients can often be split up further. This is because sound can often be made from ingredients that are not necessarily, in themselves, the individual pure tones, or *simplest* ingredients. Rather, each ingredient may be a group of pure tones, pure tones being the final "ingredients" that cannot be split further.

When sound is "split up" into ingredients, the ingredients are called *partials*. Piano tone, whether from a single string, a couple of strings, or many strings, "splits up" first of all into well separated ingredients, or *partials*, but these partials are *not* necessarily individual pure tones. Often, a single partial may itself be "split up" further, into individual pure tones. Many partials in piano tone are (with the necessary training) *individually audible* within the overall sound. When a partial is made from more than one individual *pure tone*, then these pure tones are not, in general, individually audible within the overall sound. We hear the partial, rather than the individual pure tones.

Whilst piano tone partials themselves are well separated in frequency, the pure tone ingredients of a partial, if there is more than one, are always close in frequency. We saw that close-frequency pure tones can create a beat pattern, so an individual partial may have such a beat pattern. Piano tones, however, always decay, and decay, as we shall see later, can play an important part in affecting beat patterns.

Some partials *are* pure tones, because they only contain one pure tone. However, in terms of what is *audible* for piano tuning purposes, we always consider the piano tone as made up of audible *partials*, rather than being interested specifically in its pure tones. The number of possible audible *partials* present in a given piano tone or interval, is limited. Human hearing is only sensitive to frequencies between roughly 20 Hz and 30 kHz and any pure tones present outside this range would be inaudible. Also, the initial amplitudes of the higher frequency partials are generally very small, so they are more difficult to detect by ear.

Chapter 4 - Temperament Theory

What is a musical temperament?

A *temperament* is basically a "tuning configuration" for the tuning of a musical scale, that determines the "sizes" of the intervals. The reason it is significant for practical tuning in the first instance, is that it affects both the pitch intonations *and the tone qualities* of the intervals.

The *tone qualities of the intervals* are very important here. In a very rough tuning, without any strategy, there are many different small variations of pitch intonation between pairs of notes played sequentially, especially in perfect fifths and octaves, that we could "get away with" without too much of a shock to our musical expectations. However, in piano tones, many of these intonation choices would produce interval *tone qualities* when both notes sounded together, quite unacceptable to the musical ear. A horribly mistuned perfect fifth is not perceived as horribly mistuned purely on the basis of its questionable intonation. It also has *tone qualities* very different from the well tuned fifth, that we also find musically objectionable.

All the musical intervals made between the notes of a scale, are related to each other through the fact that they share the notes of the scale. The tuning of each individual note simultaneously affects the intonation and tone of every musical interval that other notes make with it. No note can have its tuning altered, without simultaneously affecting a number of other intervals. This means that no individual interval can have its tuning *changed*, without it affecting other intervals. In other words, all the intervals are *interconnected*, in a *network*.

Long before the concept of the *network* was developed, this interconnectivity between musical intervals was known about, and it was recognised that the network of intervals in Western music posed certain special problems for the tuning of musical intervals. The process of deliberately tuning a series of intervals how one wishes, inevitably causes other intervals not yet tuned, to become tuned in certain ways. For example, we might tune C up to E, as a major third, and C down to G as a perfect fourth. The tuning of the major sixth G up to E now becomes a consequence of how we tuned the major third and the perfect fourth.

The network of intervals behaves in such a way that the attempt to tune one set of intervals, easily causes some of the remaining intervals not yet tuned, to acquire dissonant or undesired tuning. If we attempt to re-tune the resulting intervals whose tuning we do not like, this then affects other intervals, and causes them to be "mistuned", instead. Trying to eliminate the appearance of unpleasant tuning somewhere in the network of intervals, becomes, in effect, a *puzzle*.

The Pythagoreans were aware of the network's puzzle-like features, although they did not have the powerful concept and science of *networks*, that we do today. They are not known to have produced any "solutions" for the network, but they were aware that musical intervals could be represented by ratios, and they highlighted a key feature of the network in a micro-interval now known as the *Pythagorean comma* (or *ditonic comma*).

Theorists following after the Pythagoreans approached the problem

in terms of the arithmetic of ratios, and geometry, and this approach developed into the *theory of temperaments*. Different theoretical *temperaments* are in effect different ways of "solving" the puzzle, or controlling the network in some particular, desired way, through the arithmetic or geometry of ratios.

A temperament in *practical tuning*, is also a "solution", a way of controlling *in which intervals* unpleasant tuning occurs or does not occur, or which intervals sound most pleasant. It can be a way of tuning the network of intervals so that unpleasant sounding intervals are avoided altogether.

Temperament theory versus tuning theory

Many people think of *temperament theory* and *tuning theory* as synonymous. In fact, although temperament theory is perfectly precise in its own terms, it does not tell you *how* to tune the temperaments it describes, *with precision*, on stringed keyboard instruments. All it does is to describe its temperaments in terms of ratios.

With a bit of initiative and practical understanding of how the network behaves, we can of course roughly map from the mathematical models to tuning practice, and for most early temperaments used in practical music, this was quite sufficient. Getting from theoretical temperaments to practical tempering on a stringed keyboard instrument, did not need to be a precision affair. And still today, it is not really a precision operation, unless *tuning theory* is employed.

Tuning theory deals with acoustical details that temperament theory does not deal with, details that are *audible* and can, with training, be used as part of tuning technique, in order to precision tune temperaments, in practice. However, when the acoustical science of tuning theory is good enough, then the tuning theory also starts to show up weaknesses and faults in temperament theory, when one tries to apply it in practice. Temperament theory is nonetheless still used, because it is still the easiest and fastest map to follow when "zoomed out" from the acoustical territory where we have to tune. Tuning theory takes the temperament map and adds practical directions to it. Once we start "zooming in", however, we need, in effect, to change the map.

Today, especially in the context of *tuning theory*, we therefore need to keep the concept of *temperament* in its proper context, and not get carried away, as many have, with the details of its apparent precisions.

Some practical basics

A musical interval played as two notes sounding together, has two musically important features that can be adjusted in the tuning process. One is the perceived relative musical pitches of the two notes, or *pitch intonation*. The other is the perceived *tone quality* or *timbre* of the interval.

In music we often speak of musical *consonance* and *dissonance* in musical intervals, the ideas of consonance and dissonance being rooted as much in music history and philosophy, as in acoustics. In tuning, however, we now need to speak of a consonance that is more firmly rooted in acoustics, which we can call *acoustical consonance*. We saw that a major feature of the tone quality of an interval, a feature affected by tuning, is the presence of "beating" or fluctuations in the partials of the soundscape. Optimum *acoustical consonance* for an interval can be defined as the tuning condition in which its soundscape contains a minimum amount of beating or fluctuations in the partials. This "minimum" is not a mathematical point on a graph, but is a general, practical description. It refers simply to what we can hear.

Accordingly, the idea that there is a "minimum" amount of beating does not necessarily mean there is *one* precise tuning condition at which beating or fluctuations in the partials is *zero*, throughout the soundscape. There may be a small *range of tunings* in which beating or fluctuation can be considered as in a "minimum" range. Of course, if we wanted to be scientifically rigorous, we would have to define what we mean by *minimum*, mathematically. For our purposes though, this is not necessary. It is simply worth remembering that for *any* interval, the "minimum" amount of beating achievable, both in terms of how many partials beat, and what the beat rates are, is not necessarily going to be absolutely "zero".

Having said what we mean by acoustical consonance (as distinct from musical consonance, which has different meanings in different contexts), we can then say what temperament and tempering is, *in practical terms*:

> Tuning a *temperament* involves *tempering* musical intervals. *Tempering* an interval means deliberately tuning it so that it is *not* tuned to optimum *acoustical consonance*, and may be tuned considerably away from it, but doing this in a controlled way.

Historically, major thirds were at certain times considered to sound most pleasant when tuned to acoustical consonance, whilst at other times, in certain circumstances, the have been regarded as most pleasing when *not* tuned to acoustical consonance. Today, the "normal" expectation for a major third's tuning in Western music, is that it will *not* be tuned to acoustical consonance, even though most musicians may not necessarily be aware of this.

The standard tuning for a scale in Western tonal music today, is *equal temperament*, a system in which the major thirds differ considerably from acoustically consonant thirds. Perfect fifths, however, have always been appreciated for their sonority when tuned to acoustical consonance. Nevertheless, perfect fifths are often *tempered* away from acoustical consonance when tuning a temperament, as indeed they are, in equal temperament, albeit by a very small amount.

If acoustical consonance sounds beautiful in perfect fifths, why would we want to *temper* the fifths? The reason is that we cannot create a scale in which *all* the fifths plus the octave are tuned to acoustical consonance, because the laws of acoustics prevent it. This is because *all the intervals* created by a musical scale are related to each other in the complex network we have already discussed. They cannot be "separated" from each other and considered only in isolation, because they must all "fit together" in making the musical scale. No one interval can be changed without affecting other intervals. The tuning of any one note will affect *all* the intervals of which that note is a member.

Unavoidably, no tuning arrangement of the notes can be made in which *all* the resulting intervals between them, are acoustically consonant. Intervals can be tuned to acoustical consonance, but only at the expense of other resulting intervals that must then be not consonant, or even sometimes acoustically very dissonant. In general, and roughly speaking, the more acoustically consonant a chosen group of intervals are made, the more acoustically dissonant another group must become. In the modern tuning scheme, equal temperament, *none* of the intervals except the unison and the octave are in fact tuned acoustically consonant.

As it happens, a similar situation arises with major thirds. If we did prefer acoustically consonant major thirds, then we would have a problem putting three major thirds side by side to make an octave, for example, in rising thirds, as C-E; E-G# (Ab); Ab −C. If we did so, the upper C of the octave would be horribly flat (by an amount called a *diesis*).

Temperament theory deals *theoretically* with how the intervals relate to each other, in terms of their degree of tempering. In temperament theory itself, the idea of acoustical consonance does not occur. In place of the idea of acoustical consonance, is the mathematical idea of an *harmonic ratio*, which is a whole number ratio, one for each interval. The *harmonic ratio* in temperament theory represents what in practice, would be the corresponding interval tuned to acoustical consonance.

An actual musical interval on a piano, cannot have its tone and intonation properly described by just a simple ratio, because it is a whole *recipe* of components, and therefore far more complex than this. Nevertheless, in temperament theory, each musical interval is ascribed one of these simple ratios. Most of the ratios are *derived* from a basic set of ratios corresponding to vibrating *string length ratios* that produce the basic musical intervals of the perfect fifth, the perfect fourth, and the octave. We will look at this in more detail a little later.

Piano tuning practice deals with how the intervals relate to each other in terms of their soundscapes and pitch intonations. Temperament theory provides a "baseline" guide to the nature of the complex network between intervals, but it actually works as a simplified arithmetic model. As such it is useful as a kind of quantified "rule of thumb" approach. We can understand from it a good deal about the tuning behaviour of the complex network of piano intervals. However, it is not an acoustical theory capable of fully describing the actual complex behaviour of piano tone in the fine tuning process.

Tempering is not the art of piano tuning

The thing to bear in mind about this, is that whilst temperament is *essential*, it is not the sole, most important part of piano tuning. The relationship between tone and intonation, which is very important in piano tuning, is not something temperament theory itself is equipped to fully address. A "temperament" is something that can be tuned by a tuner who does not see or understand that the tuning is not stable, and who has no mastery of tone and intonation. A temperament does not make a tuning, on the piano.

On the piano, "equal temperament" in practice means something *correspondent* with the idea of temperament in *temperament theory*, but it is not exactly the same thing. This is not due to any shortcoming in pianos or tuning practice. The reason is that whatever we do in piano tuning involves adjusting tone and intonation, and temperament theory, as we shall see, is not a theory that describes piano tone and intonation, except in a highly simplified and philosophically idealised way.

The approach we need

Temperament theory is a theory with a long history. It originates from observing the relationships between proportional divisions of the sounding length of a vibrating string, and the musical intervals thus produced. In its historically "pure" or original form, temperament theory is thus a theory of *the arithmetic of proportions*, or *the geometry of proportions*, originally discovered as string length proportions, and their relationships with musical intervals.

We do not carry out practical tuning by dividing string lengths. We tune by listening to the tone properties of musical intervals. In relation to piano tuning and its required precision, therefore, temperament theory unavoidably falls into the context of the acoustics of tone, rather than the lengths of strings. This chapter therefore avoids the direct examination of string length proportions found in the history of temperament, other than showing where the ideas originally came from. Alternatively, but more appropriately, it approaches the subject more from the point of view of the acoustical analysis of tone.

The psychological idea of musical interval "size"

In practical tuning we shall be concerned with listening to notes that sound simultaneously, so it will be most useful to always think of intervals in this way, rather than as notes separated in time. It seems intuitively straightforward to think of a musical "interval" between two notes played on a keyboard instrument, as an audible pitch separation or "distance" between the two notes (this is not to be confused with the physical distance between the two keys on the keyboard that produce the notes). If we think of this "distance" as designating the "size" of the interval, then the greater the distance, the larger the interval. This way of conceiving intervals fits in with our experience of notes and intervals from the point of view of their musical pitches. The further apart in musical pitch two notes are, the larger the interval between them.

We could describe this "distance" or "size" in a number of ways. We could, for example, state the number of steps in the musical scale, between the notes. We could also express the size of the interval by stating the frequency ratio between the fundamentals of the two notes. In the first description, where we state the number of scale steps for the interval's size, we could probably alter one of the notes very slightly through tuning, and not necessarily change the label we have used to describe the interval's size.

For example, we could alter the tuning slightly on one note of a major third, without necessarily having to rename the interval as minor third or a perfect fourth. Compared to fine tuning, we would actually have to re-tune one of the notes by a relatively *large* amount, in order to change the interval from sounding like 3 steps of the scale to sounding like 4 steps. Nonetheless, in this example, the actual "distance" or difference between the two notes can still be thought of as having changed.

In the more scientific description, where we attempt to state the size of the interval in terms of frequencies, we would reasonably expect that the size of the interval, quoted as a frequency ratio, would change as soon as *any* tuning change was made to one of the notes. (Later in this book we shall see that in fine piano tuning it may be possible to make tuning changes that do not necessarily affect certain frequencies). In this case, the greater the frequency ratio, the larger the interval.

In both examples, we have intuitively used the concept that *there is a single magnitude or size to a musical interval*. In the first instance, this intuitive idea arises from our experience of musical pitch. As a perception, *pitch* seems to be a *single quality* of a musical sound. A sound may seem to contain more than one pitch, but a musical pitch, in itself, seems to be a single thing. This is how it seems. We can perceive a simple pure tone as having a single musical pitch, but we can also perceive a complex tone with many component pure tones or partials, as having a single musical pitch. We need to appreciate that:

*The fact that we **perceive** something as a single, simple quality, like the "pitch difference" between two notes, or the "size" of the interval, does not necessarily indicate that the thing we are perceiving is itself simple, or a single entity.*

The idea of a specific, single "distance" between two notes of a musical interval, or of a specific, single "size" of an interval, are concepts that arise primarily from our psychological perception of pitch *as a single quality of a sound*. These simple ideas are useful in music and in thinking about tuning. However, applied to the practical fine tuning of musical intervals, it turns out that these concepts can be misleading, because acoustically, even single notes themselves are not really *single* things. Each note is a complex mixture of ingredients.

It is appropriate to think of a musical interval as a single "distance" between two notes, or as something with a simple "size", as far as musical *pitch* is concerned, but tuning itself involves more than just pitch judgement. It involves *tone* adjustments also. A musical interval also has *tone quality*, and a change to the tuning of one note will not just affect pitch, but will also affect the tone quality of the interval.

In fact, the whole business of tuning and temperament is only of any practical consequence to us in tuning, because the tuning of an interval affects the tone quality of the interval as well as the pitches of the notes. In terms of acoustical properties and their effects on tone quality, each note, and the interval itself *as the sound of both notes sounding together*, are very complex entities, not simple ones. It is just not possible to fully describe all the tone properties of an interval that tuning affects, as a simple, single quantity. A proper description would require a very large amount of acoustical data.

There are many things that cannot be fully described by a single, simple magnitude, but which can nevertheless be usefully described in such a way, by combining, or leaving out certain information. If we wanted to fully describe of the size of a box, we would actually need to know each of its *three* dimensions. We could nevertheless multiply the three dimensions together and get just one figure representing the "size" of the box, giving its volume or capacity. This might be useful in some circumstances.

In the case of musical intervals between most musical tones, and especially between piano tones, there are *many* quantities determining the audible and tuneable qualities of the interval. This is because the tone of each note and interval, and indeed the perceived pitch of each note, is affected by many partials, each with its own properties. Unlike the box, these cannot be reduced to just one quantity that adequately describes the tone quality of the interval, or what we will hear as we fine tune it. In other words, the acoustical properties of a musical interval that are important in the tuning process, cannot be reduced to just a single quantity that adequately encapsulates all of them. For the purposes of most actual tuning, and certainly in piano tuning, a musical interval, as the sound of the interval itself, is not something that can be scientifically *defined* as a single thing like a ratio.

If we want to start describing the sound of an interval, or how tuning affects the sound of an interval, in scientific detail, then we clearly cannot just ascribe a single ratio to the interval, calling this its "size". Nevertheless, temperament theory begins, in fact, by doing just this. Each musical interval is *defined*, before any further theory is developed, by a single, simple ratio. Further interval definitions, as new ratios, are then deduced from calculations using the first ratios. The reason for this is of course that temperament theory initially developed by considering the ratios of string lengths required to produce musical intervals, rather than with a knowledge of the acoustical structure of tone.

The idea of wide and narrow

One very useful thing to arise from the simple concept of interval "sizes" and pitch "distances" between notes, is the idea that an interval can be made *wider* or *narrower* by re-tuning one or both of the notes. If we raise the pitch of the upper note, or lower the pitch of the lower note, the interval becomes *wider*. If we lower the pitch of the upper note or raise the pitch of the lower note, then the interval becomes *narrower*. This too, like the concept of "size" or "distance" for intervals or notes, is a useful simplification. The idea of *narrowing* or *widening* turns out to be very useful tool in thinking about intervals and their tuning.

This remains so, even though under scrutiny, for the finest tuning of piano tones, its meaning can break down. We shall see later, that whilst we may intuitively think of an interval as necessarily being *either* wider *or* narrower than a given "size" - but "logically" not both - in some cases an interval on the piano can in fact be both wide and narrow at the same time, and can *behave* in this way in the tuning process. This is not so difficult to understand, once we appreciate that what we are perceiving as two single pitches or one interval "size", is in fact the result of a complex array of many components.

Imagine a single marble on a table top representing a perceived musical pitch. Two such marbles can be placed a distance apart, representing a musical "interval". There is just one distance between the marbles, or just one value describing the interval between them. The acoustical reality, however, would have to be represented by two *groups* of marbles, some distance apart. What affects the sound we hear when tuning, say, an octave, is in this model the specific distances between specific corresponding marbles in each group. Imagine now just three marbles in each group, arranged in a line, from left to right, the two groups separated also from left to right.

If the spacing of the marbles in each line is identical, and we know what it is, then we can quote just one distance – the shortest distance between the two groups – and we will have all the possible information about the distances between the marbles. The interval can then be *defined* by just one value – the distance between the groups. If, however, the spacing of the marbles in each group is different, and we do not know in advance what it is, then we cannot do this.

Any single distance measurement between the groups, by itself leaves out most of the information about where all the marbles are relative to each other. To represent piano intervals, we would need large numbers of marbles, differently spaced in each group. We could say that the two groups were approximately, say, 10 cm apart, but some corresponding marbles in the two groups could be slightly *more* than 10 cm apart, whilst others were slightly *less*. In terms of tuning behaviour, which depends on specific distances between specific corresponding marbles, the interval could be both wider and narrower than 10cm, at the same time.

Tuning historical temperaments

When an interval is wider than acoustical consonance, or in temperament theory wider than its harmonic ratio, it is simply said to be *wide*. When it is narrower than acoustical consonance, or in temperament theory narrower than its harmonic ratio, it is just said to be *narrow*.

Equal temperament, the modern standard, is the tuning system in which all the semitones are equal in "size". Having equally sized semitones, results in major thirds and sixths being wide, fourths being wide, minor thirds and sixths being narrow, and fifths being narrow. In equal temperament, the amount by which every interval of any species is narrowed or widened, is the same for all those intervals. Temperaments previously in use, or *historical temperaments*, have different arrangements, in which intervals of the same species may be tempered by differing amounts.

Piano tuners may sometimes be called upon to tune early keyboard instruments, for which temperaments other than equal temperament are often required. Once the underlying principles of temperament are understood, the practical application of equal temperament will be understood better, and other temperaments can also easily be tuned. With a little experience of early keyboard instrument tuning, tuning historical temperaments on early keyboard instruments will be found to be much easier than tuning precision equal temperament on a modern piano.

Harmonic ratios

Temperament theory begins (historically and theoretically) with the concept of *harmonic ratios*, which are simple numerical ratios associated with the basic musical intervals between musical string lengths, or between the lengths of musical pipes. Harmonic ratios have been known about at least since the time of Pythagoras, and were considered important for philosophical reasons long before temperament theory in its complete form was developed. The main examples of harmonic ratios and their associated musical intervals are illustrated in Fig. (4.1), in which each ratio is expressed here, as a matter of convention, by putting the largest number first.

Fig. 4.1

Interval	Harmonic ratio
Unison	1:1
Semitone	16:15
Minor whole tone	10:9
Major whole tone	9:8
Minor third	6:5
Major third	5:4
Perfect fourth	4:3
Augmented fourth	45:32
Diminished fifth	64:45
Perfect fifth	3:2
Minor sixth	8:5
Major sixth	5:3
Harmonic minor seventh	7:4
Grave minor seventh	16:9
Minor seventh	9:5
Major seventh	15:8
Octave	2:1

What are these ratios, and where do they come from? They are all found in the old *science of harmonics*, which we will come to shortly, but let's begin with some examples of their appearance in natural phenomena. We will consider here just three instances, but there are many more.

Firstly, if the speaking length of a uniform musical string is divided into one of the ratios, then the musical interval formed between the two parts of the divided string, will be the associated interval shown in the left hand column of the table, Fig. (4.2).

Fig. 4.2

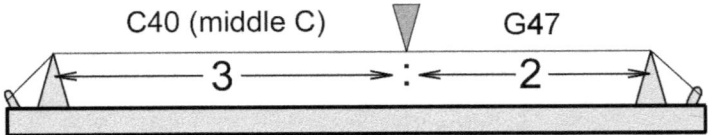

Fig. 4.2. Example of harmonic ratio division, for a perfect fifth.

In Fig. (4.2), the inverted triangle above the colon represents a stopping position on the tensioned string, for example, if there was a fingerboard and a fret at this position. The note G47 (G4) produced by the short part of the string on the right, is a perfect fifth above C40 (C4), produced by the long part of the string on the left. The ratio of the part lengths, is 3:2, the harmonic ratio for a perfect fifth. Dividing the string into any other harmonic ratio, would produce the associated musical interval between the two parts. In this example, the ratio of the full open string length to the portion on the left of colon, is 5:3. This is the harmonic ratio of the major sixth, so C40 will be a major sixth above the open string's note. The open string is therefore tuned to E-flat 31 (Eb3).

On a guitar or stringed instrument such as the violoncello or violin, stopping the string half way along its length (a ratio of 2:1, or on the 12th fret of a guitar) gives the note an octave above the open string. Early fretted instruments such as lutes and viols, whose frets were made from gut string tied around the neck of the instrument, were sometimes fretted using the semitone ratio of 16:15 to give the positions of the frets. A first fret could be tied at $1/16^{th}$ of the open string length, and the next fret placed at $1/16^{th}$ of the remaining length between the fret just tied, and the bridge, and so on. For these instruments, this approach was sufficient to produce a fretting for equal sized semitones, to practical accuracy.[19]

A second example is that two *mid-range* musical tones whose fundamental frequencies are related in one of the harmonic ratios, will (for most listeners) have pitches that appear to be separated by the associated musical interval. For example, a simple tone whose frequency is 262 Hz might be perceived as a middle C. Another simple tone whose

---

[19] On fretting and the use of equal temperament on renaissance instruments, see Lindley, Mark, *lutes, Viols & Temperaments*, CUP 1984.

frequency is in the ratio 3:2 to the first tone, will have a frequency of 393 Hz:

$$262 \times \frac{3}{2} = 393$$

The frequency 393 Hz of the second tone will generally be perceived a perfect fifth above the first tone, a pitch of G47.

It is very important to remember that it is the frequency *ratio* that is associated with the musical interval, and not the *difference* in frequencies between the two notes. Thus for two perfect fifths, one tuned above the other, we take the frequency of the lowest note and multiply by the 3/2 harmonic ratio to get the frequency of the middle note, a perfect fifth above the lower note. We then have to multiply *that* frequency by 3/2 again, to get the frequency of the upper note, two perfect fifths above the starting note. This is a "power law", and is not the same as simply multiplying the harmonic ratio by the number of intervals.

It is also worth remembering that the relationship between frequency ratios of pure tones, and perceived musical interval, is not necessarily the same for all frequency ranges, or for all listeners.

A third example is that organ pipes of the same diameter whose lengths are in one of the harmonic ratios, will in general produce notes separated by the associated musical interval.

Historical context

We shall shortly be seeing how partials of strings and pipes are arranged in a "musical" way, forming a pattern called the "chord of nature". We will also be looking at how this pattern is related to the harmonic ratios of temperament theory.

The use of the word *temperament* in the musical context can be traced back to 1496,[20] but investigations of temperaments were carried out by the ancient Greeks. The earliest reference to acoustical harmonics or the *chord of nature* appears to be in 1636, in Mersenne's *Harmonie*

---

[20] Gafurio, Franchino, *Practica musicae*, Milan, 1496, cited in Riemann, Hugo, *Geschichte der Musiktheorie*, Berlin, 1898, p. 327, cited in Murray Barbour, J, *Tuning and Temperament*, NY, 1972, p. 25.

Universelle,[21] where there is a clear but not entirely accurate description. Sauveur, however, has perhaps more often been cited as having discovered the *chord of* nature in 1701.[22]

Temperament theory originally developed through a theoretical knowledge of the ratios found in the old science called *harmonics*, whose inspiration is traceable back to Plato's *Timeaus*, and was still being mentioned as a science in its own right, in 18th century and some early 19th century texts.

In relation to musical instruments, *harmonics* was about the arithmetic and geometry of ratios associated with the division of musical string and pipe lengths, together with a practical knowledge of problems encountered in attempting to tune keyboard instruments. Geometry was sometimes considered more important than arithmetic, because there are certain ratios that can be represented geometrically, that cannot be perfectly written as a ratio of two whole numbers. For example, when we draw a circle, we can "see" the ratio, π, between the diameter and the circumference, but we cannot write it perfectly as a numerical ratio. We can approximate it to, say, 22/7, but we can never express it perfectly, because it is an *irrational number*. If we were to write it in decimal notation, we might write 3.1415926536 to 10 decimal places, but really, we need an infinite number of decimal places. The earliest theorists did not even have the benefit of decimal notation, which was not fully developed until the 17th century. They did know irrational numbers existed, but needed to express them geometrically.

Early practical instructions for musical temperaments were not necessarily based on arithmetic or geometry. They were often simply instructions to tune notes flat or sharp, using descriptions no more precise than "as flat as the ear can stand" or "a little sharp". The existence of these instructions recognises the practical need for temperament as distinct from its theoretical exposition. For our purposes, we do not need to study this arithmetic or geometry, but rather, it is convenient to think of harmonic ratios as *frequency ratios*, and to use this as a basis for the presentation of the theory.

---

[21] Mersenne, Marin, *Harmonie Universelle*, Paris, 1636-7, Tr Roger E Chapman, 1957, pp. 254-5; 263 ff.

[22] See, for example, Weiss and Taruskin, *Music in the Western World*, NY, 1984, p. 220.

Pythagoras and *the music of the spheres*

Temperaments and the inter-relationships of harmonic ratios were extensively explored by the ancient Greeks. Historically, however, it was not only musical string length ratios that made harmonic ratios important. Harmonic ratios once enjoyed *enormous philosophical status* due to the belief that they were part of the very structure and design of the universe.

The idea that the universe was Divinely constructed using harmonic ratios, is found in Plato's *Timeaus*, which following its later "rediscovery" in Europe, became a revered and highly influential book in the renaissance 'revival of learning'. The notion that harmonic ratios are divine in origin, can be traced back earlier than Plato to the *Pythagorean tradition* which influenced him. The Pythagoreans are supposed to have experimented with monochords, and Pythagoras is credited in the sources with having discovered that dividing a musical string length in the ratio 3:2, produces a perfect fifth between the two parts of the string. He is also said to have discovered the ratio 2:1 for the octave, and 4:3 for the perfect fourth.

It is worth taking some time to appreciate the profound magnitude of the context in which this occurs, and indeed in which temperament theory was developed. It would be wrong to think of Pythagoras as an acoustical scientist, and it would be wrong to think of the old *science of harmonics* or indeed temperament theory as scientific theories in the modern sense. Arguably, one of the main motivators of interest in the science of tuning musical instruments in the past, was not only the practical need for temperament, or interest in what we now call acoustics, but the already existing status of the *science of harmonics*, in whose context temperament science fell.

For a very long time, the tuning of musical intervals was not something seen as related only to practical music making, but as a lower order consequence of cosmic principles. The original Pythagoreans were a highly secret sect, but according to early Greek attestations it was Pythagoras who originated the long tradition now known variously as *the music of the spheres*, *the harmony of the spheres* or *the harmony of the worlds* (the idea often appears in Latin as *musica mundana* or *harmonia mundi*). In short, the ratios found in musical intervals were believed to be fundamental to the structure of the cosmos, and of the human soul, and it is these ideas that found their way into Plato's writings.

Plato's pupil Aristotle, writing later, said that the idea of *the harmony of the spheres* was an assertion by Pythagoras that the planets move in orbits (on crystal spheres) making a sound, and at speeds and distances that are in harmonic proportions. Aristotle himself disputed this idea, although what he was disputing turns out to be more his own interpretation of the tradition than the anything proven to be the Pythagorean philosophy itself. Nevertheless, the notion as portrayed by Aristotle was enormously influential, and the idea that the universe was physically constructed on harmonic ratios was developed further by the astronomer and mathematician Claudius Ptolemy (c.100 AD). Ptolemy's work dealt substantially with the mathematical inter-relationships of the "Divine" harmonic ratios, which in the context of musical proportion is at the root of temperament theory.

Ptolemy wrote a dense and detailed treatise linking this musical proportion and the structure of the universe, and was responsible for what became known as the *Ptolemaic universe*, a conceptualisation of the universe which placed the Earth at the centre. Ptolemy's universe was full of inter-relating harmonic proportions, and his writings on the structure of the cosmos are more than an equal in arithmetical ingeniousness for any temperament theorist's work. In Europe, Ptolemy's universe was widely accepted, and most importantly, it was endorsed and enforced by the Church, for around 1500 years. Only after Copernicus and Galileo did the ideas of the Ptolemaic universe eventually disintegrate.

The Ptolemaic model was progressively improved during the time of its influence, so that it reflected the increasingly detailed, observed motions of the planets better. Basically, this consisted of adding more and more circles or spheres to explain observed motion. But until the acceptance of the solar system with the Sun at the centre, the geocentric "design" of Ptolemy's "improving" universe, together with its harmonic ratios, dominated. (Copernicus and Galileo were by no means the first to suggest the sun was at the centre. Ironically, there are Greek attestations that the Pythagoreans themselves believed the *sun* to be at the centre of the universe).

Although there is no direct statement in Plato's *Timeaus* that the harmonic ratios in the cosmogony described there, are *musical*, the natural assumption, in the context of the tradition, was that they were. There was a strong line of influence from Pythagoras to Plato, and the Pythagorean tradition and attestations imply a direct connection between the ratios as discovered and taught by Pythagoras, and *musical* principles. Not only this, but another work, Plato's *Republic*, does contain a complete symbolic account of the *harmony of the spheres* tradition, without the mathematical quantification, in which musical harmony explicitly plays an important symbolic role in the structure of the universe, and in a cyclic system of life and death (reincarnation).

Boethius, who was largely responsible for the inclusion of the *science of music* with astronomy, arithmetic and geometry, in the medieval *quadrivium* of studies at Oxford, accepted the ratios. The aspect of music that was most important here, was not performance, but the so-called *science of harmonics* inherited from Ptolemy. It was an arithmetical discipline that continued to be respected and referred to in some sources, up to the nineteenth century, well after the so-called "scientific revolution" of the seventeenth century.

Thus, thanks to a very long tradition in academia of deference to the ancients, and not least to Plato or Aristotle, the harmonic ratios have enjoyed a status of enormous *philosophical* importance in the history of science and music. It should therefore come as no surprise that the mathematical part of temperament theory, which is also based on these ratios, has in the past enjoyed, by inference, a similar degree of tacit acceptance as being based on "divinely authoritative" sources.

Today, astronomers no longer use the harmonic ratios in the attempt to calculate the positions of the stars or planets, and no longer rely on the idea that that celestial bodies move in perfect circles. However, the harmonic ratios applied to the science of tuning, continue to work well, up to certain limits. It should always be remembered, however, that like the quantitative part of *the music of the spheres* tradition, temperament theory in its "traditional" form (and even in its contemporary, fully generalised form) remains an "idealised" theory, just like Ptolemy's universe. It is "idealised", because like the quantitative aspect of the *harmony of the spheres* tradition, it is based on idealised, perfect, harmonic ratios. Applied to intervals between actual tones, there is of course an even more important limitation: musical tones are, as we have seen, generally a *recipe* of ingredients, and that is something a simple, single ratio, cannot possibly describe.

The harmonic series

The harmonic ratios used in temperament theory themselves appear in certain phenomena in nature, where they are arranged in what is known as the *harmonic series*. To understand the place of the harmonic series and harmonic ratios in the context of modern acoustics, requires a little more explanation than the fact that they occur between string or pipe lengths. Even today, the series of proportions

$$\frac{1}{2}, \frac{1}{3}, \frac{1}{4}, \frac{1}{5}, \frac{1}{6} \ldots\ldots \text{etc}$$

is known in mathematics as the *harmonic series*. This series has special significance in relation to a single vibrating musical string, or to be more precise, to a hypothetical *ideal string*. An *ideal* string is one which is (hypothetically) perfectly flexible and perfectly uniform. Such a string, when vibrating, would theoretically produce not just one, but a whole *set* of vibrations whose frequencies are in the ratios 1 : 2 : 3 : 4 : 5 : 6 : 7... *etc*. These are called *harmonics* or *overtones* or *partials*. Real string behaviour can approximate to the behaviour of an ideal string, but not always. Sometimes a vibrating string is described as naturally, simultaneously dividing its speaking length into 2, 3, 4, 5, 6, 7... *etc*. equal length parts. We may also talk of *waves* on the string, whose wavelengths are in the ratios 1 : 1/2 : 1/3 : 1/4 : 1/5 : 1/6 : 1/7... etc.

However we approach this, the ratio set 1:2:3:4:5:6:7... etc., becomes important. Today, in the science of acoustics, a general method of approaching vibrations that produce musical tones, a method that applies not just to strings or pipes, is to analyse the vibration in terms of elemental modes of motion called *normal modes*. These are the specific, simplest kinds of vibrations that an object may be capable of producing. The normal modes of a musical string are a kind of vibratory motion called *standing waves*. Normal modes are important because they occur at specific frequencies, and because when they are all occurring together, they "add up" to the overall complicated motion of the object. In other words, a complicated vibratory motion can be considered as the addition or *superposition* of a number of simple kinds of motion, each motion being a normal mode.

Furthermore, if the overall complicated motion of the object keeps repeating itself exactly in time – a special type of motion called *periodic motion* – then it turns out that the normal modes will always have frequencies in the ratios 1 : 2 : 3 : 4 : 5 : 6 : 7... *etc*. This will be true for *any* vibrating object, not just strings or pipes, and will be true no matter how complicated the overall motion, as long as it is periodic. This is a *mathematical consequence* of periodic motion, that can be demonstrated in a mathematical process called *Fourier analysis*. The frequencies of the normal modes are then said to form an *harmonic series*, or a *Fourier series*. The vibrations of a hypothetical *ideal string* that is perfectly flexible and tensioned between rigid supports, would be periodic, and would contain normal modes or standing waves whose frequencies fell in the harmonic series. A real musical string on a musical instrument, approximates to an *ideal* string, and its normal modes will have frequencies falling *approximately* in the harmonic series.

Normal modes

Normal modes then, are motions that are the simplest kinds of vibratory motion that a "system" can exhibit. Any complicated motions of which the system is capable will then just be additions of these simpler, normal modes. In other words, the vibrating object or system, such as a string or a soundboard, cannot move in just any way. Any motion will be a mixture of motions of its *normal modes*, each one characterised by the fact that all parts of the system are moving with the same frequency. Each one of these types of motion, or *normal modes*, will be a special type of motion known as *simple harmonic motion*.

This is motion rather like the oscillation of a simple spring, or the swinging of a simple pendulum. Any other kind of more complicated motion that the system can exhibit then turns out to be a simultaneous combination of two or more of these elemental *normal modes*, in *simple harmonic motion*.[23] The particular frequencies of the normal modes are also sometimes called the *allowed frequencies* of the system.

In the case of a tensioned musical string between rigid supports, it can be mathematically shown that if the string is *ideal*, meaning that it is a hypothetically, perfectly flexible string, then its *allowed frequencies* will fall perfectly into a harmonic series. This means that if the lowest allowed normal mode frequency is f, the whole set of allowed frequencies will fall into a series

$$f, 2f, 3f, 4f, 5f...$$

which continues to infinity.

Frequencies that fall into this series are usually called *harmonic frequencies*, or simply *harmonics*. Remember that what causes the normal mode frequencies to be arranged in this way, is simply the fact that the overall vibratory motion is *periodic*, that is, it keeps repeating itself exactly, in time. Motion that is not periodic or not even close to it, is called *aperiodic*, and it will not analyse into *harmonic* mode frequencies. There

---

[23] Some systems may have normal modes that are not in perfect simple harmonic motion, but we need not be concerned with this.

are fundamental, purely mathematical reasons why this must be so. Real strings do not vibrate with perfectly periodic motion.

Harmonic tuning theory

Temperament theory is based on the idea of musical intervals being governed by harmonic ratios. The relationship between harmonic ratios and musical intervals is a strong one, at least for notes produced by strings and pipes, by the human voice, and in many other contexts. It is not, however, universal. In the case of piano tones, there is a reasonably strong relationship, and this has important consequence for the acoustical reality of musical intervals played on the piano. Harmonic tuning theory is a starting point for investigating and describing how musical intervals "fit together", and is, in effect, applied temperament theory. Because it puts temperament theory in a practical tuning context, harmonic tuning theory is a good way to approach temperament theory.

In *harmonic tuning theory*, the assertion of temperament theory that the basic musical intervals can be defined by temperament theory's harmonic ratios, is taken literally. The "text book" simplified mathematical model of a vibrating string (which assumes a perfectly flexible string with rigid supports) produces an arrangement for the frequencies of normal modes that fits perfectly with the assertions of temperament theory. The lowest frequency normal mode (with which the pitch of the note is identified) is then called *harmonic* number one, or the *fundamental*, and the other normal mode frequencies are then related in the harmonic ratios.

The *sound itself* heard as a result of each normal mode, is also called an audible *harmonic*. The rest of the audible *harmonics* or normal mode frequencies occur at particular musical intervals from each other, Fig. (4.3).

Fig. 4.3

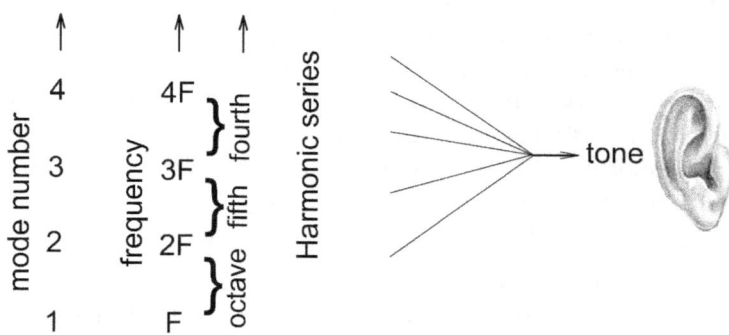

The first ten harmonics are called the *lower harmonics*. In this lower part of the harmonic series, all *adjacent* harmonics are separated by recognisable, common musical intervals, except that the 7th harmonic may sound a little too flat to the modern ear.

The 2nd harmonic is an octave above the 1st. The 3rd harmonic is a perfect fifth above the 2nd harmonic. The 4th harmonic is a perfect fourth above the 3rd harmonic, and so on. The series of intervals between the harmonics, up to the 10th harmonic is:

Octave

Perfect fifth

Perfect fourth

Major third

Minor third

Minor third

Whole tone

Whole tone

Whole tone

Putting all these facts together, what we are saying is that a hypothetical "ideal" musical string, freely vibrating, will produce, simultaneously, a whole set – a *harmonic series* – of pure tones or audible *harmonics*, each rather like the tone of a tuning fork, all nicely separated and arranged with musical intervals between them. The *frequency ratios* between these *harmonics* will be precisely those that temperament theory, through its own separate methods, also associates with the same musical intervals that appear between the audible harmonics. Temperament theory itself doesn't "know" anything about these audible *harmonics*, rather, it "discovered" the ratios by examining the sounding lengths of musical strings producing the intervals. The two approaches to the harmonic ratios and intervals are linked because of the physics of vibrating strings. If we don't look *too* closely into the details of the physics, the link between the two approaches remains perfectly "synchronised".

Looking at the chart Fig. (4.1), the harmonic ratios in the right hand column, give the numbers of the harmonics in the harmonic series, that are separated by the corresponding musical interval in the left hand column. For example, the harmonic ratio for a perfect fifth is 3:2. Looking at the table, this means that in a harmonic series of pure tones, the harmonic number 3 will be a perfect fifth above its harmonic number 2.

It should be remembered that whilst we can find musical intervals between string lengths divided in harmonic ratios, with reliable results, the relationship between pure tones and musical intervals is not always so reliable. The perceived musical interval between two pure tones may vary with pitch, from person to person, and even from ear to ear.

Demonstrating "harmonics" on the piano

Harmonics, or rather the piano string's approximation to them, can be demonstrated on the piano. Firstly lift the damper of a mid bass string, by getting someone to silently depress the key, or by wedging the damper. Then touch the string at a harmonic ratio proportion along its speaking length, with a fingertip. Typically, this is easiest to do on a bass string or bichord. The string should be touched at some whole number proportion of its speaking length – say, half way along, or two thirds along the string. Then strike the note, still touching the string. A "harmonic" (or "harmonics") associated with the harmonic proportion chosen, will ring out. In fact, this is not really a true harmonic, because the string is not an *ideal* string. It is an acoustical *partial*, which a discrete audible ingredient of the tone, that approximately correlates to the idea of the *harmonic*. The partial is still present in the tone of the string when no finger is touching the sounding string, but then all the other partials are present, and it is not so easy to distinguish it. Touching the string merely *stops other partials from sounding*, making it easier to hear that one.

Try touching the string half way along its speaking length, and experiment by moving the finger position until just the right spot is found that allows a strong, clear, bell-like tone to ring out from the partial. Then try to find the position in the same way, 1/3 of the speaking length from the bass bridge or top bridge (or agraffe). Do the same thin for 1/4 of the speaking length, and $1/5^{th}$, and so on. Each time the position is found, mark it with a piece of chalk, so it can be accurately found again quickly.

By striking the key each time, and touching the string at the marked points in order, the harmonic series of musical notes can be played. On most pianos it is easy to find the first 7 positions, but the 7th partial at $1/7^{th}$ the speaking length may be short-lived. Trying to sound the partial at $1/8^{th}$ of the speaking length, the 8th partial, may also be more difficult, because this partial is often deliberately suppressed by the design of the piano. When dividing the speaking length in this way, it does not matter whether the proportion is measured from the bass bridge or the top bridge (or agraffe) – the same partial will result.

The reason touching the string does not stop that particular partial from sounding, is because of the nature of the string's normal modes. In each normal mode the string's speaking length divides itself into smaller portions, and each portion of the string's speaking length vibrates. At the ends of each vibrating portion there is no movement relevant to that mode. Touching the string at that point, does not stop that particular normal mode vibrating, since that particular mode naturally produces no motion of its own, at that point anyway. This point of no motion is called a *node*, for that particular mode. The positions of the nodes along the string will be regular, but different for each normal mode, so as far as other modes are concerned, the finger *does* stop other modes vibrating if it is not at one of *their* nodes.

The chord of nature

The lower harmonics all sounding together, form a dominant seventh chord, but the seventh itself is a little flatter than modern ears would expect. The first *eleven* harmonics of a note C28 (C3) are shown as a chord in Fig. (4.4).

Fig. 4.4

Fig. 4.4. The first 11 harmonics of a note C

Adjacent higher harmonics are separated by smaller and smaller intervals, eventually forming a continuous semitone cluster, which higher in the soundscape's frequency range, becomes a microtone cluster, and so on.

"Coinciding harmonics"

A musical interval consists of two musical notes. As far as harmonic theory is concerned, we are interested in how the interval sounds when both notes are sounding simultaneously. Each note produces, according to the theory, a set of audible harmonics, whose frequencies are in the harmonic series (for the purposes of the theory, they are *perfectly* in the harmonic series). This means that when the two notes are tuned apart by an interval the same as one of the intervals found in the harmonic series itself, some of their respective harmonics will *coincide*, that is, their frequencies will be the same.

In other words, if the fundamental frequencies (the fundamental frequency is the frequency of harmonic number 1) of the two notes are in a harmonic ratio, then some of their respective harmonics will coincide in frequency. For example, if two notes are tuned apart by an octave, the fundamental frequency of the upper note will be twice the fundamental frequency of the lower note, because two frequencies an octave apart will be in the ratio 2:1, which "defines" the octave. The fundamental frequency of the upper note will then coincide with the second harmonic of the lower note, because the second harmonic of the lower note naturally has a frequency twice that of its fundamental, making it too, sound an octave above the fundamental. Both the second harmonic of the lower note, and the fundamental of the upper note, are an octave above the fundamental of the lower note, and both are at the same frequency.

In the case of a perfect fifth, where the two fundamentals are tuned with their frequencies in the ratio 3:2, the third harmonic of the lower note will have the same frequency as the second harmonic of the upper note, and both harmonics will sound at the same pitch.

In the harmonic series, the frequencies of the harmonics are in the ratios

$$1:2:3:4:5:6\ldots$$

and so on. Harmonics in the series are numbered, the lowest frequency harmonic (the fundamental) being harmonic number 1. Each harmonic thus has its own *harmonic number*. As we have seen, the frequency ratio between any two harmonics is the same as their two *harmonic numbers*, expressed as a ratio. So, for example, the frequency ratio between the 6th and 5th harmonics, will be 6:5. If the frequency of the 5th harmonic was 500 Hz, the frequency of the 6th harmonic would be 600 Hz. Similarly the frequency ratio between the 4th and 5th harmonics will be 4:5. As a matter of convention, when working with harmonic ratios, we will in practice always put the biggest number first, when writing the ratio.

The musical intervals between the first six harmonics, starting with the interval between the 1st and 2nd harmonic, are an octave, perfect fifth, perfect fourth, major third and minor third, as in Fig. (4.5).

Fig. 4.5

$$\underbrace{1:2}_{\text{octave}}:\underbrace{:3}_{\substack{\text{perf}\\\text{fifth}}}:\underbrace{:4}_{\substack{\text{perf}\\\text{fourth}}}:\underbrace{:5}_{\substack{\text{maj}\\\text{third}}}:\underbrace{:6}_{\substack{\text{min}\\\text{third}}}$$

We will be particularly concerned with the *interdependent relationship* between the major thirds and perfect fifths in a scale, for reasons that will become clearer later on. We can even think of a temperament as being *characterised* by the particular relationship between the major thirds and the perfect fifths in the musical scale that a particular temperament defines. Of course, other intervals are important too, but perfect fifths and major thirds are especially important in the historical context of temperaments.

To reiterate then: in *harmonic tuning theory*, when two notes are played together, each note produces its own set of harmonics, and if these two notes are tuned apart by one of the musical intervals that appear in a harmonic series itself, then some of the harmonics will coincide in frequency. The harmonics that coincide, for some of most important intervals in tuning, are shown in the next table:

Notes tuned apart by :	Frequency ratio of fundamentals (highest note to lowest note)	Lowest coinciding harmonics
Octave	2:1	2nd harmonic of lowest note and 1st harmonic (fundamental) of highest note
Perfect fifth	3:2	3rd harmonic of lowest note and 2nd harmonic of highest note
Perfect fourth	4:3	4th harmonic of lowest note and 3rd harmonic of highest note
Major third	5:4	5th harmonic of lowest note and 4th harmonic of highest note
Minor third	6:5	5th harmonic of lowest note and 4th harmonic of highest note
Major sixth	5:3	6th harmonic of lowest note and 3rd harmonic of highest note
Minor sixth	8:5	8th harmonic of lowest note and 5th harmonic of highest note

Study the table carefully. Note that there is an important pattern here. Firstly, the frequency ratio appearing here, between the fundamentals of the two notes, is the *harmonic ratio* for that interval. In the harmonic ratio, the biggest number is written first. This first number "represents" the fundamental frequency of the highest note, and the smallest number, written second, "represents" the lowest note's fundamental frequency. The numbers are not the *actual fundamental frequencies themselves*, but give the *ratio* of the frequencies. What counts is the *ratio* that the numbers express.

The harmonics that will coincide for a given interval, can be found simply by knowing the harmonic ratio for that interval. The first, biggest number of the ratio, will be the number of the harmonic from the *lowest* note, whilst the smallest number, written second in the ratio, will be the number of the harmonic from the *highest* note, that coincides. In the case of the perfect fifth, for example, the harmonic ratio is 3:2. We take the first, biggest number and "apply" it to the lowest note, so it is harmonic number 3 of the lowest note that will coincide. We take the second, smallest number, and "apply" it to the highest note, so it is harmonic number 2 of the highest note that coincides. Thus for a perfect fifth the 3rd harmonic of the lower note coincides with the 2nd harmonic of the higher note. For a major third, the 4th harmonic of the lower note coincides with the 3rd harmonic of the higher note.

For each interval, other coincidences also occur, above the lowest one. We can multiply the numbers in any ratio by any whole number, and get the same ratio, written as two other numbers. For example, the ratio for a perfect fifth is 3:2. multiplying each number by 2, this ratio can also be written as 6:4 or 9:6, and so on. These tell us where other coincidences of harmonics will occur. In this case of a perfect fifth, they will occur at the 6th harmonic of the lowest note with the 4th harmonic of the highest note, and also at the 9th harmonic of the lowest note with the 6th harmonic of the lowest, and so on.

Why "coinciding harmonics" are important

Coinciding harmonics are important because in tuning a musical interval, the fundamental frequencies of the two notes may not be tuned *exactly* in a harmonic ratio. If this is the case, then the coinciding harmonics will not *exactly* coincide in frequency. There will then occur, within the overall tone of the interval, pairs of frequencies that are close, but not exactly equal. As we saw in the chapter on sound, close but unequal frequencies are important because they can *superpose* (*i.e.* when added together) to produce a third frequency that "replaces" them, but also exhibits fluctuations known as *beats*. Fig. (4.6) is a computer generated oscillograph showing two harmonics at slightly different frequencies, and the beating waveform that results when the two are superposed, by adding them together.

Fig. 4.6

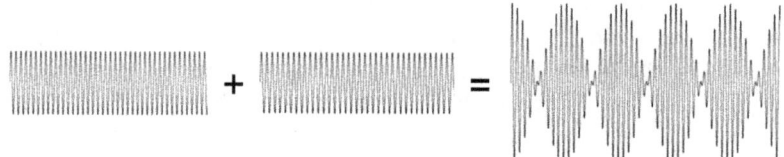

Fig. 4.6: The two waveforms on the left represent two harmonics of slightly different frequency, but constant, equal loudness. Close examination will show that this small difference in frequency appears as a small difference in their wavelengths, or a small difference in the distances that can be seen between the tips of each of the waveforms. The leftmost waveform or harmonic has tips that are closer together, representing a higher frequency than the second waveform or harmonic. The height of the waveforms represents how loud the harmonics are – the taller the wave, the louder would be the sound. On the right is the waveform that results when the two harmonics are added together. Its frequency will be the average of the two harmonics' frequencies, but its height fluctuates regularly in *beats*, from silence, to twice as high as the harmonics.

The overall tone of the interval that is sounding can be described in terms of its *spectrum*, which is the "collection" of all its harmonics. The *beating* that occurs within the spectrum of the interval when the fundamentals are not tuned in a perfect harmonic ratio, "disturbs" the perceived tone of the interval, as a form of fluctuation, agitation or movement. A slow beat rate is a little movement, and a fast beat rate is more movement. The rapidity of the beats depends on how far apart the two frequencies of the "coinciding" harmonics are. The *beat rate* measured in beats per second, will in the lowest coincidence of harmonics, equal the difference between the two "coinciding" harmonic frequencies.

Beat rate = difference in frequencies

If we were to start with a perfect fourth, for example, whose fundamental frequencies are tuned in the harmonic ratio 4:3, we would expect a zero, or practically zero beat rate. If we then raised the upper note, the beat rates in the coincidences would increase until eventually they caused so much disturbance to the tone of the interval, that it would sound musically unpleasant.

A large part of the "out of tune" quality of such an interval between strings is caused by the presence of a number of prominent beat rates in the soundscape of the interval, in the regions of coinciding harmonic frequencies. It is actually possible to take two fundamentals of certain musical tones and "mistune" them away from a harmonic ratio by a large amount, *without* this musically unpleasant tone quality appearing in the interval. This can happen provided there are either no other harmonics present in the tone's spectrum, or sometimes when the frequencies present in the spectrum of each tone do not fall in the harmonic series. This can be arranged through electronic pure tone generation, but can also occur in some instruments, such as some metallophones.

The tuned intervals on Javanese gamelan metallophones, for example, deviate considerably from harmonic ratios, without necessarily producing the kind of beating we would associate with such tuning on strings or pipes. The octave in gamelan music is divided into either seven or five unequal intervals that only very roughly correspond to Western intervals. Applied to the piano, these intervals would create very disturbed and "out of tune" sounding intervals. On the gamelan instruments, no such disturbance is audible, because of the lack of prominent close frequency pairs in the spectrum. Only the relative pitches themselves sound strange to the unaccustomed Western ear.

If we started to raise the upper note of a perfect fourth, so that the fundamentals of the notes were in a ratio larger than 4:3, eventually the beat rates would become sufficiently fast that the effect they have on the tone of the interval changes. The beats are so fast that they "blur" sufficiently and become *less* disturbing to the tone, but the pitches of the two notes no longer sound the correct distance apart for the interval. Soon we would reach a point where the interval had in fact become an augmented fourth rather than a perfect fourth. The harmonic ratio for the augmented fourth is 45:32.

This means the lowest coinciding harmonics are the 45th harmonic of the lowest note, with the 32nd harmonic of the upper note. In contrast, before, when we had tuned close to a perfect fourth, the harmonic numbers were 4 and 3. In other words, for the augmented fourth, coincidences only occur very high in the series, at higher frequencies. The perfect fourth has coinciding harmonics occurring much lower in the harmonic series.

If we continued to raise the upper note, we would eventually start to approach a perfect fifth between the two notes. The perfect fifth is a *consonant interval*, and its lowest coinciding harmonics occur low in the harmonic series, at harmonic numbers 3 and 2 (the harmonic ratio for the fifth is 3:2). As the coinciding harmonics approach each other in frequency, as we continue to tune up the upper note of the augmented fourth, the difference between them enters the range in which the beating becomes prominent. The interval stops sounding like an augmented fourth (or diminished fifth) and starts to sound like a narrow, "out of tune" perfect fifth. The tone of the interval contains much movement in the form of beating, in the regions where harmonic frequencies coincide but are not equal. As we continue the tuning still further, the beat rates eventually slow down, and the fifth becomes more and more "in tune" until the beat rates stop when the fundamental frequencies are in the ratio 3:2.

Just Intonation

In temperament theory each musical interval is ascribed a ratio, and the interval is said to be *Justly Intoned* when the ratio is the harmonic ratio for that interval. In practice, the interval is said to be Justly Intoned or *pure* when it is tuned to *acoustical consonance*, which is the condition of tuning for that interval, in which the beating is reduced to a minimum. In real piano intervals, the point at which beating is reduced to a "minimum" is not necessarily at one specific, fundamental frequency for each note. There may not be any tuning available at which all beating in the spectrum genuinely ceases.

This is because real piano tone is not made of harmonics arranged in an harmonic series, but is composed of acoustical *partials*, only approximately arranged in an harmonic series, and which differ from harmonics in other ways also. The fundamentals of the notes will be more or less in the associated harmonic ratio, but not necessarily perfectly so. An interval tuned to optimum acoustical consonance is nonetheless referred to by piano tuners as "beatless", or alternatively as "pure".

Just Intonation is a term also sometimes applied to describe a scale tuning in which each note of the scale is tuned so that it is justly intoned with the root note. It should be remembered that this "Just Intonation" tuning as it is called, is not a scale in which *all* the intervals formed between the notes of the scale are justly intoned. Justly intoning all the notes relative to the lowest scale note inevitably produces some intervals amongst the tuned notes themselves, that are *not* justly intoned, but rather, are tuned *very far from acoustical consonance*. A scale in which all intervals are justly intoned is impossible to achieve, both in terms of the arithmetic of temperament theory, and in acoustical practice, which is why temperament, or the practice of tuning intervals away from just intonation, becomes necessary. Temperament *controls* the unavoidable lack of acoustical consonance that must occur in at least some intervals of the scale, by determining which intervals it occurs in, and the degree to which it occurs in each interval.

Cents

When musical intervals are being defined as single frequency ratios, it is useful to measure musical interval size in *cents*, one cent being 1/100th of the size of an equally tempered semitone. An interval of 100 cents is an equally tempered semitone, and so there are 1200 cents in an octave. When working with intervals expressed as ratios, combining intervals requires multiplication of the ratios, whilst cent measurements can be simply added or subtracted. The relationship between a fundamental frequency ratio and the interval size expressed in cents is given by

$$C = 1200 \log_2 R$$

and

$$R = 2^{C/1200}$$

where C is the number of cents, and R is the frequency ratio.

Whilst cents are very useful for dealing with relative musical interval sizes, it should stressed that the cent is not a scientific, acoustical unit of pitch. Musical pitch itself is a subjective perception, not a scientifically measurable, physical quantity. The cent itself is not a unit of any physical quantity, but is a *dimensionless* unit used for comparing sizes of intervals that would otherwise be expressed as frequency ratios. It is simply an alternative way of measuring frequency ratio, logarithmically, so that the quantities involved can be added and subtracted, rather than multiplied and divided. This makes comparisons of interval sizes easier to think about.

The cent is, however, frequently *referred to* as a measurement of pitch, because of the intuitive *correlation* between *musical interval* "size" and *pitch*. The cent only measures interval *size*, or *change*, as a logarithmic representation of *frequency ratio*. There is no recognised SI unit of pitch.

The Pythagorean or ditonic comma

Temperament theory involves some very small ratios such as the *comma*, the *schisma*, and the *diesis*. These are all microtonal intervals. The term *microtonal* may give the impression that we are talking about musically negligible or insignificant differences. On the contrary, we only have to "mistune" a perfect fifth by a microtonal interval, for the fifth to sound horribly "out of tune", not least because of its altered, unpleasant tone properties. Any interval less than a semitone can be considered microtonal. The Pythagorean *comma* (so called because its discovery is attributed to Pythagoras) is an interval of 24 cents, or approximately ¼ of a semitone. The Pythagorean comma is particularly important in relation to equal temperament, the temperament used for tuning the modern piano. It can be discovered by studying the relationship between perfect fifths and octaves.

We begin by considering tuning a sequence of perfect fifths, as pure harmonic intervals, that is, by tuning their fundamental frequencies in the ratio 3:2. Starting with a bottom C, we would tune G above such that it's fundamental frequency was in the ratio 3:2 with that of the C. We can then continue to tune rising fifths in the same way. Using standard modern music theory nomenclature, we would end up tuning a rising sequence:

C – G – D – A – E – B – F# - C# - G# - D# - A# - E# - B#

By music theory rules, the note names have to be four steps apart in the alphabet, in order for each interval to qualify as a fifth. We might think that by the time we get to B#, we could just swap it for its enharmonic, C. In music theory, the swapping of enharmonics is a matter of convention, but in practice it presupposes some form of temperament is present. With harmonic tuning, the B# would not be the same as a C, that is, the two notes would not have the same fundamental frequencies. If, rather than tuning rising perfect fifths, we had tuned up seven octaves, we would have arrived at a top C, not a B#. This top C would actually differ from the B# by a small amount. The B# would be a slightly "higher" note than the C, the interval between the C and the B#, being the Pythagorean comma.

We have just compared an interval that is 12 perfect fifths wide, with one that is 7 octaves wide. The 12 perfect fifths are a wider interval than the 7 octaves, by the amount of the Pythagorean comma. The Pythagorean comma can thus be intuitively though of as the difference in size between an interval of 12 perfect fifths and an interval of 7 octaves.

If you have heard of "octave stretching" in piano tuning, it may be tempting to think the 7 octaves could be "stretched" to fit into the space of the 12 perfect fifths. In fact, "octave stretching" on pianos, if it is done, has nothing directly to do with the Pythagorean comma, and stretching the octaves will not solve the essential problem that the Pythagorean comma creates. If we were to "stretch" just one octave by the Pythagorean comma, it would sound grossly out of tune and would be musically unusable. For example, tuning the octave C40 – C52 wide by a Pythagorean comma would introduce numerous beat rates into the spectrum, the slowest of which would be around 7 beats per seconds. Above this would be another beat with a rate of 14 beats per second, and so on.

If, rather, we tuned each of the seven octaves (the Cs) wide by 1/7 of the Pythagorean comma, then the octave C40 – C52 would still have a minimum beat rate of around 3 beats per second, and faster beat rates above this in the spectrum, which would be wholly unacceptable. Tuning *all* the octaves wide, each by 1/7 Pythagorean comma (by tuning the remaining notes in widened octaves, starting from the notes in the sequence of fifths), would result in a gross tuning, or to put it another way, a bad mistuning. The octaves would all still be beating unpleasantly wide, and the major thirds (for example) would be so much wider than their harmonic tuning, that they too would have become unpleasant intervals somewhere between a recognisable major third and a perfect fourth.

Does a discrepancy of 24 cents really matter when it happens only over a large distance of 7 octaves? Unfortunately, the answer is yes, because it *doesn't* only happen over an interval that is 7 octaves wide. As Pythagoras is said to have discovered, a perfect fifth has another interval closely related to it, which in modern music theory we now call its *inversion*. This is the perfect fourth. If we tune two rising perfect fifths, and then down an octave, we arrive at the same note we would have arrived at had we just tuned *one* rising perfect fifth, and then tuned a *falling* perfect fourth. Similarly, we could have arrived at a B# immediately above the C we started on, not only by tuning down 7 octaves from the top C, but also by having instead tuned another sequence including some falling perfect fourths, rather than tuning all rising fifths. Fig. (4.7) illustrates this.

Fig. 4.7

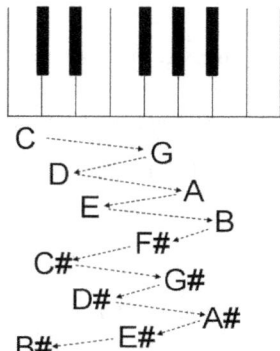

Fig. 4.7. A tuning sequence of rising perfect fifths and falling perfect fourths starting on C and ending on B# immediately above the starting C.

In the sequence Fig. (4.5) the C from which the sequence starts, is *retuned* at the end as B# a perfect fourth below the penultimate note E sharp. We then have a B sharp on the lowest white note, rather than a C. But now, because the B sharp is not the same note as the C, the interval between this lowest note and the G, is no longer a perfect fifth. Nominally, by the rules of music grammar it is a diminished minor sixth. Acoustically, it becomes a "wolf" interval – an interval that was a perfect fifth when we started, but has now become so "mistuned" that it "howls" out of tune.[24] The amount by which this interval differs from a harmonic perfect fifth, is again, the Pythagorean comma.

In terms of fundamental frequencies, this situation is dealt with using harmonic ratios. The harmonic ratio for the perfect fifth is 3:2, using the convention of writing the ratio with the largest number first. This means that if the fundamental frequency of the lowest note of a perfect fifth is f_L, then the fundamental frequency of the higher note will be

$$\frac{3}{2} \times f_L$$

---

[24] The actual origin of the term "wolf interval" is not known.

If the fundamental frequency of the highest note is f_H, then the fundamental frequency of the lower note will be

$$\frac{2}{3} \times f_H$$

The highest note of an interval must always have a higher fundamental frequency than the lowest note, so this tells us which way round to use the ratio. If the fundamental frequency of the starting C was f_C, then the fundamental frequency of the G a perfect fifth above would be

$$\frac{3}{2} \times f_C$$

The harmonic ratio for a perfect fourth is 4:3. Given the fundamental frequency of the G that we have just found, the fundamental frequency of the D, a perfect fourth below the G, must be

$$\frac{3}{4} \times \left(\frac{3}{2} \times f_C \right)$$

$$= \frac{9}{8} \times f_C$$

The D is a major whole tone above the C, and from the last result we can see the harmonic ratio for the whole tone is 9:8. If we were to work in this way through the entire tuning sequence Fig. (4.5), we could calculate that the ratio between the final B# and the starting note C, is

$$\frac{531441}{524288}$$

$$\approx 1.01364$$

This, again, is the harmonic ratio of the Pythagorean comma, an interval of about 24 cents, roughly a quarter of a semitone. It is the same number we calculated above for the intervals of 7 octaves 12 perfect fifths. It turns out that we would arrive at this number *whichever sequence* of perfect fifths and perfect fourths we tuned in order to get from the starting C to the last B#. Going up and down fifths and fourths working out ratios is quite tedious, and perhaps the easiest way to arrive at this ratio, is to use to our original sequence of 12 rising perfect fifths and 7 rising octaves. The fundamental frequency of each note following the first one in the sequence of perfect fifths, is 3/2 times the preceding fundamental frequency, and this happens 12 times in succession so the frequency of the top B# will be

$$f_C \times \left(\frac{3}{2}\right)^{12}$$

Similarly, the fundamental frequency of each note following the first one in the sequence of 7 octaves, will be 2 times the preceding fundamental frequency, so the frequency of the top C will be

$$f_C \times 2^7.$$

We can then write the frequency ratio for the interval between these two different top notes as

$$\frac{\left(\frac{3}{2}\right)^{12}}{2^7}$$

$$\approx 1.01364$$

The f_C s have disappeared because they cancel out, one being above and one being below the line. This is the same answer we had before, 1.01364, to 5 decimal places, equivalent to an interval of about 24 cents, or roughly a quarter of a semitone.

To sum up, the harmonic frequency ratio of the Pythagorean (ditonic) comma, is exactly 531441:524288, or approximately 74:73. This is the size of the interval, in Pythagorean tuning, between any two notes that in tonal music theory would be considered as enharmonic equivalents. In tuning *practice*, such enharmonics cannot be simply regarded as equivalent, and the Pythagorean comma *must* be taken into account. Without doing so, at worst, what should be a perfect fifth, may become a musically unusable wolf interval, "mistuned" by as much as a quarter of a semitone.

Thus we can think of the Pythagorean comma as the difference between 12 perfect fifths and 7 octaves, but these fifths and octaves do not necessarily have to be stacked up in one direction as all sequential "rising" intervals, over a wide distance of 7 octaves. We can actually use any tuning sequence of perfect fourths and/or perfect fifths, rising or falling. As long as we cover all the note names that we would have covered had we tuned a straightforward all rising or all falling sequence of perfect fourths or perfect fifths, then we will find that the starting note and the ending note differ by the Pythagorean comma, or by one or more octaves plus the Pythagorean comma.

On the keyboard, the last note will either need to be tuned on the same physical key we started with, by retuning that note, or the last note will end up being tuned on a key one or more octaves distance from the starting note. In the latter case, the musical interval formed between the starting note and the ending note, will not be an "in tune" octave or multiple octave, but will end up mistuned by a Pythagorean comma. In practice, it is the appearance of the comma within the intervals of a scale over just one octave's distance of the compass that is most significant.

The existence of the Pythagorean comma is one reason why it is impossible to "fit" all the intervals of a scale together so that they each are tuned according to their harmonic ratios, or to optimum acoustical consonance. There are also other commas and micro-interval differences between sequences of intervals, as we shall soon see. It is, however, that Pythagorean comma that is most directly addressed by the equal temperament tuning system that is standard on pianos today.

Equal Temperament

As was said above, "stretching" the octaves sufficiently to accommodate the Pythagorean comma is not an option. Rather, the comma *can* be addressed by making some or all of the fifths narrower than harmonic perfect fifths. Equal temperament, as used on pianos, is a tuning system in which all the perfect fifths are tuned narrower than their harmonic size, by an equal amount. The Pythagorean comma is thus particularly relevant to equal temperament. Since there are 12 perfect fifths to the 7 octaves, narrowing by an equal amount means that each and every perfect fifth is narrowed by 1/12 of a Pythagorean comma.

Since each perfect fifth is tuned narrower than its harmonic size, the coinciding harmonics in the interval will not *exactly* coincide in their frequencies, which will result in close-frequency pairs, and *beating*. If we wish to tune a perfect fourth instead of a perfect fifth, anywhere in the sequence, then because the fourth is an *inversion* of a fifth, the fourth must be tuned *wider* than its harmonic size, by 1/12 of a Pythagorean comma. Think of an octave made up from a rising perfect fifth and a rising perfect fourth, say, C1 up to G up to C2. There is a perfect fifth from C1 –

G and perfect fourth from G – C2. If we make the fifth smaller, or narrower, then because the octave stays the same size, the perfect fourth must at the same time get bigger, or wider.

Another way of conceiving equal temperament is as a tuning that divides the octave (ratio 2:1) into 12 equally sized semitones. When this is done, *none* of the resulting intervals in the scale (other than the octave itself) are at their harmonic ratios, or tuned to optimum acoustical consonance. The perfect fifths are all consequently tempered narrow by 1/12 of a Pythagorean comma.

The syntonic comma (*aka* the comma of Didymus)

The Pythagorean comma is not the only micro interval that appears in temperament theory. There are many others. As far as keyboard instrument tuning is concerned, we deal primarily with one other, the *syntonic comma*, or *comma of Didymus*. The Pythagorean comma concerned the relationship between perfect fifths and octaves. The syntonic comma concerns the relationship between perfect fifths, and the major third. Just as we said the Pythagorean comma could be thought of as the difference between 12 perfect fifths and 7 octaves (inverting intervals at will, if wish), so the syntonic comma can be thought of as the difference between two octaves plus a major third, and four perfect fifths.

Precisely why this particular comma is important will become clear as we proceed. Historically, it was theoretically more important than the Pythagorean comma, precisely because it addressed the relationship between major thirds and perfect fifths. There is in practical tuning, as a result of the comma, a conflict between the attempt to create harmonic (justly intoned) major thirds, and harmonic (justly intoned) perfect fifths. In the history of the development of Western music, this became an important issue as the use of harmony involving major thirds was introduced in addition to the use of perfect fifths, fourths, and octaves.

We will need to use the harmonic ratios for the octave, which is 2:1, the perfect fifth, which is 3:2, and the major third, which is 5:4. Starting with a fundamental frequency of f_L of the lowest note of the interval, the fundamental frequency of a note two octaves plus a major third above, will be

$$f_L \times 2 \times 2 \times \frac{5}{4}$$
$$= 5f_L$$

Here, each 2 multiplier is for one octave, and the 5/4 multiplier is for the major third at the top of the overall interval. Nominally, if we started on a C, we would have arrived at an E, two octaves and a major third above the C, at the top of the interval. If, rather, we ascended in a series of four perfect fifths, we would still arrive at an E. According to the rules of music theory then, there is no difference in how we ascend – both ascents reach the same top note, an E. But both ascents will not reach precisely the same top note, with the same fundamental frequency, if we tune each interval as an harmonic interval. The ascent in perfect fifths arrives at a top note that is sharper than the note at the top of the ascent in two octaves plus a major third. Arithmetically, the ascent in perfect fifths will be:

$$f_L \times \left(\frac{3}{2}\right)^4$$
$$= 5.0625 f_L$$

The harmonic ratio of the microtone interval between these two different notes will be

$$\frac{5.0625}{5} \approx 1.0125$$

This is the ratio of the syntonic comma. Written as an harmonic ratio, the syntonic comma is exactly 81:80. It is very close in value to the Pythagorean comma. Whilst the Pythagorean comma is about 24 cents, the syntonic comma is slightly smaller at about 22 cents. Just as we could tune and invert perfect fifths in any order over a sequence of 12 intervals, and still encounter the Pythagorean comma, so we can also tune or invert fifths over a sequence of 4 intervals, in order to make a major third, and encounter the syntonic comma. One way of doing this is illustrated in Fig. (4.8)

Fig. 4.8

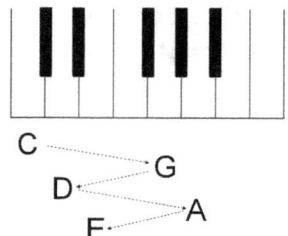

Fig. 4.8. Tuning sequence in perfect fifths and fourths (inverted fifths), from C through to E, ending in the creation of a major third C to E.

Here, if each perfect fourth or fifth is tuned in its harmonic ratio, then the resulting C to E major third formed at the end of the sequence, will be wider the its harmonic ratio by the syntonic comma. A major third so tuned, is known as a *Pythagorean third*.

A Pythagorean third is much wider than an harmonic or justly intoned one. In other words, it is tuned a long way away from acoustical consonance. It still sounds more like a major third than a perfect fourth, but on stringed instruments or organs it is very rough and coarse sounding in its tone quality, due to the presence of many fast beat rates in the interval's spectrum. In the middle ages and early renaissance, there was clearly a preference for the sound of the justly intoned major third, over any other way of tuning the interval.

There were also philosophical reasons for wishing to tune intervals according harmonic ratios, as in these times the connection between religion, philosophy influenced by Pythagoreanism, and music, was potent. The equally tempered third tuned on modern pianos is considerably wider than a justly intoned one (by 2/3 of a Pythagorean comma, as we shall see), and today this is what we expect major thirds to sound like. The Pythagorean third is wider than a justly intoned one by half as much again, so even to our ears, it sounds wide and coarse.

The Temperament Circle

In Western music we are primarily interested in scales that divide the octave into twelve "semitones". We do not wish, as far as temperament is concerned, to deviate from the harmonic ratio in the case of the octave. In temperament theory the octave ratio of 2:1 is always preserved. The practical reason for this, as already mentioned, is that mistuning the octave by the amounts normally encountered in tempering intervals, produces very unpleasant effects on the tone of the octave. Deviating the octave tuning away from acoustical consonance by even the smallest fractions of a comma used in tempering other intervals, produces an unacceptable octave. In the octave interval, half of all the partials present in the spectrum are affected by such a deviation, and will produce beating.

The sensitivity of the octave's soundscape to deviations from acoustical consonance is surpassed only by the soundscape of the unison. (The changes to both soundscapes made in fine tuning by master piano tuners, is well inside the range of changes normally made by tempering, and the reasons for such changes are beyond temperament theory).

Any of the other intervals (excluding the unison) may or may not be tuned away from their harmonic ratio, or *tempered*, depending on the temperament being tuned. However, we cannot change the tuning of any interval in the scale, in isolation from other intervals. The intervals of the scale *interconnect*, and are interdependent, because intervals share notes. Altering the tuning of any one note will alter all the intervals of which that note is a member. We need a generic and concise way of representing how the sizes of intervals in the scale are connected and affect each other, without being bound to any fixed sequence of tuning the notes in the scale. The solution is the *Temperament Circle*. The so-called *Great Circle of Fifths* of music theory is a simplified derivative of the Temperament Circle, that simply ignores the problem of temperament.

Let's begin by considering the Pythagorean comma, created by tuning a sequence of 12 perfect fifths. We can represent these fifths, as in the music theory Great Circle of Fifths, using a circle, Fig. (4.9).

Fig. 4.9

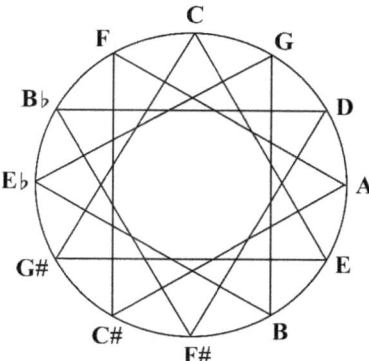

Fig. 4.9. A *temperament Circle*.

In this example, we circulate clockwise from C to G#, and anticlockwise to E-flat. There is no compulsion to do this other than as a matter of convention. We must circulate in two directions, or we stray into enharmonics of otherwise natural notes. If, for example, we circulated clockwise all the way round, we would end up with an E# next to the C, rather than an F. Anticlockwise we would reach C-flat instead of B, and end up with an A double flat next to the C. Choosing to place the meeting of clockwise and anticlockwise circulation at G# to E-flat, is, in the context of the history of music and the development of tonal music theory, one of two default choices. The other choice is to circulate one step further anticlockwise, and have an A-flat rather than the C#.

Starting from the C at the top, the Circle represents a sequence of rising perfect fifths in the clockwise direction, as far as G-sharp, and falling perfect fifths in the anticlockwise direction, as far as E-flat. Alternatively, it could represent falling perfect fourths in the clockwise direction as far as G-sharp, and rising perfect fourths in the anticlockwise direction, as far as the E-flat.

The beauty of the Circle representation is that it does not matter which way round its representation is interpreted. It simply represents the relationship of all the notes. In other words, we can go from C to G to A to E *etc.* on the keyboard, in any direction, up or down, as long as we follow the sequence on the Circle, and the Circle will be representing the sequence of notes we are following. Here, if 11 of the 12 intervals round the Circle are tuned to their harmonic ratio, travelling clockwise from C to G#, and anticlockwise from C to E-flat, then the Pythagorean comma must appear in the 12th interval between E-flat and G-sharp, which will be narrower than the harmonic ratio by the Pythagorean comma.

An inherent "rule" of the Circle therefore, is that it is not possible to have all the intervals round the Circle tuned to their harmonic ratio. If we represent harmonically tuned intervals with a zero, then we would represent the situation as in Fig. (4.10).

Fig. 4.10

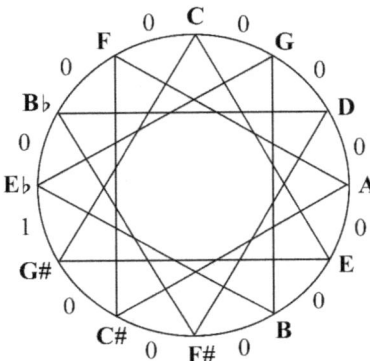

Fig. 4.10. Pythagorean tuning, in which 11 of the fifths are tuned pure, leaving a wolf interval narrow by one Pythagorean comma.

The numbers round the Circle represent the amount, in commas, by which the interval over which it appears is tempered. Note that the number 1, standing for 1 Pythagorean comma, appears in this example between E-flat and G-sharp. The zeros in the other intervals represent zero deviation from harmonic tuning, or *zero tempering*. These zero intervals, in other words, are justly intoned, or tuned "pure". The 1 between E-flat and G-sharp simply indicates this interval is narrowed from its harmonic size by 1 comma.

The convention chosen here is that a positive number means the interval will be narrowed if it is a fifth (or diminished minor sixth in the case of G-sharp rising to E-flat) or widened if it is a fourth (or an augmented major third in the case of E-flat rising to G-sharp). In most temperaments, fifths are tempered narrow if they are not tuned harmonically pure, whilst fourths are tempered wide if they are not tuned harmonically pure. There are some temperaments (notably from the French baroque) where the reverse is true, and in this case the figure would appear as negative. These are, however, the exception to the rule.

Arithmetically, the "rule" of the Circle (under this convention) is that the sum of all the figures around the Circle must remain +1, if the octaves used in the tuning are to remain harmonic (in the ratio 2:1). In principle, we are free to put whatever figures we like around the Circle, provided they add up to 1, and then the Circle will faithfully represent a tuning with "in tune" octaves that is achievable in practice. The mathematics of harmonic ratios determines that any other set of figures, that does not sum to 1, will not be possible in tuning practice, unless the octave is compromised.

Pythagorean tuning

The tuning represented in Fig. (4.10) is not, strictly speaking, a *temperament*, because a temperament is a tuning system in which some or all of the intervals are *deliberately tempered* by a specific amount. The tuning in Fig. (4.8) arises by *not* tempering 11 intervals, and then the 12th interval just ends up, unavoidably 1 comma narrow. A temperament, properly speaking, requires some intervals to be deliberately tempered in order to complete the Circle.

Fig. (4.10) in fact represents a *Pythagorean tuning*. Any tuning in which 11 of the 12 intervals round the Circle are not tempered, will be "Pythagorean". On the keyboard, Fig. (4.10) represents a tuning in which any interval that is represented on the Circle with a zero, will sound good, because it has a harmonic tuning. But in this tuning, any interval G-sharp to E-flat, or E-flat to G-sharp, on the keyboard, will sound bad, because it will be either be narrower or wider than harmonic, by 1 complete comma. A rising E-flat to G-sharp will be wider than harmonic by one comma, and a rising G-sharp to E-flat will be narrower. The physical keys used for these notes, will be ones between which we would normally expect to find a perfect fourth or fifth, on the piano.

Combining the Pythagorean and syntonic commas

The Temperament Circle is a theoretical, *heuristic model* or "rule of thumb" device for working out temperaments. It provides an easy to see representation of the nature of any temperament. As such, when using it as a "template" for practical tuning purposes, the theoretical small difference between the Pythagorean and syntonic commas is ignored, and they are replaced by one hypothetical single comma, simply called "the comma". The micro-interval interval between the syntonic comma and the Pythagorean comma is known as the *schisma*.

Interpreting the Temperament Circle

You might have wondered about the significance of the star pattern of straight lines inside the Circle. The circumference of the Circle gives information about the relative sizes of fifths and their inversion, the fourths. It shows how the *Pythagorean* comma is distributed. The set of lines making the star pattern inside gives information about the sizes of the major thirds, and their inversion the minor sixths. The division of the *Syntonic* comma for a major third, that is, the amount by which the third is tempered, is found by using the tempering values on the arc over its line. For practical use, the difference between the commas is negligible, so the figures on the circumference serve for both the fifths segments and the major thirds lines. The Circle therefore shows the size relationships between the fifths and the major thirds.

Eight out of the twelve of these lines connect notes that are a major third apart counting clockwise round the Circle (anticlockwise they are minor sixths). Each line subtends an arc of the Circle covered by 4 intervals round the circumference of the Circle. The four lines whose arc includes the interval G-sharp to E-flat, connect notes that are according to music theory, nominally a diminished fourth apart, rather than a major third. As far as temperament theory is concerned, *all* the lines can be arbitrarily considered "major third" lines, including these.

The major third lines connect notes defining "major thirds" in the temperament, and the size of the "major third" will be found from the sizes of the 4 fifths (or fourths) in the arc over the "major third" line. The "normal" arrangement in an "efficient" temperament, is that deliberately tempered fifths (as distinct from untuned "wolf" intervals) are narrow, whilst the resulting tempered major thirds are wide.

Remember that the Circumference value must add up to 1. Positive numbers on the circumference indicate by how much each fifth is narrowed. A "1" indicates a fifth narrowed by 1 comma. This would be a narrow "wolf" interval. Narrowing (or widening) a fifth by much more than ¼ comma would begin to make it musically unpleasant, and tempering it more will eventually turn it into a "wolf" interval. If a number on a circumference segment appears negative, then the "fifth" (or whatever interval it would be correct to call it) will be wide.

To find the amount by which a major third is tempered, the 4 fifths over the arc of a "major third line" must have their values added, and then subtracted from 1, and if the answer is still positive, this will give the amount by which the "third" will be wide. If some of the fifths happened to be narrow, rather than wide, then the remaining value after subtraction would be negative, indicating a *narrow* "third", but this is the exception to the general practical rule.

Clearly, there is a relationship between the tempering in the fifths and the tempering in the thirds. For the tempering over a third line to be zero, giving a harmonically pure, or justly intoned major third, the four fifths over the arc must have a total tempering of 1. We could achieve this by narrowing all four fifths by ¼ comma. This happens in ¼ comma meantone, dealt with below. The motive for doing this, is precisely to produce harmonic major thirds at the expense of the fifths (and of course their inversions, the fourths). A pure third will of course also occur if just one of the 4 fifths is narrowed by a full 1 comma, *i.e.*, if it is a narrow "wolf".

If a temperament contains a wolf interval, then the 4 thirds that fall across the wolf will be affected by it. If the wolf is a narrowed fifth, then the 4 thirds across it will generally be either pure, or much narrower than the remaining 8 thirds. This is the case in Pythagorean tuning, where 4 thirds are practically pure, whilst the other 8 are extremely wide 'Pythagorean thirds'. If the wolf is a widened fifth, then the situation is reversed. The 4 thirds falling across the wolf will then be much wider then the remaining 8 thirds.

Pythagorean tuning, in effect, has 11 useable fifths and one musically unusable, narrow wolf interval. It has 4 practically pure major thirds, and 8 unpleasantly wide ones. It is the existence of the wolf interval, that creates the four major thirds, whose lines fall across it. With only four major thirds considered musically acceptable, Pythagorean tuning limits the development of the use of major thirds in harmony. Historically, the use of *meantone* temperament, which we shall come to shortly, allowed more major thirds to be used in harmony, but it did so by sacrificing the purity of the fifths.

Equal Temperament

The unpleasant interval E-flat to G-sharp on the Circle is a "wolf" interval. One way to eliminate wolf intervals from the Circle is to use Equal Temperament in which the comma that the Circle must always contain is divided equally amongst all the intervals round the Circle, all 12 fifths (or fourths) being narrowed by $1/12^{th}$ of a comma. Equal Temperament is represented in Fig. (4.11).

Fig. 4.11

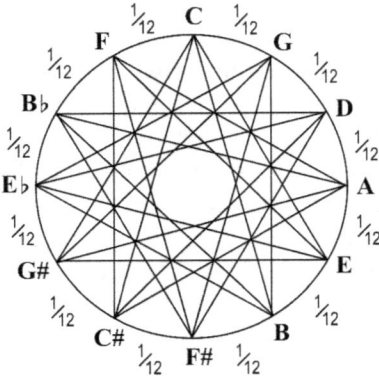

Fig. 4.11. Equal temperament.

Now all the intervals round the Circle are equal in size. Clearly, the choice of enharmonic note names – whether we choose to write G-sharp or A-flat, for example – has now become merely academic. We could just as well have called the G-sharp an A-flat, and it would be acoustically the same note. This nominal freedom of choice, however, is only true for Equal Temperament. In other temperaments, which are *unequal*, distinguishing between enharmonic note names can be very important because it may indicate an actual acoustical difference of note.

We can see from the Circle for equal temperament that the four circumference segments (fifths) over any third line add up to 1/3 comma. Using the temperament Circle rules, the equally tempered major thirds are therefore all widened by 2/3 comma.

We can also see that all the semitones will be the same size, because the 7 fifths on every arc over every *semitone line* (connecting two notes separated by a semitone or its inversion) are the same size. However, we cannot deduce the actual size of the semitone from this. The numbers currently on the circumference are divisions of the hypothetical "comma" which is an approximated syntonic/Pythagorean comma. This is only suited to indicating the sizes of the perfect fifths and major thirds. In the case of semitones, the relevant micro-interval is not similar to either the Pythagorean comma or the syntonic comma. It is the interval that occurs between 7 perfect fifths and 4 octaves plus a semitone, which is known as a *schisma*. This schisma is much smaller than a comma, being approximately 1.954 cents, which is about 1/11th the size of the syntonic comma, or 1/12th the size of the Pythagorean comma. Its harmonic ratio is 32805:32768.

Knowing the size of the schisma or of the other commas does not necessarily allow a perfect ratios to be found for all intervals. We can show that certain intervals can exist, without necessarily being able to find perfect ratios for them. The equally tempered semitone is an example. In equal temperament the octave is divided into 12 equally sized semitones, as a consequence of tempering all the fifths, each by 1/12th comma. The resulting semitone size cannot be exactly deduced as an arithmetic harmonic ratio. The size of the semitone as a string length ratio or frequency ratio is an irrational number, inexpressible exactly as a whole number ratio. The octave consists of 12 equally tempered semitones, and the octave's harmonic ratio is 2/1, so the size of the semitone can be calculated as the twelfth root of 2,

$$2^{1/12}.$$

This is approximately 1.0594631.

The complete Temperament Circle

The principle of using the circumference arc segments to suggest the effect of a given arrangement of tempered fifths on the other intervals in the scale, can be extended by drawing other lines for all the other intervals. The complete temperament Circle showing all the interval lines is shown in Fig. (4.12). Each interval has a micro interval associated with it and its inversion, relating it to the series of fifths / fourths on its arc, from which it is produced. The circle aptly illustrates how the notes are all connected as a *complex network*.

Fig. 4.12

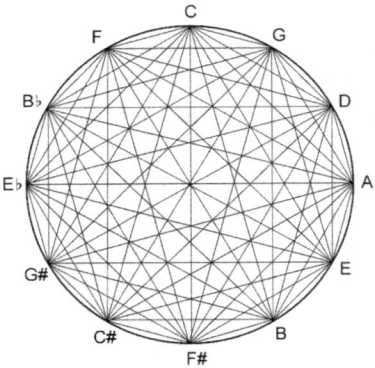

Temperament classification

Temperaments can be classified in relation to the temperament Circle. They classify into *circulating* and *non-circulating* temperaments, and also into *regular* and *irregular* temperaments.

Any temperament in which *all the fifths* round the Circle are tempered, with no untuned "wolf" interval present, is called a *circulating temperament*. If one "fifth" interval remains untempered, as a "wolf" interval, then the temperament is *non-circulating*. Thus, if the wolf interval is considered musically unusable, then a non-circulating temperament can only be used for music that avoids that interval and its inversion. In modern terms, this generally means music can only be played in certain keys, and modulation is limited. In the proper historical context in which music is mode-based, rather than tonal, this is not necessarily a problem.

If all 11 of the *tempered* "fifth" intervals round the Circle are tempered by the same amount, then the temperament is a *regular temperament*. If the tempered "fifths" are tempered by different amounts, then the temperament is an *irregular temperament*.

Both circulating and non-circulating temperaments can be irregular or regular. There is only one temperament that is both circulating and regular, and that is *equal temperament*.

A temperament that is non circulating but regular, will have major thirds that fall into two sets. This is because of the effect of the wolf fifth on the circumference, on the four major third lines that fall underneath it. These four thirds will be one size, affected by the wolf, whilst the other eight thirds on the Circle will be another size, because they have lines not affected by the wolf. Temperaments of this kind are *meantone* temperaments. It is possible for the two sets of thirds to become equal in size. At this point, the temperament is *equal temperament*, and the "wolf" interval is the same size as all the other fifths.

The mean tone temperaments

The name "mean tone" (often written *meantone*) originally refers to the size of the *major tone* being half the size of the pure or harmonic third. Some scholars consider only ¼ comma meantone to be "true meantone" in the historical context. Whether or not meantone temperaments of other comma divisions were ever actually discovered or used in the past, has no bearing on the general mathematics of temperament theory itself. It is necessary to recognise all the comma divisions in order to complete the general mathematical model that temperament theory comprises. As far as the general theory of temperament theory is concerned, the name can applied to any temperament in which 11 of the 12 fifths round the Circle are tempered by the same amount.

In equal temperament 11 of the 12 fifths round the Circle are of course tempered by the same amount, even though the twelfth fifth is itself also tempered by that same amount. Equal temperament is, therefore, mathematically a meantone temperament – in it, the major whole tone is also half the size of the major third.

Mean tone temperaments are sometimes used for early music, from medieval, through the renaissance, and sometimes for baroque and later music. They classify according to the division of the comma by which each of the fifths round the Circle is tempered. The "earliest" and most "severe" mean tone temperament is ¼ comma mean tone. The purpose of tempering fifths narrow by ¼ comma is to produce as many pure thirds as possible, the maximum being eight. The remaining four are then in effect "discarded" or not used, being, as a result, very wide.

In ¼ comma mean tone each of 11 fifths is tempered narrow by ¼ comma, with the result that the remaining "wolf interval" is necessarily wide by 1¾ commas. However, the purpose of this arrangement lies in the effect on the major thirds. By tempering the fifths narrow by ¼ comma, 4 such fifths on an arc of the Circle will total 1 comma. The value of the tempering on the arc across each major third line that does not include the "wolf" interval, is 1, so these 8 thirds are untempered or pure, since 1

− 1 = 0. The ¼ comma mean tone is thus suitable where the sound of the pure third is highly regarded, and the perfect fifths are accepted as being severely tempered by ¼ comma in order to allow this purity in the thirds.

Because ¼ comma meantone temperament is non-circulating, it has an untuned "wolf" interval between the ends of the clockwise and anticlockwise tempering sequences. The presence of the wolf on the Circle means that 4 of the 12 thirds lines will fall across the wolf interval, and will be very wide Pythagorean thirds, since they will be wide by 1 comma (the size will be 1 minus ¼ - ¼ - ¼ + 1 ¾). Any other interval whose line falls across the "wolf", will be similarly unmusical. The ¼ comma mean tone temperament therefore contains intervals to be avoided, and is therefore only suitable for playing in certain keys, depending on where circulation of the fifths starts and finishes. It produces 8 "good" major thirds that are pure, but requires that we do not mind such heavily tempered fifths.

If we were to temper the fifths by a smaller amount, say, $1/5^{th}$ comma, the temperament would correspondingly become '$1/5^{th}$ comma mean tone'. The fifths would be less heavily tempered, and the usable thirds would be wide also by $1/5^{th}$ comma (try the calculation) – quite pure by piano tuning standards, but not beatless.

Using $1/6^{th}$ comma for the tempering gives us $1/6^{th}$ comma mean tone. At $1/6^{th}$ comma mean tone the "wolf" is now wide by less then a comma (it is 5/6 comma wider than a perfect fifth), and although unpleasant, may arguably not be dismissed as musically unusable. Really, it is no longer a wolf interval. The 8 "good" major thirds will now be $1/3^{rd}$ comma wide – still purer than in modern piano tuning. The 4 "bad" major thirds falling across the "wolf" (the odd one out of the fifths, but no longer a true wolf) will be 1 $1/3^{rd}$ commas wide. $1/6^{th}$ comma meantone provides the effects and characteristics of meantone without being as extremely far removed from equal temperament as ¼ comma meantone. It is however, arguably "inauthentic" as an historical temperament.

Strictly speaking, mean tone temperament comma divisions refer to the syntonic comma, specifically, because they are temperaments concerned with the relationship between the fifths and the thirds. Equal temperament itself is in fact reached when the mean tone fifths are tempered by about $1/11^{th}$ syntonic comma, which is about the same as $1/12^{th}$ Pythagorean comma. Similarly, the major thirds in ¼ comma meantone will only be truly pure or beatless, if the fifths are tempered by ¼ syntonic comma. In practical tuning of early temperaments, using "authentic" tuning methods that do not require listening to specific beat rates, the difference between the commas is not really of practical significance.

Mean tone temperament, because it does not circulate, must define its fifths first one way round the Circle and the other. The two ends of the sequences, the clockwise and the anticlockwise, meet to define the "wolf" interval. In Fig. (4.13) there is a "nominal wolf" between G# - E-flat. This interval is, nominally, a diminished minor sixth, but in equal temperament it is the same size as all the other tempered perfect fifths. – there is no actual acoustical "wolf" to be heard. In equal temperament the choice of enharmonics is purely one of nomenclature – it has no real acoustical significance.

In mean tone temperament, on the other hand, enharmonics actually signify notes tuned to different frequencies, and one must choose which enharmonic the physical key or note on the instrument will be tuned to. The Circle as shown in Fig. (4.12), for example, circulates clockwise from C as far as G# (rising fifths or falling fourths), and circulates anticlockwise from C as far as E-flat (falling fifths or rising fourths). If the intervals round the Circle were tempered narrow by ¼ comma, the G# - E-flat "wolf" would indeed sound gross, because it would acoustically very wide. To sound good, either the E-flat would have to be re-tuned to a D#, to make a perfect fifth with the G#, or the alternatively the G# would have to be re-tuned to A-flat.

The net result of mean tone tuning is that it is suitable for playing in some keys but not others, so modulation is restricted. (Of course, in the middle ages and renaissance music was not tonal in any case, but was modal or transitional between modal and tonal). Usually, clockwise circulation is from C to G# or A-flat, depending on whether a G# or an A-flat is required for the music. Using a G# one can play in keys with up to 4 sharps or 3 flats in the key signature, or using an A-flat, one can play in keys with up to 3 sharps or 4 flats in the key signature. This is usually the choice to be made when tuning meantone. In tuning practice, one tunes up from E or C# to obtain a G#, or down from C or E-flat to obtain an A-flat. Unlike tuning equal temperament, in tuning meantone the direction of tuning determines which one of two different enharmonics will occur.

The characteristics of mean tone, whatever the tempering size, are then:

- 11 of the fifths are tempered by the same amount.

- 1 "wolf" fifth and its corresponding inversion "wolf" fourth will be produced.
- At tempering less than about $1/6^{th}$ comma, the "wolf" interval ceases to be truly a "wolf", and becomes usable.
- 8 of the 12 major thirds will be "good" (virtually pure in ¼ comma mean tone), and 4 will be wider. The difference between the size of the 8 "good" thirds and the 4 wider ones, depends on the fifth tempering amount. In true ¼ comma meantone the 8 "good" thirds are "pure" and the 4 wider thirds are unusable. As the fifths tempering is reduced, the wide thirds become less wide and the "good" thirds become wider. By the time we reach 1/11 or 1/12 comma meantone, equal temperament, the difference in thirds sizes vanishes, all major thirds being 2/3 comma wide. The 8 "good" thirds will have lines in Circle *not* crossing the "wolf" interval on their arcs. The 4 "bad" thirds will be those with lines crossing the "wolf" on the arc.
- One can always play in keys with up to 4 sharps or 3 flats in the key signature, or alternatively with 3 sharps or 4 flats in the key signature depending on the choice of enharmonics tuned. Typically, the choice is between Ab or G#. The note names must adhere to tonal grammar. Thus, for example, an Ab must be tuned as a major third below C, or as a tempered fifth below Eb. Tuning the same note as a major third above E, for example, will result in a G#. In true meantone the two are different, and are not acoustically equivalent as they are in equal temperament.
- The tempering characteristics are *equal* in all playable keys. In true ¼ comma meantone other keys cannot generally be used at all.

Equal temperament is achieved when the tempering in the meantone fifths is 1/12 comma (the generalised comma). This is approximately 1/11 syntonic comma. The "wolf" interval, by this stage no longer a wolf, then becomes the same size as all the other fifths. Meantone temperaments thus range from ¼ syntonic comma meantone to 1/11 syntonic comma meantone, sometimes called 1/12 comma meantone, which is equal temperament. The greater the tempering fraction (the smaller the denominator in the fraction), the more the "wolf" differs from the other fifths, and the narrower are the other fifths. Also, the "purer" the 8 "good" major thirds will be, but the wider the 4 "bad" major thirds will be.

1/6 comma meantone is a good compromise between ¼ comma and equal temperament, and allows playing in all keys, since at 1/6 comma the "wolf" becomes usable, as a fifth widened by 5/6 comma. The interval is not particularly pleasant, but it can be used.

Fig. 4.13

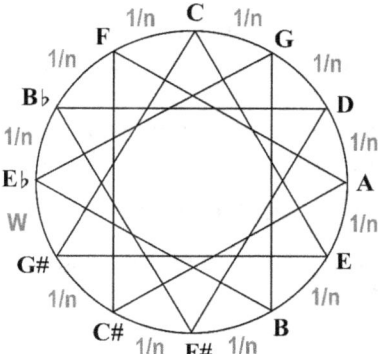

Fig.13. Meantone temperament – the general arrangement for *n*th comma meantone, placing the wolf between G# - Eb.

The "well temperaments"

In the past it was often thought that equal temperament was invented and first used by JS Bach, because, it was said, he was the first to compose in all 24 keys. In fact, compositions cycling effectively through all 24 keys (Western music prior to the 17th century was modal or transitional rather than tonal) existed earlier than Bach's *Forty Eight Preludes and Fugues*, in instrumental polyphony such as viol music. Equal temperament was used in practice on fretted instruments from the 16th century.[25] Furthermore it is not necessary to tune equal temperament in order to play in all 24 keys, or in order for an instrument to be considered "well tuned".

The name of Bach's *Forty Eight Preludes and Fugues* was of course *Das Wholtemperierte Clavier*, which means *The well tempered clavier*, not *The equally tempered clavier*. The Germans had at the time, as they do now, another word for *equal*, and there is little evidence to suggest that an equally tempered instrument would have been considered any more "well tempered" than other circulating temperaments that also allow playing in

---

[25] See Lindley, Mark, *Lutes, viols and temperaments*, Cambridge, 1984.

all keys. The term *well temperament* has thus come in modern times to refer generically to any of the circulating temperaments from the baroque period that allow playing in all keys.

The fundamental difference between meantone tuning and "well temperament" tuning is that in meantone 11 of the 12 fifths are tempered the same, the 12th being the one left out, which absorbs the tuning problem, the comma, introduced by the other 11. In the "well" temperaments, the fifths are unequally tempered (although some may be the same). In ¼ comma meantone - the only meantone universally agreed to be historically correct – a "wolf" interval results. There is no wolf in any "well" temperament, the "well" indication the absence of a wolf.

The characteristics of "well" temperaments are:

- The fifths are tempered by differing amounts, usually all being tempered.
- No "wolf" is produced.
- The major thirds will be generally different sizes.
- One can play in all keys.
- The tempering characteristics are different, in different keys. Usually, roughly speaking the "home keys" have the least widened (from pure) major thirds, and the remote keys have major thirds tempered the most wide. Correspondingly, the tempering of the perfect fifths across the raised keys tends to be greater than the tempering of perfect fifths across the natural keys.

Valotti temperament

A number of historical temperaments popular today in early music performance, follow this last principle more or less roughly. One of the best examples of the principle is the Valotti temperament. In it, the primary major thirds of the "home" keys are considerably less wide than in equal temperament, whilst the major thirds of the "remote" keys are wider than in equal temperament. The price the temperament pays for this arrangement, is naturally that the "home" key fifths are most tempered, whilst the "remote" key fifths are least tempered, in fact, they are *not* tempered. Whilst equal temperament tempers all the fifths round the Circle by 1/12 comma, Valotti tempers 6 fifths only, but by twice as much, each by 1/6 comma. In both cases the total tempering round the Circle is the required 1 comma. The 6 fifths chosen for tempering are those on the "home key" side of the Circle, so that the "home key" major thirds, are the least tempered. All the fourths and fifths between black notes only on the keyboard, are therefore tuned pure. This basis of six pure fifths makes the Valotti temperament especially easy to tune in practice.

Fig. 4.14

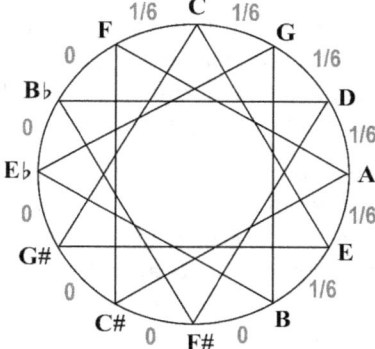

Fig. 4.14. The Valotti temperament.

Tuning equal temperament

Unlike many historical temperaments, every interval in equal temperament is tempered. There are in equal temperament no pure (untempered) fifths or thirds to act as reference points or "anchors" to build the tempering around. Tuning a pure (untempered) fifth is relatively easy for an experienced musician. Tempering a fifth narrow by precisely 1/12 comma is a different matter. For this, we need a special technique. The idea of equal temperament is for the tempering - which in turn manifests as beat rates in the intervals and in the size of the intervals - to be perfectly balanced in the sense that there is no tempering discrimination between different keys.

All scales must be the same in that all the semitones are the same size, and all the tones are the same size. All the intervals of any given kind are the same musical size. In this, equal temperament is unique in its demands for precision in the tuning of it. Any lack of precision, even in only one place, will result in deviation from the principles of equality of interval size, tempering and quality, for every interval type.

Instructions for tuning existed in 18th century England, which specified only (in effect) that the fifths should all be a little narrow and the thirds should all be wide. Whilst such instructions may have been intended to produce equal temperament, they are not sufficient to produce it. Rather, without further information they will result in a kind of *pseudo equal temperament* in which there is no particular pattern to the precise distribution of the comma and the relationships between the intervals. This may be sufficient for harpsichord tuning, but is not sufficient for tuning the modern piano. Even on the harpsichord, using these instructions with no further technique, will produce a result that a modern piano tuner will find does not resemble an equal temperament tuning anywhere near proper piano tuning standards.

For tuning of equal temperament, further technique is necessary, in the form of knowledge of beat rates, tuning by beat rates, and a sufficiently deep understanding of piano tuning characteristics, to apply the principle of *progression* of beat rates. The description of this technique and provision of the necessary instructions is the domain of *tuning theory*, rather than of temperament theory.

Temperament generalisation

The Pythagorean comma and the syntonic comma are just two examples of cases where one sequence of consecutive intervals covers an overall interval size approximating to some other sequence of consecutive intervals. It is possible to consider a finite set of harmonic ratios as an array R. Then, for two of those ratios R_i and R_j, and any two integers m and n, wherever

$$\frac{R_i^m}{R_j^n} \approx 1$$

we will encounter a microtonal interval on which a temperament, or part of a temperament, could be designed. Whilst the Greeks and many others since have explored this set of microtonal intervals arising from the relationships between the harmonic ratios, there is no longer any particular practical revelation to be found here. Consideration of the Pythagorean comma and the syntonic comma already adequately encapsulates the essential acoustical properties of any possible tuning arrangement between the twelve notes of the chromatic scale that is now firmly established in Western musical practice. In practical piano tuning it is the tone qualities of the interval soundscapes that are important, rather than any philosophy concerning ratios. These will be interdependent in a way that is unique to each instrument, but which approximately follows principles described by the temperament Circle, quite well.

Whatever approach is taken to describe the construction of the scale, using constructs like, for example, the temperament Circle, the scale can still be thought of as a set of 13 adjacent notes, "specified" by the musical "spacing" between the notes. Without regard for the tones of the intervals smaller than the octave, the notes can be "spaced" in *any* way, to form a scale, provided the outside two forming the octave, acoustically create an octave, and provided each note is consecutively "higher in pitch" from left to right on the keyboard.

We could of course have some arrangements in which many of the intervals between notes are not really recognisable as musical intervals from Western music, and in which the consecutive notes do not resemble a Western music scale. It is also relatively easy, in practice, to arrange the notes so that consecutive playing produces a series of tones recognisably like a Western scale. The main difficulty lies in the sound of the intervals when two notes are played simultaneously. The way in which the musical "spacing" between the notes affects the soundscapes of the intervals, is a direct consequence of the acoustics of the tones.

This acoustics, and its behaviour in relation to the "spacing" between the notes, is something to which temperament theory can only provide a guide, on an instrument such as the piano. The problem of tuning can just as sensibly be represented as an array of acoustical parameters, as in the form of a tuning Circle or temperament, but the array representation would allow computation of the actual audible tuning results, far more effectively.

Chapter 5 - "Traditional" piano tuning theory and elementary practice

Much of the piano's compass is designed with three strings per note, arranged in a unison group, referred to as a *trichord*. A substantial part of the bass section of the compass consists of two strings per note, as a *bichord*. A single note with more than one string is therefore called a *unison*. The unison is itself considered to be an *interval* because it involves tuning two or more strings. Only the lower bass is strung with one string to each note.

Precision aural tuning of the musical intervals is carried out by listening to *beats*. These beats are audible fluctuations in the loudness of certain *partials* in the sound recipe of the interval's tone, and the speed of the fluctuation is called the *beat rate*. In "traditional" theory beats are intended to occur in all intervals except unisons, octaves, and multiple octaves. Piano tuners usually speak of an interval having a particular *beat rate*, as though one beat rate only, is present in the tone of the interval. In actuality each interval that beats, will contain a number of different, separate beat rates.

In the case of most of the tempered intervals (the tuned intervals other than the unisons or octaves), only one beat rate is generally important. The tempered perfect fifth may typically require consideration of two beat rates. Unisons and octaves have the greatest number of simultaneous beat rates, but these intervals are said in "traditional" theory to be tuned to a condition in which all beat rates are zero.

The *scale* is a chromatic scale of 13 notes including the octave, that is tuned in the central part of the compass, at the beginning of the tuning process. Typically this will be F33 (F3) – F45 (F4), or perhaps G35 (G3) – G37 (G4). This sets up the equally tempered scale relationship between the notes. The rest of the compass is then tuned in octaves, outwards from the notes of the scale. The elementary theory dictates beat rates for the intervals of the scale, and also provides beat rates for some of the compound intervals (intervals greater than an octave) that are formed as tuning proceeds outwards from the scale in octaves.

The idea versus the reality

The original idea behind using a theoretical model to work out what the *beat rate* should be for each of the intervals, to enable precision tuning, must be understood for what it is. It starts by assuming a definition for correct tuning *as a specified set of string frequencies*, and then works out how to apply that assumed "definitive" *frequency set* using beat rates. It (a) assumes that there is a "correct" tuning defined as a set of string frequencies, and (b) assumes a simplified relationship between frequencies and beat rates.

Applying the frequency set could of course be much more easily achieved today simply by measuring the frequencies using an electronic meter. If we do this, however, the result in terms of tone and intonation is very poor compared to the standard achieved by a master tuner. In particular, the beat rates created are not generally those that the theory predicts. However, if the actual beat rates from the theory *do* appear in a scale, the scale will be good, even though the *frequency set* on the strings will then not in general be the one assumed by the theory. In other words, the function of beat rates in tuning is in reality quite different to that adopted in the "traditional" theory.

Actual tuning practice is rooted in the piano's purpose as a musical instrument. That is, what matters is *intonation and tone*. Tone is acutely affected by tuning, and not least through beat rates. Beat rates are in no way a mere tool for achieving a pre-decided set of frequencies, but are a direct part of the *tone quality* of the instrument, and indirectly, part of its intonation. In other words, in the art of tuning, beat rates are adjusted *for their own sake*, that is, for the sake of their effect on tone and intonation. To this end, in most cases, the precise beat rates quoted in "traditional" theory are only a guide. Nonetheless, it is necessary to first be conversant with the "traditional" theory and idea, if one is to truly understand how to get beyond it. The "traditional" theory remains the "rough guide", and some knowledge of it is essential in the first instance.

The "traditional" acoustical model

In "traditional" tuning theory the audible partials heard by the piano tuner are identified with the motion of *standing waves* on the strings. These are *normal modes* of vibration (we met *normal modes* in thd chapter on sound) in which either the whole speaking length of the string, or integer fractions of its speaking length, vibrate in one transverse plane (a plane at right anglds to the string's length). At each end of a vibrating section of the string, is a point of little or no motion for that particular mode of vibration, called a *node*.

In the fundamental mode, mode number 1, with the lowest frdquency, the whole speaking length of the string is vibrating, and the nodes are at the bridges at each end of the speaking length. In mode number 2, a node appears in the middle of the speaking length also, and each half of the string, each side of this central node, vibrates. In mode number 3, the speaking length divides into three equal length sections, two nodes appearing between the bridges, with each third of the speaking length vibrating. In mode number 4, three nodes appear between the bridges, an the speaking length divides into four, and so on. The nodes for any given mode of vibration, must always appear at points dividing the speaking length in *equal parts* between the two outside nodes, which themselves are at the bridges.

Between any two adjacent nodes for a given mode of vibration, only *half* the complete standing wavelength for that mode, can appear, Fig. (5.1).

Fig. 5.1

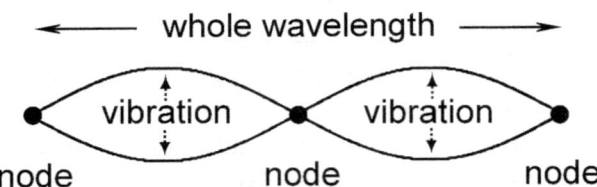

The speaking length is the longest half-wavelength that can exist on the string, and its nodes are where the bridges are at each end. All other waves must also have nodes at these same points, but can also have other nodes in between, but they must be separated by equal distances. So only standing waves with half-wavelengths that are *whole number divisions* of the speaking length, can exist on the string. This gives a set of wavelengths called *allowed wavelengths*, all of which are originally determined by the string's speaking length.

Note that the longest half-wavelength that takes up the whole speaking length, the *fundamental*, is of course only *half* a complete wavelength. The complete *wavelength* of the longest standing wave is therefore actually *twice* the speaking length. For a string of speaking length L the allowed wavelengths will therefore be wavelengths of $2L$, L, $L/2$, $L/3$ ..., etc.

A standing wave of an allowed wavelength is the result of two *travelling waves* of the same wavelength, travelling in opposite directions along the string. Whenever the string is struck this naturally produces wave motions travelling along the string. Such waves travel in both directions along the string, because travelling waves are reflected, or partially reflected back along the string, when they encounter the string *boundary* at end of the speaking length. Each bridge is a wave boundary, just like the edge of a swimming pool is a boundary for waves in the pool.

The velocity of all travelling waves on the string is assumed in elementary theory to be always the same, so the *frequency* of a given travelling wave will always be inversely proportional to its wavelength. The longer the wavelength, the lower the frequency. The shorter the wavelength, the higher the frequency. The only travelling wavelengths that can establish themselves on the string, however, are the wavelengths of the *allowed standing waves*. The frequency f of a standing wave will be the same as that of the travelling waves that produce it. This frequency is given by

$$f = \frac{c}{\lambda}$$

where c is the wave velocity, and λ is its wavelength. The allowed wavelengths are related in the ratios

$$\frac{1}{1} : \frac{1}{2} : \frac{1}{3} : \frac{1}{4} \cdots \text{etc.}$$

so because all waves travel at the same speed c, the allowed frequencies will be

$$f, 2f, 3f, 4f \ldots, \text{etc.}$$

The lowest frequency is that of the *fundamental*, the frequency of the standing wave whose wavelength is twice the speaking length of the string. The frequencies are arranged in none other than the *harmonic series*.

It is assumed in the theory, that (a) the soundboard then faithfully transmits these frequencies, unchanged, through the air to the listening ear, and (b), that different strings in no way affect one another, that is, they are acoustically isolated. In elementary theory each standing wave as described above, maps directly to an audible partial.

These simplifying assumptions provide a reasonably good model as a starting point. In actuality, the velocity of the waves on a real piano string differ with each wavelength, as a result of *frequency dispersion* caused by stiffness in the string. This leads to *inharmonicity* in which the allowed frequencies are raised above those of the harmonic series. Also, in the piano the strings are not acoustically isolated from each other, but are connected by the bridges and soundboard. Real piano strings are allowed to vibrate in two planes, rather than just one.

The important result of the elementary model, for the purposes of tuning theory, is that these frequencies of the audible partials to which the piano tuner listens, are taken to be arranged in the harmonic series, that is, they are in the ratio 1:2:3:4..., *etc.* The audible partials are simply called *harmonics*, because their frequencies fall in the *harmonic series*.

Interval ratios and beat rates

The beat rate in an interval indicates how far away from acoustical consonance the interval is tuned. Deliberately tuning an interval away from acoustical consonance by a specific amount, is to *temper* the interval. The amount of tempering in the intervals of equal temperament is determined by the principle of dividing the octave into 12 equally sized semitones. To achieve this division, the amount of tempering in the tuned intervals must be controlled by beat rates, and it is these rates that must be calculated.

For this, the theory uses the *harmonic ratios* of temperament theory, plus a *beat rate calculation rule*. The harmonic ratios are explained in the previous two chapters, on sound, and on temperament theory. There is one harmonic ratio for each interval.

We have already looked at the *harmonic series*. The theory considers the tone produced by a single piano string to be composed of a harmonic series of pure tones, simply referred to as a series of *harmonics*. The fundamental is harmonic number 1, and has the lowest frequency. The frequency of any harmonic number n will be the fundamental frequency multiplied by n.

If the fundamental frequencies of the two notes of an interval are tuned in an harmonic ratio, some of the harmonics of one note will coincide with some of the harmonics of the other, at the same frequency. As we saw in the previous chapters, coinciding pure tones of slightly different frequency will superpose to produce a *beating* partial.

The *beat rate calculation rule* is as follows:

> When two slightly different pure tone frequencies are added together, they produce a partial that *beats*, or fluctuates regularly between loud and quiet. The *beat rate*, or number of beats per second, will be equal to the difference between the two pure tone frequencies. The frequency of the beating partial itself, will be the average of the two pure one frequencies.

The aim is produce "equal temperament" by taking the octave and dividing it into 12 equally sized semitones. The size of the semitone, like the size of any interval, will be defined by a *frequency ratio*, that is, the ratio between the fundamental frequencies of any two notes an equally tempered semitone apart. Two notes a semitone apart will have fundamental frequencies in the ratio R, which we will always arrange as the frequency of the upper note, which has the highest frequency, divided by that of the lower note. Thus, as a matter of convention, the largest number in any harmonic ratio will always be the numerator. In the theory we will be using, harmonic ratios are just *frequency ratios*, so they can also be "divided out" and reduced to a single number, or a decimal fraction.

When musical intervals are put side by side, the total interval they form is found by *multiplying* their ratios together (not adding). Using the harmonic ratios of temperament theory, we can see this, for example, in the case of the perfect fourth and perfect fifth, which, side by side, make an octave. The harmonic ratio of the perfect fourth is 4/3 and that of the perfect fifth is 3/2. Multiplying these we get:

$$\frac{4}{3} \times \frac{3}{2} = \frac{12}{6}$$
$$= \frac{2}{1}$$

The final answer, 2/1, is the harmonic ratio for the octave that the perfect fourth and perfect fifth make when put together. We could also write this in decimal notation as

$$1.\dot{3} \times 1.5 = 2.$$

The equally tempered semitone

In the case of the equally tempered semitone, conceived as the division of the 2/1 octave into 12 equally sized semitones, no exact harmonic ratio that can be written as an integer ratio, is provided by temperament theory. The size of the equally tempered semitone cannot be "derived" by arranging other intervals with integer harmonic ratios. Historically, although the concept of the equally tempered semitone existed, the precise identification of its size as a ratio was an enigma.

Only after the advent of logarithmic calculation and decimal notation was it possible to calculate its size as a number, with precision. Also, prior to precision calculation of the beat rates necessary for the tuning of equal temperament, equal temperament was not a practical option for keyboard instrument tuning. Practical instructions for tuning pseudo-equal temperament on harpsichords certainly existed in the 18th century, but compared to modern piano tuning practice, these would result in what would better be described as a "random" circulating temperament. Approximate equal temperament was more successfully applied to fretted stringed instruments from the 16th century, by using relatively simple geometric rules for positioning the frets.

Today, with computers and electronic calculators, equal temperament calculations are easy. The 12 semitones put side by side together must equal an octave, whose ratio is 2/1, or simply 2. This means that R, the ratio of the equally tempered semitone, must be the number which when multiplied by itself 12 times in succession, will equal 2. In other words, R must be the *twelfth root* of 2:

$$R = \sqrt[12]{2}$$

or

$$R = 2^{1/12}$$

Thus, to 5 decimal places $R = 1.05946$. Given the fundamental frequency of any note, the fundamental frequency of another note one equally tempered semitone above can be found by multiplying the given frequency by R. This number 1.05946... is therefore known as the *semitone multiplier*. To find the frequency of a note one semitone below, the given frequency is simply *divided* by the semitone multiplier, rather than multiplied by it.

To find the fundamental frequency of a note n semitones above a note with a fundamental frequency f, we just multiply f by R raised to the power of n. In other words:

$$f_U = f_L \times R^n$$

where f_U is the fundamental frequency of the upper note, and f_L is the frequency of the lower note. For a falling interval, the formula is:

$$f_L = f_U \times R^{1/n}$$

or

$$f_L = \frac{f_U}{R^n}.$$

From this we can work out the fundamental frequencies for all the notes of an equally tempered scale, provided we know one frequency to start with.

The modern "international concert pitch" frequency standard is defined as A49 = 440 Hz. This is a standard based on a 20th century international convention. From this, we know that according to elementary theory the A37 one octave below must have fundamental frequency half that of A49, or 220 Hz, since the ratio for the octave must be 2/1. Therefore, the fundamental frequency of middle C, C40, will be given by:

$$220 \times 1.05946^3 \approx 261.6.$$

The fundamental frequency of middle C is thus approximately 261.6 Hz.

The first step in calculating beat rates for the equally tempered scale is to calculate the complete set of fundamental frequencies for the scale, using the semitone multiplier. The second step is to find which harmonics approximately coincide in frequency, for each given equally tempered interval. The third step is to find the frequencies of those harmonics. The difference in frequencies, will by the *beat rate calculation rule*, give the beat rate for that interval.

The rule for finding coinciding harmonics

A given interval will have its own harmonic ratio. In practice, we will be concerned primarily with following intervals and ratios:

Octave	2/1
Major sixth	5/3
Perfect fifth	3/2
Perfect fourth	4/3
Major third	5/4

Any harmonic ratio can be written as

$$\frac{R_U}{R_L}$$

where R_U is the numerator of the ratio, and R_L is the denominator. Following the convention we adopted, we will express harmonic ratios as improper fractions in which R_U will always be the larger of the two numbers. For an untempered interval only, this will equal

$$\frac{f_U}{f_L}$$

where f_U is the fundamental frequency of the upper note, and f_L is the fundamental frequency of the lower. The harmonics of each note are assumed to have frequencies in an harmonic series, and are numbered upwards from the fundamental, which is harmonic number 1. Thus, for example, harmonic number n of the lower note will have a frequency of nf_L.

For any interval, we take the *numerator* of its harmonic ratio and associate it with the *lower* note of the interval. The number of the *lower note's* harmonic that coincides to produce the beating partial, is the same as this *numerator*. The *denominator* of the harmonic ratio is then associated with the *upper note* of the interval. The number of the *upper note's* harmonic that coincides to produce the beating partial, is the same as the *denominator*.

For example, the harmonic ratio for a perfect fifth is 3/2. The number 3 is the numerator, and the number 2 is the denominator. This means that the 3rd harmonic of the *lower* note, will coincide with the 2nd harmonic of the *upper* note. The harmonic ratio:

$$\frac{R_U}{R_L}$$

can of course also be written as:

$$\frac{nR_U}{nR_L}$$

where n is any number, without changing the ratio. For any integer n other coincidences of harmonics will occur at the nR_U th harmonic of the lower note, and the nR_L th harmonic of the upper note.

For example, in the case of the tempered perfect fifth, with the applied harmonic ratio 3/2, we can multiply both numerator and denominator of the ratio by 2, and write it as 6/4. This tells us that the 6th partial of the lower note will coincide with the 4th partial of the upper note. Multiplying by 3, for example, the 9th harmonic of the lower note will coincide with the 6th harmonic of the upper note.

Only the coincidence of partials represented by the harmonic ratio written in its lowest terms, are used for calculating the beat rates quoted in "traditional" theory. We can easily work out higher order beat rates if we need to. The beat rates in higher coincidences work out simply as multiples of the standard, lowest beat rate. For example, if we know the beat rate for the coinciding 3rd and 2nd harmonics in the case of the tempered perfect fifth, then the beat rate for the coinciding 6th and 4th partials, will be twice the rate between the 3rd and 2nd harmonics.

Calculating higher order beat rates by the elementary method is not really useful, firstly because in practice beat patterns in these positions are weaker and not so suitable for controlling interval size and quality effectively, and secondly, because of *inharmonicity*. Inharmonicity is a natural effect of string stiffness, ignored by "traditional" theory. The effect of inharmonicity is to raise the strings' natural frequencies. The higher the harmonic, the more it would be affected by inharmonicity. Of course, this means that the natural frequencies produced by the inharmonic string do not fall perfectly in the harmonic series, so the approach to calculating beat rates taken in elementary theory, fails. It remains a useful guide for the first or lowest beat rate of each interval, but not for the higher ones.

Calculating the beat rate

We take the two coinciding harmonics, and the difference in their frequencies is equal to the beat rate. The frequency of each of these harmonics is its harmonic number, multiplied by that note's fundamental frequency. The fundamental frequencies for all the notes are calculated using the semitone multiplier.

Thus, we can write the beat rate B_R for any given interval as

$$B_R = R_L f_U - R_U f_L.$$

If the interval is *wide*, B_R will be positive. If the interval is narrow, B_R will be negative.

Revision

Let's now reiterate the process for finding beat rates. First, we calculate the fundamental frequency for every note of the scale, starting with one known frequency such as that for A37, which as a matter of international convention will be 220 Hz. We then use the semitone multiplier to find all the fundamental frequencies for the whole scale. We then take each interval and look at its harmonic ratio. Now we take the *lower* number of the harmonic ratio (the denominator) and multiply it by the fundamental frequency of the interval's *upper* note. This gives result (1). We then take the *upper* number of the harmonic ratio (the numerator) and multiply it by the fundamental frequency of the interval's *lower* note. This gives result (2). Subtract result (2) from result (1), an this will give the beat rate for that interval. If the number is negative, the interval is narrow. If it is positive, the interval is wide.

Worked example

We can use the process to work out the beat rate for the major third C40 – E44. The note A37 has a fundamental frequency of 220 Hz. We need to find the fundamental frequencies of the C and the E. C is the lower note of the interval and E is the upper note, so their fundamental frequencies will be f_L and f_U respectively. For this particular interval, f_L is the fundamental frequency of C40, and f_U is the fundamental frequency of E44.

The semitone multiplier is the twelfth root of 2:

$$2^{1/12}.$$

The note C40, whose frequency we do not yet know, is *3 semitones* above the known note A37 (which is 220 Hz). The fundamental frequency of C40, f_L, is therefore given by

$$f_L = 220 \times 2^{3/12},$$

where the 220 is the frequency we already know, the 2 on the right of the multiplication sign is always 2, and it is raised to a power which is a fraction. The denominator of the fraction is always 12, and the numerator is the number of semitones between the note we know and the one we are trying to find. If the note we were trying to find had been 3 semitones *below* the known note on the keyboard (*i.e.* if it had been F#), then the formula would have been

$$f_L = \frac{220}{2^{3/12}}.$$

The power index of the 2 remains the same, but we divide the known frequency (220 in this case) rather than multiplying it.

In our case, the note we are finding is *above* the note we already know, so we do use

$$f_L = 220 \times 2^{3/12}$$

$$\approx 261.6 \text{ Hz}$$

This completes the procedure for finding the fundamental frequency of the lower note of the interval. The major third's upper note E44 is 7 semitones above the known note A37, so the formula for its fundamental frequency f_U is

$$f_U = 220 \times 2^{7/12}$$

$$\approx 329.6 \text{ Hz}$$

We could of course have also calculated this by taking the frequency for the C40 that we just calculated and using

$$f_U \approx 261.6 \times 2^{4/12}$$
$$\approx 261.6 \times \sqrt[3]{2}$$
$$\approx 329.6 \text{ Hz}$$

where 261.6 Hz is the frequency of the C40, the number of semitones of the E44 above the C40 is 4, and so the numerator of the fraction index over the 2 is in this case 4.

Having found the fundamental frequencies of the interval's two notes, we must now find the frequencies of the coinciding harmonics above these fundamentals, and then the beat rate as their difference.

The harmonic ratio for a major third is 5/4. We therefore need to multiply the *upper* note's fundamental by 4, and the *lower* note's fundamental by 5. This will give us the frequencies of the two harmonics that coincide. The difference between them will be the beat rate. In the beat rate formula

$$B_R = R_L f_U - R_U f_L$$

R_L is the lower or smallest number (the denominator) of the harmonic ratio 5/4, and is thus 4. R_U is the upper or largest number, and is 5. Thus,

$$B_R \approx (4 \times 329.6) - (5 \times 261.6)$$
$$\approx 10.4 \text{ Hz}$$

The beat rate for the major third C40 – E44 is therefore approximately 10.4 Hz.

Working out all the beat rates, one by one, in this way, would be very tedious. In practice, a *generalised method* is useful, which can be used to calculate *every beat rate* for *every interval position in the scale*, in one operation, using a computer to do the tedious work. This is a more advanced method.

General beat rate computation for harmonic notes

The beat rates for all the intervals from the semitone to the octave can be immediately generated using arrays. We first define an array R for all the harmonic ratios from the unison to the octave:

$$R = \begin{pmatrix} 1 & 1 \\ 16 & 15 \\ 9 & 8 \\ 6 & 5 \\ 5 & 4 \\ 4 & 3 \\ 45 & 32 \\ 3 & 2 \\ 8 & 5 \\ 5 & 3 \\ 7 & 4 \\ 15 & 8 \\ 2 & 1 \end{pmatrix}$$

The equally tempered semitone size E as a frequency ratio will be

$$E = 2^{1/12}.$$

Because the fundamental frequency of a given interval's upper note is proportional to the lower note's fundamental frequency, and since all harmonics are multiples of a fundamental, the beat rate for any interval will be simply be the product of the fundamental frequency of the interval's lower note, and a multiplier M.

The multiplier for each interval size i will be given by

$$M_i = R_{i2} E^{i-1} - R_{i1}$$

where $i = 1, 2 \ldots 13$, i being the number of semitone notes on the keyboard that make up the interval. For example, for a semitone $i = 1$, and for a major third $i = 5$.

It is convenient to use the fundamental frequency f of the lowest note in the scale to generate the other frequencies. The fundamental frequency of each note j in the scale will be given by

$$F_j = f \times 2^{j-1/12}$$

where $j = 1, 2 \ldots 13$.

The beat rates for each interval size, and for lower notes at each position j in the scale (at F33 $j = 1$), is then given in an array B as

$$B_{ij} = F_j M_i$$

For a fundamental frequency of $f = 174.597$ Hz at F33, or note position $j = 1$, the complete set of beat rates for rows i and columns j is:

$$B = $$

	1	2	3	4	5	6	7	8	9	10	11	12	13
1	0	0	0	0	0	0	0	0	0	0	0	0	0
2	-18.9	-20	-21.2	-22.4	-23.8	-25.2	-26.7	-28.3	-29.9	-31.7	-33.6	-35.6	-37.7
3	-3.5	-3.8	-4	-4.2	-4.5	-4.7	-5	-5.3	-5.6	-6	-6.3	-6.7	-7.1
4	-9.4	-10	-10.6	-11.2	-11.9	-12.6	-13.3	-14.1	-15	-15.8	-16.8	-17.8	-18.8
5	6.9	7.3	7.8	8.2	8.7	9.2	9.8	10.4	11	11.7	12.3	13.1	13.9
6	0.8	0.8	0.9	0.9	1	1.1	1.1	1.2	1.3	1.3	1.4	1.5	1.6
7	44.5	47.1	49.9	52.9	56.1	59.4	62.9	66.7	70.6	74.8	79.3	84	89
8	-0.6	-0.6	-0.7	-0.7	-0.7	-0.8	-0.8	-0.9	-0.9	-1	-1.1	-1.1	-1.2
9	-11	-11.7	-12.3	-13.1	-13.9	-14.7	-15.6	-16.5	-17.5	-18.5	-19.6	-20.8	-22
10	7.9	8.4	8.9	9.4	10	10.6	11.2	11.9	12.6	13.3	14.1	15	15.8
11	22.2	23.5	24.9	26.4	28	29.6	31.4	33.3	35.3	37.3	39.6	41.9	44.4
12	17.8	18.9	20	21.2	22.4	23.8	25.2	26.7	28.3	29.9	31.7	33.6	35.6
13	0	0	0	0	0	0	0	0	0	0	0	0	0

Here, the column numbers are the positions in the scale of the lowest note of the interval, and the row numbers are the interval sizes in semitones. For example, the major third C40 – E44 is shown in row 5, column 8. The major third's size is 5 semitone notes on the keyboard, so it is in row 5. The position of C40, the interval's lower note, is the 8th note from the bottom of the scale (F33), so the interval's beat rate appears in column 8.

We can see immediately that some intervals (rows) have much larger beat rates than the other intervals. In particular, this is true of rows 2, 7, 11 and 12. These, respectively, are the rows for the intervals of a semitone, a tritone (augmented fourth / diminished fifth), minor seventh and major seventh. These intervals are consequently not suitable for tuning purposes. Those intervals whose beat rates become too fast to be practically useful in the higher scale positions, such as the minor thirds and minor sixths, may have restricted usefulness.

Only beat rates within a limited range are useful in practice. There is a limit to fast beat rate usefulness – it is not that we cannot hear fast beat rates, but rather, that the fine differences in beat rates on which equal temperament scale tuning depends, is not sufficiently reliable. This is a matter of both physics and perception. Fast beat rates are however reliable enough for use in tuning intervals beyond the confines of the scale, for example in tuning major seventeenths in conjunction with octave stretching.

Beat rate relationships

We can extract from the array B the beat rates for the intervals that can be practically used in tuning the scale. The major thirds, perfect fourths, perfect fifths and major sixths are the most useful, but the minor thirds and sixths which are also included. The beat rates for the scale area F33 (F3) – G47 (G4) are given in the tables below.

Beat Rate Table

Maj. 3rds	Beats (Hz)	Perf. 4ths	Beats (Hz)
F-A	6.9	F-Bb	0.8
F#-A#	7.3	F#-B	0.8
G-B	7.8	G-C	0.9
Ab-C	8.2	G#-C#	0.9
A-C#	8.7	A-D	1
Bb-D	9.2	A#-D#	1.1
B-D#	9.8	B-E	1.1
C-E	10.4	C-F	1.2
C#-F	11	C#-F#	1.3
D-F#	11.7	D-G	1.3
Eb-G	12.3		1.4

Perf. 5ths	Beats (Hz)	Maj. 6ths	Beats (Hz)
F-C	-0.6	F-D	7.9
F#-C#	-0.6	F#-D#	8.4
G-D	-0.7	G-E	8.9
G#-D#	-0.7	Ab-F	9.4
A-E	-0.7	A-F#	10
Bb-F	-0.8	Bb-G	10.6
B-F#	-0.8		
C-G	-0.9		
	-0.9		
	-1		
	-1.1		

Min. 3rds	Beats (Hz)	Min. 6ths	Beats (Hz)
F-Ab	-9.4	F-Db	-11
F#-A	-10	F#-D	-11.7
G-Bb	-10.6	G-Eb	-12.3
G#-B	-11.2	G#-E	-13.1
A-C	-11.9	A-F	-13.9
A#-C#	-12.6	A#-F#	-14.7
B-D	-13.3	B-G	-15.6
C-Eb	-14.1		

As before, negative beat rates indicate a narrow interval, whilst positive rates indicate a wide interval. The polarity will be otherwise ignored. We can, for example, speak of two beat rates being the same, even if one is wide and the other narrow. It is usual to refer in practice simply to an interval being a specific number of beats wide or narrow.

Some important rules begin to emerge from the results. To begin with, the tempered perfect fifths are tempered narrow, whilst the tempered perfect fourths are tempered wide. The *major* thirds and sixths are tempered wide, whilst the *minor* thirds and sixths are tempered narrow.

All the intervals of any one kind, *increase* in beat rate with rising position in the scale. This is *progression* of beat rates.

The beat rates for the tempered perfect fifths and fourths are *slow*, whilst those for the thirds and sixths, are *fast*. On any common lower note, the fifths are slower than the fourths, and the minor thirds and sixths are faster than the major ones. We may generally think of these intervals in the following terms. The fifths are the slowest beating, and the fourths are faster, but both are *slow intervals*. The thirds and sixths are *fast intervals*, the minor ones being faster then the major ones.

Major thirds – major sixths

The figures also show that there are some useful relationships between the beat rates. The first is the major sixth – major third relationship. The beat rates of the major sixths closely parallel those of the major thirds, but are staggered by one tone. In other words, the beat rate of a major sixth will be approximately the same as that of a major third whose lower note is one whole tone above the lower note of the sixth. The sixth F–D, for example, has almost the same beat rate as the major third A–F#. The beat rate for the sixth is actually faster than the corresponding third by around 1 Hz – 2 Hz, according to the calculations. It is certainly possible to hear a difference in beat rates in this order, but it should be remembered that in practice beat rates may in any case differ from those generated in the elementary calculations, owing to inharmonicity and falseness. It is sufficient to remember the beat rates as "equal" as a "rule of thumb".

Because in elementary theory we are dealing with true harmonics, there is a simple relationship between the beat rate of an interval and that of the same interval an octave higher or lower in the compass. Transposing an interval up an octave doubles the beat rate, whilst lowering an octave halves the beat rate. For example, if the beat rate of the major third C40 – E44 is 10.4 Hz, the rate for the third C52 – E56 will be 20.8 Hz, and the rate for the third C28 – E32 will be 5.2 Hz.

Fourth – fifth - octave

There are some important relationships between beat rates in intervals that together make an octave. A perfect fourth plus a perfect fifth make an octave. When the upper note of a perfect fourth is also the lower note of a perfect fifth, the fourth and the fifth will have an equal beat rate relationship provided the octave that they make is tuned "beatless". This relationship is not dependent on the tempering of the fourth and fifth, so it cannot be used to "test" the sizes of these two intervals. Whatever the beat rate in the fourth, whether "correct" or not, the beat rate in the fourth and the fifth will be the same, when the octave is "in tune". If one perfect interval is tempered wide, the other will be narrow, and *vice versa*. Similarly, for a perfect fifth whose upper note is also the lower note of a perfect fourth, the two outside notes also make an octave. In this case, the beat rate in the fourth will be twice that in the fifth.

These consequences of the calculations of elementary theory leads to the proposition that the beat rate relationship of the fourth and the fifth can be used to "test" the octave tuning. Whilst this may be true for rough tuning in the early stages of learning to tune, this prediction should be treated with caution. The change of inharmonicity over the scale area may mean that this rule breaks down, particularly on smaller instruments.

Minor third – major sixth

The minor third and major sixth within an octave also have a special relationship, and one that is important and relatively reliable in tuning bass octaves. The minor third whose upper note is the lower note of a major sixth, will have the same beat rate as the major sixth, when the outside octave formed by the two intervals is "beatless". Below the scale area the minor third beat rates are slow enough to be particularly useful. The minor third – major sixth relationship works as a "check" for the tuning of bass octaves. When the octave is "in tune" both beat rates should be the same. The minor third is narrow, whilst the major sixth is wide. Therefore, assuming tuning not to be so grossly incorrect that the third or the sixth is "on the wrong side", *i.e.* wide when it should be narrow, or narrow when it should be wide, the following holds. If the minor third is faster than the sixth, the octave is narrow. If it is slower than the sixth, the octave is wide.

Compound major thirds

A compound major third is an interval of a major third plus one or more octaves. Compound major thirds are very important in tuning the compass beyond the scale area, and in relation to tuning the octaves. For any given lower note and "beatless" octaves, the beat rate in the major third will be same as that in the major tenth on the same lower note. It will also be the same as the major seventeenth (octave-major tenth) on the same lower note. This relationship can be used as a "test" for the tuning of the octaves and double octaves in elementary tuning.

Because the beat rate halves when the major third is transposed down an octave, the beat rate of a major tenth whose *upper* note is the same as the *upper* note of a major third, will have half the beat rate of the major third. Similarly, the beat rate of a major seventeenth (octave – major tenth) whose *upper* note is the same as the *upper* note of a major third, will have a quarter of the beat rate of the major third.

Beat pitch relationships

The beating in any tempered interval occurs at a specific musical pitch above the interval in question. This is determined by the frequencies of the coinciding partials. *Any two intervals that share a note, and also share a number in their harmonic ratios, will share the same pitch for their beating partial.*

Intervals for which this is the case, include the perfect fourth and fifth arranged to make an octave, mentioned above. A particularly useful instance is the minor third - major sixth relationship mentioned above. Not only do we aim for the same beat rate in both intervals, but the fact that the pitch of the beat in both intervals is the same, makes the "test" especially quick and easy in practice.

Another very useful instance is the comparison of the beat rate of a major third, with the major sixth on the same lower note. In tuning the scale the sixth should be faster than the corresponding third, but both beat at the same pitch.

Elementary practical tuning procedure for the scale

Tuning commences with the scale, the first note tuned usually being A or C, which is tuned to the fork. By elementary theory, there is no reason why strings of the scale area should not be muted with muting felt. This is a strip of felt corrugated between the trichords in order to stop the outer strings of each trichord sounding. Tuning only the middle string of each trichord makes the task of scale tuning quicker and easier. The unisons are then tuned afterwards.

There is nothing in elementary theory to suggest that this might give a result different to tuning the scale trichord by trichord. In practice there are important differences, which is why expert tuners always proceed using a Papp's wedge, tuning the unisons as the scale tuning proceeds, rather than using muting felt. Muting felt is, however, used in the learning process, until the trainee reaches the stage where wedge tuning can be undertaken. In order to tune the scale using a Papp's wedge, firstly there must be the ability to reliably tune excellent and stable unisons, and secondly there must be some understanding of the relationship between the sound of the unison and the sound of the tempered interval.

In using the tuning fork, it must be remembered that the standard A fork has the same frequency as the fundamental of A49 (A4), and the standard C fork has the same frequency as C52 (C5). These are not notes in the usual scale area, which typically ranges from F33 – G47 (F3 – F4). In elementary theory one could tune the A49 (A4) or C52 (C5) to the fork and then tune down an octave. In practice this may not give the right result owing to inharmonicity, but as standard theory does not take account of inharmonicity, and does not deal with octave stretching, we will not elaborate on this here. These issues are dealt with later on.

The first note or "pitch note" is tuned beatless with the fork, unless there has to be a change of overall pitch of the instrument. Here, we are concerned with basic tuning procedure associated with standard, elementary theory, for which we will assume the task is one of "fine tuning" rather than overall pitch change. The rest of the scale must then be tuned using a *scale tuning sequence*.

Perhaps the simplest sequence is a "cycle" of fifths and fourths, such as the following:

Fig. 5.2

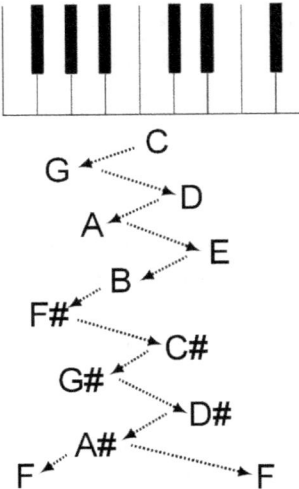

Apart from the octave on F, the only intervals tuned are fifths and fourths. The problem with such a "raw" sequence is that the other intervals, especially the major thirds, will not simply fall into place with the correct beat rates as a result. The reason for this is not entirely human error, but is also due to the many other factors controlling the physical tuning characteristics, that are dealt with later in this book.

The major thirds are particularly important because they (a) exhibit the most prominent of the fast beat patterns, and (b) determine the beat rates in the major tenths and seventeenths, which are also prominent, and contribute strongly to the tonal characteristics of the instrument over the wider compass. The first major third formed in this raw sequence is C to E, only after 4 perfect intervals have been already been tempered. If the C to E is found to be incorrect, there is no way of determining efficiently where the "error" might have occurred, or what correction strategy to follow.

It is necessary, therefore, to use in practice a sequence that allows control of the major thirds and major sixths as tuning proceeds. For the reasons that are beyond the elementary theory, it is in actuality most desirable to tune each note by both fast and slow beating intervals, as far as possible. Unfortunately the best of such sequences are "advanced", because they tend to be more demanding, requiring more experience and familiarity with the scale's complexity. There are nevertheless numerous other sequences suitable for learners.

RNC scale sequence

The next sequence illustrated here is a straightforward one taught at the Royal National College at the time of writing, that is realistic for practical use, because it includes the tuning of major thirds and sixths. It is suitable for beginners learning about beat rates and interval behaviour in scale tuning. It extends between G35 (G3) and G47 (G4) which simply avoids some of the problems due to unequal inharmonicity, that can occur in lower compass positions.

The best way to become familiar with specific beat rates is to have them demonstrated by an expert tuner, or to listen to electronically generated demonstration tones or intervals.[26] In practice it is also important to learn experientially about properties of beat patterns other than the beat *rate*, because these can affect beat rate perception. This requires electronic learning resources. Failing access to these resources, beat rates can be judged against the second hand of a clock or watch.

1. Tune middle C to the fork. Remember the pitch of the standard 523.3 Hz fork is an octave above middle C. It is actually the 2nd partial of middle C that is tuned beatless against the fork frequency.

2. Tune E above C. Being a major third, this interval must be wide. The pitch of the beating is two octaves above the E. When the interval is "pure" the beat rate will be zero. If the interval is narrow or wide, it will beat – the wider or narrower it is, the faster it will beat. The pitch of the E must be placed *above* the "pure" position, for the interval to be wide. The beat rate should theoretically be 10.4 beats per second. This beat rate has a distinctive "feel" which makes C-E a preferred starting interval for many tuners.

3. Tune the G35 below middle C. The pitch of the beat in the fourth G-C is two octaves above the G. The beat rate should be about 1 per second, and the interval must be wide, so the pitch of the G must be *lower* than its position when the interval is "pure" or beatless. Tuning the G also makes a major sixth with the E already

---

[26] For examples visit www.amarillibooks.co.uk

tuned. The major sixth is wide also. Its beat pitch is one octave plus a perfect fifth above the E, and its beat rate is 8.9 beats per second.

4. Tune the D above the G. The interval G-D is a perfect fifth, so it must be *narrow*, meaning the D must be flatter than its position when the interval is "pure" or beatless. The pitch of the beat is one octave above the D, and it should be 0.7 beats per second. Often, the beat two octaves above the D will be heard more easily. In theory, its beat rate should be twice that of the lower beat, 1.4 beats per second.

5. Tune the A, below the E and the D. The fifth A-E should be narrow, with a beat rate of 0.7 beats per second, at a pitch one octave above the E. Again, if listening to the beat at two octaves above the E, the beat rate is doubled to 1.4 beats per second. The fourth A-D should be *wide*, with a beat rate of 1 beat per second, at a pitch two octaves above the A.

6. Tune the B above the G, and below the E. G-B is a major third, so it is *wide*, its beat pitch is two octaves above the B, at a rate of 7.8 beats per second. B-E is a fourth, so it is *wide*, its beat pitch is also two octaves above the B, at a rate of 1.1 beats per second.

7. Tune F# above B and above D, and above A. B-F# is a *narrow* fifth, beat pitch one octave above the B, 0.8 beats per second. Remember if listening to the higher pitched beat, the rate is doubled. D-F# is a *wide* major third, beat pitch two octaves above F#, 13.3 beats per second. A-F# is a *wide* major sixth, beat pitch one octave and a perfect fifth above the F# (C#65), 10 beats per second.

8. Tune C# below F#, and above A. C#-F# is a *wide* fourth, beat pitch two octaves above the C#, 1.3 beats per second. A-C# is a *wide* major third, beat pitch also two octaves above the C#, 8.7 beats per second.

9. Tune G# below C#, and below C. G#-C# is a *wide* fourth, beat pitch two octaves above the G#, 0.9 beats per second. G#-C is a *wide* major third, beat pitch also two octaves above the G#, 9.4 beats per second.

10. Tune D# above G#, and above B. G#-D# is a *narrow* fifth, beat pitch one octave above the D#, 0.7 beats per second. Remember if listening to the beat pitch an octave higher, the beat rate is

doubled. B-D# is a *wide* major third, beat pitch two octaves above the D#, 9.8 beats per second.

11. Tune A# below D#, and below D. A#-D# is a *wide* fourth, beat pitch two octaves above the A#, 1.1 beats per second. A#-D is a *wide* major third, beat pitch two octaves above the D, 9.2 beats per second.

12. Tune F above A#(Bb), and above C, and above C#(Db), and above G#. Bb(A#)-F is a *narrow* fifth, beat pitch one octave above the F, 0.8 beats per second. Remember the beat pitch, and the beat rate doubling rule. C-F is a *wide* fourth, beat pitch two octaves above the C, 1.2 beats per second. Db(C#)-F is a *wide* major third, beat pitch two octaves above the Db(C#), 11 beats per second.

13. Tune G47 as a beatless octave above G35. C-G is a narrow fifth, beat pitch one octave above the G, 0.9 beats per second. D-G is a wide fourth, beat pitch two octaves above the D, 1.3 beats per second. Eb(D#)-G is a wide major third, beat pitch two octaves above the G, 12.3 beats per second. Bb(A#)-G is a wide major sixth, beat pitch one octave plus a perfect fifth above the G (D66), 10.6 beats per second.

The real scale tuning situation

The reason it is necessary to consider all these intervals, rather than just tuning each note by one interval and one beat rate alone, is far from being simply that mistakes can be made. Also, the quoting of beat rates to one decimal place does not imply that beat rates must always be tuned precisely to the quoted figures. Even inexperienced students of tuning can usually hear differences in beat rates of 0.1 Hz, when digitally isolated, but the precise beat rates as quoted are only valid where inharmonicity is zero.

In the real piano, inharmonicity is not only significant, but it varies over the scale. Furthermore, actual beat rates (as measured by digital analysis) are not in general constant or regular to within 0.1 Hz. A close correlation can usually be achieved on a good instrument, but because of the differences between the elementary theoretical model and the much greater complexity of the actual acoustical system, the beat rates quoted by elementary theory are only a guide.

Pianos are therefore not *in general* capable of a scale tuning in which all the beat rates can appear exactly as specified by elementary theory, to one decimal place. However, both in the elementary model and in the actual acoustical system, the beat rates are inescapably *related* to each other in one complex network of values. They are "connected" to each other and affect each other as one complex network. It is this *relationship* between beat rates that is most important.

The principle of excellence in temperament tuning rests not so much on specific beat rates in specific places, but on the *progression* of beat rates. The beat rates specified are a good guide to get close to progression in the first instance. Progression means the rates for a given species of interval will increase as steadily as possible with rising position in the compass, and *vice versa*. The importance of this in relation to the final quality of tuning over the whole instrument, and the final tone production, cannot be overstated. Good *progression*, not beat rates supposedly "accurate" to one decimal place, is how scale tuning succeeds. It is the hallmark of the master tuner, just as it is the most demanding thing to achieve.

This *does not mean* that "*any* beat rates can be tuned anywhere, as long as they progress", because only the "correct" set of beat rates *will* progress, and on most reasonably sized instruments this will be reasonably close to the "traditional" specified set of rates. The complex network of relationship between beat rates, both in the elementary model, and in practice, determines that each major third beat rate cannot be outside a certain very small range of values, without destroying the possibility of progression. Furthermore, it is easy to obtain progression in any one set of intervals, say the major thirds, at the expense of another set, say, the perfect fifths, but neither set should be sacrificed for the sake of the other.

Finding the correct set of rates requires a starting point. The beat rates provided by elementary theory may not necessarily be the exact solution for progression, but they are a "seed", a starting point close to the solution for progression. For this reason we must "know" the beat rates of elementary theory and be able to apply them, in the first instance.

Scale tuning is more flexible than the elementary theory suggests, but this does not make it any easier. It many ways it can make it more difficult, until we understand the actual nature of a piano's tuning characteristics. In practice, experience alone teaches the tuner how to quickly recognise a piano's true tuning characteristics, and how to work with them rather than against them, in order to bring about a scale with good progression.

Tuning the octaves

Having tuned the scale, tuning in octaves outwards from the scale should, in the elementary model, project the temperament tuned within the scale area over the rest of the compass. Octaves are therefore tuned upwards and downwards through the compass, in "traditional" theory, to a "beatless tuning". We shall see later that in fine tuning this idea is a "fuzzy" one, not very well supported by the acoustical facts.

The intervals formed between the notes outside the scale area and the notes within the scale, should contain beat rates that continue the progression of rates outside the scale area. Tuning upwards in octaves, we will soon reach notes that are a major tenth above notes in the scale. The major tenth is then regarded as a "test" interval for the tuning of the octave. When the octave is beatless, or "in tune", the major tenth should have the same beat rate (and same beat pitch) as the major third whose lower note is shared with the lower note of the tenth. Higher in the compass we will reach notes falling at major seventeenths (octave-tenths) above notes already tuned below. These will also have the same beat rate (and same beat pitch) as the major third on the shared lower note.

Because the beat rates in the scale's major thirds *progress*, the beat rates in the major tenths and seventeenths will also progress. Any lack of progression in the scale's major thirds, will be reflected in the major tenths and seventeenths if the octaves are beatless.

Tuning down into the bass in octaves, beat rates in the other intervals thus formed similarly progress decreasingly outside the scale area. As major tenths are formed below the scale, their beat rates should also decrease progressively as they are "tested" consecutively downwards into the bass. The same progressive decrease should be found in the major seventeenths below the scale.

The minor third / major sixth test discussed above can also be employed to "test" the tuning of the bass octaves. If an octave is wide, the lower note will be flatter than it should be, so the minor third at the bottom of the octave (from the lower note of the octave to a note a minor third above) will not be as narrow as it should be. Its beat rate will therefore be *slower* then "normal". The "normal" beat rate should be equal to the beat rate of the major sixth that completes the octave above the minor third. A wide or narrow octave will result in the beat rates of the minor third and major sixth being unequal. For a wide octave the third is slower than the sixth, and for a narrow octave the third is faster than the sixth. Note that ultimately, it is the *octave* tuning that is most important, not the minor third / major sixth relationship *per se*. On some octaves on some instruments (a minority) the relationship will not hold accurately true when the octave is correct.

The real octave tuning situation

In professional tuning practice, the tenths and seventeenths are not just "tests" for the tuning of the octaves. They may be used as tests as part of learning how to tune octaves in the early stages, but in actuality these intervals are important in their own right. Octave tuning cannot be compromised for the sake of the tenths or seventeenths, but later we shall see how octave tuning and the tuning of the tenths and seventeenths link together in the context of "octave stretching". Octave stretching is an essential feature of tone production in the instrument as a whole, but is often misunderstood.

In the elementary model, octave stretching is equated with octaves that are "out of tune". In actuality, this view of octave stretching is oversimplified, like the elementary proposition that an interval must either beat or not beat. Even when an attempt is made to improve elementary theory by taking into account inharmonicity, the mechanics of controlled octave stretching as practiced by the master tuner is still not properly explained. The position of elementary theory is that an interval either beats or does not beat. In the actual acoustics, this is an oversimplified description.

The physical situation in the real piano

The ways in which the actual acoustical situation in the piano differs from that assumed by the elementary theory, are numerous. Standing waves do occur on the string, but the situation is more complex. The suppositions of rudimentary theory do work for hypothetically perfectly flexible strings, vibrating in one plane, and whose speaking lengths fall between perfectly rigid supports or bridges. In the real situation, the strings are considerably stiff, rather than perfectly flexible. One end of the speaking length, at the top bridge or agraffe, approaches rigidity.

The other end, however, is deliberately designed not to be acoustically rigid. The long bridge or bass bridge must be capable of movement in order to transmit the energy from the string to the soundboard. It is the movement of the soundboard that causes sound waves to be radiated from it into the surrounding air. The fact that the bridge *moves*, means that the movement of one string can introduce movement to another string on the same bridge, or another string connected through a bridge to the same soundboard. In other words, strings *affect each other*. This is known as bridge (and soundboard) *coupling*.

The termination of the speaking length at the long bridge (or bass bridge) is by the side draft pin. The bridge and soundboard are designed to move transversely to the string's speaking length. There are, however, two transverse degrees of freedom for movement. In other words, where the string joins the bridge, the end of the string may be subject to movement in *two* planes transverse to the speaking length. Even *longitudinal* motion, motion in line with the string's length, is possible.

The effect of movement at the bridge can be acoustically measured as its *admittance*, which is a measure of how the bridge resists and reacts to movement from the string. The top bridge or agraffe, being part of the massive iron frame, has practically zero admittance, which means the bridge resists movement, and does not admit much wave energy from the string to be transmitted though it. The long bridge, however, has non-zero *admittance*, because it is designed to transmit wave energy. Looking at the design and construction of the soundboard and bridge, it seems that this admittance would not be the same in all transverse directions (e.g., on a grand piano, in the horizontal and vertical planes). The soundboard is *designed* to move, on a grand piano, up and down in the vertical, but not in the horizontal. It would seem the admittance should be *much* greater in the vertical, than in the horizontal.

In actuality the admittance values in the horizontal and vertical at the string boundary itself are not so different.[27] However, unless the value is *actually identical* in all transverse directions, the motion of the string will not, in general, be confined to one plane. A difference in admittance in different directions can cause motion in *two* planes, even when the hammer strikes the string in only one plane. Motion in two planes is often dealt with in physics in terms of *elliptical polarization*, but in the piano's case, the motion is not even confined to a fixed elliptical polarization. The string motion can thus be *much* more complicated than that described by elementary theory.

As already mentioned, the fact that the strings share the same bridge and soundboard results in motions being *coupled*. The physics of coupled motions is considerably more complicated than that of uncoupled

---

[27] See Weinreich, 'Coupled piano strings', *Journal of the Acoustical Society of America*, 1977, 1480.

motions, and can give results that the simpler system assumed by elementary theory, is not capable of illustrating. Finally, the sound heard by the tuner is not directly from the string motion, but from the motion of the soundboard itself, which complicates matters further.

From the practical point of view, all this means that the ideas of "traditional" theory fall short of describing the actual world of sound encountered in piano tuning. This sonic world is actually far richer and more complex than the theory suggests, and it is this richness and complexity that we will be exploring in the rest of this book.

Practical tuning summary

The essential information for the practical approach to tuning that the "traditional" model proposes, can be summarised as follows:

1. The strings of a unison group must be tuned beatless.

2. The strings of an octave, or compound octave, must be tuned beatless.

3. A "scale" (a chromatic scale of 12 notes) can be tuned in the centre of the compass, by tuning the intervals between the notes so that they are not beatless or "pure", but have beat rates according to the Beat Rate Table shown earlier. Only some of the intervals are actively tuned. These are perfect fifths, perfect fourths, major thirds, major sixths, and sometimes minor thirds and minor sixths. These intervals are the tempered intervals, as distinct from beatless or "pure" intervals, namely the octaves.

4. The beating partial (the adjustable partial) in each scale tuning interval, occurs at a particular pitch, in relation to the interval. These beating partials can be heard at the following pitches:

Major thirds: 2 octaves above the upper note

Perfect fourths: 2 octaves above the lower note

Perfect fifths: 1 octave above the upper note

Major sixths: 1 octave plus a perfect fifth above the upper note (2 octaves plus a major third above the lower note)

Minor thirds:	2 octaves and a perfect fifth above the lower note
Minor sixths:	3 octaves above the lower note
Major tenths:	1 octave above the upper note
Major seventeenths:	At the pitch of the upper note

1. Perfect fifths, minor thirds, and minor sixths, must be narrower than their "pure" or beatless tuning.
2. Perfect fourths, major thirds, and major sixths, must be wider than their "pure" or beatless tuning.
3. Beat rates can be *compared*. The beat rates in intervals of any species *progressively increase* or *progress* with the rising position of the interval in the scale, or in the compass.
4. Further *comparison:* The beat rate in a major sixths is approximately the same as the rate in the major third whose lowest note is one tone above the lowest note of the major sixth.
5. Further *comparison* for tuning descending octaves (in ascending octaves higher in the compass the beat rates become too fast to be useful): The beat rate in a minor third is the same as the rate in the major sixth above it, when the lowest note of the minor third and the highest note of the major sixth, forms a proper octave.
6. Further *comparison* for tuning descending and ascending octaves: The beat rate in a major third is the same as the rate in the major tenth, and the major seventeenth on the same lowest note, when the octaves are properly tuned.

Remember that all these "rules" are a *basis*. Situations arise in practice where the "rules" themselves fail to be accurate.

Chapter 6 - The soundscape, spectrum and tone

Piano tone, whether of a single note or a musical interval being tuned, can be thought of as an audible *soundscape*. In this chapter we look at the general nature of this soundscape, and how it is comprised of a *spectrum* of ingredients called *partials*. Partials are important because they provide the key to precision control of tuning, and the soundscape. In a steady state, unchanging tone, for example an steady organ tone, each separate ingredient would be uniquely[28] identified by its frequency, and the "amount" of each ingredient present in the overall recipe, would be that ingredient's amplitude.[29] The tone recipe consists of different ingredients with different frequencies, in possibly different amounts, that is, with different amplitudes.

There are *two ways* of representing the audible tone, one showing the amplitude or the amount of each ingredient in the recipe, whilst the other shows the whole recipe extended over time. If the recipe changes over time, when we show it in terms of its ingredients, we can only show the "average" or "total" value of each ingredient over time, or some other statistic, or we must actually show how the recipe changes over time.

We can illustrate this in the simplest case, where the recipe has only one ingredient. This would be the case of a single pure tone.

Fig. (6.1) shows a graph of a single pure tone as amplitude versus time. This is the *whole recipe* shown over time.

---

[28] It is also possible for different vibrational ingredients with the same frequency to occur, these being known as *degenerate modes*.

[29] For our purposes we will continue to use the term "amplitude" rather than complicate further by introducing more analysis-specific terms such as magnitude, spectrum density, or power, for frequency domain measurements.

Fig. 6.1

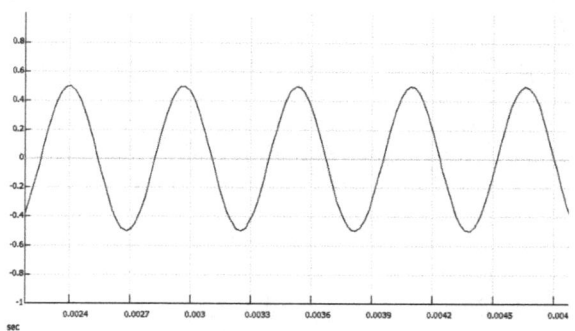

Fig. 6.1. A single pure tone represented in amplitude versus time. The wave form is sinusoidal.

Fig. (6.2) shows the same single pure tone, as "amplitude" versus frequency. This is the ingredients (there is only one) "averaged" or "totalled" over time (the statistical method can be more complicated but we need not be concerned with that here).

Fig. 6.2

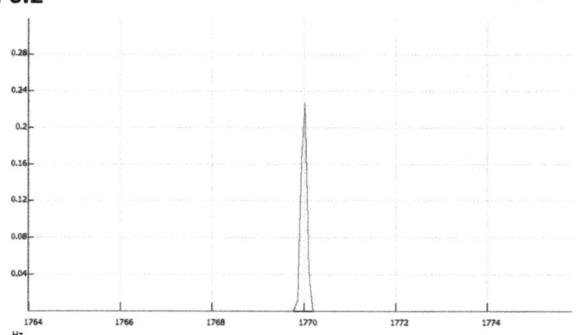

Fig. 6.2. The same tone as in Fig. (6.1), but represented in "amplitude" versus frequency. The form of the graph is generated by digital analysis, the width of the peak being a consequence of the method used to analyse digital data. The peak is a statistical representation of what is actually a single frequency ingredient whose frequency value is on the vertical centre line of the peak.

We can also do this where the recipe is more complicated and has *more than one* pure tone ingredient. Fig. (6.3) shows a more complex tone, made from four pure tone ingredients, represented as amplitude versus time.

Fig. 6.3

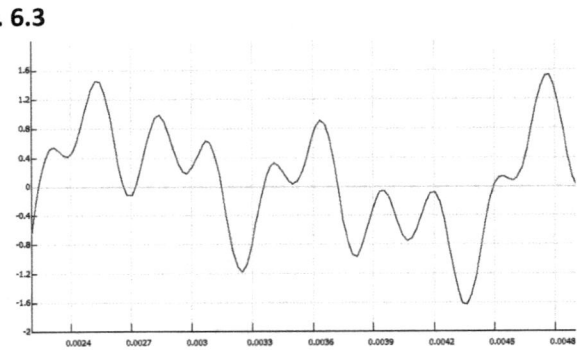

Fig. 6.3. A complex tone with four pure tone ingredients, represented as amplitude versus time. The wave form is not sinusoidal, but is a mixture of four different sinusoids.

Fig. (6.4) shows the same complex tone, represented as "amplitude" versus frequency.

Fig. 6.4

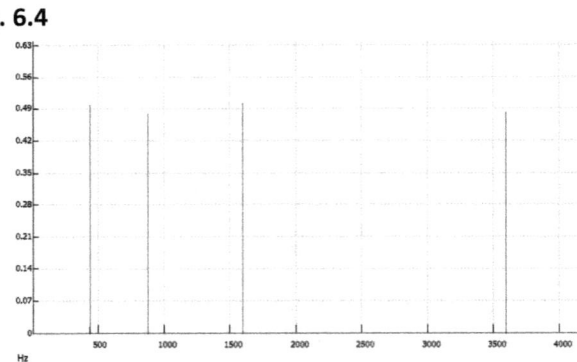

Fig. 6.4. The same tone as in Fig. (6.3), represented as "amplitude" versus frequency. Each peak represents the presence of a pure tone ingredient at that frequency value on the horizontal axis.

Where the amplitude is shown versus *time*, the representation of the tone, the whole "cake" of sound, so to speak, is known as a *time domain* representation. Where the "amplitude" is shown versus *frequency*, the representation of the tone is known as a *frequency domain* representation, showing the frequency ingredients rather than the complex wave that results from mixing the ingredients.

Each graph would in fact be a graph of a mathematical function, and it is possible to convert from the function for the time domain representation, to the frequency domain representation, using a process called *Fourier transformation*. In short, the frequency domain representation is commonly known as the *Fourier transform* of the time domain representation. The frequency domain representation is often just called a Fourier transform, or just "a spectrum".

Note that a single pure tone transforms to a single "peak" in the frequency domain, which shows the "amplitude" and frequency of that single pure tone, or ingredient. The more complex tone transforms to a number of different peaks. In other words, what makes the second tone more complex, is that it contains a number of ingredients rather than just one, and in the frequency domain or *Fourier transform*, or *spectrum*, one peak appears for each ingredient or pure tone, that the *recipe* for the complex wave contains.

A practical counterpart of this occurs in actually listening to the tone (remember that we are currently talking about a steady state tone like a plain organ note). We can hear the overall tone as a complete recipe, without awareness of the individual ingredients, and we then identify the tone by its perceived "tone quality". This is the aural (and partly subjective) counterpart to the objective *time domain* representation. When listening to the tone quality, we are, in effect, hearing a complex wave form in the time domain.

With a little training, however, it is also possible to consciously "split up" the musical tone, *by ear*, into at least some of its component ingredients, and to consciously perceive these ingredients as individual tones in their own right. When we do this, we are in effect creating in each moment, a psycho-acoustic or aural counterpart of the *Fourier transform*, and then we are in effect *hearing* each frequency peak that would appear in the transform, as a separate simple waveform that we hear in time.

To sharpen the acuity of hearing for tuning, it is good to practise listening to component ingredients of musical sounds in general, and not necessarily just piano tones. There are methods for learning how to do this, and practical help from a teacher in the early stages is invaluable. This may involve listening to the sounds of metal objects like bells and bowls.

Listening to component ingredients of sound, or *partials*, should be *enjoyable*, and often *amusing*. Many of the students I teach have great fun discerning and commenting on partials in diverse sound sources. Once you can hear partials, and particularly *beating* partials, you hear them everywhere. Your hearing changes. Your world changes. A "chink" of glasses colliding in a toast, is no longer just a chink. It becomes a couple or more clear and beautiful partials at definite musical pitches, often mixing together to produce rapid beating, indicating the frequency difference between them.

The transient

Strictly speaking, when we speak of piano "tone", we should speak of the *decay tone* after the hammer strike, rather than just "piano tone", because the decay tone, even with its recipe of partials, is only one of the components that actually determines how we perceive tone quality of a piano note. The first part of any musical tone is usually the *transient*, which is a 'noise' at the beginning of the note, before the decay tone starts. In a pipe organ, the transient will be the 'chiff' at the beginning, when the air starts to flow into the pipe, and this is generally followed by a steady tone, rather than a decaying tone. In a piano tone, the transient is largely the knocking sound of the hammer against the string.

The second section of the piano tone is then the decay tone. Both the transient and the decay tone provide important information to the ear-brain system that enables the source of the tone to be identified, and enables the sound of one instrument to be distinguished from another. Changing the transient can greatly change the perceived tone of the note. For example, if we record the piano and electronically take away the transients, replacing them with a gradual "fade-in" to the decay tone, the result can sound very unlike a piano, and more like a harmonium. In general, when we use the term "tone" applied to the piano, it is the decay tone to which we will be referring.

Changing tone - *movement*

Piano tone has another dimension to it, in addition to the idea of a recipe of frequencies and amplitudes. This other dimension is that piano tone - the tone even of a single string - *changes* in time. The recipe for the tone is itself constantly changing, as the tone dies away, or *decays*. The way a piano tone changes as it sounds, is an important part of what makes it sound like a piano tone.

Change to the recipe as the piano tone continues to sound, is therefore as important as the recipe itself. The ingredients themselves, at any time, do play an important part in determining the tone quality, but the actual *changes* to the ingredients and to the recipe as the tone proceeds, also play an important part in determining the overall perceived tone quality.

We are saying that there are two aspects to piano decay tone. One is the general *kind* of recipe that it has, whilst the other is *changes in the recipe* as the tone proceeds. The *timbre* of a note or interval depends on both the nature of the recipe at any time, and how it changes over time. As far as aural perception by the piano tuner is concerned, changes to the recipe as the tone proceeds, can all be described by one adjective: *movement*. It is changes to the recipe that are perceived as *movement*, or change taking place in the audible soundscape.

The audible soundscape can be likened to a seascape. A seascape is a much better analogy than a landscape, because the sea is always changing. Seldom is it absolutely still. A seascape typically exhibits moving waves, undulations, risings and fallings, and currents. It can be rough, calm, or more rarely, appear to be absolutely still. These features, to the trained ear, are all found in the soundscape of piano tone. Furthermore, they are largely *adjustable* in the tuning process.

Piano tone partials – revision of essential facts

In the chapter on sound we saw that a complex tone can be considered as a mixture, or a recipe, of simple *pure tones*. A pure tone is the simplest, elemental ingredient of sound, and cannot be broken down into simpler sounds of different frequencies.

We also saw that a piano tone, however, naturally "splits up" into a set of ingredients that are not necessarily, in themselves, the simplest, elemental ingredients possible. In other words, the individual ingredients that are audible in piano tone, are not necessarily single pure tones, although ultimately, of course, both the overall tone and these separate audible ingredients, can still be "split up" into individual pure tones.

Each aurally distinct ingredient of piano tone is called a *partial*, and the partial may or may not be a single pure tone. If a piano tone partial is not a single pure tone, it may be a pair, or a small group of pure tones. Each partial exists at a particular frequency, or within a particular *small frequency range*. The partials themselves, however, are much more separated from each other in their frequencies, so they are distinct from each other, and we can still approximately refer to a partial as having *a* "frequency", rather than a group of frequencies. In contrast, if a single partial is "split up" (say, digitally) into separate pure tones, the pure tones contained within a partial are always *close* in frequency, and hence they are not aurally distinct.

Harmonics

Early piano tuning theory refers to audible ingredients of piano tone called *harmonics*. As far as the natural tones of most acoustic musical instruments are concerned, harmonics can be postulated as a hypothetical kind of ingredient, rather than actual ingredients. Harmonics have special properties, and are a concept that can only properly be applied in a limited number instances, which include electronic signals, mathematical models, and other phenomena in physics. In the case of sound, a single harmonic would be a pure tone. However, pure tones do not necessarily have to be *harmonics*.

A set of pure tones are only called *harmonics* if their frequencies are whole number multiples of the lowest frequency pure tone, the fundamental. The frequencies of the pure tones are then said to be arranged in what is commonly called the *harmonic series* of frequencies.

When mid-range audible pure tone frequencies are arranged in this way, the perceived musical pitches of the pure tones are generally heard such that there are *recognisable musical intervals* between the pure tones in the first part of the series. A series of pure tone frequencies that approximate sufficiently closely to an harmonic series can also be perceived to have recognisable musical intervals between the harmonics. This can be useful when attempting to aurally identify specific harmonics.

True harmonics, such as can occur in some electronically generated sound, are pure tones that have their frequencies arranged perfectly in the harmonic series. More generally, however, musical sounds may "split up" into *partials*, rather than specifically into harmonics. Even if each partial is a pure tone, their frequencies do not necessarily have to fall in the harmonic series, *or even into a set anything like it*. Many sounds, like the tones of bells and bowls, have partials arranged in a very different way. In the case of piano tone, each partial may be a single, pair or group of pure tones, and the partials do *approximately* fall in the harmonic series. Because their frequencies *resemble* an harmonic series, their deviation from a true harmonic series is known as technically as *inharmonicity*. Piano tone partials are therefore sometimes called *inharmonic* partials.

Any complicated sound wave, or vibratory motion, will only have ingredient pure tones arranged in an harmonic series, *if the wave or motion is periodic*. We saw in the chapter on sound that to be periodic, the wave form or vibratory motion must keep repeating itself exactly in time. Piano string vibrations are not periodic, so the ingredients of the sound they produce are not arranged in an harmonic series.

Partials

The fact that piano tone partials fall only approximately in the harmonic series, and not exactly, has very important consequences, which we shall examine later. Just as important, however, is the fact that whilst a true *harmonic* is necessarily a single pure tone (this is a mathematical necessity), an actual audible *partial* of piano tone may contain *more than one* pure tone. This is very important in piano tone because two or more pure tones contained within a single partial will, as we have said, always be *close in frequency*. As we saw in the chapter on sound, pure tone ingredients close in frequency may *fluctuate* or *beat*. This means that partials, *in general*, may naturally fluctuate or beat (unless the partial also happens to be just a single pure tone), whilst pure tones, or true harmonics, do not.

Beating is therefore a natural feature of piano tone partials. The tone of even a single string could be "split up" into an approximately harmonic set of partials, some of which may beat. Beating is not something that only occurs as a consequence of the way we tune the instrument. In tuning, as we shall see shortly, some partials are *adjustable*, and we deliberately adjust some of these partials so that they contain close-frequency pure tones, which beat. Other adjustable partials we attempt to tune so that they do not beat.

Nevertheless, the potential for the existence of close-frequency pure tones exists in *all* partials, and is natural consequence of the acoustics of piano tone. From the point of view of the physical acoustics of piano tone, partials that do not contain close-frequency pure tones, certainly do exist, but are a special case, or *subset* of the more general case in which close-frequency pure tones are allowed to occur within a partial.

In sum, the tone of a single string or well tuned unison group (a single note) can be "split up" into an approximately harmonic series of *partials*, well separated in frequency. Each partial, in turn, may be a single pure tone, or it may itself be "split up" into a pair or a small group of pure tones that are close in frequency. In turn, these close frequency pure tones may cause the partial to fluctuate or beat. Some partials are adjustable, whilst others are not. We will say more about adjustable and non-adjustable partials shortly.

The musical pitch arrangement of piano tone partials

Each partial of the piano tone is in effect a musical note in its own right, at its own particular pitch, contained within the overall tone. The tonal quality of a partial itself is generally rather similar to that of a tuning fork, but because it may contain more than one pure tone, its tone is not necessarily steady. Each partial has its own initial loudness, its own rate of decay (rate of "dying away"), and its own *decay pattern*, which is how it fluctuates as it decays. It also has its own frequency, or mixture of frequencies (the frequency may also sometimes vary throughout the decay time).

The audible musical pitches of the partials for any string are arranged in an orderly series, from the lowest upwards, with recognisable musical intervals appearing between the first ten or so partials. The pitch of the lowest partial corresponds closely to the perceived pitch of the piano note itself, and this is called partial number 1, or *the fundamental*. Next above this in pitch is partial number 2, and so on.

The first six partials in the series, and certain partials above the first six, are easily identified because their pitches appear as recognisable musical intervals above the fundamental. The following table lists the first ten partials in terms of their musical intervals above the fundamental, and also in terms of the musical interval between each partial and the partial immediately below it in the series.

Partial Number	Interval above the fundamental	Aprox. interval above one below
1 (fundamental)	Unison	Unison
2	Octave	Octave
3	Octave + fifth	Fifth
4	Two octaves	Fourth
5	Two octaves + maj third	maj third
6	Two octaves +fifth	Min third
7	*Two octaves + min seventh*	* min third*
8	Three octaves	maj second*
9	Three octaves + maj second	maj second
10	Three octaves + maj third	maj second

* The 7th partial may sound flat, and the intervals it forms therefore correspondingly altered.

For the note C28, the pitches of the first 6 partials could be represented as the following C major chord:

The first 10 partials would be represented by the following dominant seventh chord with an added second:

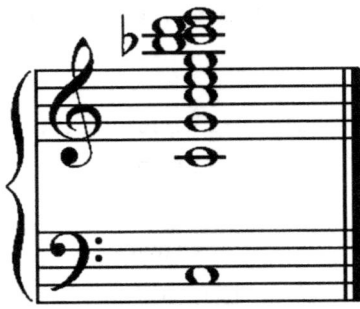

It must be remembered that the perceived pitch of individual partials, and hence the perceived musical intervals between them, is subject to psycho-acoustic factors; perceived pitch is always in part psychological, as well as being subject to physiological influences from the ear itself. This is particularly so because the pitches and intervals perceived in the partial series are perceptions of partials that are either pure tones, or partials that contain very few pure tones.

Pure tones or individual partials are often perceived differently to actual musical notes from musical instruments. In music practice, in listening to musical notes themselves, especially those of the piano, the pitches and intervals we perceive are based on the perceptions of notes that are usually each individually rich in a large number of pure tones or partials. In listening to notes from musical instruments, perceived pitch, and hence musical interval size, is very much dependent on *many* of the partials present in the tone of each note, and the relationships between them.

The perceived, apparent "sizes" of the intervals in the partial series, or the pitches of the partials, therefore only *correspond* to pitches and interval sizes found in practical music making. Whilst this correspondence exists, there is no rigorous, exact, or direct mapping from the measurable properties of partials, to the measurable properties of musical intervals used in music. If we are to fully understand the nature of piano tone in the tuning process, we must appreciate that musical intervals, *as real sounds*, cannot be rigorously defined simply by the frequency ratios found between pure tones in the harmonic series.

Partials above number 10 are called *upper partials*. The seventh partial in the table, corresponding to the B-flat in the staff notation, is asterixed. This one is an "odd one out" amongst the first ten partials, because it can seem rather flat when heard, compared to the expectation of the modern for the seventh of the overall chord that the partials form. If the fundamental was a C, the seventh partial would be approximately B flat, but it would probably sound a little too flat to many modern musicians. We could perhaps refer to it, using contemporary musical terminology, as a "B three quarter flat". Consequently the eighth partial's interval from it, is not exactly a major second, but is somewhat greater than a major second, and so it is asterixed too. Note that the intervals between adjacent partials in the series get progressively smaller as the partial numbers get higher. The higher partials above the 10th, form, in effect, a tone cluster, and partials higher still, form a microtone cluster.

Hearing partials and beats

The pitch arrangement of the partials makes them "easily identified", but this does not mean they are easily *heard* by the untrained ear. Without training, this can even be impossible. Often, it is only by listening for *beats* or other fluctuations in the partial's decay period, that the beginner starts to hear partials, because beats or fluctuations are a form of audible *movement* which is easier to notice. Partials can still be distinguished by the expert ear, even when they do not beat.

Audible beating can occur in the partials of single strings, or can be caused when two (or more) strings are sounding together. When two strings sound together, the resultant sound is composed of a series of partials similar to (but not necessarily identical to) the addition of the two series of partials from each of the separate strings. Wherever the two single string tones would each have had a partial at approximately the same frequency, if they sounded separately, there will occur in the spectrum of the two-string tone, an *adjustable partial*. Adjustable partials can readily exhibit beating, the beat rate being adjustable by altering the tensions on the strings.

It is possible to hear partials in three ways, that we could call "unconsciously", "semi-consciously" or "consciously". The ear-brain system already naturally, in effect, "hears" partials "unconsciously" by itself, when it senses a musical pitch. Although the perceived pitch of the fundamental partial *corresponds* to the actual musical *pitch* of the whole note, it is not necessarily identical to the perceived musical *pitch* of the note, as it is not just the fundamental that contributes to our perception of pitch. In fact, if the fundamental partial is entirely *absent* but a few other suitable partials are present from the harmonic series, we may still "hear" a fundamental pitch even though it is not actually there. This ability of the ear-brain system to interpret partials and their relationships in this way, is not the same thing as conscious awareness of specific individual partials, or of the *beat rates* that are required for piano tuning.

To consciously hear *beats*, we need to be at least "semi-consciously" aware of partials. Listening for beats can make us semi-consciously aware of partials, because we may hear the beating as *movement*, but not actually recognise the partial individually at its own musical pitch. We can make the perception of beats much clearer by *consciously* listening to partials, which means the beating is heard at the actual pitch of the partials causing the beating. Beating partials are often easier to hear than non beating partials. With training it is possible to hear non beating individual partials from a single string, as separate components of the overall tone quality. The ability to do this is a sign of developing acuity of hearing.

The 3-D spectrum of partials

On the one hand, we can hear the decaying tone of the piano note as one tone. On the other hand that tone is comprised of an entire *spectrum* of individual, decaying partials. The partial frequencies of a single note approximate to the harmonic series, but their individual amplitudes and decay rates usually differ in a less orderly way.

We cannot fully represent the audible soundscape of a piano tone merely by a two-dimensional graph as we did above for a steady-state, unchanging tone. *Changes* to the tone recipe, that is, changes to the ingredients as the tone proceeds, are very important features on the tone quality in their own right. These changes, or *movement* in the soundscape need to be represented in some way that not only shows what the recipe is at any time, but also shows how it changes over time. In particular, *beating* is a type of audible movement that occurs in individual partials, so we need a way of showing individual partials at their particular frequencies, as in the Fourier transform, but that also is capable of showing beating in the partial.

To graphically represent the decaying spectrum of a piano tone we need a three dimensional graph that can show this *movement*, or change to the recipe. We need the graph to show the frequency of each partial and its changing amplitude, versus time. Mathematically, there can in theory be an infinite number of partials, but in practice only the first 20 or so in the mid-compass might show up above the "noise floor" in a digital recording using inexpensive equipment. Other physical factors can limit the amount of energy available for higher partials.

Fig. (6.5) is a computer generated 3-D spectrum model of the first nine partials generated by a hypothetical single piano string.[30] It may seem strange that frequencies on the horizontal axis increase from right to left. This is because we are looking at the 'back' of the graph. The front elevation of the graph is shown in Fig. (6.6), and this can be recognised as similar to the kind of graph we looked at earlier for the Fourier transform of the tone's complex wave. Each partial is numbered.

Fig. 6.5

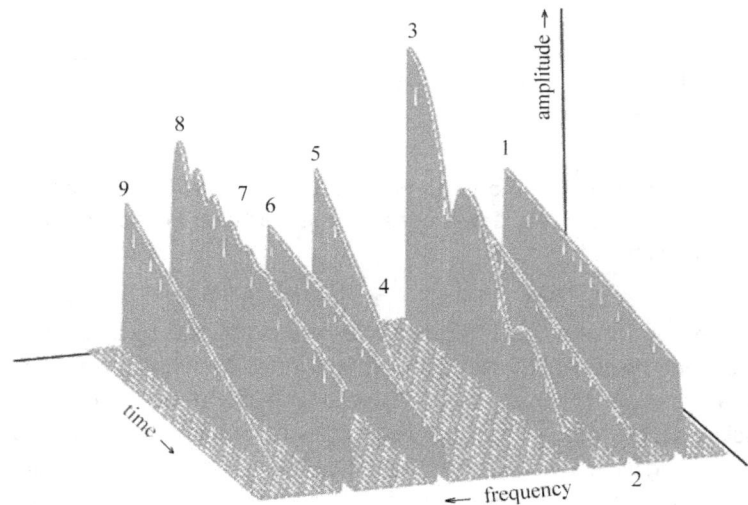

Fig. 6.5. 3-D spectrum of the first 9 partials of a single string piano tone. In this model, the natural decay of a partial that does not exhibit beating (partials 1, 2, 4, 6 and 9), gives the partial a straight line gradient, the height or amplitude of the partial decreasing uniformly as time increases.

---

[30] A hypothetical model is chosen here because it is easier to "read". Many real decays are illustrated later.

Fig. 6.6

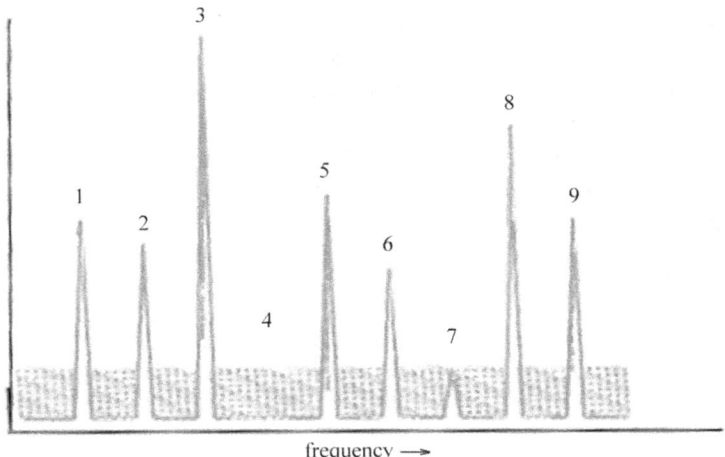

Fig. 6.6. The model in Fig. (6.2), viewed from the rear. This shows the initial amplitudes of the partials, in the frequency domain.

When tones are digitally analysed in this way into a 3-D representation, the resulting image is often referred to as a *time FFT* or *waterfall*. The term FFT stands for *fast Fourier transform*, referring to the digital process that allows a discrete transformation to be achieved quickly, at the cost of some information or precision which is lost. The term *waterfall* is a visually descriptive term for the appearance of the 3-D graph, especially when the partials are not as clearly defined as in Fig. (6.5). The beating partials found in piano tones give the impression of flowing water, and are by no means unique to piano tone.

The spectra for the tones of different individual strings or notes will differ. In this particular example there are some interesting features to note, that are not untypical of the tones of individual piano strings in general, although the specific features shown here only apply to this particular example, which has been generated for clear illustration purposes only. Each piano tone spectrum will have its own individual set of features, there being no universal rule for where particular features should appear in any given spectrum.

In this example, partial number 1, the fundamental, does not have the greatest initial amplitude. This will be true of many actual spectra of real piano strings. Secondly, partial number 4 appears to be absent, or of very small initial amplitude. Sometimes, partials in real piano string spectra may aurally seem to be "missing", because they are very diminished compared to the other partials.

Some partials have a steady, unfluctuating decay – this means that in this kind of representation they have a straight line gradient of decay in the time dimension. In other words their amplitude steadily decreases in time, as a straight line slope in the graph. (The decay of sound is actually *exponential*, meaning it could be represented as a smooth curve, but it is often helpful to graph the amplitude on a logarithmic scale, so that exponential changes of this kind appear as straight lines). Looking again at Fig. (6.5), note that the 5th partial begins with a greater amplitude than the 6th, but it decays faster than the 6th, so eventually the 6th partial is more *prominent*. Changes in prominence are themselves a form of *movement* in the soundscape.

Perhaps the most notable feature of all is that in this particular example, partials 3 and 8 *do not* decay steadily. We can see this more clearly if we rotate the graph to see the profile better (Fig. 6.7):

Fig. 6.7

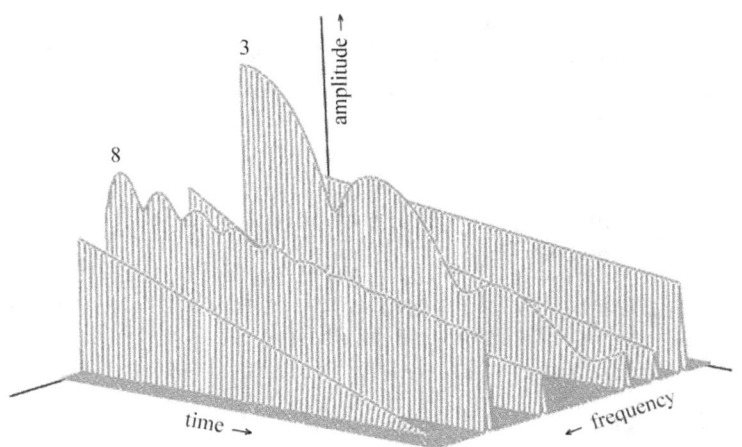

Fig. 6.7. Here Fig. (6.5) has been rotated to show more clearly the beat patterns in partials 3 and 8.

The amplitude of these partials is decaying overall, but during the overall decay the amplitude is rising and falling in a distinct *beating* pattern. These partials clearly have more than one pure tone component.

Such partials occurring in the spectrum of a single string are called *false partials*, and the beat pattern is called a *false beat*. (They are real partials but are "traditionally" called *false*, because the beating in them is occurring in a *non adjustable* partial of a single string. "Traditional" theory only accounts for beating in *adjustable* partials, which are only created when two strings are sounding together).

The beat rate of partial 3, in this particular example, is slower than that of partial 8. Note that the amplitude of the beat - as distinct from the overall amplitude of the partial – is itself decaying. In other words the height or depth of *the beat pattern itself*, gets smaller, in time, in addition to the fact that the overall amplitude of the partial is decreasing. There are two decay rates here, the overall decay rate of the partial and the decay rate of the beat, and they are not necessarily the same. These two partials also have different initial beat amplitudes. The height or depth of the beat in partial 3 begins much greater than that of partial 8. To hear, this would be a slow but prominent beat, whilst the beat in partial 8 would be faster but less prominent. Beat *rate*, beat *amplitude*, beat *decay rate*, and beat *prominence*, are all important and variable features of beating phenomena in piano tone.

All the features shown here, are the *kinds* of features found in a real piano string spectrum, in this case all packed for convenience into one illustration, but every spectrum will be different. The spectrum is like a rainbow of sound, but instead of 7 colours that can mix to produce white light, it has a much larger number of partials that mix to produce the decay tone of the note.

The most important things to remember about the spectrum of a single string tone are:

- Some partials may be absent or very small in amplitude and difficult to hear, whilst others may be particularly prominent. There is no strict rule for which ones will have which features.

- Different partials may have different decay rates.

- Some partials of single strings may be *false* – they contain a beat pattern. The false beat rate may be slow or fast, within certain limits, and the beat can be prominent or subdued, depending on the beat amplitude, and the relative amplitudes of other partials.

- The beat pattern itself may decay, at a rate different to the overall decay of the partial.

A feature that may be important in determining the *prominence* of a partial in the audible soundscape, is *masking*. In the psychology of hearing, *masking* is a relatively well known effect that involves one sound component making another sound component difficult to hear or distinguish, because it is "masked" by the first sound. This is not just a question of one sound being louder than the other. Certain partials may perhaps mask other partials making them difficult to distinguish, owing to the psychology of the ear-brain system.

Intervals

The 3-D spectrum of an *interval*, rather than of just one note or string, will be essentially the same in the *type* of features it contains, but will contain a larger number of partials, in more frequency positions. Any beat or fluctuation patterns may be more complicated, and not necessarily as regular. Unlike the case of the 3-D spectrum for a single string tone, any beat or fluctuation patterns occurring in a 3-D spectrum of an interval are not necessarily referred to as *false*. Fluctuation patterns in partials of single string tones are *always* referred to as *false*, whilst such patterns in the partials of any other tones or intervals may or may not be referred to as false, depending on the circumstances. This will be dealt with in more detail, later.

Adjustable partials

There are essentially *two* kinds of partials in the soundscape for any piano tone or interval, except for the tone produced by a single string sounding on its own. These two kinds of partial are *adjustable partials* and *non-adjustable partials*. When two or more strings are sounding together, *adjustable partials* are *always* created in the spectrum or soundscape of the overall sound, and these are the partials the tuner affects most by altering the tension on the string. Adjustable partials are adjustable in the sense that *beat rates* in the decay of the partial are critically affected by alterations to the tension on the string, within the fine tuning range.

In fine tuning, the *decay rate* of these partials may also become very sensitive to string tension changes. In addition, the soundscape heard will also contain many *non-adjustable* partials, over which the tuner has *no* real control, in fine tuning, as far as beat or decay patterns are concerned. With the exception of the notes at the bottom of the bass section, every note in the piano is created by two or more strings sounding together, so over most of the compass, every note, interval and chord that can be played, contains *adjustable partials*.

Let's summarise partials again, because it is very important:

All partials *may* or *may not* contain beating.

> *Adjustable partials* – can contain beating whose *rates* can be adjusted in fine tuning. They may additionally contain *false beating*, whose rates *cannot* be adjusted in fine tuning.

> *Non-adjustable partials* – can contain beating, but only beating whose beat rate cannot be adjusted in fine tuning. Any beating is these partials is *false beating*.

Single strings sounding on their own produce *no* adjustable partials. *Unisons*, however, in which *more than one string* sound together for the same note, have soundscapes in which *every* partial is adjustable. Thus, one of the effects of having more than one string to each note, is that some of the tone properties of the note come under the control of the tuner, just by altering the tension on the strings.

A single string or a single note well tuned, will produce a set of partials in an approximately harmonic series. If we sound a musical *interval* other than the unison, with two notes sounding *simultaneously*, then the soundscape of the interval heard, will contain a set of partials at or around the frequencies for two approximate harmonic series, one for each note. This set of partials cannot properly be considered in all its details as simply the addition of the sets that would be produced by each of the separate notes, had they been sounding alone, rather than together. Nevertheless, the *frequency arrangement* of the partials in an interval, will be like the addition of two approximately harmonic series, one for each note of the interval.

Both notes (e.g. both trichords) of an interval contribute to making the set of partials that make up the spectrum or soundscape of the musical interval, when both notes are played together. In any interval other than the unison, some of the partials in the soundscape of the interval will be generated by one note (e.g. one trichord), whilst others will be generated by the other note. (In the unison all strings contribute to all partials). Most importantly, *some* of the partials in the interval's soundscape will be generated by *both* notes. These partials generated by both notes are the *adjustable partials*.

Each note naturally attempts to *drive* the soundboard to generate a set of partials rising in frequency in an approximate harmonic series from its own fundamental. The precise nature of each of those partials, its envelope shape or decay pattern, will depend on the relative tuning between the strings of the individual note.

Finding adjustable partials

The position of adjustable partials in the spectrum of an interval can be found quite easily. They occur wherever a partial of one of the notes would *coincide* (in frequency region) with a partial of the other note. We can find where these coincidences occur, because they happen in an orderly way. We can do this in a number of different ways.

We have already seen that certain frequency ratios are associated with certain musical intervals. The ratio associated with each interval is its *harmonic ratio*. We are not in the least concerned that the interval may not conform *exactly* to its associated harmonic ratio. The harmonic ratio is merely a *tool*, useful in acoustics for finding our way around intervals, but it is not at all a definitive quantity as it is treated in temperament theory. It is merely a handy approximation. Using the harmonic ratio for an interval, we can get a rough idea of the frequencies involved.

For each interval we need to tune, the fundamentals of the notes will have frequencies approximately in the harmonic ratio corresponding to that interval. Because the partials *also* have frequencies with approximately harmonic ratios between them, the coincidences we are looking for, will occur at predictable places.

As we saw earlier, the approximately harmonic series of partials can be represented as a chord rising above the fundamental, using staff notation. Fig. (6.8) shows in staff notation form where the coincidences occur for four different intervals used in tuning, up to the sixth partial. Each notated chord represents (approximately) the pitches of the partials for one note. The lowest note shown in each chord represents the pitch of the fundamental, which also coincides with the pitch of the note that the chord represents. Placing two notes side by side, we can see where the partials coincide. Every chord is the same major chord, because every chord represents the first six partials in the harmonic series, except that each chord has a different lowest note, or fundamental for the series.

Fig. 6.8

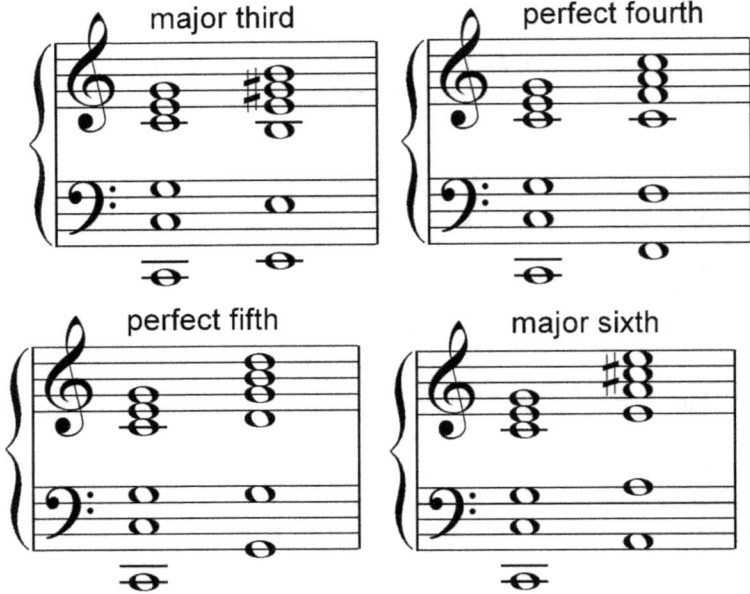

Fig. 6.8. Staff notation representations of four different tuning intervals showing the pitches of the partials for the two notes of the interval. In each case the lowest note of the interval is C16 (C2). The partials that coincide in pitch (and approximately in frequency) can be seen in each case.

For the major third C-E, the coincidence occurs at E44, *i.e.* at the 4th partial of the upper note, and the 5th partial of the lower note. For the perfect fourth, the coincidence occurs at middle C, *i.e.* at the 3rd partial of the upper note, and the 4th partial of the lower note. The same "harmonic series" form of major chord can be picked out on the keyboard using any note in the central compass, as the fundamental. To find the coinciding partial positions for an interval, the chord can be picked out for each note of the interval, seeing which notes coincide.

There is a simple rule, however, for the coincidences, if one knows the harmonic ratio associated with the interval. The harmonic ratio is expressed as a ratio of two numbers, one larger than the other. If R_U is the largest of the two numbers, and R_L is the smallest, then the first coincidence will occur at the position of the R_U th partial of the interval's lower note, and the R_L th partial of the interval's upper note. The *largest* number of the interval's harmonic ratio is associated with the *lower* note of the interval, whilst the *smallest* number is associated with the interval's *upper* note. The *largest* number of the harmonic ratio is the number of the partial of the *lower* note of the interval, at which the first coincidence can be found. Also, the same coincidence is found from the *smallest* number of the harmonic ratio, which gives the partial number of the *upper* note of the interval, at which the coincidence occurs.

This rule, rather like cross multiplication, has its origins in the mathematical expression. The fundamental frequencies of the upper and lower notes of the interval can be written as f_U and f_L, respectively. The harmonic ratio in which these frequencies fall can be written as

$$\frac{R_U}{R_L}$$

where R_U is the lowest integer numerator and R_L is the lowest integer denominator. Thus, by definition:

$$\frac{f_H}{f_L} \approx \frac{R_U}{R_L}$$

so by cross multiplication

$$f_H R_L \approx f_L R_U.$$

The term $f_H R_L$ is the approximate frequency of the R_L th partial of the upper note (whose fundamental frequency is f_H). Similarly, the term $f_L R_U$ is the approximate frequency of the R_U th partial of the interval's upper note.

Other coincidences will occur above (above in pitch, and above in the series) the first coincidence. Using either the partial number of the lowest note, or the partial number of the highest, where the coincidence occurs (or indeed using both), then either partial number can be multiplied by any whole number to find another partial number at which a coincidence will occur. For example, for the perfect fifth (whether well tuned or roughly tuned), the harmonic ratio is 3:2. The first coincidence will occur at the position of the 3rd partial of the lowest note, and the 2nd partial of the highest note. Another coincidence will occur at the 6th partial of the lowest note and the 4th partial of the highest note.

In the case of the unison, the harmonic ratio is 1:1, so coincidences occur at every partial number. Fig. (6.9) shows the case for an octave, in staff notation. The harmonic ratio for an octave is 2:1, so coincidences will occur at *every* partial position of the upper note, but only at every alternate position of the lower note.

Fig. 6.9

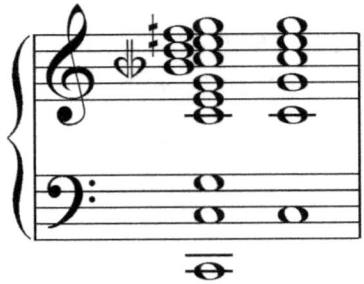

Fig. 6.9. The staff notation representation of two notes tuned an octave apart, showing the first 12 partials of the lower note and the first 6 partials of the upper note.

The accidental sign on the 7th partial, B, of the lower note, is a contemporary "three quarter flat". It indicates that this partial might sound slightly flatter to the modern ear, than expected for a B-flat. Similarly, the 11th partial has a "quarter sharp" sign, indicating a pitch rather sharper than might be expected for an F.

Proportions of the spectrum that are adjustable

Within the spectrum of any tuned interval greater than a unison, there are both adjustable partials and non adjustable partials, in different proportions for each kind of interval. A piano tuner may not necessarily even be aware of the proportions in the mixture, but it nonetheless affects tuning, because only the adjustable partials can be utilised for tuning, whilst the other, non adjustable partials, generally act to obscure the adjustable ones. Part of the aural skill of the good tuner, is therefore the ability to hear specific sounds without being distracted by other surrounding sounds.

The proportion of the spectrum that is adjustable in any interval, depends on the type of interval in question. For a unison, all partials in the spectrum are adjustable. In the octave, half of the partials in the spectrum are adjustable. In the other intervals, some of the partials are adjustable, and in practice, only one or two will actually be used for tuning purposes.

Even though piano tone is constantly changing as it decays, a Fourier transform of the tone's complex wave can be made over a given period of time, which indicates the relative strengths of partials over that time period, following a statistical interpretation, for example, by averaging over the time period. Such analysis shows that in the central compass of a mid-sized instrument, where intervals are first tuned, the strengths of partials of a single note typically ceases to be greater than the ambient noise level, roughly around the 22nd or 23rd partial.

Frequency filtering confirms this, showing that higher partials are relatively ephemeral, and are only initiated at relatively small amplitudes. Whilst the relative strengths of partials in a spectrum follows no fixed, predictable pattern in detail, the overall trend for partials to be much weaker in the higher spectrum, is marked. For practical purposes, this trend has very significant effects on their relevance to the tuning of intervals other than the unison. The reasons for this will become clearer, later.

We can divide the partials of any interval's spectrum into two subsets. Any set of partials creating the sound of an interval, will actually divide into two *intersecting* subsets, one subset associated with each note of the interval. Only the partials in the *intersection* are adjustable. Fig. (6.10) shows, schematically, the relative proportions of the adjustable partials for various interval types, that are relevant in the tuning process.

Fig. 6.10

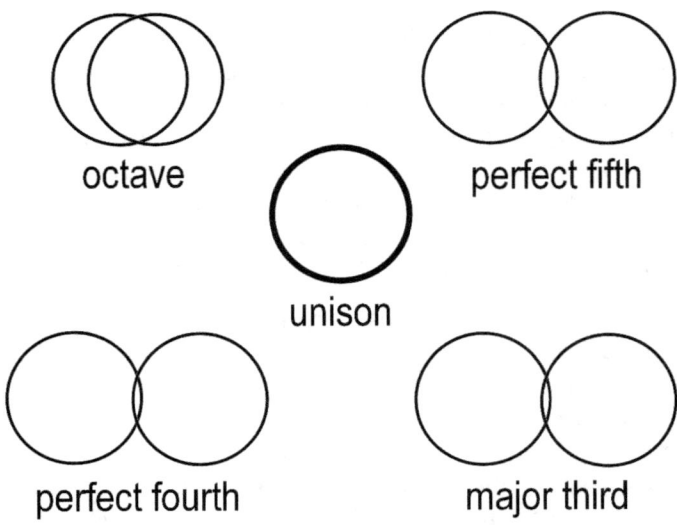

Fig. 6.10. Venn diagrams showing the proportions of important, adjustable partials to non adjustable audible partials for various tuning intervals. For clarity, the diagram is not perfectly to scale. The total number of audible partials could be around 20 or more. The intersections represent the *adjustable* partials, and the circles represent the partials associated with each note. In the octave, 50% of all the audible partials are adjustable. In the perfect fifth, two of the partials are relevant and adjustable. In the perfect fourth and major third, one of the partials is relevant and adjustable. In the unison *all* the partials are adjustable – the two circles intersect completely.

In the case of the unison, both circles or subsets coincide completely. The intersection between the circles is the whole circle. Every partial is adjustable, so the entire sound of the unison, is "adjustable". In many ways, this makes the unison "easy" to tune, but in other ways, makes it difficult, because it also makes the unison extremely "sensitive" to mistuning.

In the octave interval, only half the circles coincide, making the octave in some ways more difficult to tune than the unison, because the non adjustable half of the partial set can obscure the sound of the adjustable half, and is of course, itself non-adjustable.

The tempered intervals, that is, intervals other than unisons and octaves, have to be tuned by the effects of only one or two adjustable partials on the soundscape. In the perfect fifths, perfect fourths, thirds and sixths intervals, the number of actual adjustable partials in the spectrum is much smaller, being limited by their relatively large separation in the spectrum's frequency range, and the attenuation of partials in the higher spectrum.

A major third, for example, has the harmonic ratio 5:4, so even in theory there are only 3 adjustable partials up to the position of the 15th partial of the interval's lower note. In practice, only the first adjustable partial is generally relevant. In the central compass, the second adjustable partial, in the proximity of the 10th partial of the lower note, is invariably significantly weaker and more ephemeral than the first adjustable partial. Whilst it is usually detectable by ear, the precise tuning of the first adjustable partial, which is much more prominent, cannot in any case be sacrificed for the sake of tuning adjustments made to the second.

For all these other intervals this same principle applies. The effects of the first adjustable partial on the soundscape cannot be simply sacrificed for the sake of the tuning of a higher adjustable partial. The properties of the adjustable partials in any interval are not reliably and rigidly related to each other. In the "traditional" model, they would be, the beat rate in the second adjustable partial simply being twice that in the first. In the actual situation inharmonicity and other factors ensure that this assumption is unreliable.

For this reason, the number of useful, adjustable partials in the perfect fourth and major third, is only one. The perfect fifth, however, is different.

The partials of the perfect fifth

In the case of the perfect fifth, the principle of not sacrificing the more prominent partial for the sake of the less prominent one still applies, but both the first and second adjustable partials are potentially, equally important. The first reason is that the second adjustable partial is not always the least prominent of the two. The harmonic ratio for the perfect fifth is 3:2, and the position of the second adjustable partial in the spectrum is in the region of only the 6th partial of the lower note. No other second adjustable partial of any other tempered interval, is as low as this in the spectrum, and the fifth's second adjustable partial is, typically, relatively strong.

Digital analysis reveals that in fifths the first adjustable partial is itself typically very strong compared to the fundamentals of the interval's two notes. It is often stronger than one, and sometimes stronger than both fundamentals. The second adjustable partial of the fifth is sometimes *stronger* than the first, and if it is not, it may soon *become* stronger or at least comparable with the first adjustable partial during the part of the decay time used for tuning. Fig. (6.11) shows an example from the decay envelopes of the fifth G35 – D42 (G3 – D4) on a Steinway model M grand piano.

Fig. 6.11

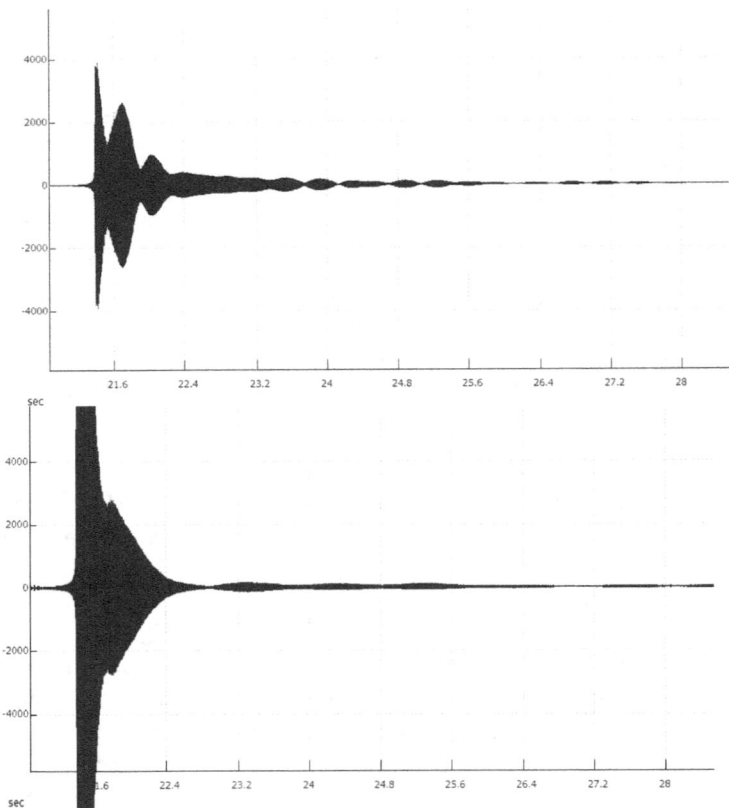

Fig. 6.11. Above, the second adjustable partial of the fifth G35 – D42 (G3 – D4) on a Steinway model M grand piano, and below, the first adjustable partial. Less than one second into the decay, the second adjustable partial is stronger than the first.

In tuning fifths, therefore, the tuner must be prepared to take account of *both* adjustable partials, since it is very common for the second one to be the *more* prominent one at some point, or points, in the early part of the decay.

Higher adjustable partials *are* generally audible in other intervals, but it is not just relative audibility that makes the fifth a special case. It is, as much as anything, the *behaviour* of the perfect fifth in fine tuning, that is subject to different principles, and leads to the need to consider both its 1st and 2nd adjustable partials on an equal basis. We look at this in detail in Chapter 9.

Chapter 7 - Partial decay patterns

Whether we are tuning unison groups, tempered intervals, octaves, or multiple octaves, the soundscape to which we listen, consists of a *spectrum* of partials - a set of many different individual sounds, each with its own pattern of changes or *movement* as it decays away. All these individual partials are sounding together, simultaneously. The individual partials are the ingredients in the overall recipe for the tone, which changes in time. Each individual partial may be a single decaying pure tone, or a pair or group of close-frequency decaying pure tones.

In adjusting the tone of a note or interval, or in adjusting the pitch of a note, it is changes to individual partials that lies behind the tone properties and qualities that we can hear and adjust. Sometimes, we should listen to individual partials, as *properties* of the tone. Much of the time, we should listen to the overall effect of the partials, as the perceived tone *quality*, or the perceived pitch.

One of the properties of partials that contributes to perceived pitch, is the *frequency*. In fine tuning, an important property of a partial that contributes to tone quality, is its *decay pattern*, which is heard as audible "movement", change, or fluctuation in the partial. We cannot simply choose what movement or fluctuation to allow in each interval, with complete freedom. Many decay patterns are interrelated with each other, and cannot be adjusted in isolation, without affecting other decay patterns. To succeed, we initially need some kind of "map" telling us how this interrelationship works.

The basic "map" that we use as a guide to movement in the partials is based on a generic description for movement or fluctuation, called *beating*. For the purposes of creating the "map", *beating* is understood as a *regular* fluctuation in specific partials. Being regular, the speed of the fluctuations can be referred to as a specific *beat rate*. The idea of *beats* and *beat rates* therefore becomes important in piano tuning. The "map" is really just a directive for putting specific *beat rates* in specific places. A fairly useful, though not strict analogy for this, might be a map of a landscape that portrays the landscape in a schematic and well ordered way. Such maps do exist, and are called *schematic* maps. The map of the London Underground train system is an example of a schematic map that bares very limited relation to actual, physical layout of the underground. Such a map can be used to find your way around, but it does not represent very well the actual, physical features. In the case of the beat rate "map", the representation is better, but still if we tried to use it as a map for the actual "acoustical landscape" or *soundscape* to which we listen, we would in many cases find it unreliable, and often *schematic* only.

Later, we shall see why in tuning intervals other than unisons or octaves, it is good to be able to discern specific adjustable partials and their actual acoustical "movement" or *decay patterns*, within in the overall sound. We shall also see why in tuning *unisons*, *octaves* and *multiple octaves*, it is necessary to listen to the *whole soundscape* without concentrating on any specific adjustable partials in the spectrum. Precisely how to hear the right partial will also be explained. Before that, however, we will take a look at the general nature of what we might be hearing. "Traditional" theory suggests that what we hear is either *beats* or *no beats* in the pattern of decay of the partial. In practice, there are many different patterns that can take place in the decay of a partial. These can easily be revealed visually by digital analysis of piano tone.

The ability to hear specific partials in any interval is an indication of the acuteness of aural perception. However, when we listen to the whole, we are hearing the *soundscape* as a whole rather than as a spectrum of discrete partials. This is rather like regarding the visible sea as a moving seascape, rather than as a set of individual waves. In order to hone the ear's perception, it is nevertheless necessary to practise hearing individual partials in all situations, and furthermore, to hear two or more individual partials simultaneously. This is like retaining the ability to see individual waves in a seascape, should we wish to, rather than just seeing the whole seascape.

The untrained ear naturally "fuses" partials into one subjective perception of the soundscape. The master piano tuner may be able to hear individual partials as individual partials, and sometimes to be able to listen to more than one individual partial simultaneously, without the ear-brain system "fusing" them. Ultimately, it is possible to hear a number of individual partials simultaneously, and at the same time, hear the psychologically "fused" result as the tone of the soundscape. This can be crudely compared to the artist's ability to see the overall effect of a complex paint colour mixture, just as an ordinary observer might, whilst at the same time perceiving the primary colour content within the overall mixture.

It is both possible and necessary to be able to consciously change the ear's perception between hearing individual partials in the spectrum, and hearing the whole as a general soundscape without focussing on any particular partials. When tuning certain intervals this provides an ability analogous to a painter paying attention to small detail in colour and form, and then stepping back a few paces to observe the overall effect in the painting as a whole, when viewed from a distance. We must be able to achieve the equivalent of this, aurally, at will, when listening to the quality of tempered intervals, octaves, multiple octaves, and chords. Proper and complete ear training enables the tuner to "dip in and out" of the spectrum at will, hearing individual partials, or hearing the effect of the whole. Fully developed aural perception is able to do this.

It is worth summing this up again, before proceeding, because the business of listening to partials is all too often regarded by theoreticians as a practical "tool" that has no real connection to tone production or the appreciation of the beauty of tone:

With experience, eventually, when listening to specific partials, it is possible to listen to them *in the context of the whole* at the same time as hearing them individually, which is rather like the painter seeing the whole canvass from a distance, as any observer might see it, but being aware of how the details determines the overall effect, which is something the unenlightened observer would not necessarily know. Even the best musicians are in this respect general observers, rather than artists, unless they also happen to be experienced aural piano tuners. The overall "effect" is as much psycho-acoustic as it is something that can be rendered in measurable acoustical terms.

The *sound quality* of an interval is something that is affected by the *decay behaviour* of individual partials. The detail on partial decay patterns that follow in the examples, does not suggest piano tuners are always concerning themselves with creating this or that specific decay pattern, but it does illustrate the true nature of the "materials" with which the art of piano tuning is concerned. In particular, it highlights the limitations in the "traditional" simplistic notions of "beats" and "beat rates" taken from early or elementary piano tuning theory. Whilst beats and beat rates are indeed an essential part of the picture describing the acoustical nature of piano tone, they are *only part* of the picture.

If we were to tune merely by attempting to apply the 19th century "rules", or "map", two things would result. Firstly, because the rules often fail to address the actuality of piano tone behaviour, we would be left guessing or approximating, without a proper guide. Secondly, the result will be "mechanical" and arbitrary, rather than an artistic result that suitably exploits the strengths and properly handles the weaknesses of the piano being tuned. In this lies the difference between the work of the true artist tuner, and the work of an acoustical technician following a map.

Piano tuning beats are not simple heterodyne beats

The beats with which "traditional" theory deals, appear in the theory as though they are straightforward "heterodyne" beat *interference patterns* occurring between *independently radiated pure tones*. Such beat patterns can arise from the mathematical superpositioning or simple addition of two slightly different simple harmonic frequencies. When listening to piano tone, we are not, however, primarily hearing interference patterns between separate sounds radiated by individual strings. We are primarily hearing sound radiated by the soundboard. In tuning actuality, we are still dealing with superpositioning or adding components, but what we are superposing are components arising from the motion of a more complicated complete system comprised of strings, bridge and soundboard.

These motions, and their consequent sound, *decay* (piano tone "dies away"). Decay cannot be treated simply as an "add on" to the idea of heterodyne interference patterns between separate sounds. In general, for piano tone radiated from the soundboard, we may be mathematically working with *complex* frequencies, that is, frequency values that involve both real and imaginary[31] quantities. This leads to a much wider range of possible decay patterns, or beat-like phenomena, than the "traditional" idea of heterodyne interference patterns would predict.

There is a definite advantage in knowing in advance what kind of phenomena we are *likely* to encounter when we start listening to partials. "Traditional" theory only tells us about regular beat patterns, or their

---

[31] An *imaginary* number in mathematics is a number with a factor that is the square root of minus one.

absence. In effect, tuners relying on "traditional" theory and looking for what the theory states should be present, will not be fully equipped to recognise what they hear, because the "traditional" theory does not provide a proper "map" to follow. Consequently, inexperienced tuners may attempt to psychologically "fit" what they hear to the "map" provided by the theory, which results in misdirection and illusion. By providing a better "map" and representation for how partials may actually behave, we provide a better understanding of the whole acoustical "territory" in which tuners work, so we can find our way around it better, have proper knowledge of it, and control it with more confidence.

More than just beat rates

"Traditional" theory specifies beat rates. Because the partials we are dealing with *decay*, any beat patterns we hear may also decay. An important consequence of the actuality of piano tone, which is only described by contemporary theory and not by "traditional" theory, is that a beat *rate* is only one of three major properties of a beat pattern. The other two properties are *beat amplitude*, and *beat decay rate*. Both can affect the aural *perception* of a beat's "rate", and *both affect tone*. In short, all three parameters, beat *rate*, beat *amplitude* and beat *decay rate*, can affect what is perceived as the movement in the spectrum due to beating. In addition to this, but very much connected with it, is the fact that *movement*, or the fluctuation pattern in a partial, may not necessarily be a true *beat* with a definite *rate*.

Two ways of representing partials

We can represent a partial as a wave *envelope*, in which the height of the envelope from its bottom edge to its top edge represents the energy or volume of the partial, against time on the horizontal axis. Fig. (7.1) shows some examples.

Fig. 7.1

A

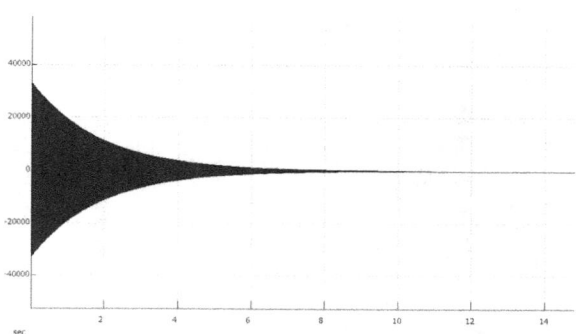

B

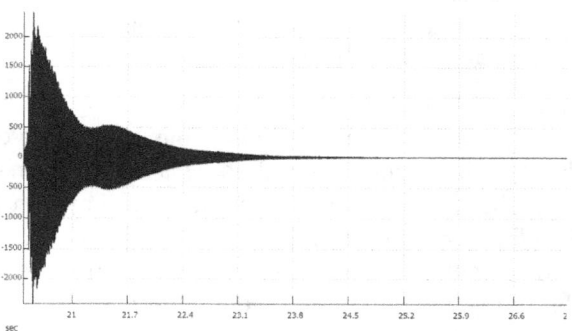

C

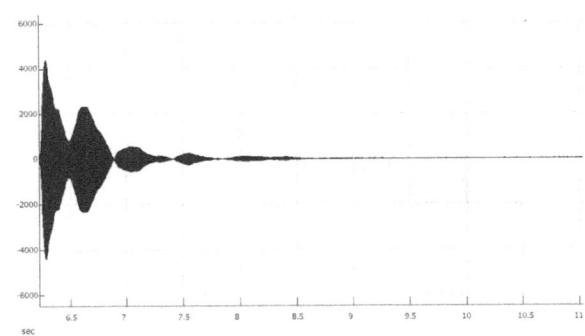

Fig. 7.1. The height of the envelope represents the energy or perceived volume of the partial, against time on the horizontal axis. A: A decaying pure tone; B: A partial of a real piano string; C: A partial of a real piano string.

The natural form of overall decay is *exponential*, illustrated in Fig. (7.1A), in which the amplitude or energy diminishes in a curve shape, the rate of decrease becoming less in time. This overall form of decay can be seen also in the overall outline of the decays shown in Figs. (7.1B) and (7.1C). Envelopes can be useful in showing the decay patterns of real partials, but it must be remembered that they represent the energy or amplitude, rather than the perceived loudness of the partial. The part of the envelope with the smallest height does represent sound that is quieter, but loudness is not actually proportional to the height.

Our hearing is approximately logarithmic – roughly speaking, we are more sensitive to differences in quiet sounds than to differences in loud ones. The decay of a partial can also be represented graphically, with its *loudness* on the vertical axis, against time on the horizontal axis. Using a *logarithmic* scale for loudness on the vertical axis, which often represents better what we actually hear, a straightforward exponential decay of the kind shown in Fig. (7.1A), would appear as a straight line, rather than curve.

Fig. (7.2) shows two envelope representations of beat patterns from actual piano tone partials, together with an example of the kind of pattern one would expect to see if the partial were represented using a line graph with a logarithmic vertical axis (a semi-logarithmic graph).

Fig. 7.2

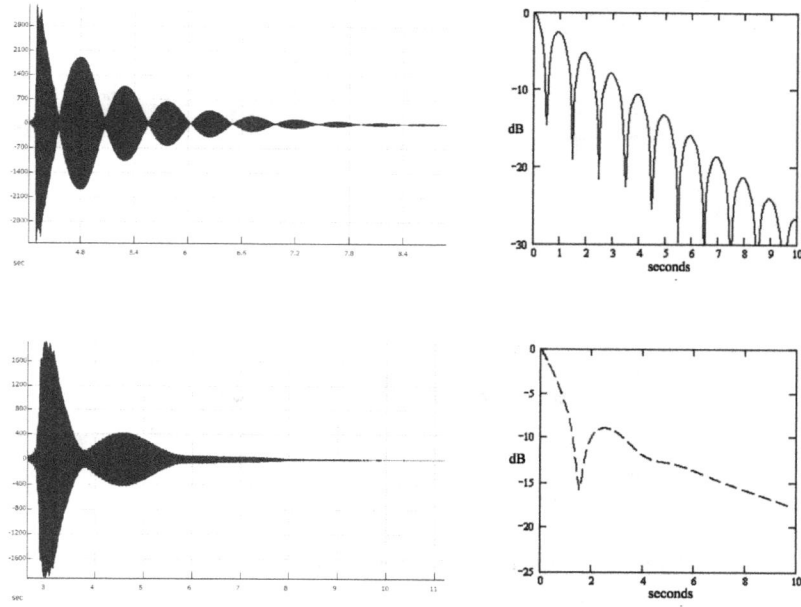

Fig. 7.2. On the left, two decay envelopes from actual piano tone partials, together with the corresponding type of semi-logarithmic line graph representation, on the right.

The simplest type of decay has no beat, no audibly recognisable decay pattern at all, except the for the fact that it is obviously decaying, or getting quieter, in time. Because we represent the pure exponential decay as a straight line on the semi-logarithmic graph, the decay will be referred to as *beatless*. The natural exponential decay of a partial's energy, when it is heard *sounds like* a "straight line" decrease in *loudness*, in the sense that its "straightness" distinguishes it from the fluctuations of a *beating* decay. If the decay does not sound "straight" but also does not sound like a regular beat, we can say it is *irregular*. When listening to a partial tone, it is *psychologically easy* to conceive the more or less pure exponential decay, as a "straight" or *beatless* decay, and to conceive a decay that beats or fluctuates in some other way, as a "non straight" or *beating* or *irregular* decay.

A beatless decay is the simplest form of decay, and although there is no fluctuation pattern in it, we can still refer to it as a beatless decay *pattern*. After all, a straight line is still a kind of *pattern*. Fig. 7.3 shows an example, in the logarithmic graphical representation, of a *beatless decay pattern*:

Fig. 7.3

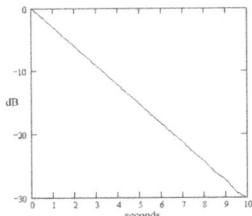

Fig. 7.3. A beatless decay - the simplest and most straightforward kind of decay pattern. The vertical axis can be thought of representing the loudness of the partial, whilst the horizontal axis represents time.

We can even refer to this line as a decay "curve", even though in this form it is a straight line, since we know it represents what would be a curve in amplitude, if we did not use a logarithmic vertical axis.

The diagonal graph line in Fig. (7.3) represents the "loudness" of the partial. The vertical scale is in *decibels*, and starts at zero, at the top, rather than the bottom. This is normal, but should not concern us here. For our purposes, we need not be concerned too much with the mathematical details, because most of the time we will only be dealing with generalities, rather than specific values. Many of the graphs, however, will show specific values because they are generated using real values in the first instance. Also, we should not worry if sometimes the units of the axes sometimes appear in different units – the principle is the same. Where the horizontal axis is marked *t*, the units are seconds.

The decay *rate*, is indicated by the gradient, or steepness of the slope of the line. A steep slope down from left to right always indicates a partial that decays more rapidly than one with a less steep slope.

Dual decays

Strictly speaking, a *beatless decay* must in this method of representing it, be a *single* straight line. Real piano note partials often have a *dual decay rate*, which means there are in effect *two* straight lines, with different gradients, and where they join, there is usually a curve. Fig. (7.4) is an example of a dual decay rate:

Fig. 7.4

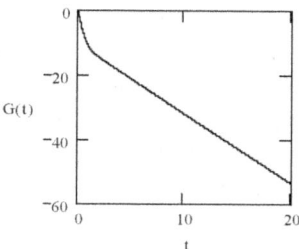

Fig. 7.4. A *dual decay rate*, typical of many piano tones.

Here, the first part of the decay is steep, with a fast decay rate, but after about (in this particular example) 2 seconds, the decay is less steep, indicating a slower decay rate. We could argue this is an *irregular decay* because it has two decay rates, or is not just one straight line. However, when heard, the dual decay rate appears *beatless* to the ear that is listening for beat patterns. Although well tuned piano notes typically display dual decays, we do not typically hear then as such.

Now we are going to be concerned with the way partials *sound*, for the specific purposes of piano tuning practice, more than the way they would actually be displayed, visually. A dual decay of the kind in figure (7.4), would in aural practice be perceived as "pure" or "beatless". For this reason, a dual decay of this kind, which strictly speaking is *irregular*, we can put into a *generic class* of aurally recognisable decay patterns called *beatless*. To the piano tuner it is just a standard, beatless decay. Both Fig. (7.3) and Fig. (7.4) represent partials that are *beatless*. Some other possible examples are given in Fig. (7.5):

Fig. 7.5

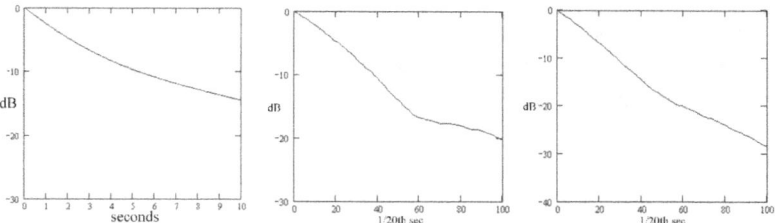

Fig. 7.5. Examples of decay patterns that are mathematically irregular, but constitute the aural *beatless decay* in a practical classification of generic decay patterns.

There is no *strict* rule about when a decay curve has a shape that qualifies it for being called a *beatless decay*. In practical terms, it is a *beatless decay* if when heard there is no appreciable pattern or movement in the decay, other than the steady decay itself. Piano tuners would simply say the partial is *beatless* or *pure*. It is important to appreciate that when we say a partial or interval is *beatless* or *pure*, there is a *range* of *different* possibilities of actual decay behaviour that we may be perceiving.

Regular beats

The next most obvious pattern to speak about is the *regular beat* pattern. The "heterodyne beat" pattern of "traditional" theory is a *regular beat*. An example is given in Fig. (7.6).

Fig. 7.6

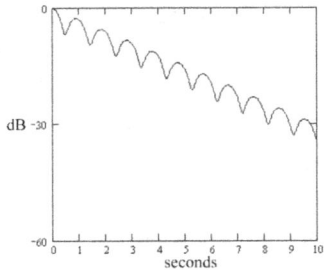

Fig. 7.6. Regular Beat decay pattern with a beat rate of about 1 Hz. The beat is relatively "shallow" or small in beat amplitude, and its beat amplitude remains constant throughout the overall part of the decay shown.

Here we can see a beat rate of about one per second – what you might expect to hear in one of the partials of an equally tempered fourth. Another example is given in Fig. (7.7):

Fig. 7.7

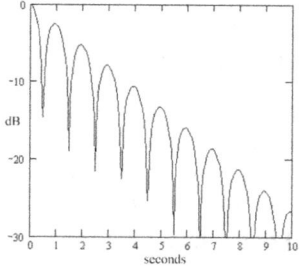

Fig. 7.7. Regular Beat decay pattern with a beat rate of about 1 Hz. The beat is relatively "deep" or large in beat amplitude. Its beat amplitude remains constant throughout the overall part of the decay shown.

The difference between Fig. (7.6) and Fig. (7.7) is that the first is a "shallow" beat, and the second is a "deep" beat. In other words, a regular beat has an *amplitude* or *loudness* that may vary from one case to another. Fig. (7.6) is relatively "unclear" or "unprominent" beat, whereas Fig. (7.7) is a "clear" or "prominent" beat, but they both have the same beat *rate* or number of beats per second. A beat may be so shallow that it is barely perceptible, or very unclear, as in Fig. (7.8).

Fig. 7.8

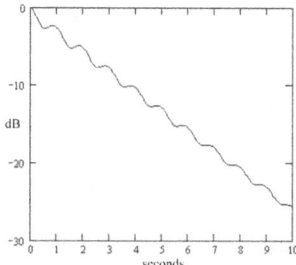

Fig. 7.8. Regular Beat decay pattern with a beat rate of about 1 Hz. The beat is very shallow or small in beat amplitude, and its beat amplitude remains constant throughout the overall part of the decay shown.

If a beat is *very* shallow, it may sound as though there is no beat, or as if the *beat rate* is zero or close to zero, when in fact it is not. It is quite easy, when listening to beats, to confuse a change of beat *amplitude* or depth, with a change of beat *rate*. At the limit of these cases, a beat in a partial can disappear either because its *beat rate* reduces to zero, or because its *beat amplitude* reduces to zero. The is a most important fact, with practical consequences for tuning, that "traditional" theory does not include.

Note that so far, there are three separate characteristics that can differ independently – (1) the *overall* decay rate, (2) the *beat amplitude* or loudness, and of course (3) the *beat rate* itself, although all these examples all happen show a beat rate of about 1 per second. Now we must look at a fourth characteristic of the *regular beat* pattern. Look at Fig. (7.9):

Fig. 7.9

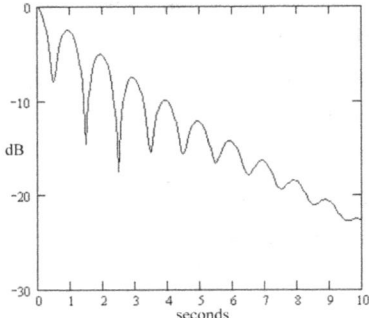

Fig. 7.9. Regular Beat decay pattern with a beat rate of about 1 Hz. The beat amplitude first grows, and then decays.

Here the amplitude, loudness or depth of the beat increases slightly at first, and then decreases. After about 6 seconds the *overall* amplitude or loudness of this partial has decreased considerably. The beat is still obviously visibly present at this stage, and although its amplitude had decreased, so has the *overall* amplitude of the whole pattern. Consequently, such a beat itself is not necessarily perceived as dying away any faster than the overall decay, so the pattern may still simply constitute a *regular beat*.

A regular beat may have an amplitude that varies in other ways, too. Fig. (7.10) shows some examples:

Fig. 7.10

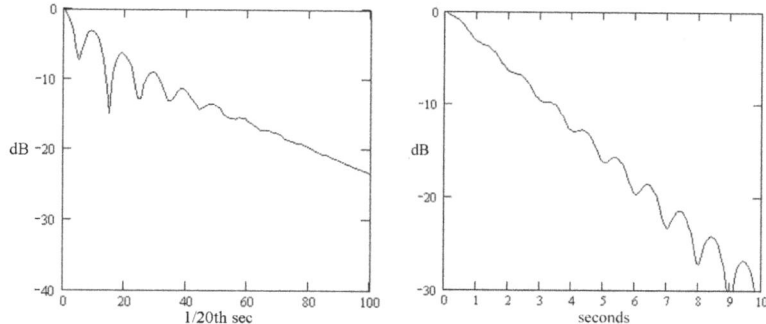

Fig. 7.10. A decaying regular beat and a delayed regular beat

Here we see on the left, the beat decays completely away by 3 seconds into the decay, and is followed by what is effectively a beatless decay section. Aurally this could sound as though the beat *rate* slows down to zero, because the beat disappears. In the second example, the beat only becomes prominent after about 5 seconds. When heard, there is a 'delay' before the beat "gets going". In all the examples the beat is nonetheless still *regular* as far as its beat *rate* (number of beats per second) is concerned.

So far, we have four types of decay pattern we might encounter – a *beatless decay*, a *regular beat*, a *decaying regular beat* and even a *delayed regular beat*, which can all be distinguishable by ear. There are other patterns we can identify by ear.

The single null

The first is the very important *single null*, which is encountered in the tone of the unison, octave, or tempered perfect fifth. Single null patterns are relatively common in piano tone partials, and are heard as a rapid "dying away" of the partial at the beginning of the decay time, followed by a resurgence back again, as if a beat is about to happen, but this is then followed not by the rest of the beat pattern but by a steady decay.

The single null happens when the beat amplitude decays so much faster than the overall partial amplitude, that a completed beat is never heard, but the beginning of the beat is nonetheless prominent. The solid line curves in Fig. (7.11) are "classic" theoretical single null patterns, shown together with the dotted line *regular beat* pattern. Single nulls always arise when the beat amplitude decays much faster than the overall decay of the partial in which the beat occurs.

Fig. 7.11

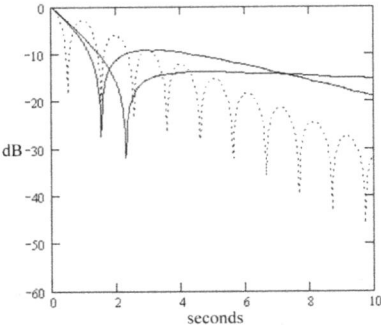

Fig. 7.11. Two examples of the single null decay pattern (solid line), compared to a regular beat decay pattern (dotted line).

The *single null*, when audible in the tone of the central compass, is a slow pattern, typically taking at least a couple of seconds from the beginning, before the sound level rises out of the null. Here we see nulls at about 1 second in one curve, and 2 seconds in the other. Rather than the '*wawawa*' sound associated with the *regular beat*, the *single null* is better described as a single '*ahhwwaaaaaa*'. After the '*wwaaaaaa*', which is the rise after the null, the sound becomes steady, or a steady decay. A second null never appears, so there is never any *complete* "beat" as such, and therefore no proper "beat rate". However, what is noticeable audibly, is the null, and the rise out of the null, which appears *like* a single beat, followed by a beatless decay. Some single null patterns are shown in Fig. (7.12):

Fig. 7.12

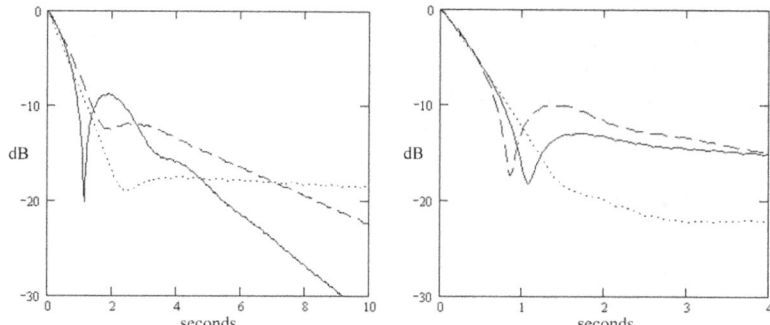

Fig. 7.12. Varieties of single null patterns. Only the dotted line decay in the right hand graph would constitute a generic beatless decay in practical terms. The varieties show how the null may be deep or shallow. The solid line decay in the left hand graph has a small "second null" and so might actually constitute a double beat pattern.

The dotted line in the second graph has no distinguishable null and would in effect be heard as a *beatless decay*. The solid line in the first graph, almost displays a second null, which tends to make the 'hump' after the first major null into a true beat. Also, there is almost a third beat detectable. In this instance the single null borders on falling into the next generic decay pattern class, that of the *double beat*. The characteristic of the double beat is that there are two distinct beats, but after these, a steady decay.

Single null decay patterns are *often* encountered when tuning slow beats. Most importantly, the null may be mistaken for a true beat, but the perceived "beat rate" is not a "rate" that relates to the tuning condition in the way that "traditional" theory suggests. Their representation in the semi-logarithmic graphs shown here, provide an intuitively easy to grasp illustration of what they actually sound like.

Progression of beat patterns in fine tuning

In "traditional" theory, as two partials are brought closer together in frequency, all that happens is that the beat rate reduces. The beat rate is said to be equal to the "mistuning" between the two partials, measured in Hz. For example, if one partial is 523 Hz when the string is sounding on its own, and the other is 524 Hz when that string is sounding on its own, then the resulting beat rate is said to be 1 Hz.

In both contemporary theory and in practice, the relationship between the resulting beat rate and mistuning is not necessarily this simple. Also, as the mistuning changes, the *decay rate* of the beat or partial may change in addition to the beat rate. This typically results in the progression of decay patterns shown in Fig. (7.13), as mistuning is decreased.

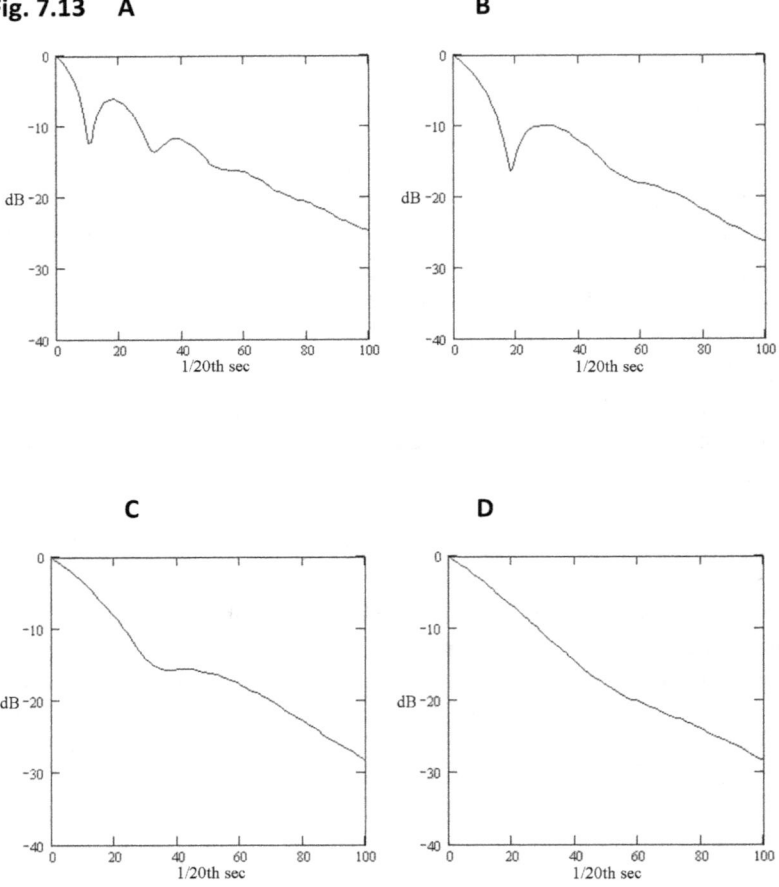

Fig. 7.13. Graphs showing a typical, natural "progression" from a beat pattern to a "beatless" pattern, in an adjustable partial. Such patterns are produced because the *beat amplitude* decays at a rate that is different to the overall decay rate of the partial.

The sequence from Fig. (7.13A) to (7.13D) represents the tuner slowly attempting to decease the beat rate. The situation here is not at all one in which we are always dealing with a regular beat that simply changes its beat rate in proportion to the mistuning between the strings. In this example sequence a true repeating beat is only present at around 1 beat per second. Trying to tune a slower beat than this, would result in decay curves that are not regular beat patterns, so in practice it might seem that the tuning passes rather easily from a beat to no beat.

Alternatively, tuning the other way, from "pure" towards a beat, the tuning readily passes from a more or less *beatless* condition to about one beat per second, without much choice being available in between. Either way, the beat disappears or emerges not just because of a change in beat *rate* (the number of beats per second), but because the beat itself, as a pattern of fluctuations or "lumps", emerges or disappears. This is called *attenuation*, which we will meet again later.

Not all tuning situations will necessarily be exactly like this, but this *kind* of sequence in which the beat amplitude decays faster, the slower the beat rate, can be considered as a typical, *general* form of progression of decay curves for very fine tuning close to "beatless". It is a generic behaviour pattern from which other or different patterns of progression, including that given by "traditional" theory, can be derived as particular cases. We shall meet this kind of behaviour in actual piano tone envelopes in the following chapters on tuning.

In practical tuning, this kind of behaviour can lead to a significantly larger *range* of mistunings that result in a "beatless" partial, than "traditional" theory would allow for. More will be said about this later.

A few more perceivable decay patterns might be identified, just to complete the picture. Practical piano tuning does not need to identify these other than as an irregular beat. These are what can be described as a *delayed null*, a *long null*, and the *irregular beat*. Strictly speaking, the *delayed null* and the *long null* are features that can be present in the *irregular beat*. In the *delayed null*, the partial appears to decay beatlessly for a while, and then suddenly drops rapidly into a null, later in the decay. In the *long null*, the partial appears to decay away entirely, and then some time later suddenly rises up again apparently out of nowhere, but in fact, out of what is literally a long null. The *irregular beat* displays a beat-like pattern, but without a fixed, regular beat *rate*. Sometimes it is like several beat rates sounding at once. Fig. (7.14) shows a couple of examples of *irregular beat* patterns. Both examples also exhibit what could be called a *delayed null*. An *irregular beat* pattern may often include a *delayed null* or a *long null*.

Fig. 7.14

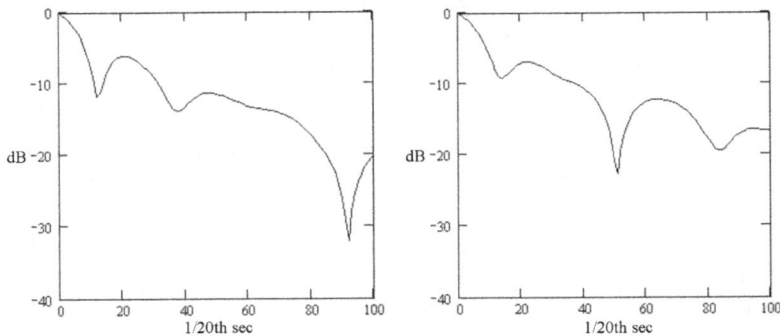

Fig. 7.14. Irregular beat decay patterns including a *delayed null*, on the left at about 4.5 seconds, and on the right at 2.5 seconds.

Summary of beat patterns

In summary, then, rather than *beat* or *no beat*, there are, somewhat arbitrarily, nine *generic decay patterns* that are recognisable aurally, that we have mentioned:

- *Beatless Decay* – The partial decays steadily without other significant features.
- *Regular Beat* - This is the "classic" beat pattern, of idealised, "traditional" theory. The partial seems to decay with a regular beat. On closer analysis, for slow beats this may consist of a beat which may grow at first in depth or amplitude, after which it may decay at a rate faster than the overall partial decay rate, so that in the later part of the decay the beat may not necessarily even be audible. Since the piano tuner usually listens to the decay for only a short time starting *after* the beginning of the tone, the beat pattern for the period of listening still appears to be a Regular Beat, even if on extended examination it may be more characteristic of a Decaying Regular Beat. Faster regular beats may have a *beat amplitude* that varies,

sometimes cyclically. We shall see later that the beat *rate* of faster beats may also vary, sometimes cyclically.

- *Decaying Regular Beat* – A regular beat rate is present at the start (or soon after), but the beat quickly decays away while the partial itself is still sounding, so the latter part of the decay may even be beatless. Although the beat *rate* does not in fact change, it may seem to reduce to zero, if the beat *amplitude* disappears.
- *Delayed Regular Beat* – The partial begins by apparently decaying beatlessly (more or less), but later in the decay a regular beat appears.

- *Single Null* – The partial drops immediately away from the beginning into a 'null' or silence, and then rises again, giving the impression of the beginning of a beat, or what could can be interpreted as a single beat. After the rise, however, it decays steadily without significant fluctuation, sometimes with a very slow decay rate.
- *Double Beat* –Two beats are heard, followed by a more or less beatless decay. This is in effect a *Decaying Regular Beat* whose beat amplitude decays rapidly.
- *Delayed Null* – The partial decays beatlessly or irregularly, but then there is a dramatic drop into a 'null' followed by a rise out of the null. There may appear to be one beat, later in the decay pattern, that is suddenly more pronounced than the other beats.
- *Long Null* – The partial may be decaying beatlessly or irregularly, and then decays into a null and remains silent for a while, and then makes a reappearance.
- *Irregular Beat* – Beat patterns and fluctuations are present in the partial, but there is no one fixed beat rate. There may be beat rates that speed up or slow down during the overall decay. Alternatively, the fluctuations in the decay may seem chaotic.

These are classes that illustrate the actual diversity of decay pattern phenomena that piano tone can exhibit. It is not just *any* partial that is, so to speak, free to exhibit any of these generic patterns. Different generic decay pattern possibilities tend to be associated with partials in different specific situations. As it turns out, the areas of tuning where generic decay patterns other than the *regular beat* are most important, are the tuning of tempered perfect fifths, unisons and octaves, and sometimes to a lesser extent, tempered perfect fourths. These are all the "slow" intervals in which we are aiming at a slow beat rate, or no beat rate, so the mistuning is small.

Unstable beat rate

Lastly, in practice, the beat *rate* itself is not always necessarily constant, but may increase, decrease, or cycle through changes. In the case of the fast beating intervals, the thirds and sixths, the *amplitude* or *prominence* of the beat typically changes as the tone decays, sometimes the beat "fading out", and other times the beat increasing in prominence. Such patterns of behaviour are commonly observable, aurally, in practice, and are easily found and measured in digitally isolated partials of real piano tones. More will be said on this in the chapters on tuning specific intervals.

The significance of decay patterns

The specific theoretical instruction to tune, say, a given tempered perfect fifth with a precise beat rate, can be misleading because a precise measurement of beat frequency can only properly apply to a regular beat decay pattern, with a constant beat rate. The ingredients of the interval's sonic quality cannot in any case be defined simply by quoting a single beat rate. In particular, the term "beat rate" applied to the tuning of the tempered fifths (and to some extent the fourths) may be only an approximate descriptor for the actual partial decay pattern. The beat may in actuality speed up during the decay, and beats slowed to around one beat per second or less, often may become a single null, rather than a repeating beat. In general, the beat may attenuate ("flatten out"), rather its actual beat *rate* reducing.

The adjustable partials

The adjustable partials when tuning equal temperament, fall into two main areas:

(1) partials that are "slow beating" (for tempered perfect fifths and fourths);

(2) partials that are "fast beating" (for the tempered thirds and sixths).

By "slow" we mean probably less than about 1 Hz beat rate, and typically, around 0.5 Hz or less in beat rate. Thus, within the scale tuning area, at least some tempered perfect fourths, all tempered perfect fifths, and throughout the compass all unisons, octaves and their multiples, are either "slow" or "pure" intervals. They are subject to all generic decay pattern possibilities, and in particular the progression shown in Fig. (7.13). Thirds and sixths, on the other hand, are always considered "fast beating" intervals, and will generally exhibit a more regular beat pattern. However, fast beating intervals can also exhibit irregularity in that the beat amplitude changes, and the beat *rate* may show cyclic variations between two limiting beat rates.

Falseness

A partial produced by a *single string sounding on its own* may itself exhibit fluctuations or beat patterns in its decay. This is known as *falseness*. More than one partial may be false in the spectrum produced by a single string. False partials usually exhibit the *regular beat* pattern. If the false beat *rate* in such a partial is very slow compared to the partial decay rate itself, it may not be noticeable as false, but the false beat pattern will nevertheless affect tuning.

An important feature of false partials, is that the false beat *rate* can be *inherited* by a partial of an interval (or unison) containing the false string. This may give the impression that in listening to the interval (or unison) it is the false partial in the single string spectrum that is being heard. We must remember that in the case of listening to an interval (two notes or strings sounding simultaneously together), the sound we hear has its *own* recipe of ingredient partials. A partial may nevertheless *inherit* characteristics of the single string's partial, in particular, its beat rate.

The *irregular beat* and the *delayed regular beat* are likely to appear where there is significant falseness in one of the strings. Where two or more strings of a unison or interval are significantly false, the *irregular beat* pattern will certainly be present in the unison's spectrum (although we may be able to change its character by judicious tuning).

For fast beating intervals, it is usually only the *regular beat* that we encounter, but often with a changing beat rate. There may also be what could be called a "dual beat rate" in the partial, or "phantom beat", where a single bate rate consist of alternately large and small amplitude beats. The significance of this will be dealt with later.

Mistuning and decay patterns

As we tune one of two strings to slow down the beat rate in an adjustable partial of the interval (including the unison or octave), we are reducing the *mistuning* between the strings. It is important to appreciate what "mistuning" here refers to. Most importantly, *mistuning* does not imply "correct" or "incorrect". It simply refers to the frequencies of partials produced by two single strings, when each string sounds in isolation. Specifically, it refers to two single strings producing partials that are close in frequency, *i.e.* one partial of one string, when sounding on its own, is close in frequency to a partial of the other string, sounding on its own. By "close" we mean much closer than the usual distance between one partial and the next in the approximately harmonic series, in the spectrum of either string. For two such partials that *coincide* in this way, the difference in their frequencies is called the "mistuning". In "traditional" theory this mistuning is equal to the resulting beat rate.

The important thing is that this *mistuning* is the difference between two partial frequencies, each partial frequency being measured when the string responsible for it *is sounding on its own*. Once we sound *both strings together*, it is not this *mistuning* that we are "directly hearing". Rather, we are then hearing its indirect *effects*. "Traditional" theory would say that the effects appear as a beat rate equal to the mistuning. However, in contemporary theory, this may only be true for larger mistunings. Once both strings are sounding together, the effects of the bridge and soundboard *coupling* the two strings can mean that the actual frequencies produced are no longer as they would have been, had the strings been sounding separately.

In the process of reducing the beat rate in an adjustable partial, when we reach a beat rate of around 0.5 Hz in the mid-compass (this value may increase higher in the compass), the decay pattern may change not only in its beat rate, but in other important features also. We may even loose the *regular beat* and encounter for example a *double beat* or *single null*. We may also hear a *beatless decay*, whilst *mistuning* is still present. We have of course, no real way of telling by ear that mistuning is still present, since listening to the interval we cannot hear mistuning directly, and when the strings sound together there is no beat anyway. The important point here, is that there is not necessarily only one position in the relative tuning of two strings that will produce what would be heard as a *beatless decay* in the partial we are listening to, but rather, there may be a small *range* of positions.

In other words, it is not necessarily only at zero mistuning, that a zero beat condition can be produced. For some partials a zero beat condition may be produced over a small *range* of mistunings. An important consequence of this, is that as the difference in tension between the two strings is gradually increased from say, zero difference, there may be a point at which the decay pattern rather "suddenly" changes from *beatless decay* to say, *single null*, *double beat* or even *regular beat*. Such rapid changes when one is attempting to bring about only a minute change, can also occur for mechanical reasons connected with string slippage over friction points, but it must be appreciated that the difficulty can also be due to the natural acoustical behaviour.

Conversely, the effect of the bridge and soundboard can mean that it is also possible that at *zero* mistuning, a beat in the decay curve is *still present*. Occasionally, a beat persists with no possibility of eliminating it, even though it is not due to inheritance of a false beat. The contemporary theory provides a possible explanation in the conjecture that the partial frequency in such a case, happens to fall close to a soundboard resonant frequency.

The coupling region

We can describe a *coupling region* as shown in Fig. (7.15), containing a range of small tension changes, within which the *regular beat* pattern may be replaced by other generic decay patterns, or has a slower beat rate than the value of the mistuning. As we reduce the regular beat rate, we enter the coupling region, and then the beat pattern tends to rapidly disappear, or be replaced by some other generic pattern. This is to be distinguished from a "jump" in the tuning due to pin or bridge friction. A little consideration of this will show that a wide coupling region could be very useful when attempting to eliminate the beat, as in tuning a unison, for example.

Fig. 7.15

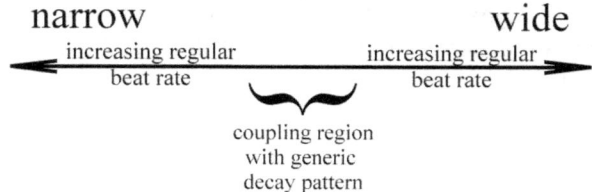

Fig. 7.15. The *coupling region* for a single adjustable partial in a tempered interval. Suppose this is the lowest *adjustable* partial in a tempered perfect fifth's spectrum. As the fifth is progressively narrowed from "pure" or "beatless", the beat rate increases. As it is progressively widened from "pure", the beat rate also increases. When it is very close to "pure", the partial enters the *coupling region*, at the centre of which it will typically decay

beatlessly. In an attempted small amount of narrow tempering, as in a perfect fifth, the partial may fall within the *coupling region*, if the region is "wide" enough.

Behaviour within the coupling region

Inside the coupling region, the normal *regular beat* pattern whose rate is equal to the mistuning, may be replaced by a slower, decaying beat, a single null, or even a beatless decay pattern. The "width" of the coupling region, if any, will depend on the individual soundboard, string and bridge properties at that position. The decay pattern inside the coupling region may depend on, and be very sensitive to changes in the precise tuning condition between the strings. In the theory, an important feature of the partial's behaviour inside the coupling region, is that its *decay rate* can critically depend on the precise tuning condition between the strings, even if there is no beat. Inside this region, the tuner may be affecting the sound spectrum and tonal qualities more by changing *decay rates* (and hence decay patterns) than by changing beat rates.

Outside the coupling region only the *regular beat* pattern generally occurs, unless there is falseness present, in which case an *irregular decay* may take place. (In fact, falseness is not really to be considered as either present or absent, but rather, its recognisable presence is a matter of degree). Inside a coupling region any of the other generic decay patterns can in principle appear.

The point to remember is that whenever two strings are tuned so that the partial in question is to contain a slow "beat rate" or no "beat rate", then the effects of a coupling region may be encountered. This means, in effect, that unisons, octaves, tempered perfect fifths, and to a lesser extent, tempered perfect fourths, are all potentially subject to coupling region effects.

Part 2 — Advanced Tuning Practice

To compare a single note on a concert grand piano with the same note on a concert grand piano of another make, could hardly be expected to reveal all the differences of tone between the two instruments, that would come to light in playing the instruments. Similarly, a comparison of the tuning of one trichord or octave, with that of another, will not necessarily reveal at all, any of the differences that accumulate to make a piano "a different instrument" as a result of its fine tuning.

Chapter 8 - Unison Tuning

For the piano tuning student, training in the tuning of unisons usually begins before learning to tune the tempered intervals between the notes of the scale, and before learning to tune octaves. Learning unison tuning usually begins on a *strung back*, a composite unit of the soundboard, iron frame and strings, without the encumbrance of the rest of the piano. Unison tuning on the strung back is achieved by plucking the strings, rather than having them struck with the action hammers. This is "traditionally" known a *chipping up*. In chipping up there is no need to wedge strings, and also "good" unison tuning in chipping is relatively easy to achieve, because the acoustical situation is different from that in the complete piano. What is actually quite crude unison tuning compared to fine tuning on the piano, produces "good" sounding unisons for the purposes of chipping up.

Each note in the central and treble parts of the modern piano compass has three strings tuned to unison, which is a "unison group" of strings, or *trichord*. Much of the bass part of the compass consists of notes each with two strings tuned to unison, known as *bichords*. All single notes in the piano that have more than one string tuned to unison, are referred to simply as *unisons*. If two strings of a trichord are being tuned in the piano with the third string muted using a wedge, these two strings can be referred to as a *unison pair*. We can also refer to a unison pair when two strings of the trichord are plucked on a chipping back, the third string not being muted. In all notes consisting of more than one string, the opportunity exists in fine tuning on the piano, for *tone control*.

The adjustable tone property with which "traditional" theory deals is a particular *movement* in the soundscape called *beating*, which is heard as a pulsating or rising and falling in the loudness of partials as they decay. In the soundscape of the unison as a whole, beating is heard as a pulsating within the tone, even if it is not identified with specific partials. Beating can occur in multiple partials, simultaneously, and in rudimentary theory is thought of as a *regular* fluctuation in the amplitude or loudness of the partial. Because it is regular, the number of these fluctuations or beats *per second*, can be quoted as a *beat rate*.[32] The only feature of beating that the "traditional", rudimentary theory treats as adjustable, is in fact this beat *rate* (the number of beats per second). In general, the beat rates in different partials will differ, at any given tuning condition, so that beating in the unison in general consists of many different, simultaneous beat rates.

The general and rough "rule of thumb" for beating between two strings of a unison is that as the tension difference between the two strings increases, the *rates* of beating in the soundscape increases. The tone of a piano unison is generally considered to be at its best when movement[33] in the soundscape is reduced to a minimum, so the reduction of beat rates is the initial aim in fine tuning the unison. We shall see shortly that in actual tuning, the use of the word "minimum" here does not necessarily imply that there will be *no* movement in the soundscape, or no beating in any partials of the spectrum.

Initially, we will look at the tuning of a unison pair (two strings), beginning where a small difference in string tensions still allows the unison tone to be perceived as a single note, but has led to the presence of beating in the soundscape. There is no universal, fixed point at which the soundscape of the unison changes from being perceived as two pitches, to being perceived as one, and it is sufficient to imagine any point reasonably approaching what appears to be a single pitch. Whether or not the tone of both strings sounding together as the unison, can still be heard as two distinct pitches, is not actually important.

A small but crude change to the tension of one string will in general either increase or decrease audible beat rates. It is of course also possible for the tension *difference* between the strings to change from positive to negative, *i.e.* for a string to be brought from flat to sharp of the other string, or *vice versa*, leaving the beat rate approximately the same. For the rough rule, it is only the tension *difference* that creates the beat rate, there being no indication in the beat rate itself, whether one string is in fact higher, or specifically lower than the other.

One can imagine a "zero beat point" at which, hypothetically, "beating disappears", when the tensions are, hypothetically, equal. Roughly, the

---

[32] Piano tuners usually refer to the beat *rate*, rather than using the more scientific term *beat frequency*.

[33] "Movement" in the soundscape is any fluctuation, beating, or *change* to the tone recipe due to unequal decay patterns or rates in the partials.

larger the difference in tension between the strings, the faster the beat rate. The sound of the beat rate itself cannot indicate which string is the "lowest" or "highest", but by *changing* the beat rate this can be determined. If it is found that an *increase* in tension results in an *increase* in beat rates, then that string can be considered, at that "tuning position", to be the "highest" (highest in tension) string. A progressive lowering of tension will then slow the beat rates, until they reach a minimum, beyond which further lowering of the tension will cause the beat rates to begin to increase again, as the tuned string now becomes the "lowest" (lowest in tension) of the two strings, and the tension difference starts to increase.

If a *decrease* in tension increases the beat rates, then the tuned string *must* be the "lowest" of the two strings. Progressive raising of the tension of the "lowest" string will then slow the beat rates, as the tension difference is reduced, until minimum beating is reached. Raising the tension beyond this point increases the beat rates again, as the tuned string becomes the "higher" of the two strings, and the tension difference increases.

For the purposes of chipping, rough tuning, or learning to tune, we can say that whether the tension is being lowered or raised, if the beat rates slow down, then the hypothetical zero point is being approached. If they speed up, then the distance of the current "tuning position" from the zero point is increasing, and the tuning condition is becoming worse. At any point where beating can be heard, the tuning position of a string may be thought of as above or below the hypothetical zero beat point, with the possibility of being moved towards the zero point. In this situation, as already pointed out, too large a movement may result in the tuning position being moved from one side of the zero point to the other, so that beating before and after moving the pin, remains approximately the same.

Fig. 8.1

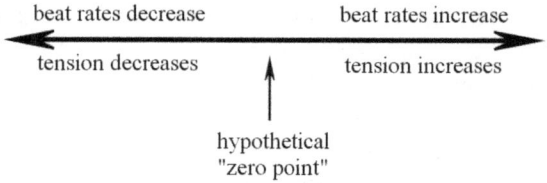

Fig. 8.1. Model for rough tuning. The beat rates in the unison pair increase, with increasing or decreasing string tension away from a hypothetical "zero point".

It is important to remember that this "zero point" is hypothetical. It reflects well the experience of tuning in the learning stages, but it is not a complete, or "scientifically correct" model. It may be possible to tune a unison so that beating appears to vanish, at a certain acuity of listening, but closer scrutiny by the expert ear, or by digital analysis, invariably reveals at least slow beating or fluctuations of partials to remain present somewhere in the spectrum.

In chipping up, the remaining presence of beating in the unison soundscape is not necessarily an indication of the need for further tuning. It may be, or it may not be. It is a question of degree and context. In the learning process an appropriate judgement requires a teacher with knowledge of the student's current stage of learning, and hearing acuity.

Reducing beat rates without crossing the hypothetical zero point, is initially difficult mainly for mechanical reasons, rather than acoustical ones. The strings are in high tension and pass at bearing angles over friction points such as the iron top bridge, under the pressure bar, under the capo d'astro or through the agraffe. The wrest (tuning) pin itself is inserted into the wrest plank (pin block) with the result that friction acts between the wood and the pin. Overcoming these frictions, and the mechanical complication that they cause, is dealt with in detail in the chapter on setting the pin. Briefly, there is not a direct and immediate connection between a change to the pin position and a change to the soundscape. Even on the chipping back on where *setting the pin* is not being considered, there remains many mechanical influences that need to be mastered, or at least tackled by the tuner, in order to take control over the soundscape, through movements in the tuning pin (wrest pin).

Listening to specific partials

It is sometimes helpful in the early stages of learning, if the beating in a specific, prominent partial, becomes the focus of attention, even though this is contrary to what will be eventually required for unisons in fine piano tuning. If, in the early stages, a specific partial does become the focus of attention, then which partial is suitable will vary from case to case, and it will not necessarily always be the fundamental. In practical tuning sessions beating is often best pointed out by the teacher, by imitating it, using the "wawawa" sound, sung at the musical pitch of the partial. In support sessions listening techniques can be developed, and hearing acuity enhanced, with the use of recorded unisons and electronic resources that can isolate individual partials.

If you are learning to tune, it is important to learn to pick out the harmonic series on the keyboard, [34] starting from any note, and to use this as a guide to help identify the pitches of the partials of any single note. This practise increase aural technique, acuity and power. However, having become efficient at this, it is then important to practice listening to the *tone* of the unison, *without* focussing on any individual partials.

Relationship of unison beat rates

The partials of the unison fall in a series of frequencies that is an approximately harmonic series. If the strings were perfectly flexible and there were no other complicating factors, the partials would have fallen in a perfectly harmonic series. Then, in rough tuning, we would always have expected the beat rate in the nth partial to be n times the beat rate in the fundamental. This follows from rudimentary theory. If this were the case, the various beat rates present in the soundscape would always remain "in step" with the fundamental beat, even though they may be beating at different rates. In practice, whilst this relationship approximately applies, *inharmonicity* (caused by string stiffness) prevents it applying perfectly, so beats in different partials are seldom "in step" with each other. It is not therefore possible to quote a single beat rate as *the* beat rate of a unison. Over most of the compass, if a unison is beating, then there are many beat rates, and they will not be perfect multiples of the slowest one. The slowest one, the beat in the *fundamental*, may also not necessarily be the strongest or most audible.

Useful principles to remember, are that for any condition of *rough tuning*, higher partials beat faster than lower ones, and in general, increases or decreases to beat rates will apply to all partials simultaneously. In the finest tuning, however, this last rule breaks down, as we shall see later. Typically, in fine tuning, it is not the fundamental but partials above it, whose beating indicates the need, or otherwise, for tuning changes. Fine tuning, as we shall see in some examples below, often involves changes that reduce beating in the higher partials, whilst leaving the lower ones unchanged.

First steps – summing up some practical points

To sum up, when a unison string pair is closely tuned, a tone will be heard, whose tonal properties – and in particular beating – are adjustable.

---

[34] The Glossary includes instructions on how to do this.

If the strings are tuned sufficiently far apart, two distinct notes or pitches can be simultaneously heard. Beating, heard as a "pulsating", can occur at different "speeds" or *beat rates*. In the initial learning process, beat rates between perhaps 1 and 8 beats per second are most easily heard. The tuner must learn how the beat rates in the unison tone vary with string tension changes. The first principle to learn is that beating slows down as the strings are brought closer "into tune", towards the hypothetical "zero beat point". In other words, the unison tone we are aiming at, is one in which the beating is slowed down as much as possible.

In particular, the beating is most obvious (to the learner) when it is not too fast, when the unison pair are close to being "in tune" but still obviously "out of tune". When the two strings are more grossly mistuned, beating is still present, but can be so fast that it may not be heard as beating. Also, two distinct pitches may then be heard, from the two strings, rather than one pitch whose tonal properties we are adjusting. The student must learn how the beat rate increases when the string being tuned is either sharp or flat of the "target" tuning, and how, roughly speaking, it more or less "stops" at the "target" tuning.

The physical mechanics of changing the string tension must be linked with what is heard. It is necessary not only to recognise beating, or how the beating *has changed* after making a turn to the pin, but also to hear *how* the beating *is* changing *as* the tuning lever *is being* moved. Turning the pin without at the same time listening to the two strings sounding, is to be avoided. Never turn *and then* listen — always turn *and* listen. However, do not expect always for there to be a change to the soundscape for every small turn of the lever. Sometimes, the tension changes in the top stringing without the speaking length tension changing, because of friction at the top bridge or agraffe (or *capo d'astro*). Also be aware that sometimes the pin turns in the wood, and other times it only twists, torsionally. Much more is said on this in Chapter 11.

The seascape analogy

It is sometimes useful to describe the soundscape as the audible equivalent of a seascape. The seascape as a whole is analogous to the soundscape as a whole, and the individual waves on the sea are analogous to the individual partials. We can either consider the sea as a collection of individual waves (the partials), or as an overall sea "picture" or seascape (the overall perceived, tonal soundscape). A seascape consists of many wave movements taking place simultaneously, and without looking at any particular wave, we can see whether the seascape is rough or calm. In tuning the unison, we are attempting to create as calm a soundscape as possible, by listening to the whole soundscape. In the learning process, it is often important for the teacher to point out, however, where beating is taking place in the soundscape, or the student may incorrectly accept a unison in which there is too much movement, erroneously believing beating to have been sufficiently "tuned out".

In the seascape analogy, the unison, like a perfectly calm sea, is not necessarily like a flat sheet of glass, although rarely, it may be. In the learning process the first aim of unison tuning is simply to achieve a soundscape that, like a seascape, is as "calm" as possible. This does not require focussing on particular individual partials, just as judging the calmness or roughness of a seascape can be judged without necessarily focussing on particular waves.

We would know if the sea was still, or if it was stormy, without having to pay attention to specific waves. We would nevertheless also be aware that its storminess or stillness was in fact due to the nature of waves on its surface, or perhaps their absence altogether. Similarly, when the tuner listens to the soundscape of the unison, the spectrum is certainly not heard merely as a musical tone, without any knowledge or awareness of the tone's internal partial structure and behaviour. The artist tuner hears the tone in far greater detail than the untrained auditor, and this is possible precisely because of the fact that the tone is comprised of a spectrum of individual, discrete partials.

Beating in fine tuning

In chipping up, and in the early stages of piano tuning, it is useful to continue to use the concept of the hypothetical zero beat point, and the general idea of "tuning out" beating. This is an important first principle, but it is only a beginning. The main characteristic of beating in rough tuning is its regularity, and it is this regularity that allows it to be described as having a beat *rate*. Actual beat phenomena in the *fine tuning* of piano unison partials may not necessarily be very regular, and can have properties other than the beat rate, that can be adjusted in the fine tuning process. In the finest tuning of piano unisons, in practice it is more generally perceived *movement* or *tone quality* factors in the soundscape, arising from the collective partial properties, that are consciously adjusted.

More generally, the minimisation of tonal movement or fluctuations in the soundscape involves, as we shall later see, *attenuating* fluctuations in the partials through the fine tuning process, in addition to reducing actual beat *rates*. This additional attenuation may include the reduction of beat amplitude, and the increase of beat decay rate. We will look at this in more detail below.

The position of "traditional" theory with regard to unison tuning, and a frequently cited practical edict for it, is simply that the piano tuner should "eliminate the beat" (eliminate fluctuations in the soundscape) in the unison. This is probably still the most appropriate guidance in the initial learning process. However, unison tuning by master tuners is always about fine control of unison tone, not simply about eliminating a hypothetical, singular or regular "beat".

Listening to beats has often been misunderstood as a tool for achieving specific, hypothesised frequencies for the strings. The master tuner does not use beating as a tool for this purpose. Beating is a tonal feature in its own right, that is adjusted by the tuner *as a property in its own right*, and not as something that is used as a "tool" for achieving particular, theoretically hypothesised string frequencies. The relationship between beating and string frequencies (and there are many different frequencies present in the tone of each individual string or note) is *much* more complicated than rudimentary theory presupposes. The master tuner is concerned with the *beating itself*, not with the frequencies that may consequently result.

To sum up then, the aspect of tone that must first become familiar in the learning process, is the movement in the soundscape called *beating*, which is a more or less regular fluctuation pattern in the partials. Good tone for a unison is always associated with a "minimum" of beating, at every level, but this does not mean that an optimum tone is always identifiable with complete absence of beating throughout the entire soundscape. Such an absence is not a natural feature of piano tone, as expert tuners are aware, and as digital analysis easily reveals. This does not, however, mean that the expert tuner deliberately leaves movement in the soundscape of the unison when it could have been reduced.

Saying that the tone of a unison is generally considered to be at its best when movement in the soundscape is reduced to a minimum, also does not imply that there is necessarily only a single tuning position that constitutes a single "minimum" movement. Rather, *movement* or beating in the piano tone soundscape is not a single thing, but is the result of changes taking place in *many different partials*, and these may be behaving in different, independent ways, in relation to tuning. The movement in the unison soundscape as a whole is the result of all these separately changing or fluctuating partials. The precise tuning position (the precise tensions on the strings) required to minimise the movement in any one partial, is not necessarily the same for all partials in the unison spectrum, as we shall later see.

All partials in the unison are nevertheless affected simultaneously by any tuning change, because all the partials in the unison's spectrum are *adjustable* partials. Making a change to the tuning of a unison can thus readily affect its *entire soundscape* by potentially altering *every* partial's decay pattern, simultaneously. In the finest tuning some partials may be affected more than others, but always the smallest tuning (tension) changes in the unison can readily produce relatively large and obvious audible soundscape changes, at least to the expert ear.

Unisons and beat inheritance

The unison soundscape's sensitivity to the unison's precise tuning condition, affects the entire tuning of the piano. Although it may seem obvious to experienced tuners, many students of tuning in the early stages do not "instinctively" appreciate the dependence of the tuning condition of other intervals, on the unison tuning. There is little point in attempting to fine tune a tempered interval or an octave, if the unison against which the octave or tempered interval is being tuned, contains avoidable beating. The interval containing the beating unison will, in general, *inherit* this beating from the unison. The most effective way to reduce such beating is through the tuning of the unison itself. The unisons are the building blocks of the tuning as a whole, and it is necessary for the tuning quality of every unison to be excellent in order for the tuning of the piano as a whole to be successful.

Two beat species - *false beats*

There are in fact *two* distinct kinds of beating in the unison tone (not just the one that rudimentary theory proposes). Beating of the first kind has a beat rate that is indeed very *sensitive* to changes of string tension, as rudimentary theory predicts, and so can be slowed or eliminated through tension changes to the string. Beating of the second kind, however, which includes what piano tuners recognise as "false beats",[35] is beating whose beat *rate* is very *insensitive* to changes of string tension. The soundscape

---

[35] The term "false" here is historical. It does not imply the usual modern meaning of the word "false", except perhaps in relation to deception. Inexperienced tuners may be deceived by false beats.

of any single string, or any unison, may include beating of both kinds, simultaneously.

Typically, the pitch of a string can be raised by a large amount such as a semitone or more, without actually affecting the *rate* of a "false beat". Beating of the second kind that is present in the spectrum of a single string, will in general be *inherited* in some way by the soundscape of the unison. Whilst a change to the tension of one string of a unison pair can slow the rate of a beat of the first kind, it will not generally slow the rate of a beat of the second kind (false beating). The presence and nature of false beating, its effect on the soundscape, and how it is affected by tuning, is not recognised at all by the idealised, rudimentary theory.

Beating of the second kind, false beating, is found in the tone of strings throughout the compass, but occurs with different properties, depending largely on the position in the compass. In a very good instrument the fundamental partial is rarely false in the notes of the central compass, but there are almost always some higher partials in the spectra of central compass notes, that are false. Above the treble break, however, it is not unusual on a good instrument for fundamentals to exhibit falseness. In the bass, fundamentals are less likely to be false, but the spectrum of a bass note typically has *many* false higher partials.

The effort by the tuner to reduce movement in the soundscape to a minimum, *must take into account beating of the second kind*, as well as beating of the first kind. Although beating of the second kind, false beating, has a beat *rate* that is insensitive to changes of string tension, this by no means makes it entirely unadjustable, as we shall shortly see. This is because in beating of *both kinds*, both the *beat amplitude* and the *beat decay rate* are adjustable within a fine tuning range, through a process of beat *attenuation*.

False beating, as beating present in the partials of a single string, was recognised in early texts and investigations into piano tone,[36] and today its presence is very easily demonstrated through digital isolation of partial envelopes. Braid White commented on false beats as being 'the worst single enemy the tuner has',[37] and the meaning of this would be

---

[36] See, for example, Cree Fischer, J, *Piano Tuning / Regulating and Repairing*, (1907), reprinted as *Piano tuning. a simple and accurate method for amateurs* (Dover, NY, 1975), pp. 160-161; Schuck, O H, and Young, R W, 'Observations on the Vibrations of Piano Strings', *JASA*, 15, 1, 1 – 11 (1943); Wolf and W. Sette, S K, 'Some application of modern acoustic apparatus', *JASA*, 6, pp. 160 – 168 (1935).

[37] Braid White, W, *Piano tuning and allied arts*, Boston Mass., 1917, 1972, p. 106.

recognised by any experienced tuner. He is referring to false beating in single strings, that is then inherited by the soundscape of the unison, and severely impedes the attempt to reduce beating in the unison soundscape, since tension changes to the string will not in general reduce a false beat rate. However, Braid White's comment as it stands, also unfortunately suggests that single strings either "normally" produce no beating partials, or may abnormally produce beating partials, *i.e.* false beats, of such a magnitude that tuning is severely impeded. Whilst false beating severe enough to be considered an "enemy" to the tuner does indeed exist, it is by no means the only instance of false beating.

In fact, beating partials in the tone of a single string are a "normal" part of piano string tone behaviour, but are not always severe enough to fall into the class that Braid White is describing. There is no specific point at which natural false beating phenomenon becomes the "enemy" to tuning that Braid White describes. It is a question of degree. Most false beating or fluctuation goes relatively unnoticed, because it does not severely impede the attempt to reduce beating or fluctuation in the unison soundscape. Nevertheless, it is present, can in fact easily be heard with proper aural attention, and can be easily seen and heard in digitally isolated partials.

Braid White specifically identified false beating in its severe form, as a *fault*, and associated it with specific supposed physical faults, described as "uneven strain" in the string, "uneven thickness" of the string, or a twist in the wire.[38] However, the contemporary acoustical theory for the vibration of the piano string *in situ* indicates that such faults are not necessary for false beating to arise. The theory predicts beating partials can be present in the soundscape, as a natural part of the behaviour, even for a string that is perfectly uniform and untwisted. The theory shows that some false beating should be *expected* to be present, and whether or not the theory is accurate, actual observation shows that it is indeed present where there is no reason to suspect a physical fault. One of the theoretical reasons for its presence lies not in the string itself, but in the boundary of the speaking length, specifically, the soundboard bridge.

The soundboard is designed as a two-dimensional shallow arch, the curvature being known as "crown" or "buck". The soundboard arches towards the strings at its centre, and in doing so elastically resists the collective downbearing force of the strings pressing against the arch. The structure is such that motion in the string in a plane perpendicular to the

---

[38] Braid White, *op. cit.*, p. 106.

surface of the soundboard, causes the soundboard to move in the same plane. The purpose of the soundboard is to act as an amplifier, draining energy away from the strings, and radiating sound from its much larger surface area.

The structural design of the soundboard is such that its elastic properties in the direction perpendicular to its surface would be expected to be different to those properties in the direction parallel to its surface. Roughly speaking, it should be much easier to move the soundboard in the direction perpendicular to its surface, than in a direction parallel to its surface. In the acoustical theory, this ability of the soundboard to move elastically, is included in a parameter called the soundboard's *reactive admittance* (or alternatively, its inverse, its reactive impedance). We should expect, then, from the structure of the soundboard, that at the end of the string's speaking length where it connects with the soundboard bridge, the *reactive admittance* at the bridge would be much smaller in the direction perpendicular to the soundboard, than in the direction parallel to its surface.

In fact, the reactive admittance has been shown to be about the same in both directions.[39] One possible explanation for this finding is the side draft pin. The side draft pin terminating the string's speaking length is cantilevered out of the bridge surface, and any movement of the string parallel to the soundboard surface must act on this pin. Even though the string should pass the pin at its base, the side-draft force of the string is resisted by the pin, so that string and pin are in elastic equilibrium, under considerable force. It is possible that the net effect is that the reactive admittance at this point *can* be equal in both directions, at least partially because of the influence of the side-draft pin.

We cannot, however, simply assume perfect equality always to be the case. The theory demands that any sufficient inequality can lead to the string producing false beats, depending on a number of other parameters. The degree of inequality can determine the false beat rate, so if it is small, a slow false beat can still result, the limiting factor being whether the beat amplitude decays faster than the beat itself develops.

An easy practical way to demonstrate the effect of the bridge (side-draft) pin on false beating, is to locate an obviously false string, and sound the string both with and without a screwdriver providing additional support to the pin, against the side-draft force of the string. When additional support is provided, a false beat rate will clearly be slowed, or

---

[39] Weinreich, G, 'Coupled piano strings', *JASA*, 62, 6, 1977, pp. 1474-1484.

even eliminated. The facts suggest that less than optimum rigidity of the side-draft pin will increase the admittance in the direction parallel to the soundboard surface. What constitutes "optimum" will of course depend on the other parameters, determined by the wood, the dimensions of the pin and string, the tension, and the side-draft angle. An inequality arising where the conditions are not "optimum", can cause false beating, even if not necessarily severe enough to be noticed and labelled as such, by the tuner.

The salient point is that the construction of the pin in the wood need not necessarily be faulty in order to allow for *some* inequality, and hence false beating that is not severe. Where faults are obvious, such as the string passing the pin above its base (where its elastic admittance of the pin will obviously be greater), or where the wood supporting the pin is obviously compromised by splits, large inequality might be expected, and in such cases more severe falseness is indeed frequently found in practice.

There are other mechanisms also, in contemporary theory, that might explain observed false beating where there is no physical fault, as such. One such possibility, shown by Morse and Ingard,[40] is longitudinal motion acting as a coupling mechanism between transverse modes. Irrespective of how accurate or otherwise the attendant theory might be, false beating in numerous degrees, from being aurally insignificant, to being an outright "enemy" of tuning, is in any case an observable fact of piano tone. As a normal part of general behaviour, the phenomenon in itself cannot be considered an anomaly or fault in string behaviour or piano tone. It can only properly be labelled a fault in the context of tuning where it is both particularly severe (fast and prominent) and could be reduced by removing clear physical faults.

It is probably worth emphasising here, that false beating in partials is a phenomenon quite distinct from *inharmonicity*. Inharmonicity is the raising of the frequencies of partials, principally due to string stiffness. Early investigations into inharmonicity in piano tone were made by Schuck and Young,[41] and today it is recognised as a particular case of a more general and widespread phenomenon known in acoustics and engineering as *mode frequency dispersion*. Inharmonicity could perhaps be colloquially

---

[40] Morse, P M, and Ingard, K U, *Theoretical Acoustics*, McGraw-Hill, 1968, pp. 861 - 863.

[41] Schuck, O H, and Young, R W, 'Observations on the vibrations of piano strings', *JASA*, 15, 1, 1943, pp. 1-11; Young, R W, 'Inharmonicity of plain wire piano strings', *JASA*, 24, 3, 1952, pp. 267-273.

described as an effect in which "harmonics are thrown out of tune", and this phrase appeared originally in Braid White's *Piano tuning and allied arts* as an explanation of false beating.[42] This has sometimes lead to a confusing of inharmonicity with falseness, and *vice versa*. The two phenomena are quite distinct.

Visual conceptualisation of falseness in the soundscape

Fig. (8.2) illustrates parts of two 3-D spectra of single string piano tones, one of which we met earlier in Chapter 6. On the top is represented part of a spectrum containing false partials, in which the beat patterns can be seen. Below is a part of a spectrum closer to what might be expected from "traditional" theory, in which there are no false partials. The illustration is a visual conceptualisation of how the part of a soundscape containing the lower partials might be composed of a mixture of both steadily decaying partials, and false partials that beat.

---

[42] Braid White, *Piano tuning and allied arts*, 1917, 1972, p.106.

Fig. 8.2

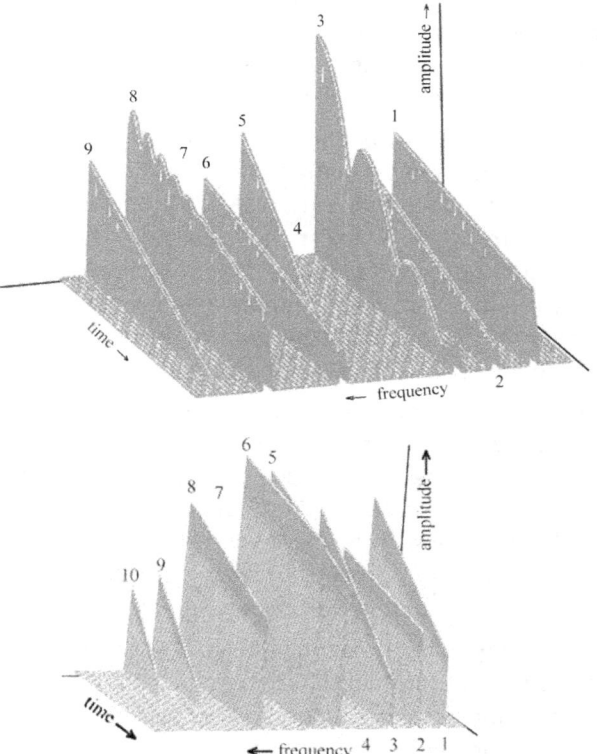

Fig. 8.2. 3-D graphical representation of part of an audible spectra from single strings, showing the individual partials. Above shows a spectrum containing some false partials, exhibiting beat patterns in the decay. Below shows a spectrum without any false partials in this section of the spectrum. Each shows only the lower part of the spectrum, including only the first 9 or 10 partials. Time is along the horizontal axis running towards the point of view, and the vertical axis represents loudness of the partial.

Falseness and context

Despite the fact that a high degree of falseness is often associated with poor quality instruments, there may be falseness widespread across the compass of some high quality instruments. It is certainly not the case that high quality instruments, by virtue of their name and quality, can be assumed to be inherently free from false beat behaviour. Falseness is contextual, a natural part of the way vibrating piano strings *in situ* behave, so rather than being regarded as a "fault" *per se*, it must be understood and incorporated into theory and practice. As we shall see, false beating within certain limits, despite its resistance to having its beat rate reduced, is certainly not necessarily an obstacle to tuning aspirations, and to the production of a beautiful tone.

Falseness is heard as beating – however slow - in the spectrum, a beating that originates in the individual string itself, and is then inherited in the unison (and indeed other intervals). In a unison trichord on at least a reasonable instrument, falseness *can* usually be dealt with in a way that is not necessarily detrimental to the tone. Much of it is reduced to the point where it is not noticeable, simply as a consequence of the expert tuner's ability to reduce beating and movement in the soundscape, without the tuner necessarily having to be specifically aware of it.

In these cases, as we shall see, the reduction is generally by *attenuation* rather than beat rate reduction, although the overall effect is very similar. Where the tuner is specifically aware of its presence, the tuner may refer to this as "hiding" falseness or "pushing it into the background", or other such terms. As we shall see, there are acoustical mechanisms by which this process is made possible, which allow far better results than merely "masking" false beats with other sound.

Falseness - as the phenomenon in general rather than in a form severe enough to be very detrimental to tone – can result in unavoidable movement remaining in the soundscape of the well tuned unison, but it is the degree of movement in terms of the number of partials that are false, the false beat rates, and the just as importantly the beat amplitudes, that can make the difference between ugliness and beauty of tone.

Tonal beauty need not be analysed

We have said that some form of movement is generally always present in unison tone, even if only in higher partials. It is true that in most cases, even a small amount of *readily eliminatable* movement in the soundscape can be considered undesirable, or ugly, to the ear of the expert tuner. It is not the case, however, that movement in the soundscape is ugly, *per se*. Movement in the soundscape is a natural and ubiquitous feature of piano tone, and can also be part of tonal beauty. The degree of movement or fluctuating changes in the tone recipe, that is a part of beauty in *piano tone*, is, however, very small. That fact that it may still be present in beautiful unison tone is not a convenient excuse for poorly tuned unisons.

After the hammer strike, there are features of the tone recipe - apart from partial frequencies - over which we can have control, which are scientifically measurable properties. It would nevertheless be folly to try to define good tuning, or to try to carry out fine tuning by considering these properties separately. They collectively contribute to the *perceived tone quality*, but the latter is both sufficient to tune by, and, after all, the point of fine tuning. It would be scientism to suggest that tonal beauty can be entirely scientifically definable, and even if it were, it would not be necessary to deal with difficult or complex "scientific" criteria when the perception of *tone quality* itself, is already easily available to the expert ear.

Judgement

If different partials require different tuning positions (different tensions on the string) for their movement to be minimised, then how is one to know which tuning position is the "right" one? Also, if some movement is inevitable, perhaps due to falseness, how is one to know *which* movement, or how much movement is acceptable, and which movement, or how much, is not? The answer to this lies, again, in *tone quality*. There is no established, neat theoretical answer, nor does their need to be.

It is not a question for science, it is a question of *art*. In scientific terms, the perceived tone quality of the unison, including any movement in the soundscape, is not just dependent on any one partial. The whole partial recipe - the whole spectrum – plays a part in determining the tone. Fine unison tuning is not generally carried out with conscious regard for movement in this or that particular partial, but with regard for movement in the soundscape *as a whole*, in the context of the *tone quality* of the soundscape *as a whole*.

Knowledge of tone quality, as the awareness of what is or is not possible for a given tone, and the ability to control it with finesse and mastery, comes only from experience. The judgements made by the master tuner with respect to the tone of a unison cannot be encapsulated in a straightforward formula. This does not, however, mean that unison tuning becomes a freestyle game in which any tuning can count as good. There is a solid consensus for reducing movement to a minimum, against which any proposed defence for "leaving in" readily eliminatable fluctuation or beating, is seen by those who are capable of eliminating it, as merely an excuse for poor tuning ability. Notwithstanding this, the fact still remains that beating or fluctuation is seldom entirely eliminatable for the reasons already mentioned, so reducing it to a minimum, never in general means removing it entirely.

Above all, even on poor instruments, beauty in the instrument cannot be fully realised without the tuner's conscious awareness of tone *as musical tone*, in addition to the more objective manipulation of it, through the awareness of its acoustical structure and behaviour. This is very much a *musical* perception. The difference between one tuning and another, when some movement is unavoidable, is a matter of art that may seem arbitrary or insignificant in the individual unison. However, in tuning the final results on the piano may be more than the sum of the parts - see the comments on the first page of the *Tuning in practice* section.

Some scientific studies of unison tuning

According to rudimentary theory, two or three strings tuned to a "perfect" unison would have the same frequency fundamentals. The unison would then be beatless. Martin and Ward[43] in 1954, however, noted that expert tuners did not necessarily tune the fundamentals of the unison strings to equal frequencies. In 1959 an experiment by Kirk[44] seemed to indicate that musicians actually preferred the sound of unisons with a small frequency difference between the fundamentals of the strings, to those with no frequency difference. Kirk's results also suggested, therefore, that piano tuners might be deliberately tuning in a way that resulted in small frequency differences between the fundamentals of unison group strings.

Another perhaps unexpected result appeared in investigations into the decay rates of piano tones. Martin, Chase Hundley, and Benioff found that unison partials typically exhibit a dual decay rate in which the first part of the overall decay may have several times the decay rate of the later part.[45] Benade[46] hypothesised in 1976 that the decay rate of the unison should initially be three times that of the single string. He reasoned that when all three strings vibrate in phase, as they should initially, the force

---

[43] Martin, DW, and Ward, WD, *JASA*, 26, 1954, 932(A); 33, 1962, pp. 582-585.

[44] Kirk, RE, 'Tuning Preferences for Piano Unison Groups', *JASA*, 31, 1959, pp. 1644 – 1648,.

[45] Martin, DW, 'Decay rates of piano tones', *JASA*, 19,4, 1947, pp. 535-541; Chase Hundley, T, Benioff, H, and Martin, DW, 'Factors contributing to the multiple rate of piano tone decay', *JASA*, 64,5, 1973, pp. 1303-1309.

[46] Benade, AH, *Fundamentals of musical acoustics*, NY, 1976, 1990, p. 336.

exerted on the bridge is three times that of a single string alone, the bridge motion is tripled, and the rate of energy loss is thus tripled. After a little time, due to slight mistuning the strings lose their phase relationship and the decay rate falls to equal that of a single string. This explanation seemed to lend weight to the notion that frequency discrepancies deliberately tuned between the fundamentals of the individual trichord strings, were an advantage. The strings of a unison "too perfectly tuned", so to speak, would not lose their phase relationship very quickly, so the unison would simply decay too fast.

The argument was that expert tuners introduce "mistunings" between the strings of the trichord in order to improve the decay time – "perfectly tuned" unisons, it was argued, die away too rapidly and sound too dull. (It should be noted, however, that whilst dual decay rates are observable, this intuitively appealing explanation nevertheless assumes all three strings vibrate in the same single plane only, and that only vibration in one plane contributes to the radiated sound. In fact string motion naturally takes place in two planes, despite the string being struck in only one plane, and motion in both planes contributes to the sound radiated by the soundboard).[47]

In 1977 Weinreich[48] showed that in a two stringed unison, the *decay rate* of a partial is strongly dependent on the mistuning between the strings when the mistuning is small. Furthermore, at small mistunings, the beat rate itself is not as dependent on the mistuning as "traditional" theory supposes. This is because the soundboard-bridge couples the string motions. The frequencies and decay rates produced by the unison with both strings sounding, as a coupled system, are not necessarily the same as those produced by the individual strings, sounding alone. Weinreich's work suggested that in the finest tuning region, piano tuners can be making changes that primarily affect partial decay rates, rather than beat rates. Importantly, zero mistuning does not necessarily produce a beatless unison, and conversely a small mistuning does not necessarily mean the that beat rate predicted by "traditional" theory must be present.

The decay rate of a partial would also be affected by the *phase relationship* between the strings, and Weinreich pointed out that the latter can be affected by small irregularities in the hammers, causing strings to begin vibrating at different times, or with different amplitudes.

---

[47] Weinreich, G, 'Coupled piano strings', *JASA*, 62, 6, 1977, pp. 1474-1484.

[48] Weinreich, *op. cit.*

Small mistunings might, Weinreich suggested, be being deliberately introduced by expert tuners as a means of altering the decay rates of partials, as part of a process of adjusting tone, in order to compensate for hammer irregularities.

Summary of the difficulties in the scientific study

All these developments illustrate that there is more to fine unison tuning than simply tuning all strings so that their fundamental frequencies are the same. The complications met in both scientific investigation and tuning practice can be summarised:

1. The decay behaviour of one partial on one piano string or unison, is not representative of all partials and strings in general. The behaviour can vary significantly from one partial to another. Demonstrating partial behaviour in one partial does not necessarily imply all partials will behave in the same way.
2. The behaviour of many partials can simultaneously contribute to audible movement in the soundscape.
3. The string motion driving the soundboard to create the audible sound, is complex. Piano strings in general may move in two transverse planes, as a result of being struck in one plane. The resulting elliptical motion is not, in general, necessarily of a fixed polarisation, and movement in two orthogonal planes can contribute to the radiated sound from the soundboard.
4. The frequency of an audible partial may not be constant.
5. Many audible partials of even single strings exhibit beating, indicating more than one component frequency to be present in the single partial.
6. The vibrations of the unison strings are in general coupled by the bridge and soundboard – one string affects another, especially in unisons. Coupled vibrations can lead to much more complicated behaviour than uncoupled vibrations.
7. Fine tuning can alter not only beat rates, but also decay rates and other specific features of decay patterns in the partials.

***Mistuning* - the technical meaning**

It is convenient to talk of *mistuning* between the strings of a unison group or pair, or more specifically, the mistuning at a particular partial.

This will be necessary in order to understand how unisons behave in practice, in the finest tuning. The word "mistuning" in contemporary theory[49] is a technical term that does not necessarily mean there is any "incorrectness" in the tuning condition between the two strings. Rather, we take "zero mistuning" between two strings to mean that at that particular partial number, the partials of each string *sounding alone* would have the same frequency. We need to refer to the partial frequencies measured with just one string sounding, because as soon as both strings are sounding, a *different* set of partials is created, which may have different frequencies.

When we say, for example, there is a mistuning of 0.2 Hz between two the fundamental partials of two strings, one from each string, we mean that if each string was sounded on its own and the partial frequency was measured, we would find a difference of 0.2 Hz between the two measurements. When the strings sound together, there may be a unison tone partial in the same frequency range, with two frequency components within it. However, we may not necessarily find the same 0.2 Hz difference between these two frequency components.

In "traditional" theory, a beat rate is simply equal to the mistuning. Thus, in "traditional" theory, a mistuning of 1 Hz will yield a beat rate of one beat per second, and beating will vanish when mistuning is zero. In contemporary theory this relationship holds well for sufficiently large mistunings, but for smaller mistunings, in the range of fine tuning, different rules apply. In contemporary theory regular beating can vanish when the mistuning is not zero, and a beat can persist when the mistuning is zero, depending on the conditions. This is primarily due to bridge and soundboard coupling. We will look at this in more detail later.

Movement as tuning changes are made

One type of soundscape *movement* occurs naturally in the spectrum, when it is not being altered by the piano tuner. Another distinct type of soundscape movement arises additionally as a direct consequence of the piano tuner deliberately *changing* the mistuning. In a sense, one could say this second kind of movement is a change or movement, of the acoustical movement in the spectrum. The perceivable nature of movement in the soundscape when no change to string tension (in the speaking length) is taking place, is different to the perceivable nature of soundscape

---

[49] The definition of *mistuning* is taken from Weinreich, *op. cit.*

movement as tension change takes place. In other words, even though the soundscape is already moving, we can hear it change, as a tuning change takes place. Fig. (8.3) illustrates this simply.

Fig. 8.3

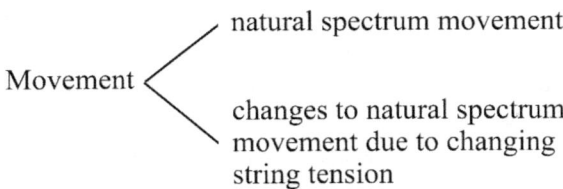

The relationship between these two aspects of audible *movement* in the spectrum are important in connection with the art of "setting the pin". We can tell from it not only where the tuning condition is at any time, but how it is changing in relation to the changes of tension we are introducing through tuning. This, in turn, informs us about the stability of the tuning. In the process of fine tuning, we must be able to distinguish between one type of movement and the other, and must be able to "feel" the connection of the latter to the lever movements. We will deal with this in more detail in the chapter on setting the pin.

How partials are mapped from normal modes

In the chapter on sound we saw that just as complex sound "splits up" into simple pure tone ingredients, so, similarly, complex vibrations can also be shown to be a sum of Simple Harmonic motions, known as *normal modes*. In the methods of modern acoustics, a complex vibration like that of a piano string or soundboard, is typically analysed as a set of normal modes, or Simple Harmonic motions that together constitute the overall complicated motion. The normal modes found when two connected (or *coupled*) parts of a complete vibrating system are both in motion, are not necessarily the same as the modes for each part if it were in motion by itself. For example, when two pendulums are connected by a spring, they can both be capable of motion that neither pendulum on its own could achieve (in fact, the resultant motion can be quite counter-intuitive).

In the case of vibrating piano strings, two or three strings of a unison group vibrating together, can produce bridge motion that no one string on its own would produce. This leads to the situation in which the tone of the unison must be considered as a recipe of partials in its own right, rather than simply the sum of the recipes for the individual string tones. It is easy to imagine that in listening to a unison, one is simply hearing the sounds of individual single strings, simultaneously. This, however, is not the general case, and as a concept can create difficulties in understanding what is actually heard. What one hears in the finely tuned unison is a soundscape arising from the two or three strings behaving as a connected (coupled) system.

The normal modes of piano string vibration in "traditional" theory, are the standing waves on the string. In "traditional" theory one audible partial is associated with one normal mode of motion (standing wave) on the string. The string's normal mode frequencies in the simple model are arranged in the harmonic series – all the normal modes are well separated in frequency. In contemporary theory, and as can be easily demonstrated through digital analysis of piano tone, real piano tone partials differ from this picture in two major ways. Firstly, partials fall only approximately in the harmonic series, being actually sharper or higher in frequency than the corresponding true harmonic would be. Nevertheless, the individual audible partials, being in an approximately harmonic series, are well separated in frequency. Secondly, a single audible partial *may* be generated from a close-frequency *pair* or *group* of normal modes, rather than just one normal mode. This is evidenced by the fact that some single audible partials demonstrably contain two or more close-frequency components (Fourier component), and typically exhibit beating or fluctuation as a result.

When we adjust the string tension with the tuning lever, we are adjusting the recipe of normal modes responsible for producing the resulting complex sound. It can be helpful to understand how the audible partials of a unison that we listen to, are mapped from the *normal modes* that we adjust.

Let us take a look at the mapping for a single string. The complex motion of the string, soundboard and bridge, creating a piano tone, can be divided into a spectrum (or a *recipe*) of normal modes, each moving with simple harmonic motion. The soundboard motion then radiates into the air the *sound* that we hear. Sometimes, two normal modes are very close in frequency, and create sound that will be heard as one audible partial.

Fig. (8.4) shows 6 normal modes of the system being mapped to 5 audible partials.

Fig. 8.4

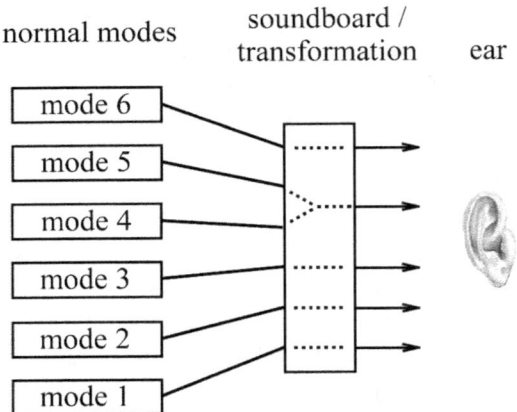

Fig. 8.4. Example mapping of normal modes of motion to audible partials, for a single string. The modes themselves, are approximately in an harmonic series of frequencies, except that in some cases two or more modes close in frequency appear in place of one. In an harmonic series of frequencies, or even an approximately harmonic series, all the frequencies are well separated. The series of partials radiated from the soundboard, falls in an approximately harmonic series, some partials being created from two or more close-frequency normal modes.

The normal modes are approximately arranged in the harmonic series, except that sometimes two or more close-frequency modes appear in place of one. In this example modes 4 and 5 are close in frequency. The final set of audible partials emitted from the soundboard (on the right hand side) are always in an approximately harmonic series, so we still hear the "harmonic series" of partials familiar to piano tuners. In this example, the six normal modes on the left, are transmitted by the soundboard as the five audible partials in an approximately harmonic series, on the right.

The normal mode series does not therefore correspond directly to the audible partial series. In the case of a single string sounding, the six modes shown on the left are normal modes of a combined system that is composed of the string plus the bridge-soundboard system. Modes 4 and 5 combine to make one partial because they are close in frequency, and if they have slightly different frequencies, the partial may display a *false beat* (but not necessarily, depending on the mode decay rates).

If we were representing to two or three strings of a unison sounding *together*, rather than a single string, the same *kind* of diagram could still apply, but the modes on the left hand side would then not be normal modes of a single string vibrating on the bridge and soundboard. Rather, they would be the normal modes of a *different* system that consisted of more than one string, plus the soundboard-bridge system. In terms of the audible *soundscape* to which the piano tuner listens, the soundscape of the unison as a whole is not simply the sum of the soundscapes of the single strings in the unison group. It is a soundscape mapped from a *different set* of normal modes to those of the single strings. However, it will have much in common with the soundscapes of the single strings, and may inherit many of their properties.

Inheritance of false beats

When noticed aurally, beating of the second kind, *i.e.* beating whose rate is not sensitive to changes of string tension, is referred to by tuners as falseness, false beating, or false beats. False beats can be easily identified when they occur in a partial of a single string when sounding on its own, simply because *any* beating in a partial of a *single string* will be false beating, by definition. It cannot be "tuned out" in the single string, by altering the tension in the string (within the normal tuning range). Increasing or decreasing the tension so as to cause a pitch change of a semitone, for example, invariably does not change the false beat rate significantly.

False beating in partials of a single string is very important, because the false beating can be *inherited* by any interval containing the false beating string. In a unison pair, inherited false beating is recognised by its insensitivity to tension changes. An inherited false beat in a unison pair usually exhibits essentially the same beat rate as the corresponding false beat in the single string from which the beating has been inherited. Where more than one partial is false beating, and/or more than one string is false, which is often the case, multiple inheritance takes place, sometimes leading to complicated fluctuation patterns in the envelope of the unison.

The *way* in which false beating is inherited, and the results of this inheritance on the soundscape of the unison, however, can be affected by the tuning condition. The effects of false beat inheritance on the unison seldom produces fluctuation in the unison soundscape the same as the false beat in the single string, except for the beat *rate* itself. The tendency for the rate to be inherited more or less unchanged, if offset by the fact that, as we shall see below, a beat can also be removed or reduced by reducing its amplitude, and increasing its decay rate.

To begin with, it is useful to appreciate the principle of inheritance. Each single string has its own individual spectrum, that when sounded on its own, constitutes a *parent* spectrum of the spectrum for the unison group as whole, when all the strings are sounding together. Thus in the whole trichord, the three individual string soundscapes are three parents of the same unison child soundscape, which can then inherit some of the parent properties, especially false beats, Fig. (8.5):

Fig. 8.5

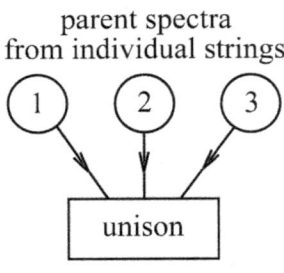

In other words, if one string is false, that is, if one parent string's spectrum includes a false partial – *i.e.* a partial that beats on its own in the single string - then the unison with all strings sounding together, may inherit false beating.

We should not, however, think of a false beat in the unison spectrum as still "belonging" uniquely to the string from whose spectrum it has been inherited. With more than one string sounding, it belongs to the unison spectrum as a whole. In fact, it is *created* by the unison trichord as a whole, because the strings, though separate, are all members of the same, coupled, acoustical system. Similarly, if just two strings are sounding, then the spectrum of a unison pair is still a *child* of the two single string parent spectra, and will inherit some of their properties, as in Fig. (8.6).

Fig. 8.6

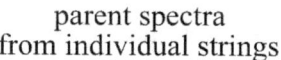

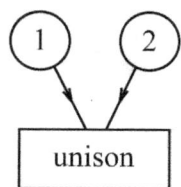

child spectrum

An important consequence of false beat inheritance, already mentioned, is that the inheritance may be dependent on the tuning condition. The main tuning characteristic of a false beat is that its beat rate is insensitive to changes in string tension. This is true whether it is a false beat in the spectrum of a single string, or an inherited false beat in the spectrum of a trichord or bichord. However, a false beating partial inherited in a unison spectrum, *may be changed* by altering the tuning between the strings, often, to such an extent, that the inheritance of the beat is substantially reduced or even eliminated. This, as already mentioned, is not generally a process of beat reduction by beat *rate* reduction, but a process of beat reduction by *attenuation*. In attenuation, it is changes to the beat amplitude and decay rate that bring about the reduction.

Fig. (8.7) shows an example of false beat inheritance for the 2nd partial of F#46 on the Steinway model M. In the example, the unison pair inherits a false beat from one of the strings. The right string of the trichord is wedged (muted), the left string exhibits false beating in the 2nd partial (a beat rate of around 1 Hz), and the middle string is "pure" ("beatless") at its 2nd partial.

Fig. 8.7

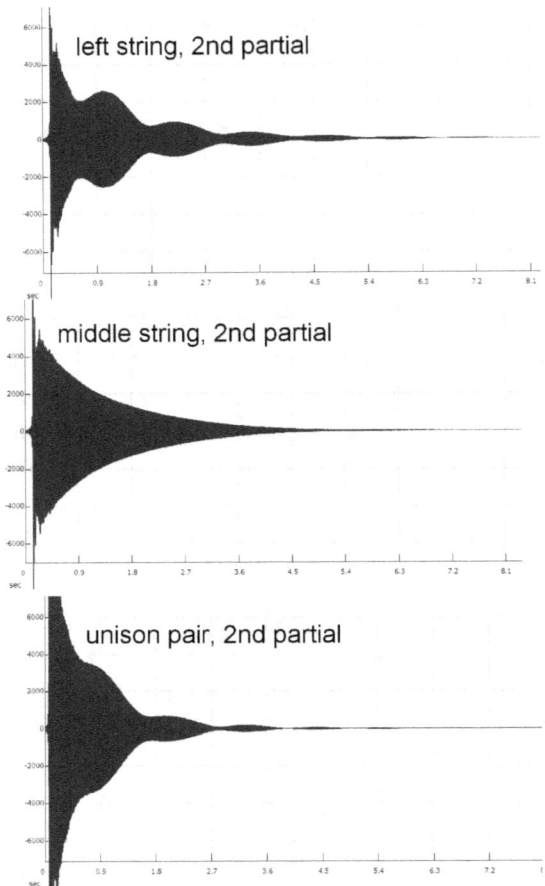

The 2nd partial of the unison pair sounding together, inherits the false beat rate from the false 2nd partial of the left string, but attenuates it. It may be tempting to think, looking at the envelopes in Fig. (8.7), that the envelope of the unison pair will always be simply a mixture of the envelope shapes for the two individual strings. In the actual situation, bridge coupling effects can cause the envelope of the unison pair to differ from that obtained by merely adding together the envelopes for the single strings. However, even for the simple addition of the single string envelopes, the situation is not quite that simple, for the following reasons.

We must remember that the false beating partial necessarily contains at least two frequency components, in order to produce the beat pattern. One might be tempted to argue that the partial could be a *single* frequency, with its amplitude modulated in a beat pattern. There are indeed ways we could create such a beat pattern on a single frequency without deliberately introducing more than one frequency in order to achieve it. Nevertheless, any such beating partial, no matter how produced, is still mathematically equivalent to two or more close-frequency components that are not beating, so the situation can be viewed in either way.[50]

The false 2nd partial of the single string, therefore, necessarily contains more than one frequency component, no matter what the actual physical mechanism producing the beat. Now unless the frequency of the middle string's 2nd partial happens to be the same as one of these frequencies contained in the beating false partial of the left string, mixing the two envelopes of the single strings' partials may produce *additional* beating, because *three* (or more) close-frequency components can be present.

Even simple addition of the single string envelopes will tend to produce attenuation of the beat. However, the unison pair must be finely tuned in order for the envelope of the unison to simply inherit the false beat from the false string, without additional beating being present. In such fine tuning, the unison pair may then also be subject to coupling mechanism effects through the bridge, which can alter frequencies and decay rates. In this example, the frequency of middle string's "pure" second partial, when it sounds alone, is 737.85 Hz. This falls *between* the two frequencies present in the false partial of the left string sounding alone, which are 737.18 Hz and 738.02 Hz. The difference in the latter two frequencies, producing the false beat in the single string, is 0.84 Hz, the false beat rate. When the unison pair sound together, two frequency components are present in the 2nd partial of the unison's spectrum, which are 737.09 Hz and 737.93 Hz. Their difference is also 0.84 Hz, the false beat rate (in each case the FFT frequency precision is 0.08 Hz).

The scales of all three envelopes in Fig. (8.7) are the same. Some attenuation or reduction of the false beat can be seen in the unison, compared to the left string, in terms of the amplitude of the beat pattern. This is not, however, an example in which the false beating inherited from the left string is very much attenuated. We will meet some examples of more significant attenuation a little later.

---

[50] Beats created by imposing beat modulation on the amplitude of a single frequency wave generally cause the new modulated waveform to resolve into three frequency components.

Sequence of tuning unison strings where one is false

It does not follow from any currently established theory that if one string is false, the falseness inherited in the unison should necessarily be adjusted by adjusting the false string. Nor does it follow that the adjustment should necessarily take place specifically on one of the strings that is not false. Furthermore, in a trichord, one, two, or all three strings can be false. Although the falseness in a unison soundscape may be inherited from a particular string, the inherited false beat in the unison itself, is created by the unison working as a whole. The number of variable physical factors affecting how the unison behaves is large, so there is no fixed rule for tuning strings in a particular order, that applies best to all unisons.

The physical system is one of three coupled strings, and in the contemporary theoretical description this means there are up to six component oscillators per partial. It is therefore necessary to discover the best sequence by "trial and error", when success appears to be elusive. However, inexperienced tuners may find choosing the least false string as the first one tuned, is the best strategy. We will examine the sequence of tuning strings in the trichord in more detail, later.

The skill – general beat attenuation

Inherited false beating is easily identified by the experienced tuner because it has beat *rates* that are noticeably *insensitive* to changes in mistuning, or string tension. However, false beating does not necessarily even have to be specifically *identified* by the tuner, for its *attenuation* to come into play as part of the tuning process.

The presence of false beating inherited by the soundscape of a unison, leads to an important consequence. Each adjustable partial in the spectrum that inherits false beating, will during the tuning process contain a *mixture* of beating of the first and second kinds, *i.e.*, beating whose beat *rate* is sensitive to changes of string tension, and beating whose beat *rate* is insensitive to changes of string tension. Excellence in unison tuning is dependent on the fact that beating *of any kind* in the partials, can in general be *attenuated*, which is a process quite distinct from reduction of beat rate. The skill in tuning unisons lies in the *combination* of both reducing beat rates, *and* attenuating beats or other fluctuations. This is especially the case in the treble and upper treble.

There are *three* important aspects to tuning beyond the ability to slow beat *rates*. These are:

The ability to select the best tuning result for beauty and best consistency of tone, where audible fluctuation or movement must remain in the soundscape, due to false beat inheritance. Often, fluctuation must remain, because there is no one tuning position at which all fluctuation ceases, even though this may be slow, and not necessarily prominent. Both inharmonicity and false beating can cause this to be the case, but other than in the bass, it is the latter that is most obvious. The choice of fluctuating soundscape is then a matter of the tuner's empirical knowledge and experience.

1) The ability to select the best tuning result in such a way that the tuning *remains stable* (this is *setting the pin* which is the subject of a separate chapter).

The ability to select the best tuning result for beauty and best consistency of tone, where the only audible movement is due to differing partial decay rates, or decay itself. This can be affected, as we shall see, by tuning causing changes to the decay rates of certain partials affecting the tone quality.

False beat attenuation – an overview

The initial aim in unison tuning is to reduce beating, as far as possible. The fact that inherited false beat *rates* are insensitive to tuning changes, does not mean that the tone of the unison is necessarily doomed to be as poor as that of the single false string, in terms of the amount of audible fluctuation in the partials. False beats should certainly not be ignored simply because their beat rate is insensitive to tuning changes. False beats, as we shall see, *must* in general be *attenuated*, rather than having their beat *rates* reduced. Attenuation, however, is something that applies to *all* beating, not just false beating, and it arises primarily from the fact that all partials *decay*.

False beat attenuation, and its incorporation into tuning strategy in general, is one of the most important, but most understated parts of the art piano tuning. As already stressed, false beating is not a rare phenomenon, nor is it necessarily a "fault". Many piano tuners only usually declare a string to be false when false beating is prominent and fast enough to be particularly noticeable, and to be very obviously spoiling the tone. In actuality the phenomenon that appears as false beating is a perfectly normal part of piano tone behaviour. Whilst it is by no means present in all partials, digital analysis shows that it is *ubiquitously* present in piano tone spectra, in varying degrees.[51] When false beating is less severe and so is not specifically noticed or acknowledged by the tuner, it will nonetheless be dealt with by the skilful tuner in the normal course of events, as a natural result of the process of attempting to reduce movement. The tuner does not *necessarily* have to be aware of attenuation as distinct from beat rate reduction, in order to be carrying it out. In fact, in unison tuning, the adept tuner is unlikely to be listening to any one particular partial, but rather, to the tonal effect of the whole spectrum, on the audible soundscape.

Eliminating the beat: rate reduction and attenuation

It is important to appreciate the possible effects of the bridge and soundboard on the tuning of unisons. The soundboard bridge boundary of the speaking length is designed to cause decay, by draining energy away from the partials. The fact that a partial decays has important consequences on the behaviour of the beat pattern as beating is slowed down. In addition, because the bridge is designed to move in order to transmit motion to the soundboard, the motions of the strings may be *dissipatively coupled* by the bridge and soundboard.

What this means for the piano tuner is not especially difficult to appreciate, although understanding the mechanics of dissipative coupling in the actual situation can be a complicated business. Essentially, if strings are *bridge coupled*, then the motions of the strings affect one another, with the result that frequencies and beat patterns can be altered. *Dissipative* coupling, however, means that decay rates also, are affected, which has further consequences in relation to the tuning condition. We will shortly look at these in detail. The effects range from being negligible, to having significant effects on fine tuning.

One of the factors determining the potential effects of bridge coupling, is the mistuning between the strings. The mistuning can affect whether or not potential coupling plays an important role. For any given partial of a unison pair, a sufficiently *large* mistuning between the strings, such as occurs in the obviously "out of tune" condition, the effects of bridge coupling will be relatively insignificant. At larger mistunings, regular

---

[51] For early references to falseness see Cree Fischer, J, *Piano Tuning / Regulating and Repairing*, (1907), reprinted as *Piano tuning. a simple and accurate method for amateurs* (Dover, NY, 1975), pp. 160-161; Schuck, O H, and Young, R W, 'Observations on the Vibrations of Piano Strings', *JASA*, 15, 1, 1 – 11 (1943); Wolf and W. Sette, S K, 'Some application of modern acoustic apparatus', *JASA*, 6, pp. 160 – 168 (1935).

beating will of course occur, and the adjustable partial is typically divisible into two (or more) acoustical components more or less the same as the equivalent partials of the individual strings sounding alone. At these larger mistunings, the fact that the bridge and soundboard is shared by both strings, therefore has relatively little effect on the motions of the strings.

However, if coupling is potentially present, then at the smallest mistunings, as in the final fine tuning of the unison, the effects of bridge and soundboard coupling between the strings can be much more pronounced. In effect, there is a small *range* of fine mistuning over which coupling between the strings may become significant, and this is what we met earlier as the *coupling region* in tuning the unison. The *coupling region*, for practical purposes, can be thought of as a small range of mistunings over which coupling may *potentially* be taking place, whether or not it is actually confirmed. The reason for this is that *as far as the practical consequences on tuning are concerned*, bridge coupling effects are an "extension" of the *kind* of effects that would already be present due to decay, even without coupling.

Even where coupling may not be taking place, the behaviour of a partial as the beat rate is reduced, can still differ significantly from that portrayed by rudimentary "traditional" theory, simply because of the decay mechanism itself. The rudimentary theory, even if inharmonicity is included, predicts a beat whose rate is simply equal to the mistuning, at any time. Thus, a beat in a partial would be expected to vanish at zero mistuning, measured at that partial. In actuality, because of decay mechanisms, a beat may vanish, by decaying, even if its beat *rate* is not zero. This principle is readily observable in digitally isolated partials, and can apply even where other signs of bridge coupling are not clearly observable. As far as consequences on practical tuning are concerned, bridge coupling itself generally intensifies the effects of decay that are in any case present.

We saw earlier some theoretical decay patterns showing the *attenuation* of beats in partials. Now let us take a closer look at some actual piano partial envelopes, and how "beat" patterns change as a unison pair is tuned. Fig. (8.8) shows some sequences of partial decay patterns taken from a unison pair (on a Steinway model M) in the central compass, as the unison is brought into tune.

Fig. 8.8

A - fundamental

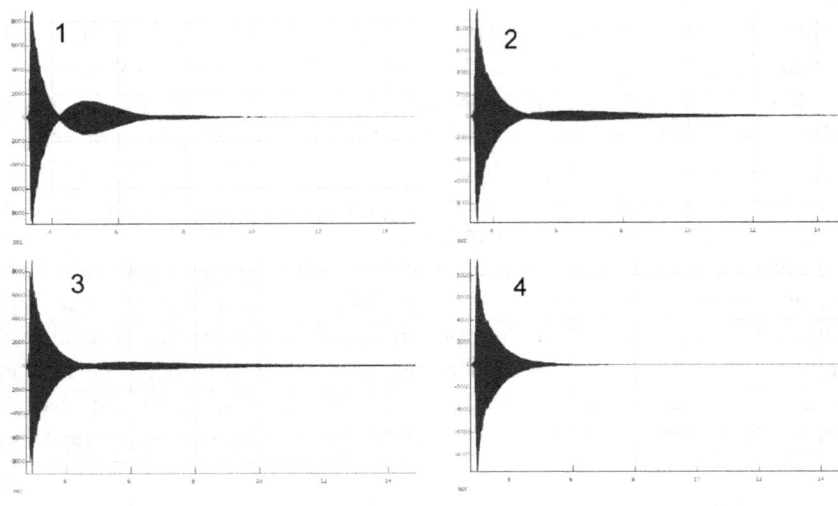

B – third partial

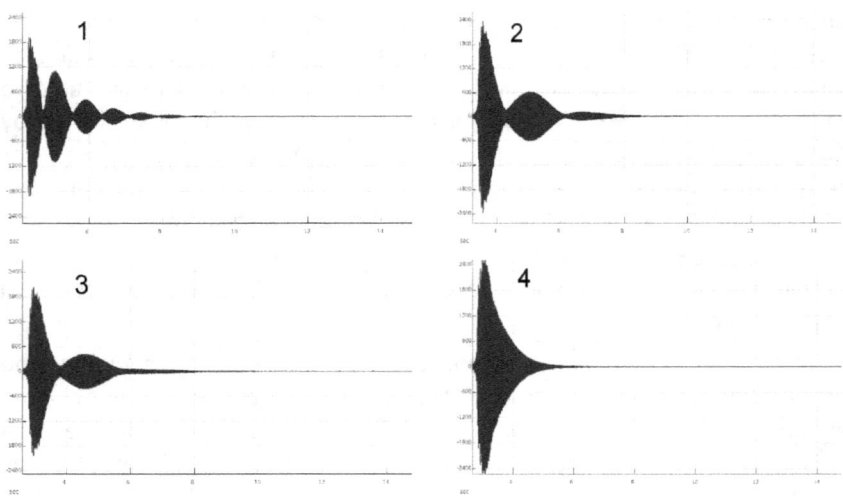

C – fifth partial

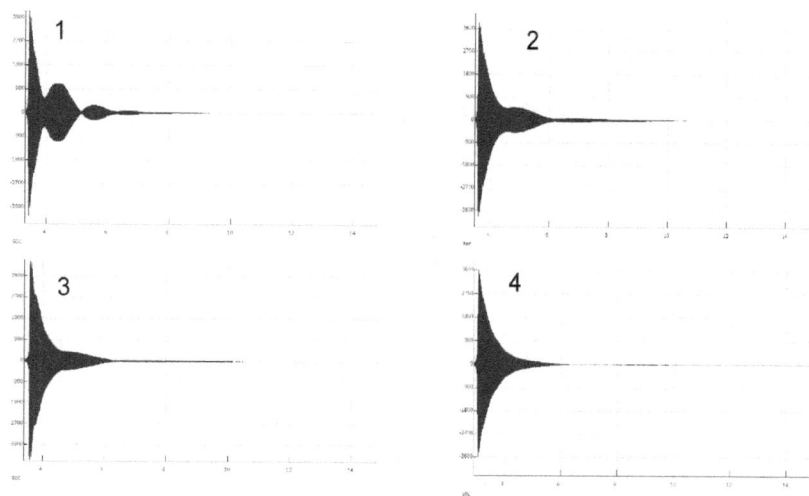

Fig. 8.8. Decay patterns of three partials at four stages of tuning a unison pair in the central compass of the Steinway model M grand piano. A: The fundamental; B: The third partial; C: the fifth partial. The attenuation of the beat's amplitude by its increasing beat decay rate, is as at least as significant as the beat *rate* reduction, especially, here, in the first and fifth partials.

This kind of behaviour is typical of that found in the central compass, and could even be considered a "classic" form of progression of beat patterns in the process of reducing beats or movement. As the beat rate is reduced to less than around 0.5 Hz, the *attenuation* of the beat due to the fact that the beat amplitude is decaying, starts to become more significant than any reduction in beat *rate*. The *beat decay rate* (as distinct from the overall partial decay rate) begins to increase, so the beat is decaying faster than the partial in which it appears. In other words, approaching the beatless state, there is not so much a gradual reduction of regular beat rate as the unison is further brought into tune, but rather, a rapid increase in beat attenuation. In other cases, where false beating may be present, a final *single null* is sometimes the *minimum* amount of movement possible in a partial.

A little theory – Weinrich behaviour

Fig. (8.9) shows the behaviour assumed by rudimentary, "traditional" theory, for the beat rate in a partial of a unison pair, in relation to the mistuning between the strings (at that partial).

Fig. 8.9

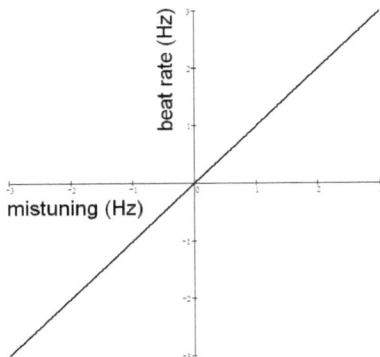

Fig. 8.9. Graph showing the "traditional" theory relationship of the beat rate (in Hz), to the mistuning (in Hz), for any partial of a unison pair. The beat rate in the partial is simply equal to the mistuning (at that partial).

The beat rate in the partial is simply equal to the mistuning (at that partial), and is quite independent of the partial decay rate, which is in fact ignored. At zero mistuning, the beat rate is zero. At a mistuning of, say, 0.3 Hz, one beat would arise about every three seconds, and if the partial was audible for 15 seconds, five equally distinct beats would be heard (this assumes no false beat phenomena to be present). In practice, such a beat is often only heard if it is due to false beating, the behaviour of adjustable partials following a different pattern of behaviour, for the reasons that follow.

The prominence, depth, or technically, *beat amplitude* of a beat, depends on the relative amplitudes of the two frequency components producing it. When the two amplitudes are the same, the beat will have its greatest beat amplitude. Two components with very different amplitudes, will produce a beat with only a small beat amplitude.

In rudimentary theory, in a unison pair, these two components are assumed to be produced from the two strings, one from each string. It might be reasonable to assume that both strings, being more or less identical, would have more or less equal decay rates for the partial in question. This would mean that although both components are decaying, so their amplitudes are constantly reducing, their amplitudes at any time would remain equal. The beat produced, would therefore only decay at the same overall rate as the partial in which it appears.

However, in contemporary acoustics, the acoustical system is comprised of *both* strings plus the bridge and soundboard coupling them, as *one* system. As a complete system, this produces its own frequency components responsible for the beat. These components cannot necessarily be identified only one with each string, as they would be if the strings were uncoupled. A consequence of this, is that the *decay rates* of the two frequency components producing the beat, would not necessarily even be *expected* to be the same. Often, in consequence, the beat has *its own decay rate*, different to the overall decay rate of the partial.

Fig. (8.10) shows the kind of relationship expected, between the decay rates of the two frequency components, and the mistuning.[52] This additionally assumes no false beat phenomena to be present.

Fig. 8.10

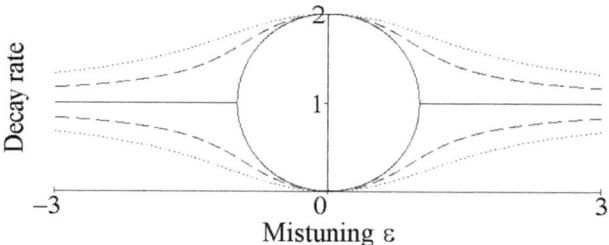

Fig. 8.10. The decay rates versus the mistuning, of the two frequency components producing the beat in a partial, in contemporary theory (from the Weinreich model). The solid, dashed, and dotted lines are for various bridge coupling parameters (the solid line with the circle being a hypothetical extreme, where the bridge is purely resistive). The mistuning units are "natural" units based on the uncoupled decay rate of one string's partial. In the central compass the range would be over about 0.5 Hz. The decay rate units are equal to the uncoupled decay rate of one string's partial.

The solid graph line represents a hypothetical extreme that would occur if the bridge and soundboard had no "elastic" or inertia-like properties. Outside the central circle there is only *one horizontal line* on each side. A real bridge and soundboard does of course have these properties, so the situation is one in which, in general, the decay rates *always differ*, and there are always *two lines*. This means that the *beat amplitude* of the beat produced, can *change*, as the two decay rate values change. The greater the difference, the smaller the beat amplitude. From

---

[52] The theory was first presented by Weinreich. See Weinreich, G, 'Coupled piano tones', *JASA*, 62, 6, 1977, pp. 1474-1484.

the graph, it can be seen that the difference in decay rates increases as the mistuning decreases. Thus, the beat amplitude decreases as the mistuning decreases – the closer we are to zero mistuning, the more the beat will be attenuated.

Contemporary theory also shows that bridge coupling can change the relationship of the actual beat *rate* and the mistuning. Fig. (8.11) shows how the frequency components change in relation to the mistuning, the beat rate at any mistuning being determined by the difference between the frequencies at that mistuning.

Fig. 8.11

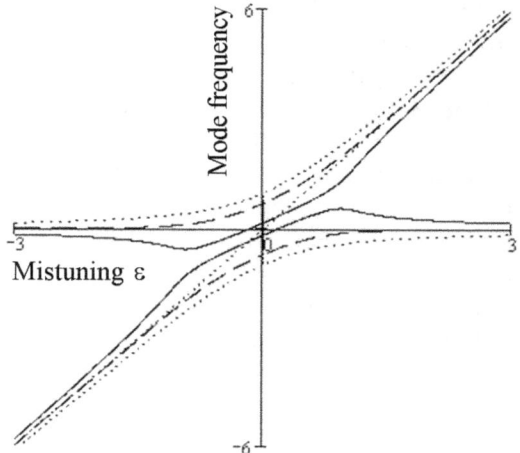

Fig. 8.11. The two frequency components responsible for the beat in an adjustable partial, versus mistuning. The diagonal dotted line represents the behaviour of one component in the absence of coupling, as in "traditional" theory (the other component runs along the horizontal axis). The other dotted, dashed, and solid lines are for various bridge coupling parameters. Where there is sufficient resistive component present (the two solid lines), as zero mistuning is approached, the difference between the two frequency components and hence the beat rate, is less than it would be in "traditional" theory (the lines are closer together). The beat rate therefore decreases faster than "traditional" theory predicts, as the unison is tuned towards "beatless". However, the frequencies never become equal (the lines remain separate even at zero mistuning), so the beat *rate* never actually becomes zero. If the beat disappears, it does so due to attenuation. At larger mistunings the behaviour approaches that for "traditional" theory.

Where there is sufficient resistive component in the bridge admittance, represented by the solid lines, then slowing or eliminating the beat becomes "easier". As the mistuning is reduced close to zero, the difference between the two frequency components, and hence the beat rate, will be less than "traditional" theory predicts, because the lines are closer together. However, *very close* zero mistuning, and at zero itself, the lines are further apart. In this region, and at zero mistuning, a very slow beat might be expected to remain, because there is still a difference between the two frequency components. However, at this point the difference in decay rates is at its greatest, so the beat is, in general, attenuated. The beat vanishes, not because its *rate* has reduced, but because its *beat amplitude* has been eliminated.

Beating is thus "eliminated" or "reduced" not by reducing the beat rate to zero, but by reducing it substantially, and then attenuating it through the decay mechanisms. Close to zero mistuning these mechanisms attenuate the beat by reducing the beat amplitude. A fuller picture is obtained when false beat phenomena is included, however, the situation then becomes much more complicated. In general, expansion of the Weinreich theory suggests that a minimum fluctuation in the partial will not always necessarily occur at zero mistuning.[53] Thus we would expect that tuners would sometimes naturally leave mistunings in the unison as a result of the effort to minimise fluctuation or beating, where falseness is present.

The coupling region

First let us consider a single unison tone partial (although in practice we would not focus on a single partial, but listen to the whole soundscape). What happens is that as we reduce the beat rate, we may at some point enter the *coupling region* for that partial, *before* its beat rate is zero. Once inside the coupling region the regular beat pattern may disappear or change to a more generic pattern. A very "wide" coupling region would cause the regular beat to change to a more generic pattern, or perhaps even to disappear, as soon as the mistuning was reduced from a point at which the beat rate is still relatively fast. A very "narrow" coupling region or no coupling region would cause the unison to behave just as "traditional" theory suggests. Both types of behaviour can occur. Wide coupling regions would be expected to make unison tuning easier.

The following diagram (Fig. 8.12) illustrates the coupling region for one string being tuned flat or sharp, when sounding against another unchanging string.

---

[53] Capleton, B, 'False beats in coupled piano string unisons', *JASA*, 115 (2), 2004, 885-892.

Fig. 8.12 – coupling region for an individual unison partial

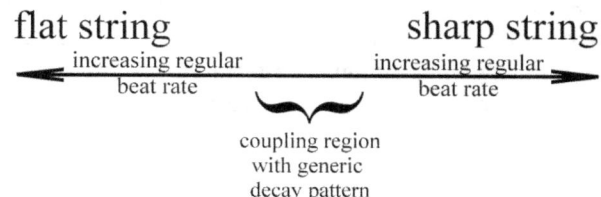

The actual "width" of the coupling region – and the beat rate in the regular beat pattern at the point where we enter it - depends on the individual partial and unison; there is no fixed rule. It may be effectively zero, as in rudimentary "traditional" theory. The only general rule that contemporary theory suggests, is that it will tend to be wider, the faster the decay rates of the individual string's partials at this position in the spectrum.

Thus for say, the fundamental of a long stringed unison with a long decay rate, the coupling region may be quite narrow, but for higher unisons with shorter strings and faster decay rates, it may be wider. Since there is no fixed rule for the partial decay rates, this still does not give any fixed rule for the "width" of the coupling region. However, an important consequence is that as a rule of thumb, the coupling regions can be expected to be relatively wider for partials of the higher strings in the piano. The net effect of this would be that, apart from the obvious increased difficulty in hearing acuity and control of the string tension through the lever, the higher strings would actually be *easier* to tune. The additional ease is likely to be offset by the other considerations, until mastery of technique is attained.

Inside the coupling region we can encounter decay patterns other than the true regular beat pattern with a properly definable beat rate. Furthermore, the nature of the decay pattern will still generally vary with the degree of "mistuning". Typically, as we bring the strings "closer together" we might for example move from a *regular beat* towards a *single null*, and then possibly to a *beatless decay*. As we tune the strings "apart" again, we would then move in reverse order, from *beatless decay* to *single null* to *regular beat*. Digital analysis shows this kind of behaviour to be common.

Remember this is only for the one partial in question. At any point of "mistuning" different decay patterns are to be expected at different partials, for the same unison. While some may be a slow *regular beat*, other partials in the spectrum might at the same point of relative tuning between the strings, be a *beatless decay* or *single null, etc.* Each partial may have a different width of coupling region, and the tuning position between the two strings may create a different situation for each partial, in relation to its coupling region. The final choice of tuning will take into account the whole picture, not just one partial.[54]

Listening to the whole

In general, unisons are not tuned by concentrating on specific partial decays. In fact, concentrating on any one partial is a recipe for poor tuning. The unison as a whole is like a whole painting – we must "step back" aurally and observe the whole, which means to listen to the whole without concentrating on any particular partial. This does not necessarily mean we do not *hear* particular partials. The painter observing the whole painting will still be aware of individual colour grades and their colour components in particular parts of the picture. "Stepping back" aurally and listening to the unison as a whole, the properly trained ear still continues to "pick up" individual partials and their decay patterns, which is essential to the task of unison tuning.

As far as the individually distinguishable partials are concerned, they are registered automatically by the ear and do not become a subject for conscious decision making – the conscious decision making happens elsewhere, as we will discuss in a moment. The untrained ear that cannot hear partials and their decay patterns individually, will not be able to tackle the whole picture anyway. It is possible to train the ear to hear individual partial decay patterns, and then to hear a number of individual ones simultaneously. This hones the aural sensitivity in a way that nothing else can. But after sufficient experience, the observation of individual decay patterns can go back to an almost subconscious level, but now the sensitivity is much finer. Once trained to hear a complex tone as a spectrum of partial decays, we are always aware of partials in the spectrum, even without trying.

Listening to the whole, the decay tone of the unison is determined by the relationship between all the partial decay patterns in the spectrum. This involves the relative *prominence* at any time of each individual partial to the whole, and the *movement* (anything other than *beatless decay*) in

---

[54] Capleton, B, 'Piano unison tuning', www.amarilli.co.uk/academic, 2006-.

each partial. As a whole, the decay tone of the unison in any case *changes with time*.

The perception of *movement* in the decaying spectrum largely depends, of course, on the sensitivity of the ear listening to it. This may depend on how "hard" we are listening. Most unisons that are well tuned appear, superficially, to have *no movement* whatsoever, but this is deceiving. Invariably even the best unisons reveal partial movement somewhere in the spectrum. This may be in the high partials only, but the high partials play a very important role in the tuning and tone. Typically, differences in behaviour means it may be necessary to leave a lower partial beating slowly, for example, in order to attenuate or slow certain higher partials.[55]

Unison tone and the *vowel* or *cat* in the unison

Tuning the unison should not involve concentrating on listening to specific partials. The overall effect of partial behaviour is one on *tone*, and *movement* in the soundscape. This *movement*, in the case of the unison, is not necessarily to be perceived as beating. In fact, in fine tuning unisons, the movement *should* be perceived as *tone changes*. Let us look at this in detail.

One of the best known and most straightforward ways in music technology generally, to change the *tone* of a sound with great control, is to use a *graphic equaliser*. Simple amplifiers for music reproduction were once often found with a treble and bass control included. The treble control could increase or decrease a broad higher frequencies band, whilst the bass control increased or decreased a broad lower frequencies band. Turning up the treble control would make the sound more treble biased, "brighter", and "clearer", but would also show up *hiss* more, because *hiss* tends to occupy the higher frequencies. Hiss can be hidden by attenuating higher frequencies, but this also shuts out higher frequencies from the wanted sound, making it less clear or crisp, and more dull.

Boosting the bass makes the sound more bassy, but can overwhelm the treble, making the sound seem unclear again. It can also highlight *hum*, because *hum* tends to occupy the lower frequencies. Turning down the bass can suppress hum, but will also suppress the lower frequencies of the wanted sound, making it seem "hollow" or without tonal substance, rather like sound heard through the tiny speaker of a mobile phone.

---

[55] Capleton, B, 'Piano unison tuning', www.amarilli.co.uk/academic, 2006-.

The graphic equaliser is rather like a much more advanced version of the old treble and bass controls. Rather than just controlling treble and bass, it controls right across the frequency range, splitting the entire frequency range into controllable, narrower bandwidths. It typically consists of a (usually large) set of sliders, each one controlling a specific frequency band, the whole set of sliders covering the whole bandwidth for the sound. Using a graphic equaliser, the settings of the sliders can be used to "fine tune" in detail, the tone quality of the sound. The chosen positions of the sliders are not usually juxtaposed, but rather, a smooth line with bends and curves can be traced from slider to slider right across the whole width of the set, like a graph line. Hence the term "graphic".

Now let's see how this relates to the unison soundscape. The soundscape of the unison is *already naturally split* into numerous discrete frequency bands that contain "signal", with nothing much in between these bands. These bands are the partials. In effect, the amplitude of each partial at any moment in time, is like the slider setting on some imaginary graphic equaliser, for that frequency band. In the case of the fine tuned unison, these partial "slider settings" are constantly changing as the unison decays. Firstly, the decay of each partial is, in effect, a continual lowering of the "slider" that that particular partial would represent. Unless all the partials begin at the same "slider setting" and decay at the same rate, then the overall "graph" on the "slider settings" will constantly change. This is one kind of *movement* in the tone of the unison, as it decays. Different partials begin at different magnitudes, and decay at different rates, causing a continual change in tone.

Another type of audible movement arises from beating or fluctuations in individual partials. In the finely tuned unison, any such beating or fluctuation will be relatively slow variations in the partial amplitude. Different partials beating or fluctuating at different rates, and by different amounts, are in effect somewhat like moving "sliders" in the soundscape up and down relative to one another. The tone quality, in terms of the dominant frequency regions present in the soundscape, steadily changes from one condition to another. This can be *perceived* not just in terms of its component changes, but more naturally as a *changing vowel sound*.[56]

This changing vowel sound in the tone, is sometimes said to be not unlike the meowing of a cat, and hence is sometimes called "the cat" in the unison. In fine tuning it is this change or movement of vowel sound that can be adjusted or sometimes eliminated, rather than any particular

---

[56] A demonstration of the mechanism may be found at www.amarilli-books.co.uk

beat. Options that occur in various unisons, range from the complete removal of the *cat* (the changing vowel sound), to the slowing of it, to the placing of the most noticeable part of the change, later in the overall partial decay. It is also often possible to select which part of an otherwise changing vowel sound is tuned to be static and dominant in the tone of the unison. Thus we can often change the *tone quality* of the unison, by altering the dominance of certain frequency partials, and we may be able to do this as something distinct from altering *movement* in the soundscape.

Stopping a change of vowel sound altogether is not necessarily the end result of the fine tuning, as some change may remain due to false fluctuations, and even just changing relative partial amplitudes due to decay. Once learned, it is much *easier* to listen to the vowel sound, than to any particular beat. The cat, like beating, is not to be regarded as an abstract "tool" for tuning. It should be listened to *in the context of the tone quality* of the unison, as a dimension of that quality.

To emphasise again, movement *per se*, in the very finely and closely perceived tone quality, *is not necessarily ugly*, but can in fact be part of the beauty. Some "cat" may be unavoidable, as false beating may be creating it. The "traditional" theory implies, in contrast, that the most desirable tone quality of the unison is entirely featureless. In fact, this is not true. This does *not* mean a unison soundscape should to be tuned to less than one of its *most still states*, but it does mean that stillness in a beautiful unison tone is not necessarily devoid of natural fluctuations, or slow changes of vowel sound. The false beats or fluctuations that cause this are natural, and are only a negative feature if present beyond a certain degree. Beauty of unison tone does not simply equal zero fluctuations in the partials, or zero movement in the soundscape. It is much more a question of *what*, *how much*, and *where* fluctuation is present.

There is a fine difference between a changing vowel sound that is acceptable, and one that the master tuner would not accept. Every string pair and trichord is different, and every piano is different. The final tuned result is a very important feature of the tuning, that collectively adds up to a major part of the tone of the instrument on completion of tuning. Adjusting the tone of the unison as *tone quality* is therefore a major part of the art of tuning, and requires considerable experience and knowledge of piano tone, down to the finest nuances.

Magnified partial envelope shapes

Before we look at some examples of false beat attenuation on a Steinway piano, there are a couple of observations worth mentioning, that are sometimes seen when partial envelope shapes are digitally magnified. As can be seen in some of the illustrations included here, some partials decay with fast beat pattern in the initial part of the decay, whilst the overall amplitude is dropping rapidly. But this is then followed by a *much* slower beat or single null pattern. In these decays, it is generally only the later slower beat or null that is heard as the movement in the spectrum as a whole (all the visual illustrations of envelopes in this book come from audible data). The "beat rate" in the transient period, is often actually *much* faster than the order of beat rates that would be associated with finely tuned unisons.

In cases where the partial decay shape is a single null or beat, digital magnification of the later part of the decay period, sometimes reveals very shallow subsequent beats to actually be present, but in general these are not audible, even in the electronically isolated partial. In such decay shapes, the beat rate is in general not constant, and typically may increase over the overall decay time.

Examples of false beat attenuation

Example 1 – Steinway A49 (A4)

Fig. (8.13) illustrates one example of false beat attenuation in the second partial of a two string unison A49 (A4) of the Steinway model M, the third string being wedged.

Fig. 8.13

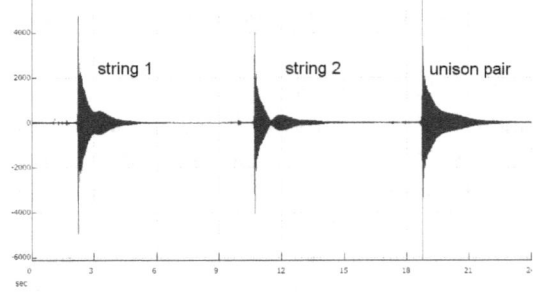

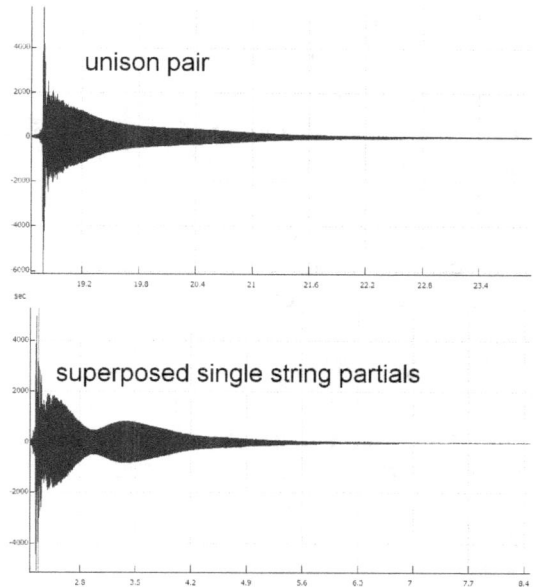

Fig. 8.13. Envelopes of the second partial for each of two strings of the A49 (A4) unison on a Steinway model M, and of the unison pair. The actual envelope for the unison pair is also shown (middle) magnified. At the bottom is the envelope resulting from the digital addition of the envelopes for the single strings. In terms of reducing beats or beat-like movement in the partial, a better result is attained in the actual situation where both strings are sounding, than can be synthesised by simply adding the two envelopes from the single strings sounding alone.

Both string 1 and string 2 show false beating, the beat in string 2 being more pronounced. However, when both strings sound together, the decay envelope of the partial for the unison pair, is beatless. This is not the same result as the two envelopes for the single strings sounding alone simply being added together. Addition of the envelopes for the two separate strings, such as would be expected if the sound of the unison was simply the sounds of the two individual strings added together, produces an inherited beat pattern. The actual decay produced when both strings sound is beatless.

It is important to not that this "elimination" of the false beating, or the prevention of it being inherited into the soundscape of the unison, is only achieved at a particular tuning condition. Contemporary theory based on an extension of the theory first presented by Weinreich (by considering transverse string motion in two planes) suggests that the mistuning for this to be achieved is not necessarily always zero.[57] In the following we shall see some examples supporting this.

In aural tuning of the unison it will be beat patterns in more than one partial that determine the total amount of movement in the soundscape, since the soundscape consists of many partials. Nevertheless, the mechanism of attenuation will remain as important as beat rate reduction, and there seems to be little theoretical ground for the assumption that beating in the soundscape a whole necessarily must occur at zero mistuning. This is consistent with the observation made in 1917 by Braid White in, when he suggested false beats could be "neutralized" by tuning the strings "slightly off".[58]

This could give the impression that all one needs to do in order to "neutralise" inherited false beating is to introduce an arbitrary mistuning. Of course, in practice, the optimum attenuation of false beats requires tuning of the highest precision and stability. Elimination of beating (even very slow) altogether from a soundscape is often impossible, firstly because of the presence of specific false beats (which may be very slow) that may resist total attenuation, and secondly because the mistuning required to maximise attenuation is not necessarily the same for all partials.

Example 2 – Steinway A73

Fig. (8.14) shows the fundamentals of a unison pair of A73 (A6) on the same Steinway. The string pair is the right and middle strings, the third string of the trichord being wedged. The middle string is clearly false, exhibiting a false beat rate of about 5 beats per second. The right string is "pure", showing an exponential decay. With judicious tuning, when both strings sound together, the envelope for the unison pair shows, overall, an attenuation of the obvious regular beat pattern in the false string.

---

[57] Capleton, B, 'False beats in coupled piano string unisons', *JASA*, 115, 2, 2004, pp. 885-892.

[58] Braid White, W, *Piano Tuning and Allied Arts*, 1917, 14th reprint 1972, p. 106.

Fig. 8.14

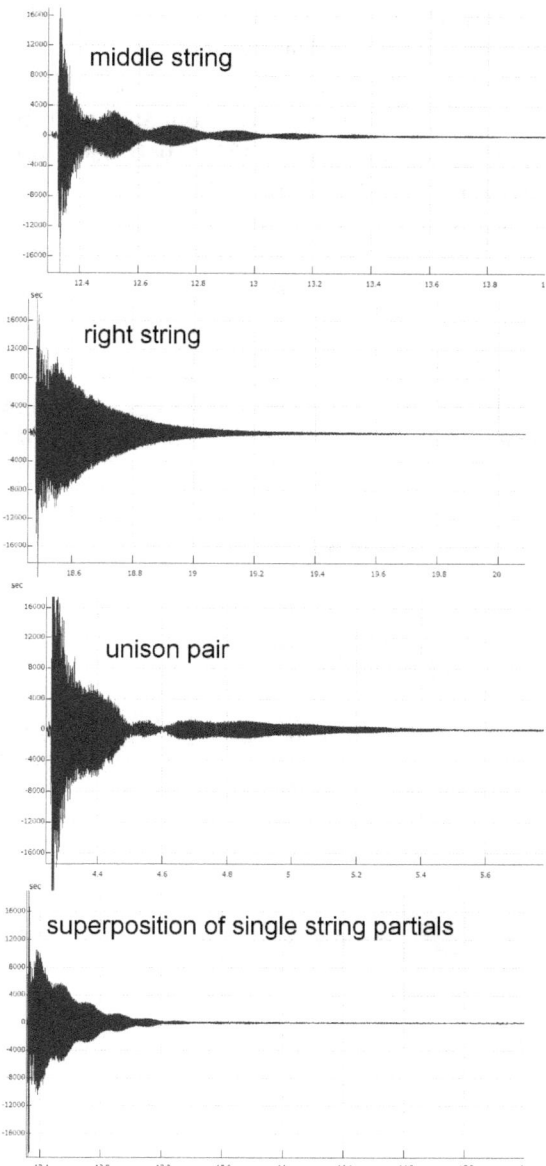

Fig. 8.14. Fundamental partials from a Steinway model M A73 (A6) unison pair. The middle string is false. When the pair sounds, the envelope contains less obvious regular beating than appears in the synthesised superposition of the two single string partials. Rather, there is an initial decay into a null, followed by one fairly prominent beat,

followed by a more sustained aftersound with only very shallow beating. Aurally, this does not really sound like a regular beat pattern, but more like a single null and aftersound.

The false beat rate is aurally still just discernable in the soundscape of the unison pair, but the beat depth is attenuated over most of the decay. Aurally, the partial "sings" with an "aftersound" (sustained tone after the initial fast decaying "prompt sound"), in which false beating is much less obvious than in the single middle string. The synthesised superposition of the individual string partials differs from the actual unison pair partial. There is some suggestion of dissipative coupling mechanisms at work, but this is not clear.

Example 3 – Steinway A61

Fig. (8.15) shows the fundamentals of a unison pair of A61 (A5) on the same Steinway.

Fig. 8.15

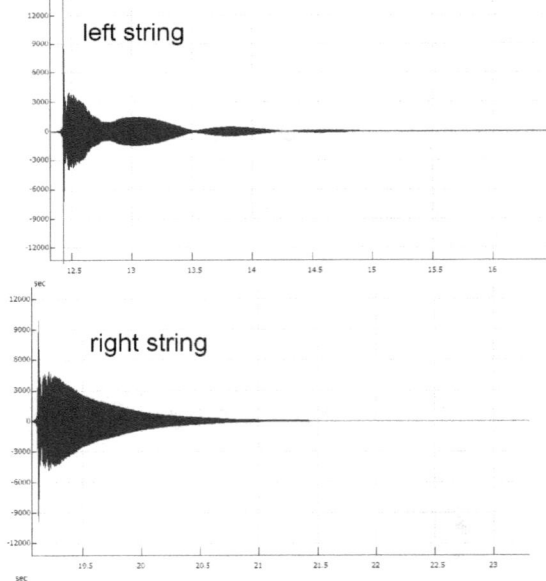

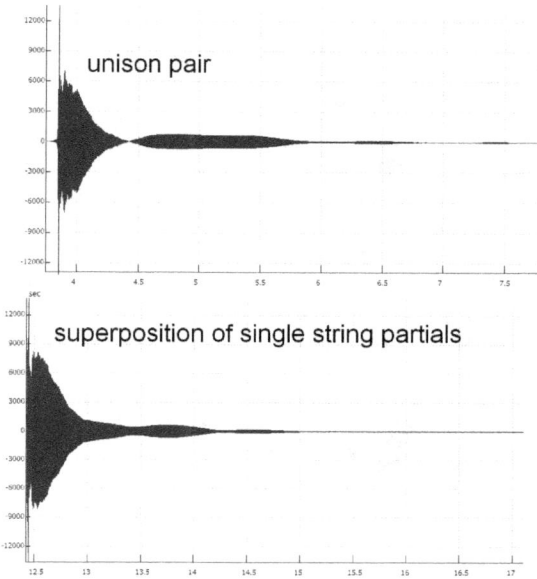

Fig. 8.15. Fundamental partials of the A61 (A5) unison pair on a Steinway model M. The left string is false. When both strings sound (unison pair) the envelope shows the same characteristic null and "aftersound" in which the false beat presence is attenuated, but there is a prolonged aftersound.

This time, the left string is false, with a false beat rate of about 2 beats per second. A similar result to that for the higher A occurs. The envelope for the judiciously tuned unison pair contains an initial "prompt sound" followed by a sustained "aftersound", the regular beat pattern of the false beat being attenuated throughout. Again, the shape of the unison pair envelope suggests dissipative coupling mechanisms at work.

Example of the coupling region – effect on beat rate

Where falseness is not prevalent, the Weinreich theory suggests that for small mistunings, not only does beat attenuation occur, but the actual beat *rate* may be slower than expected by "traditional" theory. In one sense, this would appear to make unison tuning *easier* than expected from "traditional theory", because one could reduce the beat rate more readily – leaving in a small mistuning would not necessarily leave in a beat equal to the mistuning, as "traditional" theory states. Fig. (8.16) shows an example in a unison pair from B-flat 74 (Bb7) on the Steinway model M. Here both strings are "pure".

Fig. 8.16

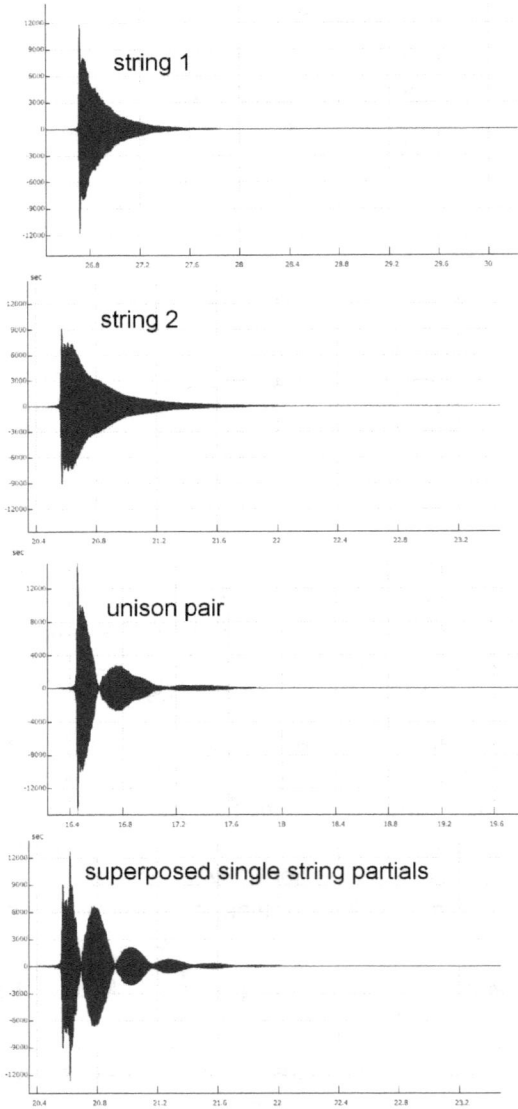

Fig. 8.16. Fundamental partials from a unison pair from B-flat 74 (Bb7) on a Steinway model M. Neither string is noticeably false. The beat rate predicted by "traditional" theory would be equal to the mistuning, approximately 4 Hz, which can be seen in the superposition of the single string partials. The actual prominent beat rate in the unison pair is around half this. A shallow modulation of around 4 Hz is visible, but not audible. The results "mimic"

Weinreich behaviour, but the precise mechanisms are not clear from this data alone. These effects are much more noticeable in the upper treble.

The FFT precision in this example is ± 0.09 Hz, and the frequency measurements for the two individual string partials are 1868.96 Hz and 1864.75 Hz. The superposed single string partials produce the beat rate that would be expected from the difference between these frequencies, whilst the actual beat rate in the unison pair is only around half this. The actual unison pair partial shows two prominent frequency peaks at 1868.774 Hz and 1866.760 Hz corresponding to the actual, observed beat. Thus, the system with two strings sounding together actually generates frequency components in the adjustable partial, that are closer than those generated by the two strings sounding separately.

String 2 shows some deviation from an exponential envelope shape, and a shallow modulation of around 4 Hz can be seen in the unison pair envelope. Neither are audible. The behaviour "mimics" that predicted by the Weinreich one plane model, but its mechanisms will be more complicated as motion in two planes will be present. Effects as clear as those in Fig. (8.16) are only generally found in fundamentals of notes in the upper treble.

Whatever the actual underlying mechanisms, the results show that judicious tuning can produce results "better" than that predicted by "traditional" theory, in terms of the manifest beat rate versus the actual mistuning.

Practical effects of bridge coupling

Essentially, ignoring false beat complications, bridge coupling can have both "friction-like" (*dissipative*) properties that contribute to decay, and "spring/inertia-like" (*reactive*) properties that contribute to frequency changes. The relative "balance" between these determines the precise behaviour, and this, in turn, will depend on the physical properties of the bridge and soundboard. In sum, there are two consequences:

1) Beat patterns and fluctuations can be reduced in tuning by *attenuation* as well as by *beat rate reduction*.

2) At small mistunings, the beat rate is not necessarily proportional to the mistuning (contrary to "traditional" theory).

It is important to remember that there does not have to be *coupling effects* in order for beat amplitude *attenuation* to occur. This can occur simply by superposing beating partials with other non-beating partials. But *coupling effects* can produce pronounced attenuation in their own right.

The effect of the beat rate being smaller than the mistuning, however, does imply actual *coupling effects*.

The coupling range can help in the attempt to reduce beating, but it can also hinder. The beneficial effects are largely dependent on the dissipative part of the bridge properties. According to the theory, it would be possible in some instances for the reactive part of the bridge properties to be large enough compared to the resistive part, to make beat elimination *impossible*, rather than easier, in other words, a beat will always remain, because there is no mistuning value at which it disappears.

The outer dotted lines in Fig. (8.11) illustrate such a situation. Although the frequencies are drawn together approaching zero mistuning, they remain apart by a substantial distance. If the resistive mechanisms are insufficient to allow the beat to effectively attenuated, then a beat will always remain. This could be experienced in practical tuning in much the same way as an inherited false beat, except that it could occur where no false beat is present in the single strings. Fortunately, this theoretically predicted situation is only likely to arise in practice where a partial frequency happens to fall close to a soundboard resonant frequency (in resonance reactive effects outweigh resistive ones).

A Steinway trichord

Fig. (8.17) compares partials of the fine-tuned A49 trichord on a Steinway model M grand piano, with the corresponding false partials of the trichord's left hand single string. The single left hand string can easily be heard to exhibit very slow false beats, whilst the other two strings appear pure to the ear. The vertical scales in each pair of graphs is equal, but the scales are not the same for all pairs – if they were, the shapes of the smallest amplitude partials would be difficult to discern. The time axes along the horizontal are all equal (0.7s units), so that the beat rates or speeds of fluctuations can be compared.

Fig. 8.17

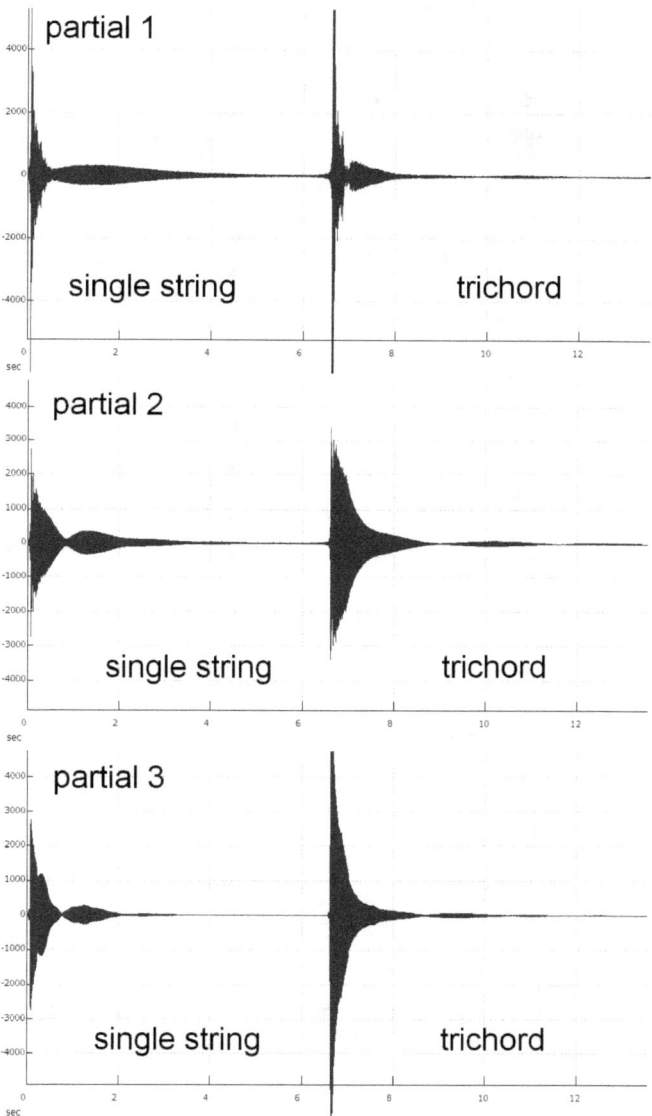

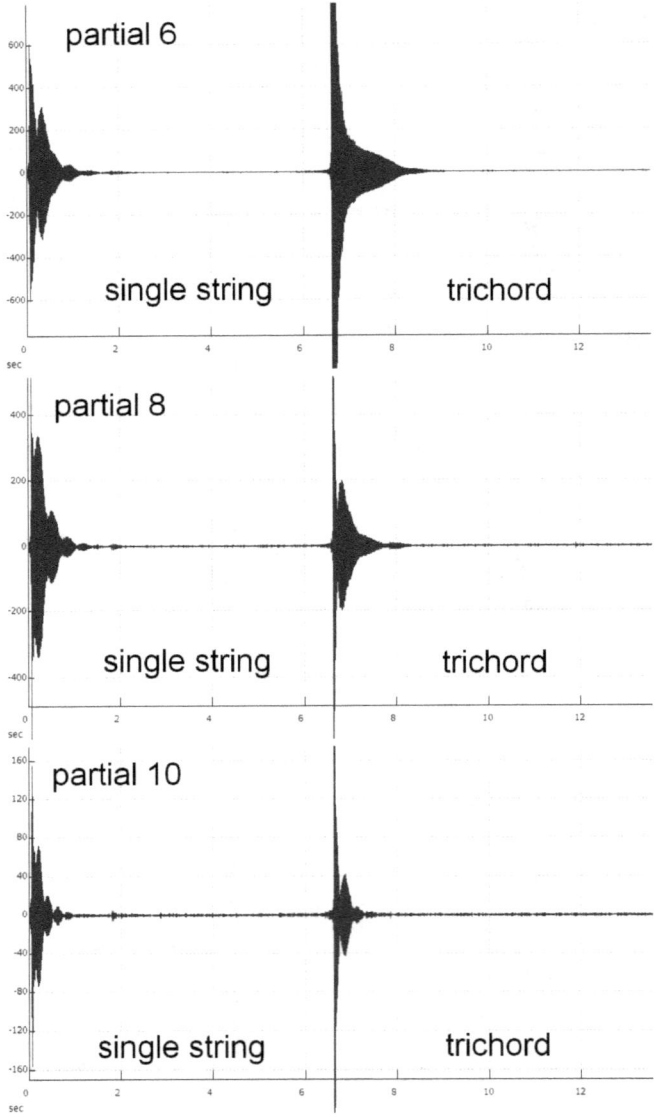

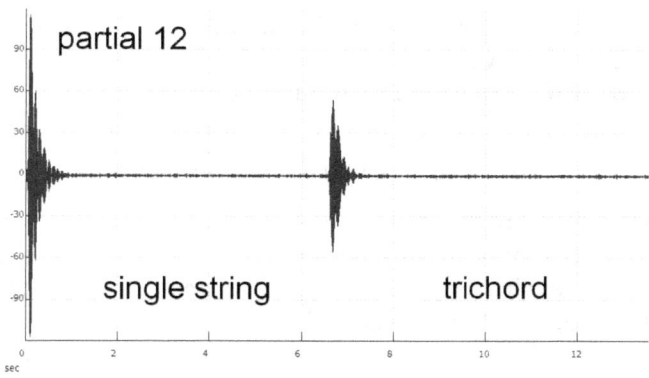

Fig. 8.17. Comparisons of partial envelopes of the left string of the A49 trichord of the Steinway model M, with the envelopes for the full trichord, finely tuned.

In most cases the complete, finely tuned unison's partial shows less audible fluctuation overall, than the equivalent partial of the single string, for at least one three reasons:

The remaining beat rate in the unison is slower then the false beat rate in the single string (the remaining beat is not necessarily false);

The beat amplitude (prominence) in the unison is less than in the single string;

The first null comes later in the decay, in the unison.

In the case of partial 12, the beating is not reduced.

Aurally, the unison is tuned to reduce the amount of *movement* in the soundscape to within a "minimum range", and then to optimise the tone without introducing further unwanted movement. Measurably, many of the resulting partials of the finely tuned unison contain less movement or fluctuation than the corresponding partial of the false string, in terms of rate and beat amplitude over most of the decay time. In many of these partials of the finely tuned unison, false beat attenuation can be seen as distinct from beat *rate* reduction.

This is just one example of course. There is not necessarily any "typical" prevalence of false beating in individual trichord strings. The example illustrates that although false beat fluctuations from a single string can be inherited in the unison's soundscape, the final soundscape of the properly tuned unison can also be "better" than that of the single string, when the aim is to "reduce beating" or fluctuation. This is not achieved merely by masking false beating in the single string by an array of further beat patterns in the unison. Rather, the false beat fluctuations that would otherwise be inherited, can themselves be eliminated or reduced in the final spectrum, through judicious tuning.

The singing quality – *single nulls* and *aftersound*

The tone of an individual piano is a property of the instrument and its condition, but there are other aspects of its tone that are not determined purely by the inherent quality of the instrument, and the condition of its action, soundboard and bridges. The tone of an instrument is something that is perceived in relation to the music being played, and the acoustical environment in which the instrument is heard. Only when the instrument is performed upon, can all the tonal characteristics that the instrument is capable of producing, be appreciated. In addition to the contributions of the player and the acoustical environment, and by no means least in importance, is the quality of the tuning.

Very crudely speaking, musical tone quality can range between "mellow" and "bright", or "soft" and "harsh", or "rounded" and "nasal", etc., to use just a few of the popularly used adjectives. Given that the transient[59] is fixed, then a somewhat more objective way of describing a range of tonal qualities, would be to say, for example, that the *spectrum* of the *decay tone* is more or less prominent with high or low frequencies. Enhancing high frequencies makes the tone "brighter" or more "brittle" or even "harsh". Suppressing high frequencies makes the tone more "mellow", "rounded" or even dull. Suppressing low frequencies can make the tone "weak" or "tubby". This has already been discussed in some detail in relation to graphic equalizers.

In tuning the unison we can alter the shape of the partial envelopes, and one of the things we can do is create a "single null" decay pattern shape, rather than a regular beat, or a "beatless" decay. One of the reasons we may *want* to do this is that the partial may be sustained longer, overall, if the decay pattern is a "single null", than if it is "beatless" (for reasons that will be explained below). Certain partials, if sustained longer than others, may become more dominant and alter the tone in a desirable way, providing a better "singing" quality in sustained notes. The "single null" pattern is itself not always undesirable, as most regular

---

[59] The short period of noise at the beginning of the note, which in the piano is the sound of the hammer knocking against the string, and which critically affects the ear's assessment of tone.

beating would be, and after the one null it is in any case effectively beatless.

In the "single null" pattern, the part of the decay at the beginning that rapidly falls into the null, is called the *prompt sound*. The part after the null, which is typically much more sustained, is called the *aftersound*. There is the possibility of deliberately producing a sustained *aftersound* occurring after an initial *single null*, in certain partials. By making some partials more sustained than others, we would change the sustained tone of the note. In a moment we will look at the possibility of *regularising the tone* from one note to the next, through this process, where irregularities in the hammer strike exist.

In my own view, an area of tuning where affecting the envelope shape is clearly important, quite apart from hammer irregularities, is in the tuning of the upper treble, where decay times are naturally short. In notes of the upper treble this general principle can in theory at least, affect the fundamental[60] as part of the process of attenuating falseness, which results in both *less obvious beating* in the note, and a better "singing quality" produced by the more sustained *aftersound*. Let us now look at some theory behind the idea of adjustable *prompt sound* and *aftersound*.

"Hammer irregularities" and tone – the Weinreich model

Small variations in the condition of the hammers can be called hammer *irregularities*. Hammer irregularities can be responsible for tonal irregularities from note to note. Because in fine tuning we are adjusting the tone of the unison, we can in theory compensate for hammer irregularities in the fine tuning. Weinreich[61] proposed an actual mechanism through which such compensations could take place. In Weinreich's model the motion of the *bridge coupled* string pair, for any partial, has two *normal mode* ingredients in one plane – a *symmetric mode* and an *antisymmetric mode*.

---

[60] Digital analysis easily shows that in the upper treble the fundamental is much more dominant than any other partial, often almost to the point of being, in effect, the only partial present.

[61] Weinreich, G, *op. cit.*

The symmetric and antisymmetric modes

Let's first review what normal modes are. In the chapter on sound, we met the *principle of superposition* for waves, in which a complex wave can be considered as a mixture (a superposition) of simple sinusoidal waves. As far as vibratory motion is concerned, a complex vibration can be considered as a mixture (a superposition) of *normal mode* vibrations. In the case of just one string, and for motion in one plane perpendicular to the soundboard bridge, a normal mode of the string would be a mode in which the string was moving perpendicular to the bridge, in simple harmonic vibratory motion.

The normal modes in this case happen to be standing waves on the string. When we are dealing with a system comprised of two strings *coupled* by the bridge, however, we are dealing with a *different* kind of system. The strings may still have standing waves, but each string's standing wave is not, in itself, a normal mode of the system as a whole. The system as a whole includes the bridge, and it has different normal modes to those of individual strings.

Remember that normal modes can be thought of as modes of motion of a system, in which all parts of the system are moving in simple harmonic motion, with the same frequency. When we are dealing with two strings sounding together as a pair of oscillators coupled by the bridge (but still for motion in one plane only), there are *two* normal modes of the system as a whole, for each *single* adjustable partial. One is called the *symmetric* mode, and the other is called the *antisymmetric* mode.

In one of these modes, the *symmetric mode*, the two strings at the bridge position are moving back and forth (or up and down) together, "in phase", that is, always in the same direction as each other. In the other mode, the *antisymmetric mode*, they are always moving in opposite directions – "out of phase" with each other. By the *principle of superposition*, any other kind of motion of which the system is capable, will be a superposition (mixture) of these two modes, possibly in different quantities. The two modes may or may not have the same frequency, depending on the properties of the bridge, and the mistuning.

The theory of hammer irregularities proposes that when two strings are struck perfectly equally by one hammer, they will begin motion together, *in phase*, with equal amplitude, for a given partial. This type of motion would be the *symmetric mode* of motion. However, it is also possible for one string to be set into motion, not directly by a hammer strike, but by the fact that the string next to it is in motion first. Any string already in motion will cause the soundboard bridge to move, and if the bridge boundary of another unstruck string is moving, that string will tend to be set into motion, because it is being *driven* from the bridge motion at that end of its speaking length.

This would happen, for example, if one string of a trichord is struck (the other being wedged), whilst the *una corda* pedal is applied. The unstruck string starts to move in *resonance* to the struck string, because of the motion of the bridge. When this happens, the motion of the unstruck string always starts *out of phase* with that of the struck string, so that its motion is always in the opposite direction to that of the struck string. This might seem counter-intuitive, but it is indeed the case, and is due to the nature of reflection of waves at a boundary. This type of motion would of course be the *antisymmetric mode*.

The theory can then be roughly outlined as follows. If a unison pair is struck with an *irregular* hammer surface, it is possible for one string to start its motion with a slightly different amplitude or phase, to the other string. Being only slightly different, the two strings nevertheless initially move in a more or less synchronised manner on the bridge, forcing it to move more or less twice as far as one string on its own would do. In this state, the motion of the strings is largely *symmetric*, but also contains a small amount of *antisymmetric mode*.

Remember that the bridge is effectively a *resistance* to incoming waves from the strings. As it moves, it drains energy away from the strings, in a way similar to friction. The further it moves, the more energy is drained away. The mostly synchronised motion of the two strings "working together" on the bridge, causes the bridge to move further than either string alone. The result is a rapid initial decay, as the larger movement of the bridge drains energy almost twice as fast away from the strings, allowing the soundboard to radiate the initial, loud, but fast decaying *prompt sound*.

If the string motions are not perfectly equal, however, due to the hammer irregularity, then one string will actually approach zero motion before the other, where upon it starts to become *driven by* the motion of the bridge, more than it is *driving* the bridge. It now begins to behave more like a resonating aliquot string. At this point, what was initially a largely symmetric mode of the two strings, then "converts" to a largely antisymmetric mode, with the motions of the strings now acting *out of phase* or "against" each other. The *net force* on the bridge is now very small, because as one string "pushes" the other "pulls", so to speak. This results in much smaller bridge motion, a much slower energy drain, and a sustained *aftersound* radiated by the soundboard.

The precise decay shape of the partial will thus depend on the original admixture of symmetric and antisymmetric modes, determined by the irregularity in the initial hammer strike. However, any *mistuning* between the strings *also* introduces some antisymmetric mode in the initial, mainly symmetric mode. Fine tuning can therefore be used to regulate from note to note, the amount of antisymmetric mode initially present, which can both compensate for hammer irregularities, and ensure a sustained aftersound in certain partials.

It can be seen from the actual decay envelopes shown in this chapter that partials of finely tuned unisons often do display an initial rapid decay into a "null", followed by a more sustained (less rapidly decaying) "aftersound". In the real situation, in trichords there are three strings, and the strings move transversely in *two planes*, even though they are struck in just one plane. The Weinreich model is therefore still a simplified one, and the real situation is necessarily more complicated. In particular, the tuner is primarily engaged in reducing fluctuation movement in the unison soundscape, and in doing so, attenuating false beat patterns.

Regardless of this, the proposition that unison tuning may compensate to some degree for the effects of hammer irregularities, remains sensible, given that factors of tone other than beat rate are deliberately addressed in fine tuning by expert tuners. It is of course not necessarily only hammer irregularities that may cause variations from note to note. Whatever the causes, and whatever the mechanisms, the tuner is certainly engaged in a process of imposing more uniformity of tone on what to begin with, may be a lack of uniformity.

Movement in the spectrum – some important points highlighted

Bringing a unison to a "beatless" condition does not actually mean all partials will be simultaneously beatless, or free of fluctuation. Different partials may require different tuning positions of the strings for their own *beatless* condition to be reached. Tuning the unison to make one partial enter a beatless state, may cause another partial to leave its beatless state and start beating, or fluctuating with some other pattern.[62] There are no universal rules, except to say that close to zero mistuning many partials will be beatless, or as close to beatless as possible.

Any very high frequency band giving a sharp or unpleasant "edge" to the unison must still be eliminated, or made as stationary as possible. If

---

[62] Capleton, B, 'Piano unison tuning', www.amarilli.co.uk/academic, 2006-.

movement has to remain, then it should involve the least unpleasant sound, or should occur as late in the decay as possible. The skill lies not just in the mechanics of achieving the most beautiful sound, but in combining this with a condition of *stability*, such that the chosen soundscape does not rapidly deteriorate in any case, due to tuning instability. Stability is dealt with in the chapter on *setting the pin*. The spectrum of an excellently tuned unison can often be shown to contain fluctuations in individual partials, but nonetheless, *as a whole*, the unison may be *in effect*, absolutely "still".

The way partials in a spectrum combine to form a tonal *whole*, is in part psychological. However, the *broad differences* between excellent, reasonable, and poor tuning of the unison, considered in terms of *movement of the tonal whole*, is quite easy for tuners to agree upon. When dealing with fine differences, often the only way to demonstrate the existence of unnecessary movement in a unison, is to eliminate the residual movement and show that the unison can be made stiller. This is not really "beat elimination". Rather, it is more accurate to describe it as an adjustment of partial decay patterns and decay rates.

Adjusting tone through decay rates

Whether or not we have a "cat" or vowel in the unison to tune, in the fine tuning region we will be adjusting partial decay rates. Inside the coupling region a unison partial may exhibit some audible decay curve other than a *regular beat* pattern. This might typically be, for example, the *single null* pattern. Fig. (8.18) shows for illustration, three theoretically generated decay curves for the same unison partial, at different points of relative tuning between two strings as we finalise the unison towards its optimum tuning. Whilst these curves are not recorded from actual piano tone, they are not merely fictitious illustrations – they are mathematically generated from the associated theory, using "realistic" parameters. Examples of actual envelopes displaying the same kind of behaviour can easily be found in piano tone, but the theoretical semi-logarithmic graphs are "tidier" and much easy to superpose for comparison.

Fig. 8.18

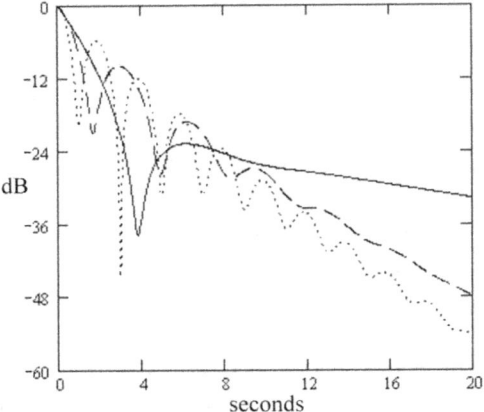

Fig. 8.18. Three stages of tuning in a unison partial as tuning approaches the optimum. At the dotted line the unison is beating with a regular beat. At the dashed line the beat rate has reduced. At the solid line the beat has converted to a single null, with a prominent "aftersound" after the single null at 4 seconds.

Remember this is not the whole unison – it is just one partial. The dotted line is a *regular beat* pattern occurring when tuning is well outside the coupling region. Now as we finalise the unison, the "mistuning" between the two strings decreases, and the beat rate decreases, giving the dashed *regular beat* pattern. As we finalise the unison still further, we enter the coupling region and the partial's decay curve becomes a *single null*, shown as the solid line. Note that after the null, the decay rate is much slower than in the *regular beat* decay patterns, giving a sustained "aftersound".

Once inside the coupling region, *i.e.* once the regular beat has converted to its new form, in this case a single null, further changes to the relative tuning condition between the two strings may produce a large effect on the aftersound *decay rate*.

Fig. 8.19

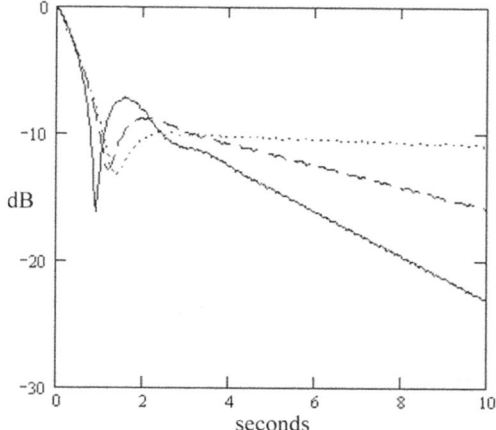

Fig. 8.19. The decay curve of a partial with further fine tuning. The mistuning is reduced, from solid, to dashed, to dotted lines. The smallest mistuning (dotted) produces the most sustained "aftersound". If false beating were present, the situation would be much more complicated, the main criteria being the attenuation of fluctuations.

Fig. (8.19) shows some theoretical decay curves for a single partial, at three different stages of "closing" a unison. The solid line is the curve just inside the coupling region, the dashed is a little further inside, and the dotted line is the curve occurring close to the centre of the coupling region, where the decay rate is the slowest. These graphs are for generally long decay times – specific curves and decay rates may differ for different situations.

At any point of relative tuning between the two strings, each partial in the spectrum will be at a different point, relative to the coupling region for that partial. Consequently, different partials will have different decay rates, and the arrangement of decay rates will depend on the exact tuning condition between the two strings. Which partials dominate the spectrum at the beginning of the decay, and later in the decay, will thus also depend on the tuning condition between the two strings.

Thus in tuning the unison we may often be listening to the changing tonal "colour" of the unison as determined by relative aftersound decay rates. It may be that there is not a single point of tuning at which audible *movement* is minimised, but a *range* of tuning positions that all qualify for having "minimised" the overall amount of movement. Different tuning positions will then give a different tonal spectrum, in terms of relative partial amplitudes, decay curve shapes, and *aftersound*. We do not, of course, think of it in this complex way whilst tuning, but rather, "instantaneously" interpret the soundscape as a whole.

Unisons in the high treble

Ease of tuning

Complete success in tuning the upper treble is elusive to most students of tuning, for a number of reasons including the aural difficulties of working with high frequencies. Apart from the aural difficulties, actual mechanical and acoustical factors work to both increase and decrease the difficulty of tuning in this region of the compass.

In the higher treble compass, the upper partials of a note become progressively less important in the tuning process, the higher in the compass one tunes (although they may of course affect other aspects of tone perception). Fig. (8.20) compares the 2-D spectra of a note from the mid-compass, A37 (A3) on the Steinway model M, with a high treble note, A73 (A6).

Fig. 8.20

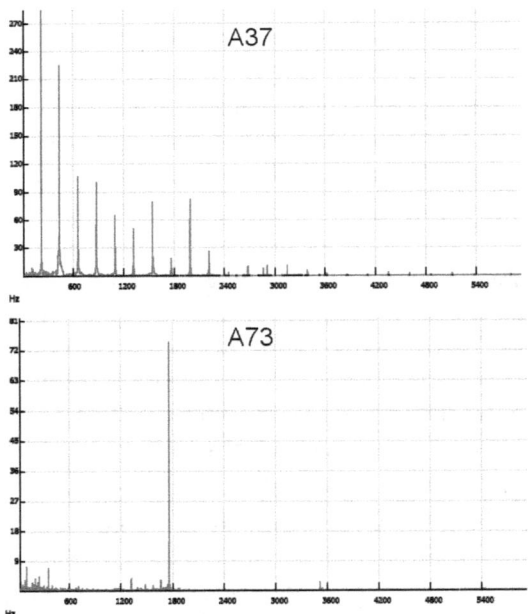

Fig. 8.20. Comparative 2-D spectra of two notes, A37 (A3) in the central compass, and A73 (A6) in the top treble. The tall peak in the spectrum of A73 is the fundamental, the other peaks lower in frequency being noise.

In the case of the high A73 (A6), only the fundamental is significant for tuning purposes. The tuning pins (wrest pins) are (on a properly strung instrument) the same diameter across the compass, but the speaking lengths of the strings in the upper treble are of course much shorter than those of the mid-range, so a given degree pin turning causes a greater change to string tension in the upper treble than it does in the mid range. The upper treble strings thus require in theory, greater finesse in tuning lever control. This apparent difficulty is offset by three factors.

Firstly, the shorter decay times of partials (including fundamentals) in the upper treble can be associated with a "wider" *coupling region*, making it relatively easy to enter a tuning condition where beating stops or some decay pattern other than a *regular beat* takes place. Secondly, we are more concerned with partials only at the lower end of the spectrum, and perhaps only the fundamental. Thirdly, as we saw above (Fig. 8.16), notes in the higher treble can exhibit more pronounced drawing together of the frequency components of an adjustable partial, as zero mistuning is approached. Thus the beat can be slower for a given mistuning, than elementary theory suggests. This effect seems to be consistent with the proposition that there is more dissipative coupling in the upper treble, due to the faster decay rates. Such effects on the beat rate are not generally found in fundamentals of notes lower in the compass.

Falseness in the high treble

Treble tuning is typically made more difficult, however, by a much increased incidence of falseness, compared to the mid-range. The ability to tune fine unisons right up to the top treble requires considerable command of unison tuning as an art, and is very dependent on experience. Most, if not all strings will be false,[63] even if only with relatively slow false beat rates, so the minimisation of fluctuation in the soundscape through false beat *attenuation* becomes very important.

Typically, in the high treble it is adjustment of the decay shape of one or two partials only (the fundamental and 2nd partial) that is being

---

[63] Remember that what is "traditionally" acknowledged as falseness is an acute subset of the much wider phenomenon.

specifically adjusted, and this is a different experience to the tuning of a mid-range unison, where there are many more partials contributing to the choice of soundscape adjustment. In the higher treble, it is not so much "cat" or vowel movement in the soundscape that is important, but the primary consideration is the reduction of fluctuation and beating in the fundamental, which is often false, combined with the production of a good "aftersound".

The ability to tune the top treble well, takes years to learn, and is not generally achieved by the end of a formal course of training. The top treble makes the highest demands on the hearing, on the knowledge of tone, on the lever control, on tuning for stability, and on the ability to effectively attenuate falseness. The ability to attenuate falseness may well be by far the greater part of the art in tuning this part of the compass. The top treble is not necessarily the most difficult part of the compass to tune, once it is mastered, but it is a part of tuning that takes the longest to master.

Unisons in the bass

Bass strings invariably exhibit a high incidence of readily audible false beating in the spectrum. This is not necessarily to be considered a fault – it is a natural consequence even when the piano has been constructed to the highest standards. It means that the spectrum of a single bass string can be *expected* to exhibit beating in many of its higher partials, and that this is a *natural feature* of bass note tone in the piano. There is, however, a "limit" to this feature beyond which the presence of false beating *will* become detrimental to tone. If pronounced false beating is inherited in the adjustable partials of the octave's spectrum, then the tone of the octave may well be compromised.

If the two strings of a bass bichord were truly identical then we might expect a perfect match between the string's two spectra, provided their boundary conditions at the bridge were also identical. The real situation is not necessarily like this. We often find, in practice, that in tuning a bichord, the simultaneous elimination of all *regular beat* patterns in the adjustable partials, is impossible. Usually, we can eliminate a prominent beat x somewhere in the spectrum, but another prominent beat y will remain somewhere else in the spectrum. The beat y will only be eliminated at a tuning position that leaves x still beating. There may be *no tuning position* at all at which *both* the beat patterns x and y are eliminated simultaneously.

Wherever the unison is finalised, some beating must remain. This will of course be expected to happen if there is an inharmonic mismatch between the spectra of the individual strings – they cannot always combine or superpose to create a new unison spectrum in which *none* of the partials beat. We can choose to eliminate beating in this or that partial, but not in both, or all, simultaneously.

Like the treble unisons, a bass unison must generally be tuned *as a whole soundscape*, not deliberately with regard to this or that specific partial, exclusively. The decay times of the partials are relatively long in the bass, so we would in theory expect either relatively narrow, or no effective *coupling regions* for the partials. The behaviour of a partial with changing tuning condition will tend to be more like that predicted by "traditional" tuning theory. A *coupling region* can still in theory exist where the partial decay rate is small, but its effects will be somewhat different to that in a treble unison.

In practice, in the bass unisons we do not find quite the same sense of co-ordination and co-operation between the partials of the spectrum as we attempt to finalise the unison. We will not necessarily have the sense of a *physical* region into which the soundscape can co-operatively drop as a whole, as we do in the case of the treble unison. The region may still be "there", offering a sense of satisfaction on entry into it, but the "feel" of it is less positive than it is in the treble unisons. There may still be a "cat" or "vowel" produced by the high partials whose movement we may wish to minimise, but this is not the only consideration.

For the spectrum of any note above about the 16th partial, *i.e.* for partials whose pitches are from about three and a half to four octaves above the fundamental, the partial pitches form a musical cluster, rather than having well separated pitches, and this is still true for bass unisons. This continuous cluster of pitches of partials is important in determining the tone of the unison. A small change x of tuning pin (wrest pin) position may not produce as much effect on the fundamental of a bass unison, as it would in a treble unison, but it can nevertheless affect the higher partials significantly, so fine lever technique is still required for bass tuning.

The precise tuning of the bass unison cannot be ultimately be dictated by a formulaic approach, but will be determined by the individual soundscape, together with the judgement and experience of the tuner. Bass unisons are never "beatless" in their spectrum, but generally the lower partials would be expected to have beating eliminated or reduced as much as possible.

Three string unison groups (trichords)

We have seen that two strings tuned to unison, constitute a different acoustical system to each of the strings separately. The kinds of motion that are possible for the two stringed system, are not necessarily the same as those for each of the individual strings. Consequently, we cannot think of the sound of the unison merely as the sound of both individual strings, simultaneously heard. Of course, many of the characteristics of the individual strings are "inherited" by the two stringed system, but the latter is a new system in its own right, with its own laws of motion, and consequently its own properties of sound.

We must consider the third string, since most of the unisons in the piano are trichord groups. Again, when all three strings are sounding, the trichord is an acoustical system in its own right. After tuning the first two strings of a trichord, we must regard the third string not just as an "add on" to the unison already tuned between the first two strings, but as having a more useful potential to do two things. Firstly, it *may* improve any unresolved problem in the unison between the first two strings, due to falseness or some other decay feature, so that the result in the trichord is *better* than the result possible in only the first two strings. Secondly, it may exacerbate or highlight any problem between the first two strings that was otherwise not really noticeable.

In the theoretical modelling of unisons that takes into account the possibility of false beats, the number of variables affecting the generic fluctuation behaviour of a single adjustable partial, increases as the square of twice the number of strings involved in producing the unison. This means that whilst there are 16 variables for the unison pair, adding in the third string increases the number of variables to 36. In other words, roughly speaking, adding in the third string of a unison can make the acoustical behaviour of the unison more than twice as complicated as it was with the unison pair. This has some bearing on unison tuning technique, as far as the sequence of tuning the strings is concerned.

Trichord string tuning sequence

When tuning a trichord unison, there are two alternative procedures that can be taken, using a Papp's wedge or grand wedge, and assuming one is not using listing (muting) felt inserted between the trichords:

Tune the middle string to the first string, and then tune the third string to the middle string, with the first string wedged.

Tune the middle string to the first string, and then tune the third string to the first two strings, unwedged.

Starting, for example, with the left string, these procedures can be represented:

Fig. 8.21

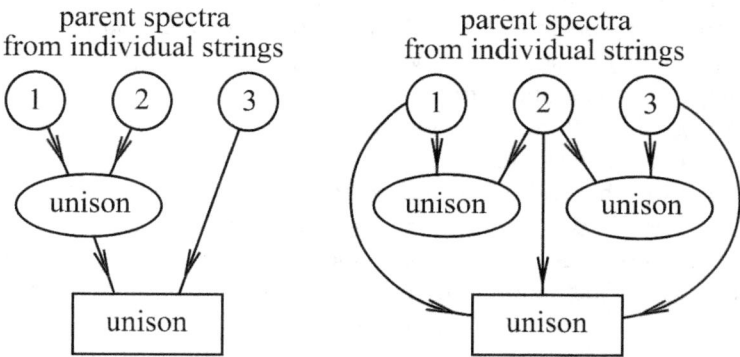

Fig. 8.21. Schematic representation of the two sequences for tuning a unison trichord.

In the first procedure of Fig. (8.21), on the left, the final string, number 3, is tuned with all three strings sounding simultaneously, in the final unison shown as the rectangle. In the second procedure, on the right, the final trichord unison in the rectangle is never actually *tuned* with all three strings sounding simultaneously. This unison between the three strings arises subsequently as a consequence of the tuning of the two bichord unisons, shown in the ellipses. All three strings, however, contribute to the final unison.

The question is: How does an "optimum" tuning condition of the final unison (the rectangle) relate to the tuning condition for each individual pair of strings? Will two "optimised" pairs of strings produce an "optimised" trichord, or must we tune the last string with all three strings sounding in order to get an "optimised" trichord?

Two strings are not necessarily tuned to zero mistuning (*i.e.* zero tension difference) in order to "optimise" the soundscape of the two string unison. Remember that in contemporary theory a "mistuning" does not imply an "out of tune" unison, but rather, merely indicates a difference in partial frequencies *when the strings are sounded separately*. Intuitively, we might think that tuning the trichord in two separate, "optimised" unison pairs (procedure 2), will leave the mistuning between two strings (usually the outside pair) "unoptimised". In other words, we tune string 2 to string 1, and then tune string 3 to string 2, which leaves the tuning of string 3 to string 1, "untuned", as far as placing a "deliberate mistuning" between them is concerned. The situation, however, is not necessarily as it might seem.

A trichord has three string pair combinations: left string and middle string; middle string and right string; and left string and right string. Computer modelling[64] suggests that, for non false strings, as far as *reduction of fluctuation* in any unison partial is concerned, *one* of these string-pair combinations only, *critically affects* the tuning of the whole. If *only one* of the three string-pair combinations in the trichord has zero or close to zero mistuning, then the other *two* pairs can be much more "mistuned" than otherwise expected. The mistuning in either of these other two pairs can be of a size that would produce considerable fluctuating or beating if the pair sounded alone. But when sounding in the trichord, the fluctuation is not produced, *provided* there is zero or close to zero mistuning in the third string-pair.

For example, in the central compass one might expect mistunings between a unison pair (the third string of the trichord wedged) within 0.3 Hz to be inside the coupling range, but in the case of the full trichord the mistuning can be twice this or more, and still result in the elimination of a regular beat pattern, providing just one pair of strings has zero or close to zero mistuning.

The situation is more complicated if the strings are not "well behaved", that is, if they are considerably false. In this case, the definition of "optimisation" as the elimination of beating, or its replacement with a single null, itself fails, and what constitutes an "optimised unison" becomes more obscure. The practical side of this is that where the strings are particularly false and not "well behaved", it may be *imperative* to complete the trichord tuning with *all three strings sounding*, in order to achieve the best results. The presence of moderate or slow falseness, on the other hand, is not shown by the mathematical model to necessarily be an obstacle to tuning an "optimised" trichord as two unison pairs with a common string.

So should one tune the trichord as two pairs of stings, or as one pair, followed by all three sounding together? Perhaps the best answer is that *in the absence of significant falseness*, any difference between the two methods is more likely to be a matter of psycho-acoustical considerations.

---

[64] *i.e.* generating decay curves from the 3X3 single plane dynamical matrix extended from the Weinreich model. From unpublished research by the author.

One such psychological consideration is the fact that the *changes* of decay pattern in a partial of the full trichord, as one string is altered, are in general *much more complicated* than they are in the two-string unison.

There is certainly an argument that as long as falseness is not causing a detrimental effect on the tone, then *tuning stability* is in fact the main issue determining the answer. First, string 2 is tuned to string 1, for optimum tone and stability. The aural flag for any slight instability is a *change* to the soundscape decay characteristics over the next few strikes of the note. If a string is slipping minutely, one can hear the change it produces. Time and speed of tuning considerations dictate that one cannot simply keep striking the unison pair, once it is tuned, without moving on.

If the third string is then tuned with the first unison pair, so that the whole trichord is sounding, one cannot necessarily always tell from the soundscape of the whole trichord, if some of the audible movement has been introduced due to instability in the first pair, as the whole trichord is being struck. Movement resulting from a change in the first pair will be inherited by the trichord in much the same way as false beating would be, and this could result in the attempt to attenuate the movement through the tuning of the third string, when in fact a better result might be achievable by retuning the first pair.

Having tuned the first pair, one already knows the resulting soundscape. Tuning the third string to just one string of the first pair, one knows the resulting soundscape for this pair also. *Subsequently* listening to the whole trichord immediately reveals any instability in the first pair, and if instability in either pair is suspected, the offending pair is easily identified using the Papp's wedge.

Where falseness is an issue, however, it may well be the case that a better result is attainable in the trichord (all three strings sounding) than in either string pair alone, due to the effects of coupling and beat attenuation. In this case there is, both according to the theory, and certainly in my own practical experience, *a distinct advantage to tuning the third string with all three strings sounding*. The behaviour in the trichord is too complex, and the parameters too variable, however, for a single "best" sequence to be determined. The first string of the trichord will be tuned as part of another interval, so it is generally sensible to choose the least false string as the first one tuned. Even so, as long as a satisfactory result for both the interval and the trichord is elusive, it may be necessary to change the sequence in order to arrive at the best result, even if this means tuning the most false string first.

Inharmonicity, stringing errors and unisons

If the three strings of a trichord have the same coefficient of inharmonicity, then inharmonicity, however great, does not present an obstacle to excellent unison tuning. If, however, the three strings are not identical in their inharmonicity, for example if one string has been replaced with an incorrect gauge, then the *difference in inharmonicities* can present problems. Often there may be no tuning position of the string pair with the gauge mismatch, at which beating can be sufficiently reduced. There is no *one* position in which *all* the partials have their beating sufficiently reduced or stopped. This effect is quite distinct from falseness, but can easily be mistaken for it.

Utilising unison tuning to fine tune other intervals

Bridge coupling in the unison itself can have an effect on other intervals, of which the unison is a member. This means that if we tune the first string of the unison relative to another note, as a tuned interval, then by the time we have tuned the whole unison trichord, the nature of the interval between the unison and the other note, may have changed. In other words, the beat rate in an interval when only the first string of a trichord forming one note of the interval has been tuned, may not be the same as when the whole trichord has been tuned.

Adding in the second and third strings of the trichord may change the beat rate in the interval. This will of course happen if tuning is not stable, but it may also happen when the tuning is stable. It can be verified by checking the interval beat rate with just one string sounding, of the trichord in question, the other two strings being wedged, and then checking the interval beat rate with the whole trichord sounding.

The fact that an interval between two trichords may behave differently to the same interval when only one string of one or both trichords is sounding, is the main reason why expert tuners tune the scale in full trichords, using a Papp's wedge, rather than using muting felt to isolate the middle strings. If we tune the whole scale by the centre strings first, using muting felt, and then tune the unisons to the centre strings, the character of the scale can change, even if the unison tuning is excellent. In tuning a unison after tuning a tempered interval, it may be found that the tempering has changed, and yet if two strings are wedged and the tempered interval tested, its tempering remains as originally set.

One must sometimes re-tune a unison in order to "utilise" this effect. For example, suppose on completing the unison, the tempered interval is found to be too slow. The procedure is to tune the unison again, tempering the interval on the first string correspondingly too fast, so that when the rest of the unison is then tuned, the tempering is as required. Typically, the effect is associated with irregular beat rates in tempered intervals. Some tuners *use* this effect to alter the tuning in the tempered interval *by altering the unison*. For example, if the tempered interval is just a little slow, the first string of the unison can be altered in the direction that would speed up the tempered interval, by "spoiling" the unison slightly, rather than by tuning the tempered interval itself. On re-tuning the trichord, the tempered interval is then improved.

Tuning Blüthner unisons with aliquot strings

Many Blüthner grand pianos have aliquot stringing. Aliquot strings are additional, unstruck, fourth strings to each trichord over most of the treble, each having its own bridging. Some of the aliquots are tuned to unison with the struck trichord, and some are tuned to aliquot intervals above the main trichord, in octaves and fifths.

The aliquots are "sympathetic" strings, each with its own wrest pin (tuning pin). They must initially be tuned by "chipping", *i.e.* by plucking. The aliquot strings tuned to unison with the struck note have a relatively strong effect on the tone of the unison. After tuning the struck trichord, the aliquot string can be fine tuned by striking the note in the normal way, and adjusting the tone of the unison. In much the same way that fluctuation attenuation allows the full trichord to produce a better result in the unison where falseness is present, so the fourth string in the Blüthner can improve the trichord tone. Falseness is very prominent in the aliquot strings, and fairly prominent in the struck trichords. However, this is not a fault. Blüthner grand pianos have a beautiful tone, and they highlight the difference between handling falseness in the process of tuning for tone and intonation, and the purely technical approach to tuning by beats that "traditional" theory encourages.

Chapter 9 - Tuning the Scale

The learning curve

In this chapter we look at the ways in which actual scale tuning is more involved than the "traditional" theory suggests. Chapter 5 on "traditional" piano tuning theory, especially the *Elementary practical tuning procedure for the scale* should ideally be read, and perhaps even tried in practice, first.

The "scale" is the chromatic scale of thirteen notes (including the octave) first tuned in the central part of the compass. Even if tuning initially ranges over a part of the compass wider than this, the scale is still any suitable complete chromatic scale in the central part of the compass. We shall see that there should not be in any case any demarcation between a chosen "scale" and the adjacent parts of the compass, in terms of the tones of the intervals, and their beat rates. Even if initial tuning takes place over, say, two octave's width of the compass, any octave's width within that, must still contain good *progression* of beat rates, and could in principle have been tuned first. We shall see that the critical thing in scale tuning is *progression* of beat rates, seamlessly flowing through the scale and beyond it.

Typically the tuned scale ranges from F33 – F45 (F3 –F4) or G35 – G47 (G3 –G4), but may be tuned over other note ranges. Once the scale is tuned, tuning can proceed outwards through the compass in octaves and other intervals. The characteristics of the scale, in particular the good "progression" of beat rates in the thirds, compound thirds, and sixths, should never be treated as being confined to the scale itself. The scale should "integrate seamlessly" into the compass so that there is no obvious boundary indicating the particular range of notes over which it was tuned. In this respect, it should not matter which range of notes is chosen for tuning the scale – the end result should be equally good.

Learning to tune the scale usually begins with the technique of muting the outer two strings of each trichord using a strip of muting felt, or perhaps in the case of a grand piano using a gang wedge. This isolates the centre strings and removes the problem of unison tuning and its effect on tempering, simplifying the process for learning purposes. It should be remembered that acoustically a trichord unison is by no means merely a triplicate copy of a single string. Once the unisons are tuned, the nature of a tempered interval between the unisons can change, *even if the tuning of the strings is entirely stable*. In other words, the required tuning position of a single string being tuned to another complete trichord, may be different to its required tuning position when it is tuned to just one string of the other trichord.

Student tuners often complain that after completing the tuning of a trichord, the tempered interval changes. This can of course happen if tuning is unstable, but it also happens even where tuning is stable. For this reason, expert tuners construct the scale in complete unisons during the scale tuning process, and do not first tune the whole scale in single strings only, followed by unison tuning. Each first string of a trichord being tuned, is tuned to another *complete trichord*, rather than just to another single string. This produces much better control over the results, by eliminating the changes in the scale that can otherwise take place if it is tuned on single strings only, the unisons being completed afterwards.

Tuning a "muted scale", in which a strip of felt is inserted between the trichords, is a helpful part of the learning process, because it allows the student to concentrate on tempering issues without the complication of tuning unisons. It also allows the student to begin tackling tempering intervals and tuning a scale, even before mastering unison tuning. Tuning of the scale professionally, however, should *always* be done by using a wedge, completing the unisons note by note, as scale construction proceeds. The drawback of this, for the learner, is that it requires excellent unison tuning skills as a pre-requisite. Any poorly tuned unison trichord will have its movement or weakness inherited by every tempered interval of which it is a member. It will especially show up in the perfect fifths.

The approach

Firstly we must appreciate the purpose of scale tuning. The common error, as already said, is the assumption that it is just to get the first 13 notes at the "right" pitch or the strings to the "right" frequency. The reality of its purpose lies jointly in *tone* and *intonation*, not just in the scale area, but *throughout the compass*. Tone and intonation are inextricably connected.

Every interval has a tone quality as affected by the beating in its partials. All the beat rates are interconnected in a complex *network*. Correct pitch intonation *follows from* the correct tone quality as affected by beating. But beating *is not* merely a tool to achieve the right pitch intonation. The tone qualities of the intervals, as affected by beating, *are of great importance in their own right*.

No interval's tone quality can be changed without changing the tone qualities of other intervals to which it is connected. Tuning the scale is a setting up of the whole network of interval tones so that it is uniform, even, balanced, regular. In terms of *beat rates*, this uniformity and balance manifests as the steady *progression* of beat rates in the fast beating intervals, *i.e.* the thirds and the sixths. In the slow beating intervals it manifests more usually as *similarity* of beat rate. Certainly, if we *increased* the beat rates in the fifths with rising position in the compass, as fast as "traditional" theory dictates, they would be beating far too fast in most of the treble. Inharmonicity, of course, dictates a different pattern.

❖ To achieve the necessary uniformity one must understand *how the network works* !

"Traditional" theory provides a "ready made" solution for an assumed idealised piano, as a mathematically specified set of fixed beat rates for the intervals, which would work without the need to understand the network, provided the piano's tuning characteristics were close to the ideal assumed by the theory. Since all pianos differ from the theory's assumed piano in some respect, and many differ substantially, mastery of scale tuning is not about applying "traditional" theory's proposed solution with mathematical perfection, but is more about understanding *how the network works*, and being able to manipulate it at will. The traditional" theory's fixed beat rates are a good starting point.

To be able to do scale tuning really well *on any piano*, you have to see the beauty and rightness in tempered piano tone intervals, and be free of any cherished theory that piano intervals are somehow "compromises" or "deviations" from some other "ideal" such as "Just Intonation". The latter is an attitude, a conditioned response that will not easily lead to an appreciation of the beauty of piano interval tone. Remember (from the chapter on temperament theory) that interval tuning has a very long history from which such ideas arise, a history full of partial observations and spurious oversimplifications, that to this day, because they have now penetrated the psyche, continue to inspire but to mislead. "Pure" intervals with a minimum of beating are indeed beautiful. But so are equally tempered intervals when they appear in piano tone, with their shimmering partials. There is no more or less divine in the "Just" interval than in the tempered one, except in a Pythagorean belief system. There is beauty in both, just as there is beauty in the fact that, contrary to the Pythagorean tradition, the planets *do not*, after all, move in perfect circles, the cosmos *is not*, after all, constructed on harmonic ratios, and piano tone is not, after all, truly constructed on the harmonic series.

The basic principles

To reiterate a most important point: Beating is important in aural tuning because it constitutes a part of the tone quality of the interval that is precision-adjustable. This is contrary to the assumption that beating is merely a utilitarian tool that is used to attain the "right frequency" or "right pitch" for each note.

As is emphasised throughout this book, a musical interval has *two* essential musical qualities that are affected by its tuning condition. The *pitch* relationship between the two notes, or *pitch intonation*, is one. The *tone quality* of the interval is the other. In natural piano tone, the two are related to each other, because generally, as the intonation changes, so does the beating in the tone. But they are still two *different* dimensions of the sonic, musical quality of the interval.

Beating, in itself, is essentially a feature of tone quality, rather than a feature of pitch, even though it may well play a part in the psycho-acoustic perception of pitch. With digital manipulation it is perfectly possible to create two tempered "piano" intervals with exactly the same *pitch intonation*, but one which contains beating in its soundscape, right where you would expect to find it in natural piano tone, whilst the other does not. It is also possible to create two "piano" intervals with distinctly *different* pitch intonation, but with *identical* beating in their soundscapes. Such examples are artificially produced, but they nonetheless demonstrate how pitch intonation and beating are two *distinct* sonic qualities, each with their own implications for the sonic quality of the interval, even though in *piano tone* they are inevitably interconnected.

The fact that they are interconnected in piano tone does not mean that we only need to set the beat rate *just in order to get the pitch intonation right*. In fact, although pitch intonation matters, it by no means takes the prize as the biggest issue in fine tuning. The real difficulty lies in the *combined* issues of beating in the tone, and interval intonation. Both are important.

Let us take an extreme example of the tonal effect of beating. The infamous "wolf fifths" that occur in Pythagorean tuning or quarter comma meantone temperament, are intervals somewhere between a recognisable augmented fourth and a diminished minor sixth, that are so "out of tune" they are normally considered musically unusable. They contain very fast beating adjustable partials. It is quite true that the pitch intonation of a wolf "fifth" sounds different to the pitch intonation in a normal, perfect fifth. But what really clinches the tone of the wolf interval as being musically appalling, is the *beating* in the soundscape. The untrained ear may of course not recognise this aspect of its tone quality as being at the root of its dissonance, but to a tuner it is obvious.

Again, if this beating in the soundscape of a wolf interval is digitally "ironed out", so that the adjustable partials remain in the soundscape, but the beating is taken out of them, then the tone of the "wolf" interval is altered drastically. Subjectively, it sounds rather heavily tempered, but the characteristic "wolf" quality is gone. Even at this gross mistuning, its tone quality no longer makes it entirely musically unusable, and in the right context, it could even be mistaken just for a rather heavily tempered fifth in an early temperament. Put the beating back in the partials, and the interval "howls" again. In other words, the effect of beating in the tone of the interval is an important consideration *in its own right*, distinct from its functional relationship with the pitch intonation.

The wolf is an extreme example, but even in the usual piano intervals, beating needs to be adjusted in the tuning *for its own sake*, as a tone property, and not just as a means of adjusting pitch. Musicians generally do not hear beating specifically, but tone in the piano is cumulative and holistic. The beauty of the piano's tone and intonation lies very much in the way it is "distributed" over the compass. Piano designers go to great lengths to achieve smooth tone and intonation transition over the compass, and the last thing the piano needs is a tuner who cannot achieve the necessary complimentary effect through the tuning.

What about the pitches of the scale notes? The reality is, that despite the ear's undoubted sensitivity to pitch, the *precision changes* to beat rates made by the expert tuner in normal *fine tuning* adjustments – and I do mean the *finest tuning adjustments* by the *expert tuner* - in the central scale, *do not even produce changes of pitch intonation that the musician will discern*. The precision adjustments made, are simply not made for this reason. Let me clarify further, so there is no doubt what I mean by such a radical sounding statement.

The "traditional" difference in beat rate between one major third in the scale and the next, is roughly around 0.5 Hz, and no more. This means, for example, that at A37 (A3), changing the beat rate in the third A37 – C#41 (A3 – C#4) by 0.5 Hz, the fundamental frequency of the A must be changed by around 0.1 Hz, since the adjustable beat happens at the position of the 5th partial. This corresponds to a microinterval change of just over three quarters of a cent. An isolated microinterval of this size might well be discernable in a pitch sensitivity test, but a deliberate change of this size to a tempered interval on the piano itself, *will not be noticed by a musician as a pitch intonation issue*. If it were, then master tuners would constantly have problems with pitch intonation, and clients complaining about it, for the reasons that follow.

It is not a question of whether such a change *can* be detected by the human ear as a pitch intonation change, but a question of whether or not it *is* noticed in the normal course of events, when listening to *piano tone* or playing the piano. The fact that the tuner is theoretically attempting to graduate beat rates from one major third to the next by around 0.5 Hz, belies the nature of the sounds to which the tuner is listening. Beat rates in the thirds (even on some of the best instruments) *can naturally vary or fluctuate by 0.5 Hz* during the decay (as we will be seeing), with no assistance from the tuner. This is easily demonstrated by digital extraction of the beating partial, and direct measurement. This means that *variations in at least the same order* are often deliberately made by the tuner, in order to produce overall progression of beat rates. On many instruments beat rate variation over the first few seconds can be much larger than this.

The equating of precision beat rates with "precision pitch" is therefore a mis-reading of the situation. The fuzziness of the relationship between beat rate and pitch is not a problem in practice, precisely because the expert tuner engages in precision beat rate adjustment *for its own sake*, at a level that has relatively little meaning as far as the pitch intonation of the interval is concerned. In short, the "right pitches" do of course naturally enough occur during the process of good beat rate tuning, but beat rate tuning goes well beyond this point. It goes *much* further than its ability to provide good, equally tempered pitch intonation in the central scale.

Precision beat rate adjustment beyond the need to attain satisfactory pitch relationships between the notes of the scale, is seen in the tuner's considerable efforts to attain the perceived *progression* in the beat rates of the thirds and sixths. Outside the scale, however, precision beat rate adjustment and progression is all a matter, simultaneously, of both *tone* and *pitch intonation*.

Progression means that adjacent thirds or sixths will have very similar beat rates, but the lower one will never have a faster beat rate than the higher one, or the higher one a slower beat rate than the lower one. This sounds very straightforward, and is in any case surely just what "traditional" theory's beat rates do in any case, isn't it?

Well, yes, except that simply trying to apply "traditional" beat rates is not they key to progression in practice. Also, in "traditional" theory beat rate progression is presented as just a side effect of the effort to produce a certain "axiomatic" set of string frequencies. In practice, progression is for the sake of progression itself, which in turn is for the sake of consistency of tone and intonation, and the smooth transition of tone and intonation over the compass.

It has a real purpose in its own right, that the prescription of "traditional" theory itself merely hints at. It is critical not just for the beauty of the scale itself, but even more importantly for what happens over the rest of the compass, where the characteristics of the scale will be magnified through the octaves and multiple octaves. As in most aspects of piano tuning, the final, overall result, as the tone and intonation of the instrument, is very much determined by the collective effect of many small nuances.

The "traditional" prescription for beat rate values is a starting map, but it does not take into account *inharmonicity* (which varies from note to note, and is different on each instrument), *bridge and soundboard effects*, or *false beating*. Achieving *progression* is a question of intelligently *finding* it on each instrument. The bigger and better the instrument, the easier it is, if you follow the idealised map, and the better the possible result, but even then there is always the possibility of variation. The skill is required in the fact that the intervals are *not* just a set of intervals in which we can physically tune whatever beat rate we like. As a set, they form a *complex network*, in which whatever is done with one interval, will inevitably affect the beat rates in other intervals at the same time.

The situation is rather like an acoustical equivalent of the Rubik's cube, except that it is actually more difficult. There are in fact fixed "solutions" for the Rubik's cube that apply to every cube, but each scale tuning puzzle has its own variations demanding a slightly different "solution" for each scale. It is not just a task, it is certainly a puzzle to be solved. On some instruments, the "traditional" beat rates may provide, more or less, the solution, but the puzzle, if excellently solved, is still different on every instrument.

The puzzle and the solving of it is what expert tuners are involved in when tuning the scale. Each solution depends on factors (not least inharmonicity) that are not known in advance, and the solution is therefore in no way guaranteed to provide any particular set of absolute string frequencies, even if we were interested in this. Hence the solution is not really linked to any particular set of pitches, other than that the resultant pitches will of course fall within bounds where they are perfectly satisfactory to the musician.

In aural tuning the true behaviour and relationships between the beat rates must be ascertained during tuning, and in electronic tuning the best ETDs solve the puzzle by *measuring* the inharmonicity that will affect the beat rates, on the individual instrument, before tuning commences.

Beat behaviour

Beat rates are thus an essential part of tempering and tone control, and are an important component of tone in their own right. "Traditional" theory proposes that there is a set of specific beat rates, one for each interval tuned, that can be quoted with precision to several decimal places, and that the "correct" scale will exhibit these rates. The reality is that the acoustical behaviour of piano tone, and hence the properties of the scale, are far more complicated than the "traditional" theory assumes. There are a number of ways in which actual beat behaviour varies.

Firstly, as already mentioned, an interval in practice will not necessarily beat with a regular, fixed beat rate. Beat patterns may change, as the note decays. The beat rate in a major third or tenth, for example, may *often* vary (as we shall see), sometimes cyclically, by around at least 0.5 Hz during the audible decay time. It is therefore meaningless to make a generalised statement that an interval such as C40 – E44 (C4 – E4) for example, has, or should have a beat rate of precisely 10.38 Hz. The degree of variation in the beat rate of a major third or tenth, for example, is very often in the same order of magnitude as the "traditional" difference in beat rate between one third and the next.

Secondly, beat *rate* is only one factor to take into account in creating a good scale tuning. Beats also have their own *amplitude* or "depth", distinct from the overall amplitude of the partial in which they occur. The beat amplitude may *typically* vary by both decreasing and increasing over the decay time. Beats, in other words, can decrease or grow in depth or prominence, as a separate feature to the beat *rate*, and as a separate feature to the overall decay rate of the partial in which they occur.

The beat amplitude, as we shall see, is important in influencing the quality of an interval, and can even easily influence the aural judgement of beat rates. A beat's amplitude may reduce so much over part of a partial's decay time, that during this time it becomes practically inaudible, and hence aurally equivalent to a zero beat rate, even though technically the beat *rate* has not slowed. Conversely, a prominent (large amplitude) beat may sound faster than a shallow (small amplitude) one, even though the actual beat rates are the same.

These complex natural features of beat phenomena in themselves preclude the reducing of good scale tuning merely to the notion of applying a set of precise uniform beat rates only. Nevertheless, beat rates and their relationships *are* important in piano tuning! Even though a real beat pattern is not in actuality a single, simple thing or number, the idea of a "beat rate" remains useful. The reality is that good scale tuning is still characterised by uniform progressions of "beat rates" in the thirds and their compound extensions, and in the sixths, and by carefully controlled "beat rates" in the fifths and fourths.

If we were to just abandon the simplified and idealised framework of "beat rates" suggested by "traditional" theory, we would have no starting map. We must, however, understand the nature of this framework, as a guiding map, rather than as a kind of photographic representation of some hypothetical perfection. The expert aural tuner takes *all* the complications of beat behaviour into account in scale tuning, as part of the art, without necessarily quantifying them. It is a natural result of experience and expertise in the art. These include factors of beat amplitude, beat decay, beat rate instability, other complications we will deal with later, like dual beat patterns.

False beats

The tone of every string and every interval consists of numerous partials. Some of the partials in a tempered interval are *adjustable*, which means their beat *rate* can be readily adjusted by altering the string tension. Nevertheless, beating is not confined to adjustable partials. When beating that is non adjustable is obtrusive, piano tuners refer to this beating as *false beats* or *falseness*. The actuality is that "false beat" phenomena, *i.e.* non adjustable beat patterns or other fluctuations in partials, is a ubiquitous factor in *all* piano spectra. Beating partials do not always decay with regular beat patterns as "traditional" theory supposes. Beating does not just occur in the *adjustable partials,* and, confusingly enough, beating in adjustable partials themselves, is not always *adjustable* in the sense that its beat rate can be altered.

The idea that a truly "pure" or truly beatless interval is even *theoretically attainable* on a piano, is another myth. In the acoustics of the piano, it is a fact that beats are constantly, a natural part of the acoustical phenomena, even in what piano tuners would refer to as a "pure" interval, and *even in the tone of a single string*. Piano tuners only *need* to notice this "beating" that cannot be so easily controlled by tuning, when it is beyond a certain severity, and then they refer to it as "false beating". In practically every piano tone spectrum there will be *some* partials that do not have a smooth, exponential decay, which fluctuate, or which actually beat, even if their beat rate is only slow. False beat rates or fluctuations are not always slow, however. Many false beat rates are quite fast, especially in the treble.

Furthermore, in the context of modern acoustics, what piano tuners call "false beats" are no more "false" than any other beats. They are only "false" in piano tuning practice in that they do not, in general, have adjustable beat rates. Acoustically, all beats are the real result of superpositions of what are usually referred to as *eigenfrequencies* of the vibrating system as a whole. The only thing that really distinguishes "false" beats from beating in general, is therefore the lack of adjustability in their beat rates.

False beating cannot be simply ignored in scale tuning. It is not good tuning to simply allow an interval to be tempered such that beating spoils its tone, on the grounds that it is the false beating causing the excess. What counts is the musical *tone* of the interval, and if necessary, even the overall tempering of the scale may need to be adjusted to this end.

"Tempering" the scale is therefore much, much more than anything *temperament theory* itself can describe. It takes into account beat rates, variations in beat rates, variations in beat amplitudes, decay rates, false beating, and *tone*. None of these are addressed for the sake of temperament theory ideals. They are all addressed for the sake of *pitch intonation* (which is complex and in part psycho-acoustic), and *tone*, (which is also complex and in part psycho-acoustic).

Aural scale tuning - background

Learning aural scale tuning begins by learning to hear beating partials, and precision tuning of tempered intervals is achieved by adjusting the *adjustable partials* (those with adjustable beat rates) of the intervals in question.

In tempered intervals these are *specific partials only*, and the beat rates quoted by "traditional" theory as guidance, apply *only to the specific adjustable partial for which the beat rate was calculated*. For a given interval, partials other than those for which the published beat rate was calculated, will in general have *different* beat rates. In listening to beating in a tempered interval, attempting to "apply the right beat rate" is therefore misdirected unless you are sure you are listening to the correct beating partial. This is surprisingly widely misunderstood.

Tuning the scale at a professional standard also involves tuning unisons. In fine tuning, unisons are best tuned by listening to *tone*, and *not* listening to any specific partials. There is a good reason for this difference. *Every* partial in the unison is adjustable, and will change its decay shape as the string tension changes. In the mid-compass, typically at least 15 to 20 partials or so, and often many more, will show up above the noise floor in acoustical analysis of a unison tone. With so many partials all simultaneously adjustable, it is the *whole soundscape* to which one should listen when tuning unisons. The unisons are fine tuned in terms of *tone* and *tonal movement*, not in terms of specific beat rates in specific partials. The tempered intervals themselves, however, need to be tuned with regard for both tone *and* the beat rates in specific partials.

Fig. (9.1), as an example, shows the spectrum of a unison tone (E44) recorded from a Yamaha grand piano, using standard equipment. *Every* peak shown is an adjustable partial. The height of the peak is a (very often crude) representation of the audible prominence of the partial.

Fig. 9.1

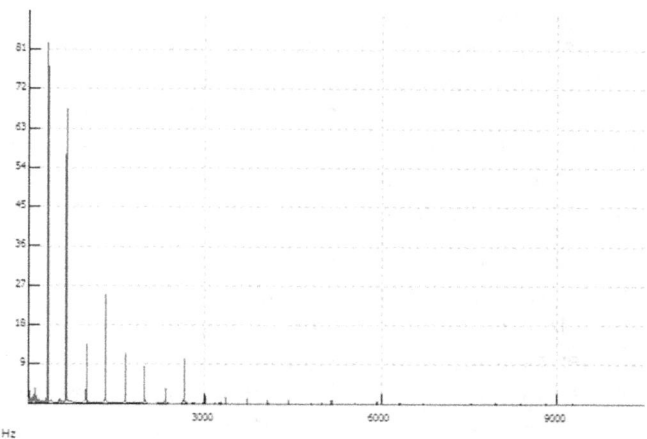

Fig. 9.1. Spectrum of the E44 unison on a Yamaha C3 grand piano. All partials are adjustable.

Fig. (9.2) shows by comparison the spectrum of a tempered interval, in this case, a major third (C40 – E44) on the same piano, in which the adjustable partials are marked.

Fig. 9.2

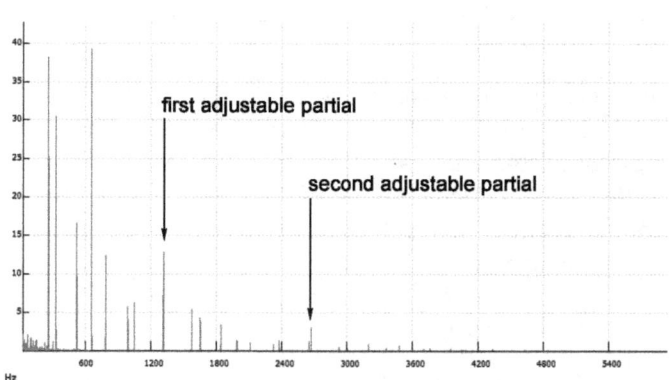

Fig. 9.2. Spectrum of major third interval on the same instrument.

When *most* of the partials of an interval are *not* adjustable, as in Fig. (9.2) for the major third - and as is also the case for all other *tempered* intervals - then it is important to be able perceive the specific partial that *is* adjustable. With a little practice, the ear can indirectly perceive a beating adjustable partial through its (usually quite large) *effects* on the soundscape, without necessarily perceiving the individual partial itself responsible for the beat. In other words, the tuner may be well aware of the partial's beating, but without necessarily hearing the beating as associated with that specific partial. Typically, the beating may be (psycho-acoustically) perceived as though it were in the fundamental, or at some other pitch or in some other partial. This is a psycho-acoustic illusion, of which there are many in listening to tones, and perception based on psycho-acoustics sometimes varies unreliably.

In learning, the secret of clarity and complete control lies in listening *at the correct pitch, to the correct partial*, so that you are aware of the specific partial, and its beat pattern. You then perceive what is actually there, rather than just your subjective impression. This banishes doubt, and eliminates confusion that can otherwise be caused by the fact that other adjustable and *non* adjustable partials may also be beating.

Once the ear is trained to immediately focus on the correct partial for a given interval, then after sufficient practise, hearing the beat phenomena with accuracy and reliability is automatic whenever that tempered interval is heard. The ear should be trained so that eventually no effort is required to work out where to listen, but rather, the ear just automatically knows "where to go". Only the minority of tuners in the learning process seem to put in the time and effort to practise this, but without such proper self-training, the result is a working practice rather like that of the pianist who never bothered to learn scales with the correct fingering.

There will always be a skill barrier beyond which the tuner's art cannot then pass. Once the ear is trained to this degree, it is easy to also psychologically "step back" and "take a listen" to the overall sound of the interval "from the musician's point of view", rather like a painter stepping back from the canvas to see the overall effect. The difference is that with the formal training, both the musical tone and the influencing factors, namely, the behaviour of the *specific adjustable partials*, can be heard and controlled *simultaneously*.

Looking at the spectrum diagrams above, being able to pick out the adjustable partial in a tempered interval may look like a daunting task, but in reality this is not so. It merely looks that way when visually represented. Fortunately, partials are arranged in a musically recognisable pattern, roughly, the *harmonic series*, which makes picking out the correct partial *easy* if you know what you are doing.

How to "ear train" and find partials

Just like learning to play a musical instrument, or learning a sport, or learning many of the tasks of practical musicianship, hearing beating partials accurately and reliably is simply a matter of practise, practise, practise, in the early stages of learning.

Start by learning to quickly find which note on the keyboard would be roughly the same pitch as the correct beating partial – we will see in a moment how to do this. It can be just learned by rote, or it can be "worked out" easily enough. To learn it by rote, the pitches for the adjustable partials of each interval are given in the section for each tuning interval discussed in this chapter. Just memorise these. Alternatively, to work out where to listen for adjustable partials, is not so difficult, provided the *harmonic series* is learned first. "Learning the harmonic series" means first learning how to find the sequence of notes on the keyboard, that represents the harmonic series starting on any given note, up to the sixth harmonic in the series. Here's how to do it.

The harmonic series for any note, will consist of a sequence of rising musical intervals, starting on that note. Start by choosing a note. This pitch of this note itself, is the first, or lowest pitch in the harmonic series we are about to find. In other words, the pitch of this note itself represents the pitch of the 1st partial, or *fundamental*, being produced by each vibrating string of this note. Not surprisingly, the strings of this note, as they vibrate, are each producing a (usually strong) partial at the perceived pitch of the note itself. But these strings are also producing other partials (sometimes called *overtones*) at higher pitches, whose pitches are approximately the same as the pitches of certain other notes that can be played on the keyboard above (we are assuming the piano is in reasonable tune). The next note above to find, representing the pitch of the originally chosen note's 2nd partial, is an octave above the chosen note itself. The 3rd partial's pitch is the same as the note a perfect fifth above this.

We are saying that starting with a chosen note, whose pitch represents the pitch of the *fundamental* or 1st partial of that note itself, we can find a practical listening guide for the pitches of its other partials by playing other notes above it, on the keyboard. If we wanted to hear the third partial of a note, for example, we would need to know what pitch to listen for, within the tone of the note, as it is sounding. So we would play a note "representing" the third partial's musical pitch, an octave and a fifth above the original, and use this mew note as a preliminary *pitch guide* for the ear, *i.e.* to tell the ear what pitch to listen for, before listening for an actual adjustable partial at this pitch.

Let's reiterate the sequence. We need to go up an octave from the starting note, to find the 2nd note (to give us a pitch guide for the 2nd partial). Then up a perfect fifth from this second note to find the 3rd note (to give us a pitch guide for the 3rd partial). Then up again a perfect fourth to find the 4th note. Then a major third to find the 5th note. Then a minor third to find the 6th note representing the 6th partial. This is as high as we need go, ordinarily. Try this sequence on the keyboard, starting from C. The sequence should be C, C (an octave above), G, C, E, G. Now try from A. The sequence should be A, A, E, A, C#, E. Try it from F. The sequence will be F, F, C, F, A, C. Practise finding the sequence starting from *any* note in the scale area (central octave). Always use the same series of musical intervals. Practise until you can find the series from any note, and recognise, musically, the sound of the series, which forms a root position major chord.

For a more complete picture, the full sequence of intervals up to the 10th harmonic is: octave; perfect fifth; perfect fourth; major third; minor third; another minor third; a whole tone; another whole tone; yet another whole tone. This brings us to the 10th harmonic. Playing the notes on the keyboard, starting on C, we would have C, C an octave above, G a perfect fifth above that, C two octaves above the starting note, E a major third above that, G, B-flat, C, D, E. These are the notes of a dominant seventh chord. If we only want to think about the first 6 harmonics, which is all we need for the commonest tempered intervals, then the notes form just the major chord, root position. Practise finding the series on the keyboard, playing the notes, one at a time in rising sequence, just by reproducing the major chord by ear, starting from any note.

Once the harmonic series can be found starting on any note, the pitches of the adjustable partials for any *tempered interval* can easily be found. They will occur where the harmonic series for each of the notes making the tempered interval, share a common note. For example, suppose we are tuning the fourth G35 – C40 (G3 – C4). The 4th harmonic position starting on the G35 (G3) will be at G59 (G5). But the 5th harmonic position starting on the middle C, will also occur at G59 (G5). The pitch of the first adjustable partial for the fourth G35 – C40 (G3 – C4), is therefore the same as the pitch of G59 (G5). There will be other coincidences above the G59 also, but as we shall see below, in practice, there is only one tempered interval for which we really need to consider an adjustable partial above the first one, and that is the tempered perfect fifth.

Try this for the fifth A37 – E44 (A3 – E4). The 3rd harmonic for the A37 (A3) will coincide with the 2nd harmonic for the E44 (E4), at the pitch of E56 (E5). You will also find the 6th harmonic of the A37 (A3) will coincide with the 4th harmonic of the E44 (E4), on the note an octave above the first coincidence. It is good to discover these coinciding harmonics by yourself, through the harmonic series played out on the keyboard. This learning is better than merely being told where the coincidences fall.

However, *there is a pattern*. Each type of tuning interval has its own *harmonic ratio* associated with it, which can be used to find which harmonics in the series would coincide, for that interval. The principle harmonic ratios are as follows:

Perfect fifth - 3:2
Perfect fourth - 4:3
Major third - 5:4
Minor third - 6:5
Major sixth - 5:3
Minor sixth - 8:5

Harmonic ratios have been known about for thousands of years (and arise because of natural *standing wave* motion on the string, and its *allowed frequencies*). If you have not already done so, to find out more about the origins of harmonic ratios, you can read the earlier chapters. Here, we just need to learn them.

Here is how to use the ratios. First learn the harmonic ratio for each interval, given in the list above. The *largest* number in any harmonic ratio (the number quoted first in each ratio in the list above), gives the number of the first coinciding partial from the *lowest* note of the interval. The *smallest* number in the harmonic ratio (the one quoted second), gives the number of the first coinciding partial from the *highest* note of the interval. Thus, the first number is the partial number of the lowest note, and the second number is the partial number of the highest note. These two partials will coincide. Having learnt how to find an harmonic series from any note, test these ratios out in practise, on the keyboard, and see if they are correct.

For any harmonic ratio, we can also *multiply* each number by any given integer (any whole number), to get another ratio expression, telling us where higher coincidences, and hence, higher adjustable partials, will fall. Thus, the major third (5:4), in theory has higher adjustable partials at the coincidences of the 10th and 8th harmonic positions, and the 15th and 12th positions, and so on.

There will also be some coincidences outside harmonic ratio relationships, especially given that *actual* partial frequencies produced in piano sounds *do not* fall perfectly in harmonic series, due to *inharmonicity*. The harmonic ratio method of finding adjustable partials is reliable and complete for the lower partials used in tuning practice. It is not, however, an exhaustive or complete scientific method.

Playing notes on the keyboard to represent the harmonic series in this way, serves as a training technique. The technique obviously requires a piano in reasonably good tune. "Working out" where to listen for adjustable partials is not necessary for the expert tuner, as eventually, the appropriate partials can be heard immediately, without the aid of reference pitches provided by other notes on the keyboard, to tell us where to listen. Through training, one should get to the stage where the ear automatically goes to the correct adjustable partial for any tuning interval heard, but can also, at will, focus on any other partial chosen.

All you need to do is to practise listening to adjustable partials by playing the intervals. With practise the ear will learn to automatically go to the right partial when any interval is heard. This is the foundation of solid, reliable, aural technique.

The student of tuning should explore the harmonic series in as many practical ways as possible. Practise playing the harmonic series on the keyboard from any starting note. Practise listening to, and analysing by ear, the tones of individual notes, and hearing the individual partials in the (approximately) harmonic series. Practice finding the points on a piano string, or ideally on a bass bichord of a grand piano, where touching the string lightly whilst playing the note, produces a clear partial with the other partial suppressed. Touching ½ way along the speaking length with produce the 2nd partial, 1/3rd of the way along (or 2/3rd of the way along) will produce the 3rd partial, ¼ of the way along (or ¾) will produce the 4th partial, and so on. Mark the points on a bass string with a piece of chalk or tiny piece of tape. Practise touching the points and playing the note, in sequence, to produce the sequence of partials in the harmonic series. And then practise listening for the same partials when the string is not touched.

Learn where to listen for the adjustable, beating partials of tempered intervals, and then practise listening to them by just playing the intervals on the piano. Preferably do this on as many different pianos as possible.

All pianos are different in their tonal and tuning characteristics, and some instruments will be found to be "easier" than others, for hearing partials. Eventually, with practise and experience, partials will always be found to be easily audible, but still more so in some instruments than in others.

Finally, it is possible to train without a piano, using headphones (or good speakers) and electronic resources. Piano tones can be recorded and frequency filtered, and the partials can then be heard in isolation. Using such resources it is possible to compare the perception of the partial when listening to the whole tone, with the actual filtered partial, in isolation.

What we have been talking about is training for *aural technique*. Good technique is essential for any art (or sport, for that matter). Practise the technique sufficiently, and the art of hearing partials accurately and reliably will become second-nature. Without the technique of aurally "filtering" individual partials from the spectrum, the perception is subject to all kinds of subjective aural illusions, which can mislead. For the musician, this is not necessarily so important, but the piano tuner has a technical, precision task of sound manipulation to perform, and needs to penetrate with greater depth into the structure of the sound.

The technique can be likened to the master painter's technique of analysing colour content in paint on the canvas. It is often not a merely question of making a colour "better" or "brighter". The artist, with knowledge of colour theory, can see, for example, that the purple needs, say, more blue, and less red. The artist can see specific colour content in the overall colour. So it is with tempered intervals, where the "colours" are the partials. Just as the painter can also see the overall colour mix and its effect, as one colour, so the piano tuner can also hear the overall interval quality, as a musical quality.

Solid aural technique opens up the world of piano tone. It enables one to hear what is actually going on in the soundscape. It eliminates uncertainty, guessing, and imagination. It is not something that will come without effort. If you want to have really good aural skills, you have to put in the time and dedication.

The scale tuning sequence

The tuner's "datum point" for the entire instrument is the "pitch" note, the first note "tuned to the fork". This note is tuned to the tuning fork, and then tuning expands outwards from this note. Details about tuning the first note to the fork are given in the chapter on *setting the pitch*. We can only tune one note at a time, so there has to be a *tuning sequence* of notes tuned in order to tune the scale (this has sometimes in the past been called "laying the bearings"). There is no single "correct" *tuning sequence* – different tuners use different tuning sequences, and some tuners will use different sequences on different occasions.

The master tuner will be able to use any sequence, but will probably have a favoured sequence. The tuning sequence will cover at least all 13 notes of a central octave in the piano, typically F33 to F45 (F3 – F4), but not always. We might, for example, tune the scale between G35 to G47 (G3 –G4). The starting note is typically A or C, but an expert tuner would be able to start from any note. The standard tuning fork notes are A440 Hz or C523.3 Hz. The C fork is pitched one octave above middle C, at C52 (C5) and the A fork corresponds to A49 (A4). It is of course also possible to tune a piano imprecisely without initially setting a scale, but only by first setting a complete scale in the most frequently used and most exposed part of the compass – the centre - is it possible to reach the required level of tempering precision for all the scale intervals within the scale, and throughout the rest of the compass.

Some example sequences are given further below, but all sequences require a pre-requisite knowledge of the "target" beat rate for each interval, which is the starting point. These are provided by the "traditional" theory calculations, and can be found in the table in Chapter 5. The rough "rule of thumb" that is sometimes useful, for A 440 Hz or C 523 Hz "pitch standard" is as follows:

1) The perfect fifths beat *narrow* at around 1 beat every two seconds.
2) The perfect fourths beat wide at around 1 beat per second.
3) The major third F33 (F3) – A37 (A3) beats *wide* at around 7 beats per second, and each major third above beats wide 0.5 Hz faster for each semitone step in the scale it lies above F-A. Thus F#-A# is 7.5 Hz, G-B is 8 Hz, and so on.
4) The major sixth F33 (F3) – D42 (D4) beats *wide* at around 8 Hz, and each major sixth above beats wide 0.5 Hz faster for each semitone step in the scale it lies above F-D.

This is a *rough* guide that is *easy to remember*, and the purpose of this chapter is to illustrate all the ways in which fine tuning goes beyond this rough guide. Going beyond the rough guide is not a question of saying that we need to tune beat rates to greater accuracy than the 0.5 Hz the above mnemonic works to. The rounding off to 0.5 Hz steps is not the mnemonic's weakness. The weakness in *any* set of specified beat rates lies in the false idea that good tuning can be ultimately defined by *exactly* prescribed beat rates, and also in the equally false idea that beat rates in natural piano tone are always constant, and therefore specifiable as a precision rate.

Specified beat rates are an invaluable *starting guide*, but not an absolute solution to the problem of fitting beat rates together in their complex network. They are a pretty good map for the solution, but they do not show much "detail on the ground". There is no reason to suppose real piano tone should somehow not be what it is, but should aspire instead to emulate the hypothetical tone properties of *ideal* strings, and there is equally no reason for any piano tuner to feel obliged to exactly reproduce the beat rates associated with *ideal* strings. The task of tuning beat rates in real piano tone intervals is more sophisticated than this.

The tuning sequence takes us through tuning all the notes, one at a time, so that the intervals formed between the tuned note and previously tuned notes are all good. This is quite a difficult task mainly because of the way the intervals and their beat rates are all interconnected in a *complex network*. If you are reading this and do not already have a good grasp of this essential principle, then now would be a good time to break off and read *Piano tuning – the basics*, in Chapter 2.

Fast and slow intervals

A scale contains two types of tempered intervals that are actively tuned – "fast beating", from around 7 Hz to 14 Hz and "slow beating" from around 0.3 Hz to 1 Hz (and sometimes a little faster). Perfect tempered fourths and fifths are called "slow" intervals, whilst thirds and sixths (both major and minor) are "fast". Fifths are regarded as slower than fourths. We may tune both major and minor thirds and sixths, but we do not generally use augmented or diminished fifths and fourths as tuning intervals. Usually, of the intervals that are used for tuning the scale, minor thirds and minor sixths are the least used, and most of the scale creation arises from the relationship of tempered perfect fourths and fifths, major thirds and major sixths. Certainly, the major fast intervals cannot be compromised for the sake of the minor ones, because of their pronounced effect over the rest of the compass once the octaves are tuned.

Beating partials in the tuning intervals

The scale intervals are tuned by adjusting beat rates in the *adjustable partials* of the intervals. Consider just two strings, tuned a tempered interval apart. Some of the partials of one string, when the string is sounding on its own, will have approximately the same frequencies as certain partials of the other string, when that string is sounding on its own. These are at the positions of the coinciding notes in the harmonic series, that we dealt with above. These shared frequency regions can be called frequency *meetings* of partials of the two strings. When both strings are sounding together, a new spectrum from the tempered interval is created, but it will have *adjustable partials* at frequencies that are in the same regions as these *meetings*. Adjustable partials carry the beat rates that we can adjust by altering the string tension.

Every tuning interval has more than one such adjustable partial in its spectrum. The *first* or lowest adjustable partial (the one closest to the fundamental of the interval's lowest note) has the lowest musical pitch, and the lowest frequency of all the adjustable partials. This is the adjustable partial whose beat rate "traditional" theory attempts to quote. The second adjustable partial will sound an octave above the first, and it will have, according to the model adopted by "traditional" theory, twice the beat rate of the first one. In practice, the ratio between these beat rates can generally be greater than 2:1, due to inharmonicity and other factors. But we could still find the second adjustable partial, if we wanted to, at a pitch an octave above the first, with practical accuracy.

As already mentioned, further adjustable partials exist above the second, but fortunately this does not complicate things too much. In *all* tempered intervals we are concerned with the *first* adjustable partial, but we only really need be also concerned with the *second* adjustable partial, when tuning the tempered fifth (and very occasionally in the fourth). The second adjustable partial of the tempered fifth is usually prominent, and can often be *more prominent* than the first, so it is very important to take it into account. It can easily have a detrimental effect on the tone of the fifth if it beats too fast. On the other hand, the tempered fifth's third adjustable partial is invariably much less prominent than the least of it's first two, and its psycho-acoustic effect is any case less severe. In the tempered fifth, then, priority is not necessarily always given just to the first (lowest) adjustable partial.

In all the other tempered intervals, the second adjustable partial is considerably higher in the interval's total spectrum than it is in the tempered fifth, and it is generally much less prominent than the first adjustable partial, and so is not generally significant in tuning. Priority must always be given to the first (lowest) adjustable partial. Occasionally, the second adjustable partial of a tempered fourth *is* about as prominent as its first adjustable partial, however, when this is the case, it is still not generally wise to give it priority. Both adjustable partials can be electronically filtered from recorded fourths, and heard simultaneously in isolation. Within the fine tuning range, the decay patterns of the second adjustable partial in fourths generally contain significantly more irregularity than those of the first adjustable partial.

We can examine some spectra of tuning intervals on a 1930 Model M Steinway grand piano, to see some characteristics. Fig. (9.3) shows, for comparison, the fifth A37 – E44 (A3 – E4), the fourth A37 – D42 (A3 – D4), and the major third A37 – C#41 (A3 – C#4). Each vertical peak line represents a partial of the overall sound. The first and second adjustable partials are indicated in each case. The vertical axis represents "spectrum density", the height of each peak line giving an indication of the audible prominence of that partial. It does not indicate how fast the partial decays, but in this region the partials have roughly the same decay rates. Frequency is represented along the horizontal axis, so each peak indicates the presence of a partial at the frequency given at that point on the horizontal axis.

Fig. 9.3A

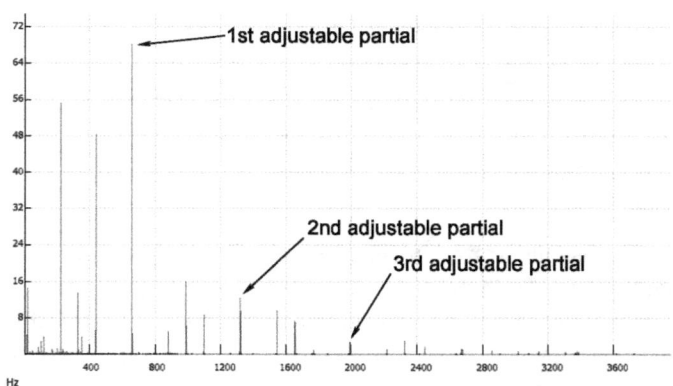

Fig. (9.3A). The spectrum for the fifth A37 – E44 on a model M Steinway.

Fig. 9.3B

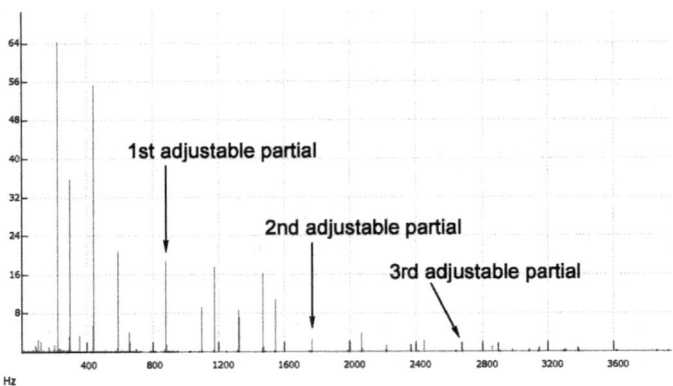

Fig. 9.3B. The spectrum for the fourth A37 – D42 on the model M Steinway.

Fig. 9.3C

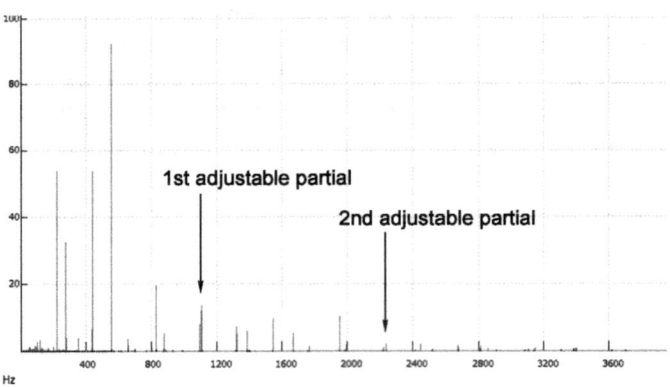

Fig. 9.3C. The spectrum for the major third A37 – C#41 on the model M Steinway.

In each case illustrated in Fig. (9.3), the first tall peak (on the left) is the fundamental of the interval's lower note. To its left, in a lower frequency range is small group of frequency components associated with the *transient*, or noise as the hammer hits the strings, and other ambient noise sources. These give a rough indication of "noise" level. To be clearly distinguishable and useful for tuning, a partial peak needs to rise considerably above this height. The 3rd adjustable partial in each of these examples does not significantly achieve this, and in the case of the tempered fourth and the major third, the 2nd adjustable partial also fails to

345

achieve this. This will not necessarily *always* be the case, but the examples shown are fairly typical.

We can also see some typical characteristics from the decay curves of the adjustable partials, in the example Fig. (9.4), which shows the first and second adjustable partials for the fifth A37 – E44 (A3 – E4) and the fourth A37 – D42 (A3 – D4) on the Steinway.

Fig. 9.4

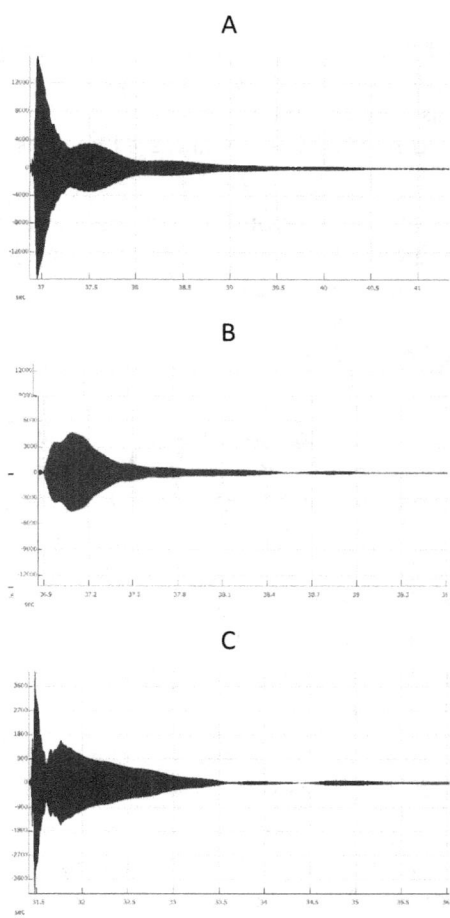

D

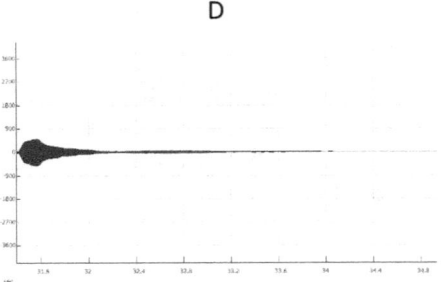

Fig. 9.4. A and B: First and second adjustable partials for the tempered fifth A37 – E44 on a model M Steinway grand piano. C and D: First and second adjustable partials for the tempered fourth A37 – D42 on the same instrument. The second partial of the tempered fourth is much weaker. This is typically the case.

The first adjustable partials of the both the fifth and the fourth (A and C) begin at larger amplitude than the second adjustable partials (B and D). Of the four, the second adjustable partial of the fourth (D) is significantly weaker than the other three. The first and second adjustable partials of the fifth, and the first only of the fourth, are about equally strong.

The perfect fifth, of all the *tempered* intervals, is the interval to which the musician's ear will be most sensitive, as far as its tuning or intonation is concerned. Many musicians are used to tuning perfect fifths in ensemble, or on stringed instruments of the violin family. An increase over the normal beat rate, of say, 1 Hz in *an isolated* major third or sixth in the central octave of the compass, would probably not be noticed by a good musician. The same change in a fifth could wreck its tuning quality, and might well result in a beat rate in the second adjustable partial of 4 or more beats per second. On an otherwise well tuned, quality instrument, good musicians might well feel it was "out of tune".

Beat rate relationships

The canonical aim in tuning the scale, in terms of beat rates, is to obtain good beat rate relationships between the intervals. When this is achieved, all intervals of any kind (semitones, whole tones, minor thirds, etc.) will exhibit the same musical tempering quality. This means (a) all intervals of any kind will have the same relative pitch intonation, and (b) their *tone qualities* as affected by beating will graduate smoothly across the compass.

Correct beat rate relationships will produce *progression* of beat rates. This is especially important in the thirds and sixths, because of the effect this will have on the rest of the compass when the octaves are tuned.

The scale is comprised of 12 sizes of interval ranging from the semitone (the smallest) to the octave (the largest), and each kind of interval appears at 12 different note name positions, that is, it appears with its lowest note on 12 different note names. If we were to consider all 12 of these, we would have to consider more than one octave's worth of the compass for tuning the scale. The minimum number of beat rates that comprises the entire system for a tempered scale is therefore 144, and these intervals involve notes necessarily extending beyond the central octave's length of the compass.

Piano tuning in practice generally involves "configuring" this system not by considering all 144 beat rates, and attempting to get all their relationships "correct", but by concentrating on, typically, say 27 to 30 tempered intervals within an octave's length of the central compass, and then considering up to around, say, 20 further tempered intervals in the process of tuning octaves outwards from the first 13 note chromatic scale tuned.

For example, a scale of 13 notes may be tuned between F33 – F45 (F3 – F4), inclusive, by tempering all major thirds, perfect fourths, perfect fifths, major sixths, and perhaps some minor thirds and sixths, that fall within this area. Then in tuning outwards in octaves from the initial scale, the tuning of the tempered intervals consequently formed, e.g. fifths, fourths, major thirds, major sixths and major tenths, is also considered. Tuning downwards into the bass, minor thirds can be also be important. Different tuners will have different approaches.

In scale tuning, tuners do not generally consider the beat rate relationships between the semitones, whole tones, tritones, sevenths, *etc.* One *could* do so, but not at the expense of attention to, for example, perfect fifths, perfect fourths, and major thirds. What generally happens is that the tuning of the initial scale of 12 notes (13 including the octave) is achieved by tempering, say, 35% of the 78 intervals that "define" the overall tuning condition. The reason this works adequately is that the 78 intervals are not independent of the rest of the scale but are *inter-related* with each other and all the intervals that are not directly tuned. The entire temperament system, in practice, is an acoustical *network* of beat rates. Once the correct, initial 35% of beat rates are properly related to each other, the rest of the intervals tend to "fall into place", because of the nature of the network. Less than this percentage will tend to produce rougher results.

In tuning the scale, there are some essential beat rate interrelationship rules to be appreciated, in addition the beat rates themselves, *as specified by "traditional" theory*. The most useful of these are:

- Beat rates should *progress*, that is, a beat rate especially in a major third or sixth should not be slower than the beat rate in the next lower adjacent interval of the same kind, and should not be higher than the beat rate in the next higher adjacent interval of the same kind. This should, however, take into account that beat rates themselves may be ambiguous, and change throughout the decay time.

- Beat rates should double (or more than double with inharmonicity and controlled octave stretching) over an octave, that is, for example, C52 – E56 (C5 – E5) should have twice the beat rate of C40 – E44 (C4 – E4).

- In "traditional" theory, a major sixth should have about the same beat rate as the major third whose lower note is one whole tone above the lowest note of the sixth.

- In "traditional" theory a minor third should have the same beat rate as the major third whose lower note is a perfect fourth above the lowest note of the minor third.

- An approximate beat rate guide for the major thirds can be obtained very easily by remembering that G35 – B39 (G3 – B3) has a beat rate of (approx.) 8 Hz. An estimate of the length of a second is easy to learn, and the division of a second into 8 to obtain 8 Hz is relatively easy, by dividing a second into four, and then doubling the speed. Beat rates in the major thirds then increase by (approx.) 0.5 Hz for each rising semitone position of the major third in question, or similarly decrease by 0.5 Hz for falling thirds. For example, G#36 – C40 will be 8.5 Hz, and F#34 – A#38 will be 7.5 Hz. These figures are not precisely those specified by "traditional" theory, but they are easy to remember and close enough. In practice, beat rates in the thirds even on quality instruments typically vary by around 0.5 Hz during the decay period, so citing figures to 2 decimal places is relatively meaningless. Inharmonicity, sound board effects and falseness can also cause significant changes to the beat rates.

- The "rule of thumb" for the tempered perfect fifths and fourths is that each fourth should have about 1 Hz beat rate, and each fifth around 0.5 Hz beat rate, at the lowest adjustable partial. This may seem at variance to "traditional" theory. The whole issue of beat rates in the fourths and fifths is more involved, and is dealt in detail later. High levels of inharmonicity change over the compass will mean that the fourths will in general need more than 1 Hz beat rate, and on smaller pianos perhaps as much as 3 Hz. The fifths may need to be faster or slower, depending on the relative prominence of the two adjustable partials, and depending on inharmonicity, bridge coupling effects and falseness (discussed below). The beat rates tuned in all the fourths and fifths should be psychologically regarded as necessary only in order to avoid excessive beat rates appearing in any one.

- Major thirds are destined to have fast beat rates if they are to be equally tempered. This cannot be avoided, because of the influence of the comma, as set out in temperament theory. Thirds that are too fast will always accompanied by thirds that are too slow, and *vice versa*. This too, is a consequence of the basic theory. Correct progression of beat rates is therefore of paramount importance, if the tempering is to be *equal*, and the tones of the thirds consistent. This then later ties in with consistency of tone in the tenths and seventeenths, when tuning the octaves. Never listen to a beat rate as an abstract thing. The beat rates in the major thirds, provided they are not faster than they should be, should be psychologically regarded *as having aesthetic merit in their own right*.

Tuning tempered perfect fifths: harmonic ratio 3:2

<u>Musical Pitch relations of adjustable partials</u>

First adjustable partial – Listen a perfect fifth above the upper note.

Second adjustable partial – Listen an octave above the first adjustable partial.

The first adjustable partial of the tempered fifth occurs where the 3rd partial of the lower note meets with the 2nd partial of the upper. This has a musical pitch a perfect fifth above the upper note of the fifth. In the second adjustable partial the 6th partial of the lower note meets with the 4th partial of the upper. This has a pitch an octave above the first adjustable partial. These are the pitches at which the adjustable partials will be heard.

Tempered fifths are likely to deviate noticeably in their beat behaviour from the predictions of "traditional" theory. The beat rates for fifths given by "traditional" theory range from 0.6 Hz to 0.8 Hz in the scale F33 to F45 (F3 –F4), and these beat rates are calculated for the *first adjustable partial* only. The theoretical beat rates in the second adjustable partial will be twice these values, or more in practice, because of inharmonicity.

There are two broad areas to consider in the tuning of fifths, relating to how their actual behaviour can differ from that suggested by "traditional" theory. The first concerns the rates we should employ. The second concerns the nature of so-called "beats" themselves.

Beat rates for the fifths

In the tempered fifths, both the first and the second adjustable partials are generally prominent. Sometimes the second adjustable partial is *more prominent* than the first. With no inharmonicity or other complicating effects, the beat rate in the second adjustable partial would be twice that in the first. This is simply due to the fact that the frequency difference between the component meeting partials will be twice as great in the higher position. The beat rate in "traditional" theory is calculated as the frequency difference between the two component meeting partials. The calculation is based on the assumption that the partials are in an harmonic series, so that the 6th partial of any string will be twice the frequency of its 3rd partial, and the 4th partial will be twice the frequency of the 2nd.

Inharmonicity, caused by natural string stiffness, causes the partials to have progressively higher frequencies, the higher the partial number. Taking into account inharmonicity, the beat rate at the second adjustable partial of the tempered fifth will be *more than twice* the beat rate in the first adjustable partial. Depending on the degree of inharmonicity, the beat rate in the second adjustable partial may even be three times or more that in the first adjustable partial. If, however, the fifth is tuned so that it is *wide* at the first adjustable partial, it is possible for the beat rate in the second adjustable partial to be *slower* than that in the first, depending on the inharmonicity and degree of widening.

It is questionable whether "traditional" beat rates in the first adjustable partial as fast as 0.6 Hz to 0.8 Hz, are actually acceptable in the fifths of a good instrument, and questionable whether any expert artist tuners really apply such rates. Many tuners I have observed believe they are applying the "traditional" theory's 0.6 Hz to 0.8 Hz to a fifth, when in actuality they are often applying this rate to the second adjustable partial, not the first. This means the actual rate being applied to the first, is considerably *less* than 0.3 Hz – 0.4 Hz, given the "traditional" theory's ratio between beat rates in the first and second adjustable partials. This is particularly common amongst students of tuning.

Does this mean we are making a mistake when we do this? It is definitely a mistake as far as applying "traditional" theory is concerned, but as far as what *ought* to be done is concerned, I believe the answer is "not necessarily". Applying the range of beat rates 0.6 Hz to 0.8 Hz to the first adjustable partial, will result in beat rates of *at least* 1.2 Hz to 1.6 Hz in the second adjustable partial, and possibly 1.8 Hz to 2.4 Hz or even more, taking into account inharmonicity. This may or may not be acceptable, depending on how prominent the second adjustable partial is, but I think in most cases on good instruments which a clear second adjustable partial, it is *clearly* unacceptable, *and is in any case entirely unnecessary*. The reality is that on good instruments beat rates as large as this are not necessary in the fifths, in order to create a successful scale with properly progressing thirds.

Certainly, to my ear, both as a tuner and a musician, beat rates as fast as those dictated by "traditional" theory *and applied to the correct partial for which they have been calculated*, are, on a good instrument, unacceptably fast, precisely because they produce unacceptably fast beating in the second adjustable partial. Which is often more prominent than the first. They make the tempering in fifths far too unpleasantly obvious. As I have said, good tuners whose work I have observed often believe they are applying the "traditional" rates when in fact the true "traditional" rates are much faster than those they are actually tuning. This is relatively easily established by recording the tuned intervals and isolating the partials electronically. The actual beat rate can be established by direct measurement of the resulting envelope.

Beat patterns in the fifths

The second area to be considered concerns the nature of the beat phenomenon itself. In "traditional" theory, a beat pattern is regular, and it has a fixed beat rate. In practice, not only is the beat pattern not necessarily one with a regular beat *rate*, but the beat pattern itself may exhibit its own *decay pattern*, meaning, the *beat amplitude* (and hence its audible prominence) may grow or diminish over the course of the partial decay.

Rather than expecting to always hear a beat that remains equally prominent throughout the partial decay time, we should *expect* variation. "Traditional" theory proposes that as a beat rate is reduced, the beat remains a regular beat, and is not affected in any other way. In contrast, Fig. (9.5) shows the basic, simplest, theoretical prediction for typical beat patterns, using contemporary theory. The graphs show the typical, generic type of behaviour, when the natural dissipative effects of the soundboard are taken into account, as *decay*. These effects are ignored by traditional theory, even though the whole point of the soundboard is, in effect, to cause decay.

"Traditional" theory, then, does not take into account that partials *decay*, and when decay is included in the theory, there are certainly different results. The Fig. (9.5) graphs show what might typically be expected as the beat rate is reduced, based on theoretical calculations considering the strings, soundboard and bridges as one system.[65] The graphs have a logarithmic vertical axis, so the normal exponential decay of a beatless partial would appear as a descending straight line. The graphs are therefore a good visual representation of how we might psychologically *hear* the pattern. The sinusoids that make up the beating partial in this illustration have constant decay rates and frequencies. In a real piano, these frequencies and decay rates may not necessarily be constant, introducing more complication, and possibly an *irregular* decay pattern.

Note that in **A**, the beat amplitude (depth of the beat, corresponding to its audible prominence) *grows* before it decays. In **B**, the beat amplitude decays slightly faster than in **A**. In **C**, the beat amplitude decays so fast, that only one real beat appears, followed by a more or less "beatless" remaining decay. The illustrations show how it would be difficult to slow the beat rate to much less than about 1 beat per second, without it disappearing altogether. The effect in practice would be that the partial would be relatively easy to tune beatless, and relatively easy to

---

[65] The method for generation of the graphs is to find the complex eigenfrequencies of the Weinreich dynamical matrix, or an extended version of it, representing the non-conservatively coupled oscillator system. Because the matrix is normalised, the whole generic range of decay curve possibilities for *relative* parameter values can be found without the need to know specific values.

tune to tune, say, with 1 beat per second, but a beat rate in between, of one beat every two seconds, would be elusive, or unattainable.

Fig. 9.5

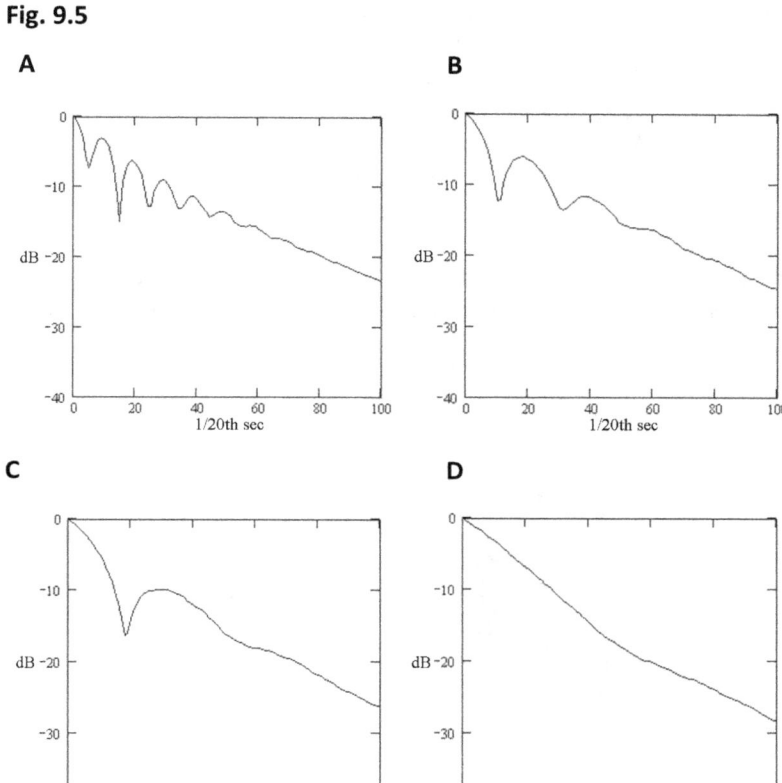

Fig. 9.5. Theoretical decay curves, assuming "realistic" bridge coupling for the first beating partial of a tempered fifth as tempering is reduced. The bridge parameters in this example are such that a *regular beat* pattern does not persist as tempering is reduced. In **A** the beat rate is 2 beats per second and there is a *regular beat* pattern decay. In **B** the beat rate has been reduced to 1 per second. There is still what would be perceived as a *regular beat* decay but the beat rate is too fast for a fifth in the scale area. Reducing the tempering still further, at **C**, results not in a slower *regular beat* but in a *single null* pattern – there is what appears to be a distinct "beat" reaching a maximum at about 1.5 seconds which is followed by what amounts to a *beatless decay*. Listening to the first part of the decay we might conclude a beat rate of about 1 per second, but listening further into the decay time the

interval will seem "beatless". At **D**, the mistuning between the components is 0.5 Hz, and the beat has all but vanished.

The patterns in Fig. (9.5) are of course theoretical, rather than actual. Actual beat phenomena in the adjustable partials of a fifth follow very similar "rules" but can show further complications, notably, ambiguity of beat "rate". Fig. (9.6) shows the behaviour of the first adjustable partial from the fifth A37 – E44 on a model C3 Yamaha grand piano, as the beat rate is reduced (the A37 is a full trichord, and one string of the E44 is being tuned to it). Here, and in all the illustrations of actual partials, the vertical scale is not in dB, so an ordinary beatless decay would appear as a smooth exponential curve, rather than a straight line. Time, along the horizontal axis, is in seconds. The displays were obtained from analysis of studio recordings.

Fig. 9.6

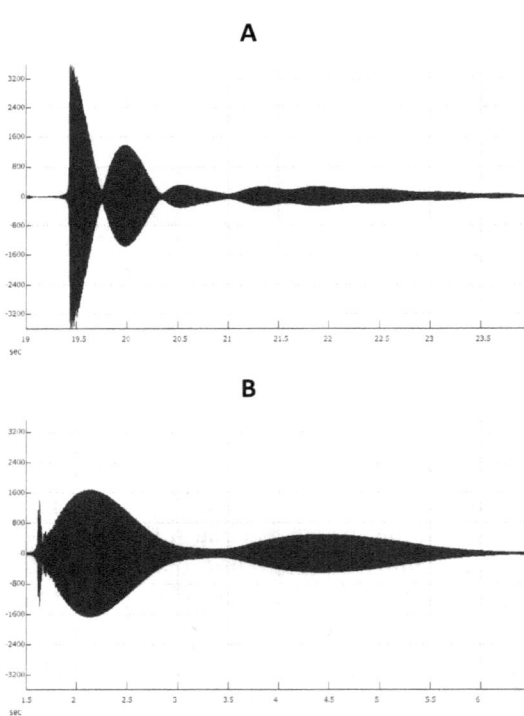

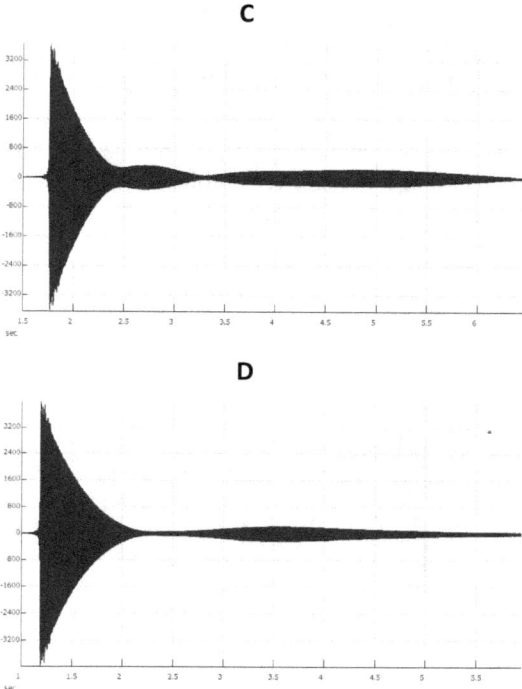

Fig. 9.6. The first adjustable partial of the fifth A37 – E44 on a Yamaha G3 grand piano, as the interval is tuned from beating (A) to as "pure" as possible (D) in four stages. The beat pattern is not regular except at the larger mistuning at A, when the beat rate is 2 Hz.

Notice in Fig. (9.6) that at B there is a "short" beat followed by a "long" beat. Notice also that at C there appears to be a beat rate somewhat less than one per second, followed by a much slower (longer) one of about one beat in three seconds. This is arguably due to the first beat rate having decayed, the second, longer beat being a persistent false beat, which remains present also in D. The E44 string sounding on is own produces a decay curve almost identical in shape to D. The audible beat *rate* is not regular except at A, where the beat rate is relatively fast (about 2 beats per second). Variations from the regular beat pattern are most likely to occur at slower beat rates.

Fig. (9.7) shows the same adjustable partial, but on an older upright piano (manufactured by Jermyn),[66] for various tuning conditions of the

---

[66] The piano is a full iron frame, overstrung and underdamped instrument.

fifth, playing only single strings. In this case notice that rather than the beat rate being always "regular" as it is in Fig. (9.7A), the beat amplitude in B, C and D, varies, causing irregular decay patterns. Where there is a very thin "neck" in the pattern a clear beat is defined. Elsewhere, the beat is shallow and not so clearly defined. B, C and D all show patterns in which an underlying regular beat is discernable, but the decay may sound like a long beat followed by a shorter one.

Fig. 9.7

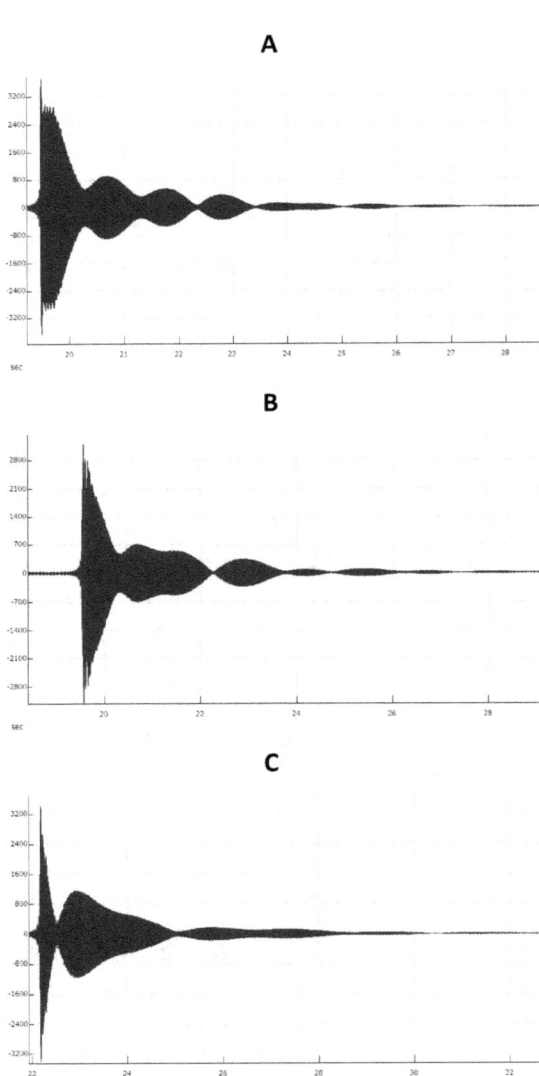

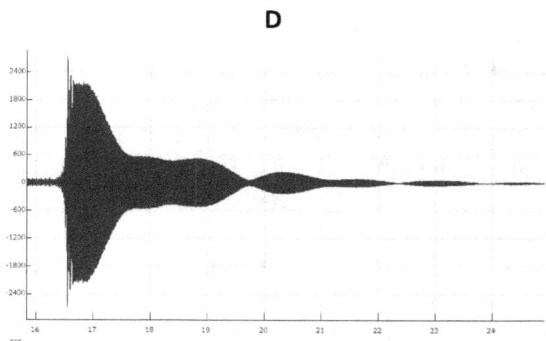

Fig. 9.7. The first adjustable partial of the fifth A37 – E44 on an older upright piano (Jermyn) for various tuning conditions. Only at the A, when the beat rate is about 1 Hz is there a regular beat pattern. At smaller mistuning the "beat rate" is ambiguous due to variations in both beat lengths and amplitudes.

Irregularity of beat rate and amplitude in the beat, may in part be due to falseness, even where strings would not normally have been regarded by the tuner as "false". False beating in a single string within the interval may not necessarily have been previously detected by ear. Even a very slow false beat not immediately noticeable in the course of tuning can be enough to "upset" the decay pattern of the interval. The presence of falseness can mean that the decay curve will have more movement in it than "traditional" theory predicts, or that the decay curve is likely to be *irregular* or to have a dual beat rate. Any apparent "beating" may be faster than we would otherwise expect for a given degree of "mistuning" between the strings, and it may be impossible to obtain a *beatless decay*. In this case, a beat in the partial that is due to falseness, can be mistaken for a beat due to tempering, and actual tempering by the normal amount may lead to an unacceptable degree of beating in the interval.

In theory a rarer soundboard effect emulating false behaviour can occur if the frequency of the partial is close to a resonant frequency of the bridge-soundboard system. In this case, it may be impossible to obtain a *beatless decay* in the interval's adjustable partial. As we "close" the interval from its narrow condition towards expected purity, the beat rate slows, but never disappears, and then increases in speed again as we pass through the point of expected purity, and the interval begins beating wide. This is similar to what happens when one of the strings is false.

The actual behaviour of partials in fine tuning the fifth

As we already said, the first adjustable partial is not necessarily more audibly prominent than the second. It may be that the second, higher adjustable partial is much more prominent than the first. We also saw that these first two adjustable partials are, however, generally more prominent than still higher adjustable partials. The relationship between the beat rates of the first and second adjustable partials is not always a straightforward, fixed ratio, when we are tuning very slow beat rates. For very small mistunings, which are what we are dealing with when we fine tuning a fifth, the relationship of the beat rates can change with the tuning condition or amount of tempering.

Fig. (9.8) illustrates the changes in the first and second adjustable partials of the fifth A37 – E44 on the Yamaha C3 grand piano, as the fifth is tuned from an obviously beating condition, to a best "pure" or untempered condition. The top note of the interval, E44, is a single string in each case, the other two strings of this upper trichord being wedged in the normal way.

Fig. 9.8

1st adjustable partial   2nd adjustable partial

A

0.22 Hz narrow   2.19 Hz narrow

B

0.84 Hz wide   0.17 Hz narrow

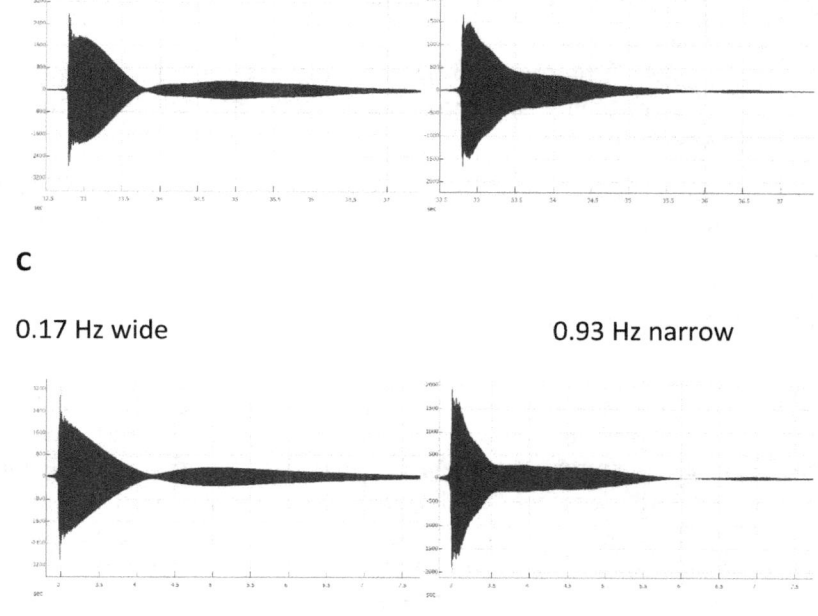

Fig. 9.8. The simultaneous 1st and 2nd adjustable partials of the fifth A37 – E44 on a Yamaha C3 grand piano, for different tuning conditions of the interval. The tension on E44 is raised between A and B, and then lowered a little to C. Each stage, A, B and C, is for one tuning condition, or one tension on the string E44, showing the first adjustable partial on the left, and the second adjustable partial on the right. The "mistunings" in the lower adjustable partial (on the left) are as follows: A = 0.22 Hz narrow; B = 0.84 Hz wide; C = 0.17 Hz wide. The "mistunings" in the higher adjustable partial (on the right) are as follows: A = 2.19 Hz narrow; B = 0.17 Hz narrow; C = 0.93 Hz narrow.

In the left column of Fig. (9.8) is the usual *first adjustable partial* produced by the meeting of the 3rd partial of the A37 with the 2nd partial of the E44. On the right is the *second adjustable partial* produced by the meeting of 6th partial of the A37 with the 4th partial of the E44.

Spectral (FFT) analysis of the tones from the individual notes played separately, reveals the "mistuning" for each of the three cases. "Mistuning" is the frequency difference between the two partials of the separately sounding notes, that meet to produce the adjustable partial of the interval. (The adjustable partial itself will analyse into two or more sinusoids that do not necessarily have the same frequencies as the meeting partials of the individual notes). In terms of this "mistuning", which is actually *defined* as the partial frequency difference when the notes are sounded separately, we can say whether the interval is technically wide or narrow, and by how much, at each of the adjustable partials. This mistuning of course has an influence on other intervals sharing the same notes.

For the first, lower adjustable partial (in the left column), the situation is as follows: At A the interval is narrow by 0.22 Hz, at B it is wide by 0.84 Hz, and at C is it less wide, being wide by just 0.17 Hz. At the same time, according to the mistuning at the second, higher adjustable partial, the interval is *always narrow*, at A by 2.19 Hz, at B by 0.17 Hz, and at C by 0.93 Hz.

In other words, the sequence Fig. (9.8) A, B C, begins at A with the interval narrow at both adjustable partials. At stage B the upper note E44 has been raised, so that the interval is now *wide* at the lower adjustable partial, but is still *narrow* at the higher partial. Finally, at C, the interval is still wide at the lower partial, whilst narrow at the higher one, but the E44 has been lowered a little from its position at stage B. The interval is still wide at the lower partial, but a little less so. Having lowered the E44 now makes the interval a little more narrow at the higher partial, than it was at stage B.

In this example, we *cross* the "zero mistuning point" going from narrow to wide (going from A to B) at the lower adjustable partial. This reduces the beating in the upper adjustable partial from A to B, but it does not cross its zero point, but instead remains narrow. At B and C the fifth remains *narrow* at the upper adjustable partial, but is simultaneously *wide* at the lower one.

In practice we can only *know* a fifth is wide from the fact that the adjustable partial *increases* its beat rate as we raise the tension on the upper note's string. It turns out that in this example, in the interval's *lower adjustable partial*, this increase in beat rate *does not happen* even though we increase the tension. Look at the decay shapes in the left column. At C, we would not even *know* that the fifth is wide at its lower adjustable partial, because the "beat rate" *does not change* between the mistuning of 0.17 Hz wide at C to 0.84 Hz wide at B. We cannot be sure it is wide or narrow unless we have heard a beat rate change in relation to a change in the tension. No such beat rate change happens. We can move from stage A to stage B to stage C or *vice versa* without the lower adjustable partial changing its beat rate. Even if there is some point of tuning in between the recorded stages, at which the "beat rate" or decay curve shape *would have* changed, had we recorded it, the lack of change over the whole range fails to represent the change that "traditional" theory says *should* take place over that range of mistunings.

The shape of these first adjustable partial decay curves, over the mistuning range 0.22 Hz narrow to 0.84 Hz wide, is not the regular beat pattern with a beat rate from 0.22 Hz to 0.84 Hz, that "traditional" theory predicts. In fact, if we gave the vertical axis on these displays a logarithmic scale in dB, we would obtain decay curve shapes very similar to the "single null" shape predicted by the theoretical model for strings coupled by a dissipative bridge. Fig. (9.9) illustrates two such theoretical *single null* curves, together with an example regular beat pattern (dotted) of "traditional" theory. Fig. (5), earlier, also shows how, as the mistuning decreases, and the interval is tuned towards the "pure" or "beatless" condition, the regular beat pattern may gradually transform into a *single null*.

Fig. 9.9

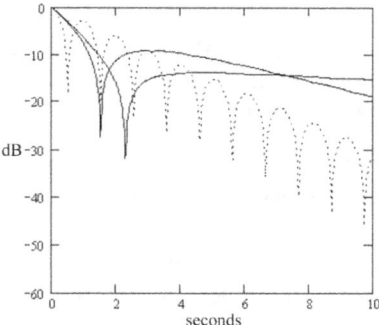

Fig. 9.9. Theoretical decay "single null" curves (for a tempered fifth) together with a "classic" beat pattern of 1 beat per second (dotted line) as predicted by "traditional" theory.

These *single nulls* are the result of the beat amplitude decaying much faster than the partial itself, so that after the first beat null, the beat has disappeared. A similar pattern could also be produced where the "beat" after the null is inherited from a false beat in one or more of the strings. In Fig. (9.5 D), the decay is beatless, but the mistuning is still 0.5 Hz. Perhaps in Fig. (9.8), in all three stages, the lower adjustable partial remains in itself beatless over the range 0.22 Hz narrow to 0.84 Hz wide, the "beat" being due only to a false beat in the E44.

Whatever the case, this illustrates an important principle that sets the actuality of practical tuning well apart from the image of it that "traditional" theory presents. In this example we have actually changed the mistuning at the lower adjustable partial from 0.22 Hz narrow at A, to 0.84 Hz wide at C, *with no significant change to its beat rate*. This in turn means there is a mistuning variation range from zero mistuning to at least 0.84 Hz wide, over which no significant beat rate change takes place. At the same time, the beat rate in the upper adjustable partial *does* change, over part of this range. Even then, between stages B and C, in the upper adjustable partial, there is a *mistuning change* of 0.76 Hz over which, as can be seen, there is not an equivalent beat rate change. By "traditional" theory we would expect a 0.76 Hz greater beat rate at B, than at C. This does not happen.

Clearly, within the fine tuning range, the "beat rate" in either adjustable partial of the fifth does not always simply equal the difference in the coinciding partial frequencies, as "traditional" theory supposes. Qualitative parallels can be seen in this behaviour, with the kind of behaviour associated with coupled unison strings, first shown by Weinreich in 1974. The physics of the situation is, however, somewhat more complicated by the facts that full trichords and notes separated by relatively large bridge distances, are involved.

Even where such effects do not occur, that might perhaps be caused by bridge coupling, the presence of inharmonicity alone means that perfect fifths do not *as a general rule* become truly "beatless" at some fixed tuning condition. As a fifth is made less narrow, the lower adjustable partial will tend to become beatless before the higher one, and can even cross over into a wide condition whilst the at the higher partial the interval is still narrow. If a crossover is made in one tuning movement, which is entirely possible, then the beat rate in the higher partial will change whilst the rate in the lower one may remain unchanged, if it goes from n beats/sec narrow to n beats/sec wide.

How to tell a wide fifth

In scale tuning practice, it is often necessary to ascertain whether a fifth is beating narrow or wide. It is not always necessary to alter the tuning in order to determine this. Firstly, with sufficient experience, one may be able to tell from the pitch intonation. Secondly, because of the effects of inharmonicity just discussed, there is a difference between the beat patterns if it is narrow, and if it is wide. When narrow, *approaching pure*, the rate in the higher adjustable partial is invariably faster than that in the lower. When just wide of being pure at the lower adjustable partial, *i.e.* when the lower adjustable partial just begins to beat wide, the upper adjustable partial will not be significantly faster, may be the same, and may typically be beating *slower* than the lower partial at this point.

To sum this up:

- A tempered fifth can be narrow at one adjustable partial, whilst simultaneously wide at the other.

- Within the fine tuning range of the tempered fifth, one beat rate can sometimes be changed without the other necessarily changing noticeably. This is not always merely due to one beat rate crossing its zero point whilst the other remains on one side of its zero point.

- The tension on the string can be changed over a certain range *without necessarily affecting the beat rate in an adjustable partial*, the interval remaining wide or narrow throughout, at that adjustable partial.

- Within this range, (in the particular example shown) the actual change in mistuning over which there is no change to the beat rate is around 0.8 Hz. This is approximately the same for both adjustable partials.

- Such a change to single string partial frequency, *could* (in the example) produce a 0.8 Hz change to the beat rate in another interval involving that partial, for example, another tempered fifth, or a major third, whilst leaving the fifth practically unaltered in its beat rate.

Plasticity in tuning fifths

The example illustrated is just *one* example. Different instruments, and even every different interval on the same instrument, can be expected, potentially, to behave differently. However, a most important consequence for the tuning of the scale, emerges from this general *kind* of behaviour. The *range* of mistuning over which the beat rate remains more or less unchanged, is in this example around 0.8 Hz. This range is itself *greater than the difference in prescribed beat rates between two adjacent major thirds in the central scale*. It is also *comparable to the beat rate prescribed for a tempered fifth or fourth*.

This means that in tuning the scale so as to obtain all the right beat rates in both fifths and thirds, there is a very significant *plasticity* or "sponginess" in the beat rate relationships between the various intervals. Sometimes we can change one interval without it necessarily having a proportionate change on another sharing the same note, and sometimes even the slightest change on one interval may cause a disproportionate change to another. "Traditional theory" gives no indication of this aspect of scale tuning behaviour, leading to the erroneous idea that we simply need to know in advance what the required beat rates are, and then apply them. The reality of the situation, on the other hand, demands that scale tuning is approached as a kind of sound sculpture, the tuner responding to the medium according to how it behaves.

We will deal with the scale's plasticity in more detail below. We shall also see that the relationship of beat rate and mistuning in the case of the major thirds also, can sometimes deviate in a similar way, from the "traditional" theory.

Practical approach to tuning fifths

There is little point in attempting to tune specific "traditional" beat rates in fifths in the belief that this will automatically result in a good progression of beat rates in the thirds. There is also no justification for leaving a fifth tuned with a desirable beat rate in the lower adjustable partial, if it is at the expense of a prominent beat that is too fast, in the higher adjustable partial. It would be better to have the lower adjustable partial even technically "pure", if necessary, in order to slow an upper adjustable partial that is otherwise too fast and prominent.

What is important in the final analysis, is the overall tone quality of the fifth, and that it allows a good progression of beat rates in the thirds. The only way to reliably and consistently satisfy the criteria of progressive beat rates in the thirds and sixths, and the avoidance of any fifths that are uncomfortably narrow, wide, or noticeably different from the rest, is to work with all the criteria at once, in the process of tuning. We cannot, for example, approach tuning by fifths and regard thirds as "checks" or *vice versa*. Rather, fifths, fourth, thirds and sixths all need to be given equal status in the tuning process, because they are all important musical intervals, and all need to be actively tuned, as well as used as checks against other intervals.

On some instruments with compromised scaling design or other problems, where the rate of change of inharmonicity over the scale is too great, the "separation" between the lower and upper adjustable partials of a fifth may be so severe that in order to reduce the rate in the upper partial sufficiently, the lower adjustable partial has actually to be tuned *wide*, according its beat rate. Even in the example of the Yamaha illustrated in Fig. (9.8), which is a well designed, good quality instrument in good condition, this is indeed what happens, except that the degree of this effect on the Yamaha grand is very small. The effect on many other instruments can be very marked, so much so, that unlike the case of the Yamaha grand, it may be very obvious that we are tuning the interval *wide* at its lower adjustable partial while it is still *narrow* at the higher adjustable partial. It turns out that this is *typical* behaviour on smaller instruments, and on many upright pianos towards the bass end of the long bridge.

Let us just clarify this in a practical description. What we are saying is that, starting from a very narrow fifth, as we slow down the higher partial's beat rate by raising the tension on the upper string, making the interval less narrow, the lower beat rate also slows down. But by the time we have completely "tuned out" or eliminated, as far as possible, the beat in the lower partial, the higher partial is still beating far too fast. By continuing to tune in the same direction, the upper partial's beat rate slows down sufficiently, but the rate in the lower partial has then started to increase again. It is now wide, whilst the at the higher partial it is narrow. Is the fifth then tempered narrow or wide?

The answer depends on what criteria you use to define whether the interval is narrow or wide. It turns out, in other words, that the notion of an *interval* "beating wide" or "beating narrow" is a generalisation that does not always have a very specific meaning, especially in the case of fifths. More accurately, it is *individual adjustable partials* that "beats wide" or "beats narrow", not the interval as a whole. Only to the degree to which the partials happen to *agree* on this, so to speak, is the *interval*, *i.e.* the whole spectrum, narrow or wide.

In practice, on a good instrument we first aim to temper the fifth narrow by giving the first adjustable partial a beat rate of about one every two seconds, or less. The whole picture for the tempered fifth consists of:

- The movement in the first adjustable partial, and its prominence.
- The movement in the second adjustable partial, and its prominence.

- The combination of the two determining the overall "effect" of movement in the tone of the interval.

If the beat in the higher adjustable partial is prominent and too fast we must attempt to slow it. If the lower adjustable partial contains falseness, we may want to reduce the beat rate in the higher one to less than one per second or even no beat, if this is what is necessary in order to reduce the overall movement taking place in the first partial. On any instrument we can allow the higher partial to beat relatively fast, say, three beats per second, provided it is not too prominent, and provided its speed is not due simply to the fact that the lower partial is set too fast, the interval being tempered too narrow.

We must be aware that there is a difference between movement at the beginning of the decay and movement towards the end of the decay. There may be, for example, a beat at the beginning of the decay that seems to have a beat rate of 1 per second, but only one actual "beat" that occurs. After this the interval may seem "beatless", due to the beat having decayed faster than the partial itself. The pattern is not really a beat, but a *single null*. How then, do we know how to leave the tuning of a fifth if there is no single, specific regular beat rate to which we can conform?

In terms of the *art*, the answer lies in the question "How does it sound, musically?" In terms of the technique, the answer lies in the *tuning sequence* and in the principle that a fifth should *never* be tuned in isolation. The decision on how to tune a fifth should not be based purely on the fifth itself, but also on the other intervals that share its notes, *i.e.* fourths, thirds, and sixths. There is also the principle that *movement* (genuine regular beat patterns or otherwise) in either adjustable partial of the fifth should be viewed as "undesirable but necessary", which will force us to always *minimise* it as far as is consistent will allowing progression.

The solution also lies in the *plasticity* of the scale already mentioned. We must have tempering in the fifths, and hence beats, or equal temperament will be impossible to achieve, but we do not need to *want* movement for its own sake. I would tend to regard with suspicion the comments sometimes heard, that the sound of equally tempered fifths, rather than pure fifths, is desirable in its own right. In general, even "pure" fifths do actually beat slowly anyway, due to ubiquitous falseness and other effects, as acoustical analysis invariably shows.

Maintaining the principle of introducing narrowing to the fifths because it is necessary, rather than desirable, we can gauge what is possible in any given fifth, through the use of the *tuning sequence*, because in the scale no one interval exists in isolation, but must "fit into" the whole. In the scale, all the intervals, both the fast and the slow, are jostling for their own "perfection", but each one must consider its place in the whole, and its effect on the other intervals with which it shares its notes. The principle to apply to fifths is one of the least possible movement that does not result in too much movement in some other fifth, or lack of progression in the thirds and sixths.

The perfect fifth is even today a *special interval* in modern music, as far as its tuning is concerned. Fifths are for modern musicians a base interval used in tuning not just by string players tuning their open strings, but also in ensemble, usually with no regard for temperament at all. We *cannot* tune many fifths that are "too pure" and get away with it on the piano, because the network of beat rates will ensure that if we do, other fifths will be beating too fast, and the fast intervals will never progress. So we *do* have to temper them. But we do not have to literally apply the rather extreme directives of "traditional" theory, which would make many fifths much too fast. The pianist should certainly have no feeling that the fifths are anything other than "perfect fifths".

The *tuning sequence* enables us to find how to do this, how to find the best possible relationship between the fifths and the other intervals, for the particular instrument in question. The "traditional" beat rate guide is a pretty good guide, but the tuning sequence itself shows us how each individual piano responds, and shows us *precisely* where to go.

Inharmonicity mistaken for falseness

A very inharmonic tempered fifth may seem false. As we have described above, beginning with a narrow fifth, we may find we can slow the beat in the lower adjustable partial until it stops, but the higher partial's beat continues too fast. If we slow the lower beat to zero, the higher beat may still be too fast, in some cases. We can slow the higher beat further only by-re-introducing a beat in the lower adjustable partial, *on its wide side*. This way we can arrive at suitably slow beats in both partials, provided we understand this relationship between the upper and lower partials' beat rates. The relationship is now contrary to "expected normal" beat behaviour, and contrary to "traditional" theory. If we are not focussed on the specific adjustable partials, but only on the *quality* or the *beat effect* of the interval, we may erroneously conclude one of the strings is false, because there is an aspect of the beating that *seems* uncontrollable and inescapable.

If we are *expecting* the higher beat to always get slower as the lower one gets slower, as per "normal theory", then our apprehension of what is going on gets confused – and we conclude falseness. In fact, the situation is probably much more controllable than it would be if it were just due to falseness, but to take control we must be *aware* that decreasing one partial's beat rate may be *increasing* the other's. To really control the tuning, we need to listen to the specific adjustable partials and set the "best" compromise position. We need to hear both adjustable partials simultaneously, but as separate partials. Then, when we "back off" with the ear and listen to the overall interval *quality* or *beat effect*, we will invariably be pleasantly surprised at the result.

Understanding the behaviour and controlling it is much more difficult if we *only* listen to *quality* or *beat effect* as we are setting the tuning, as the feedback between how the overall "sound picture" is changing with changes in the lever, is confusing. Clarity comes from analysing the picture into its component partials – that way we can understand the dynamics of what is going on.

Of course, the strings may be false as well an inharmonic! In this case the approach is still much the same. Firstly test to find what is the relationship between the beats in the lower and upper adjustable partials. Knowing this we can put both rates where we want them, within the range of possibilities. We must also "back off" with the ear to hear the interval *quality* and *effect* of the beat rates with various relationships between the lower and upper beats. Every case will be different.

Tuning tempered perfect fourths: harmonic ratio 4:3

<u>Musical Pitch relations of partials</u>

First adjustable partial – Listen two octaves above the lower note.

Second adjustable partial, not usually used – three octaves above the lower note.

The first adjustable partial for the interval of a fourth occurs where the 4th partial of the lower note meets with the 3rd partial of the upper. This has a musical pitch two octaves above the lower note. In the second adjustable partial the 8th partial of the lower note meets with the 6th partial of the upper. This has a pitch three octaves above the lower note. These are the pitches at which the main adjustable partials will be heard.

Fourths are always wide, and the "traditional" beat rate for the fourths is around twice as fast as the fifths. In general, the kind of behaviour described above for the fifths, which occurs at small "mistunings" only, can be regarded as *potentially* applicable to the fourths also, but perhaps less noticeably. In particular, the second adjustable partial is not generally prominent or regular enough to be considered equally useful (see the example, Fig. 9.4). Sometimes it is prominent, but generally is not as important a factor as the higher partial in the case of the fifth. It may be *easy to hear*, but that does not make it *prominent*, or a critical feature of the overall tone quality.

It is acceptable to tune to the 2nd adjustable partial *in addition* to tuning to the first, but not *instead of* the first. The first adjustable partial cannot be simply sacrificed for the sake of the tuning in the second partial. Falseness, inharmonicity, and attenuation, preclude a reliable fixed relationship between the adjustable partials, and because the intended "mistuning" in the fourth is larger than that in the fifth, the relationship is not the same as in the case of the fifth. The same issues, particularly of simultaneous narrowness and wideness, do not apply to the fourth.

The main hazard with the first adjustable partial of the fourth is therefore falseness, rather than its relationship with the second adjustable partial. In other respects, however, the considerations can be similar. The general rule of thumb for fourths (which are always wide) where acute inharmonicity variation is a potential problem, is to temper with a beat rate of the fourth *somewhat more than* one per second, even if this does not seem necessary in the early stages of the scale tuning sequence.

As is the case with the fifths, the answer to the question of precisely what tuning to apply to a fourth, lies not in the fourths themselves, but in the tuning sequence, in the relationship between all the scale's beat rates, and in the scale's plasticity. No fourth's tuning should be considered in isolation. The required tuning quality of a fourth depends not only on the fourth itself, but on the other intervals with which it is connected and compared. The default aim, however, is still for adjacent fourths to be as equal as possible in their beating characteristics. Beyond this default, will be considerations of beat prominence, and tone.

Tuning tempered thirds and sixths

Major third: harmonic ratio 5:4

Minor third: harmonic ratio 6:5

Major sixth: harmonic ratio 5:3

Minor sixth: harmonic ratio 8:5

Musical pitch relations of partials

Major third, *first adjustable partial* – two octaves above the upper note.

Major sixth, *first adjustable partial* – one octave and a fifth above the upper note.

Minor third, *first adjustable partial* – two octaves and a fifth above the lower note.

Minor sixth, *first adjustable partial* – three octaves above the lower note.

Thirds and sixths, major and minor, are "fast" beating intervals. The mistuning in these intervals is large enough that we tend to encounter *regular beat* decay patterns, or "modifications" of *regular beat* decays, such as the variable beat rate that we will now illustrate. We will not encounter anything like a *single null*, provided the interval is somewhere close to its properly tempered condition and not very close to being "pure" or "beatless".

The beat rate may not be constant. Even on good instruments with excellently tuned unisons, acoustical analysis often reveals beat rate variations over the decay of a fast beating interval of around 0.5 Hz or more. This is as much as the change in beat rate from one major third to the next, specified by "traditional" theory.

The beat rate can increase as the overall decay takes place, or it can decrease, or it can "cycle" through increases and decreases. Secondly, as is the case with all beating intervals, the amplitude (or "depth") of the beat can be great or small, and can change during the decay time. In the case of the "fast beating" intervals the amplitude may rise and fall within the overall decay of the interval. In the "slow beating" intervals this kind of behaviour would tend to be perceived as an *irregular decay* or modification to a generic decay curve. In the case of the "fast beating" intervals it can be heard as a "fading" and "reappearing" of the beat rate. Fig. (9.10) illustrates some decay curves, generated from contemporary theory, for the adjustable partials of "fast beating" intervals whose beat *rate* changes when falseness is present. They illustrate in a clear, easy-to-see way, how a beat rate may convert or change over the decay.

Fig. 9.10.

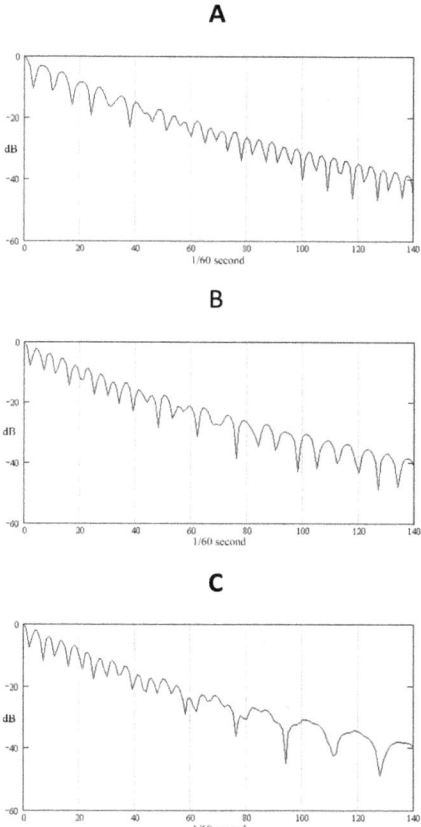

Fig. 9.10. Theoretical illustrations of possible major third beat behaviour. At A, the beat rate increases, whilst at B, it decreases, over the decay time. C illustrates a scenario where the third's proper beat decays away, and is eventually replaced by a much slower beat.

Fig. (9.11) shows some actual decay curves for the adjustable partials of various major thirds in the scale on an upright piano.

Fig. 9.11

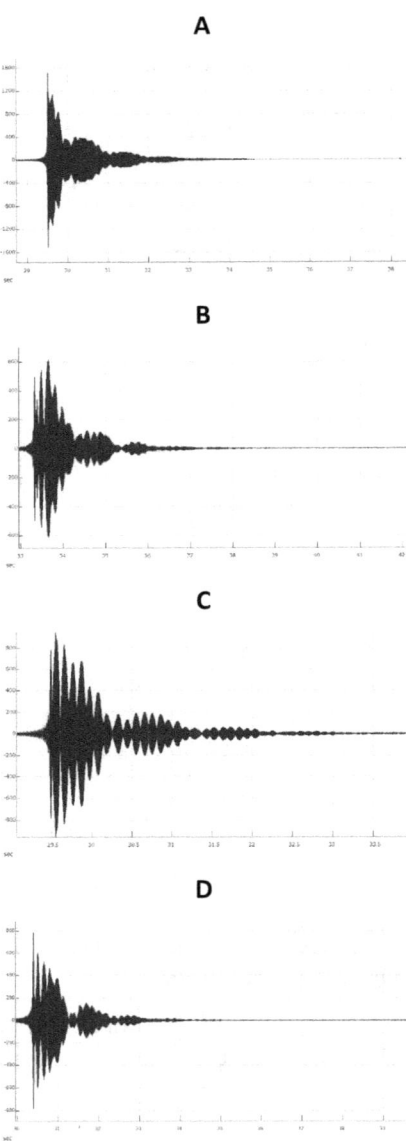

Fig. 9.11. The adjustable partials from four different major thirds in the scale on the upright Jermyn piano. At A, the third's beat rate is very obscured, the slower amplitude modulation (a false beat?) in the partial being much more prominent. The same is true for parts of the decay at B. At C, the beat is

clearer, but still modulated by the slower beat. At D, after the initial rapid decay, the beat becomes very ambiguous before re-establishing itself.

On better quality instruments, the beat rates are generally clearer and more regular, but still not always necessarily with constant beat rate or steadily decaying amplitude. Fig. (9.12) shows the adjustable partials for the major third C40 – E44 on the Steinway model M, and the major third A-flat36 – C40 on the Yamaha C3. The beats are much clearer, but in the case of the Yamaha, an irregularity is immediately obvious. In both cases, there is an additional, slow beat modulation pattern present, and in neither case is the beat *rate* perfectly constant.

Fig. 9.12

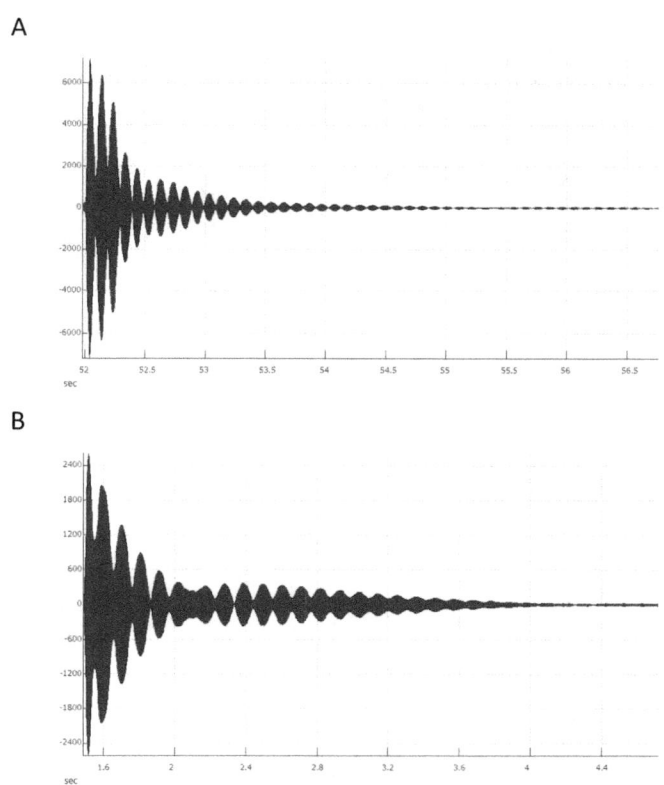

Fig. 9.12. Beat patterns in the adjustable partials of (A) the major third C40–E44 (C4-E4) on the Steinway model M grand piano, and (B) the major third A-flat36–C40 (Ab3-C4) on a Yamaha C3 grand piano. Both exhibit an additional slower beat pattern that modulates the amplitude. In the case of the Yamaha, at (B), the beat amplitude almost vanishes after the initial rapid

decay, but re-asserts itself with the rise of the slower beat. In neither case is the beat rate perfectly constant.

Irregularity of beat rate will of course increase when the unisons are not well tuned. Imperfections in the unison tuning will always be inherited in some way by the tempered interval. On most good instruments with excellently tuned unisons the beat rate will in general be regular and sufficiently close to constant that aurally it will indeed appear to be constant. However, on any instrument natural string behaviour is capable of contributing to irregularity no matter how good the unison tuning may be. Fig. (9.13) shows the adjustable partial of the third $G35 - B39$ ($G3 - B3$) from an older, overstrung, upright piano, in which only single strings are sounding – issues of unison tuning have been excluded. The beat *rate* varies by as much as 1 Hz. Such behaviour is not uncommon in smaller, upright pianos.

Fig. 9.13

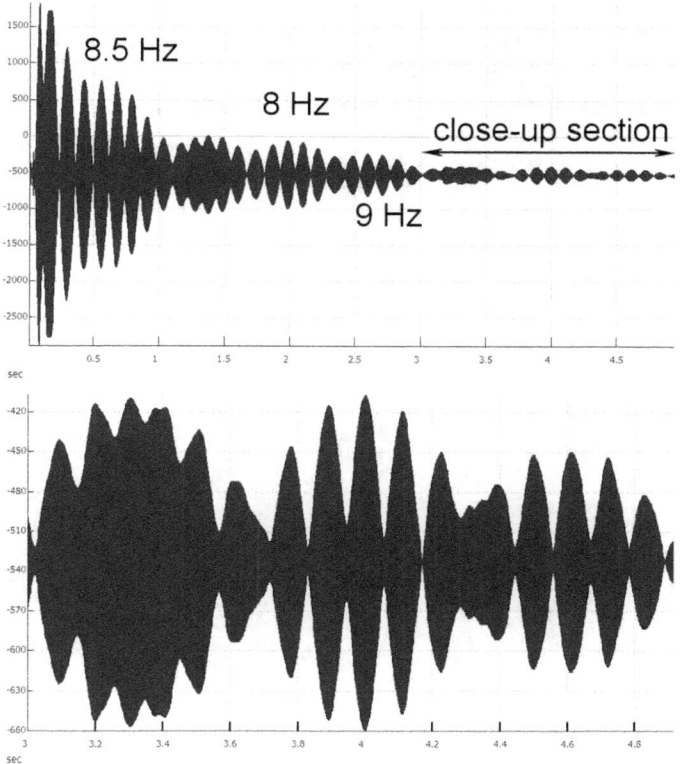

Close-up section

Fig. 9.13. The adjustable partial of the major third G35 – B39 (G3 – B3) on an older, overstrung upright piano. The beat rate varies by around 1 Hz.

Why are the beat rates and beat patterns not stable? The simple answer is that the adjustable partial is not merely composed of two pure tones, sinusoids, or "harmonics" as "traditional" theory supposes. An audible partial *may* consist of only one sinusoid or pure tone, but in general, an audible partial consists of a pure tone *group*. By "traditional" theory we would expect two component pure tones to combine in the one adjustable partial of the major third. Fig. (9.14) shows the actual situation in the example. The FFT shows *four* peaks in the spectrum. It is the frequency distance between the lower *pair* and the upper *pair* of peaks that produces the beat rate proper. We can only *hear* the mix of all four, but each close pair of peaks also contributes the partial patterns illustrated in Fig. (9.14), as ingredients in the overall sound we hear.

Different pianos and different intervals will of course behave in different ways. An adjustable partial of a tempered interval will always consist of at least two component pure tones, but may consist of more. Even an apparently perfectly regular beat pattern that on analysis shows only two peaks in the partial, *may also exhibit variation in the beat rate.*

377

Fig. 9.14

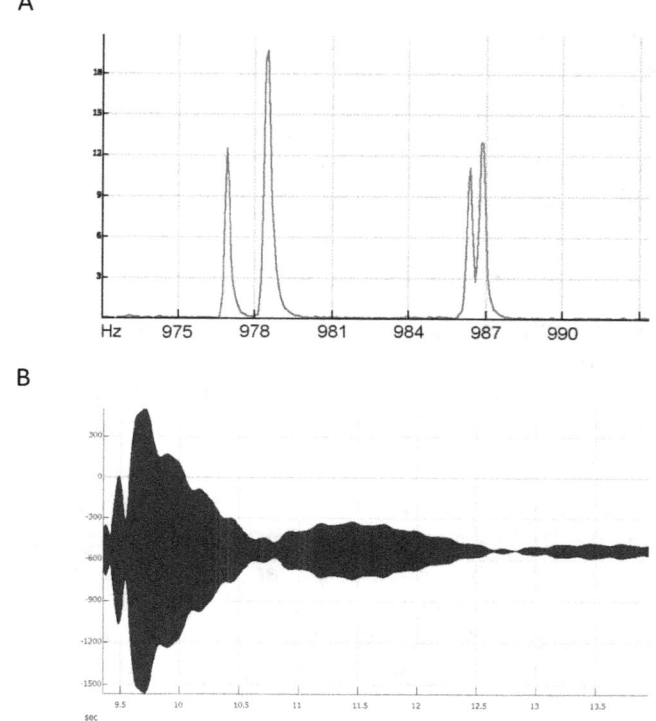

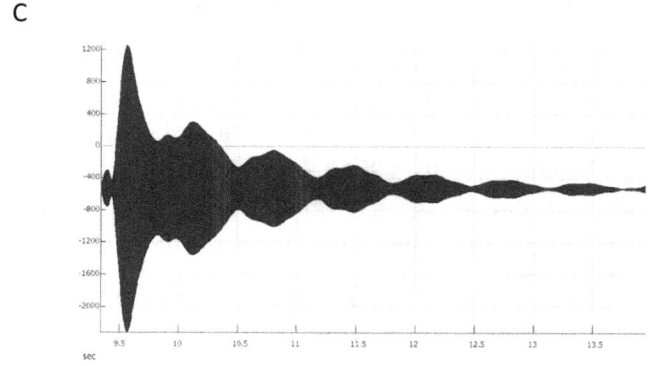

Fig. 9.14. A: Spectrum analysis of the adjustable partial of the major third G35 – B39 (G3 – B3) on the older, overstrung upright piano, shown in Fig. (9.13), showing the "elemental" Fourier components. All four peaks are contained in the one audible, adjustable partial. B: The partial component produced by the lower pair of Fourier components. C: The partial component

produced by the upper pair. The addition of these two leads to the audible partial decay pattern, Fig. (9.13).

Once the complete unisons are tuned and included, some of the characteristics of the single string's partials may be *inherited* by the tempered interval's adjustable partial, but the latter will have its *own* component sinusoids quite distinct from those of the individual strings. In general, including the unison produces a *better* overall result, because of beat *attenuation*, provided the unisons are excellently tuned.

Judging beat rates for thirds and sixths

One can become familiar with beat rates through mentor-guided experience in tuning them, and through access to appropriate teaching resources, as may be available during a formal course of training in tuning. Even without such resources, estimating the beat rate for G35 – B39 (G3 – B3) is not too difficult.

1. Learn to judge seconds. This is *not* an *ad hoc* or "unreliable" technique. There are verbal tricks for this. One example is to count "one Mississippi, two Mississippi", *etc.* at a normal speaking rate. "Elephant" is another suitable word. Practice this against a stopwatch or clock, until the speed of speaking is familiar and reliable. With a little practice judging seconds should become sufficiently good that it becomes possible to judge the length of a minute with precision. You should be able to get to within 1 second per minute. This means you can judge seconds to within less than 2% error. This means your beat rate judgement for, say, the third C40 (C4) to E44 (E4), will be altered by around only 0.2 Hz, by this error, which is generally less than the natural beat rate variation in this interval (which can be around 0.5 Hz or more).

2. Learn to recite the "one, two.." etc., at the same speed, without the accompanying "Mississippi" or other chosen word.

3. Divide the rhythm of counting into four, so that there are now four "beats" to every second. This can be done mentally or verbally using a consonant sound such as "da".

4. Double the rhythm speed again, so there are now eight "beats" to a second. Verbally, one can "flutter tongue" or use a double consonant such as "dubba" can be used. In one second, we would have "dubba dubba dubba dubba".

5. The rhythm of these eight beats to one second is the approximate beat rate for the major third G35 – B39 (G3 – B3), at normal A440 Hz concert pitch.

It is possible to practise with other beat rates using a clock or watch with a second hand (and a little musicianship) or devise similar techniques. If two major thirds such as G35 – B39 (G3 – B3) and C40 – E44 (C4 – E4) are used a reference points, the other thirds can be graduated in between. Remember that as a rough rule of thumb, you can start with F33 – A37 (F3 – A3) with a beat rate of 7 Hz, and increase by 0.5 Hz for each major third a semitone higher in the scale. Thus C40 – E44 (C4 – E4) becomes 10.5 Hz (its "traditional" value is 10.38 Hz).

The generic beat form

What we have seen is that the overall beat patterns in the thirds are often a *mixture* of more than one beat pattern, typically being a mixture of fast beat rates and slow beat rates. If the beat pattern is thought of in this way, its true nature will be easier to appreciate, aurally. Fig. (9.15), the envelope of the adjustable partials of the scale third F-A on a Steinway model M grand piano, illustrates the *typical* generic features in a straightforward way.

Fig. 9.15

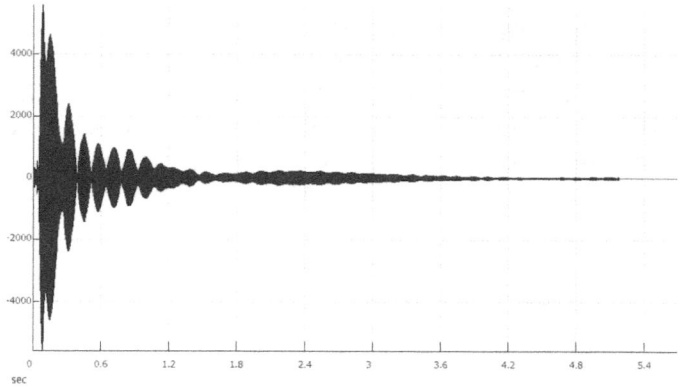

Fig. 9.15. The adjustable partial of the scale major third F-A on a Steinway model M. The beat amplitude decays at a rate different to the overall partial decay rate. The beat amplitude decays irregularly. The overall amplitude of the partial is modulated by a slow beat pattern. This general pattern is very common.

As the partial decays, its amplitude is modulated in a slow beat pattern. The beat amplitude itself decays at a different rate to the overall partial decay rate, and it decays irregularly. A slow beat pattern overall will occur if falseness is present.

The dual or phantom beat rate.

The "best" size of a third with a variable beat rate can often be judged from its "tonal context". The tonal *quality* of tempering in a third (or sixth) as distinct from the beat rate itself, is a combination of both the beat rate and other tone quality factors. On some instruments, a faster actual beat rate will produce the same tonal quality of tempering as a slower beat rate on other instruments. Many tuners will take this other tonal quality factor into account in setting the actual beat rate of the first third. Knowledge of the required "tempering quality" comes from years of experience on a wide range of instruments, enabling the tonal properties of a given piano in its scale region, and its requirements for "tempering quality", to be more or less instantly recognised by the tuner. This "tempering quality" may be taken into account even where the beat rate appears to be perfectly stable.

Occasionally (more rarely), on smaller instruments of lesser quality, we may encounter a "double beat rate" or "phantom beat rate" in the third, sixth, tenth or seventeenth. It is possible for the partial to exhibit a beat pattern in which the individual beats have alternately large and small beat amplitudes. Listening to such a pattern, the ear tends to follow the beats with the large amplitude, and to perceive a rate about half the actual beat rate. Fig. (9.16) shows an envelope illustrating this pattern. The rate still seems ambiguous, aurally, and a change of tuning may cause the envelope to loose this feature.

Fig. 9.16

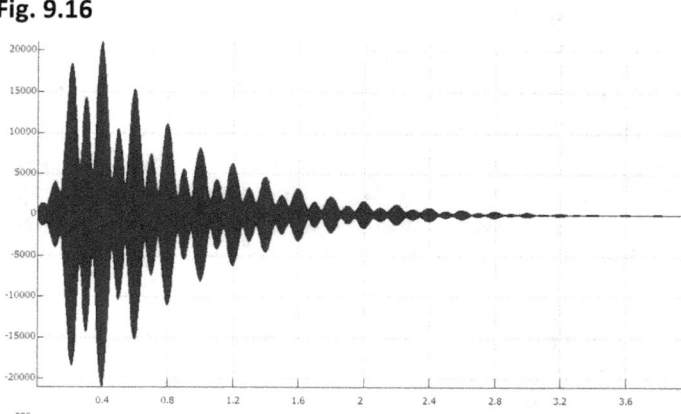

Fig. 9.16. A double beat pattern that is heard as a "confusing" mixture of two beat rates, one twice the other.

Because the dual beat feature can come and go with changes in tuning, it is possible, for example, to raise the tension on the upper string of a tenth, and yet move from a beat rate that is too fast, to one that is acceptable (normally the beat rate would just carry on increasing).

Sometimes, rarely, there also may initially appear to be no *clear* or *definite* beat rate at all in the third, even though the "tempering quality" clearly shows that the third is not "pure" - it *is* wide by approximately the right amount. On closer scrutiny there may be the suggestion of a beat rate that *is* present, but of very small amplitude or prominence. Widening the third further, may show that this beat is indeed present, and that its beat rate increases over the decay time to well beyond what is normal for a third. However, widening still further, introduces another beat that starts to become more prominent, whose beat rate is in the correct order.

There may also, of course, appear to be *two* simultaneous beat rates, one much faster than the other. This can be due to a dual beat, but not exclusively. Sometimes the faster of the two is actually at the interval's second adjustable partial, however, its apparent *pitch* may not seem to confirm this. This can be a psycho-acoustic effect. Situations with such double or phantom beats require careful judgement of "tempering quality", a degree of trial and error, and attention to how the note in question "fits into" the structure of the rest of the scale.

Tuning the scale – the construction principles

It is still best to begin by understanding the "traditional" beat rate model for the equally tempered scale. This is given in the chapter on "traditional" tuning theory. The scale is *basically* tuned by applying the "traditional" beat rates to the intervals, but taking into account the considerations already dealt with above. Whatever the tuning sequence used, one should be aware of the three basic, scale beat rate relationship guides. *These are part of "traditional" theory, neglecting inharmonicity and other complicating factors*. These are:

1. In an octave comprised of an adjacent tempered perfect fifth and a tempered perfect fourth, in any order, the beat rates of the fifth and fourth will be the same if the octave is in tune.
2. A major sixth rising from a given note, will have about the same beat rate as the major third rising from the note one tone above the sixth's lower note.
3. A minor third rising from a given note, will have about the same beat rate as the major third rising from a note a perfect fourth above the minor third's lower note.

We must also take into account the following factors that arise from the *actual behaviour* of partials:

- In a slow interval there may not be a *regular beat* with a clearly definable *beat rate*.
- In a fast interval the beat rate may not be constant.
- The beat rate of the second adjustable partial in the tempered fifth will need to be considered, especially if it is both too fast and prominent.
- Inharmonicity means that that "traditional" beat rates may not apply very precisely, even to *regular beat* decay patterns.

Thus, rather than imagining a scale as an arrangement of notes with absolute "tuning positions" for each note, we consider the *relationship* of each note relative to all the others in the scale. This relationship is in terms of beat rates and decay patterns, and the quality of the interval arising from these. The scale will be successful if all the relationships are "right". What is "right" is still based on the "traditional" model's beat rates but the approach is different. If the physics of the scale perfectly accorded to the "traditional" model of it, we could arrive at a correct final result, through any *tuning sequence*, provided each interval we tuned had the correct beat rate. The purpose of the "check" intervals would then be simply to detect any errors in our judgement of the beat rates we had set.

In actual tuning, the "check" intervals are not really "check" intervals at all, but intervals that are to be actively tempered at the same time as the other intervals. There is no "priority" of intervals, of those that are actively tempered and those that are "checked".

In the finished piano, the tuning "position" of every note tuned in the scale is determined by the tempering in *all* the intervals of which the note is a member. We can represent this diagrammatically as in Fig. 9.17.

Fig. 9.17

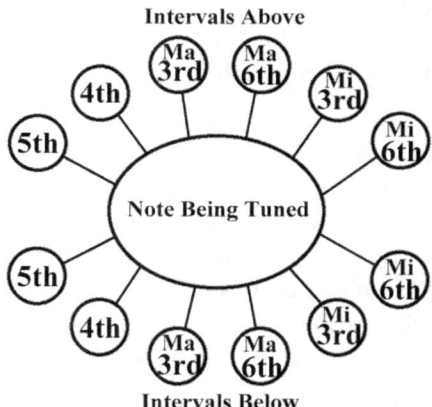

In tuning the scale, not all of the intervals in the diagram will apply to a given note being tuned – only those that are actually already available from other notes already tuned. However, any note being tuned must provide satisfactory tempering in *all* those intervals that do apply. Some tuners consider the minor intervals to be relatively unimportant, and these may not be even be referred to.

Starting from any "pitch note" in the scale area, the number of intervals that apply to a given note being tuned increases with each note tuned. In any completed scale of 13 notes a maximum of 7 tuning intervals and a minimum of 6, within the scale area, can be associated with any one note.

We can list again the ways intervals may deviate from "traditional" theory:

- The tempered fifth may not necessarily exhibit a *regular beat* decay when it is only very slightly narrow or wide, but may exhibit a *single null* decay pattern or other generic patterns that do not have a well definable "beat rate".

- Inharmonicity may result in a situation where the fifth can be narrow at one beat rate, whilst simultaneously wide at the other.

- The effect of falseness in the slow intervals will be to increase the rate of a *regular beat*, or more often to produce an *irregular decay* pattern that does not really have a well definable "beat rate".

- The effect of falseness in the fast intervals will be to modulate the beat amplitude of a *regular beat*. As we discussed above, the beat *rate* may not be constant, and also the amplitude of the beat may vary, other than gradually decaying. The beat rate may cycle from one limiting value to another.

- In the fast intervals there is also the possibility of the *simultaneous dual beat rate*. Look at Fig. (9.10 C). There is a fast beat rate that "converts" into a slower beat rate. But 3 to 5 seconds into the decay (the middle of the graph) both beat rates are present, the faster one "impressed" so to speak, on the slower one. In this particular curve this happens only for a short while, but it is perfectly possible for this kind of effect to last substantially throughout the overall decay.

When there is not a single, genuine, *regular beat* decay pattern in an interval, how should we judge the interval, and how should we use it to construct the scale? Many students of tuning must have been confronted with this question in practice, whilst attempting to understand what is going on in terms of a "traditional" tuning theory. Experienced tuners know the answer to this in empirical terms. The answer is as follows, but we are now in the realms of tuning *art*, so one cannot expect a scientific description. We must have recourse to the language of art. There are three things to consider:

1. Beat rate *effect*
2. Interval *quality*
3. Scale *logic*

Beat rate *effect*

In the fast intervals the "speed" of an irregular or modified beat rate provides an overall *effect* of a "beat rate" from which a judgement can be made. In the slow intervals the same is true, even though we may be dealing with an *irregular decay* pattern rather than an irregular but continual beat. In both cases it is necessary to hear the actual beat phenomena and also to "step back" with ear and hear the overall effect.

Interval quality – coarseness as distinct from beat rate

The beat rate or movement in an interval's decay plays an important part in determining the *quality* of the interval. In the case of a tempered fifth, and to a lesser extent a tempered fourth, the combined effect of the *movement* in the first and second adjustable partials will contribute to the interval's tone *quality*, which is all about the total amount of prominent movement in the interval's spectrum. In the case of a fast interval, the *quality* of the interval is more a question of "coarseness" (undesirable) or "shimmering" (desirable), than the actual beat rate or movement in the spectrum. To illustrate more what this means, we could describe a pure (beatless) major third, for example, as having zero "coarseness", and a Pythagorean third as being extremely coarse.

The overall coarseness in a fast interval is *not fixed entirely by the precise beat rate*, but rather, by the way that beat rate "sits" in the context of the interval's other tonal properties. For example, a beat rate in a major third that is considerably faster than its "traditional" value may sound more "coarse" on a very small instrument than it does on a large instrument. The fast beat rates of the thirds and sixths should "shimmer" but the interval should never be coarse. The difference between shimmering and coarseness depends on where the interval is in the compass, the "beat rate" causing the shimmering naturally becoming greater, the higher in the compass an interval of any given kind falls. A faster beat rate will also seem coarse, if it is out of place, *i.e.* surrounded by adjacent intervals of the same species that are slower. This will happen if beat rate progression is absent.

Scale *logic*

Combinations of notes forming given intervals have a logical relationship between their beat rates. This is true even if the "beat rates" are not true regular beat rates but are determined by beat rate *effect* and interval *quality* as described above. We can use this logic to our advantage in building the scale, once we are experienced enough to know what the "beat rates" of all the intervals should sound like, whether as actual rates or as a combination of *effect* and *quality*.

As far as possible, we tune any note from at least two others, "simultaneously". Each interval of which the note being tuned is a member, tells us through its "beat rate" something about that note's "position" in the scale. The interval's "beat rate", if it is not correct, tells us that the note needs to be higher or lower, for that interval. If we are unsure because of irregularity in the decay pattern, then we rely on the fact that there are *other* intervals of which the note is also a member, also telling us where that note should be positioned. We listen, so to speak, to the demands of all possible intervals and make a "democratic" decision based on the severity of the demand from each interval.

In tuning the scale, we thus use a *tuning sequence* to deliberately create a situation where as far as possible each note being tuned can be set by such a "democratic" decision, rather than in isolation. Furthermore, because of the differences in the way fast and slow intervals behave, we try to tune each note on the basis of demands made by at least one slow and one fast interval.

Scale plasticity; three note "closed box" relationships

There is a mutual relationship between any three scale notes forming tuning intervals. We should be aware of how this relationship may differ from the predictions of "traditional" theory. In fine tuning the scale, any three notes with standard tuning intervals between all of them, form a "closed box" that can be dealt with as a unit. Diagrammatically this is of course is a triangle, but "box" is appropriate because we can use the relationship to "box in" notes.

No one interval between the notes can be altered without potentially affecting the other two. If one of the notes is already "fixed" by previous tuning, then the other two must be tuned with regard for all *three* interval formed between the notes. Each new note tuned can also be tuned with regard to at least *two* intervals of which it is a member. "Boxing in" notes in this way provides a fine degree of "error trapping". The "error" here, is not necessarily tuning error, although it can be. The "error" can also just be problems inherent in the scaling or acoustics of the instrument.

Knowing which intervals are wide and which are slow, and testing more than one interval's beat rate simultaneously as we tune, we can find optimum tuning positions for the notes from the logic of their relationships. For example, consider tuning G – C – E in the scale. Let us say C is fixed, being tuned to the fork. Logically, if G – E is too fast, then either E must be lowered, or G must be raised, or both. If C – E cannot take any slowing, then the G must be raised. Otherwise, both must be compromised a little.

To the greatest extent possible, each note tuned should be tuned to at least two other notes, preferably those that form *both slow and fast beating intervals*. In deciding whether the note should be raised or lowered, at least two intervals then make demands for either raising or lowering the note being tuned. If both or all the intervals agree, then generally this is the required direction in which to alter the string, unless the reference notes from which we are tuning are themselves incorrect. If there is disagreement, say, if one interval demands the tuned string is raised, whilst the other demands it is lowered, this is either a cue to re-examine the previous tuning of the reference notes, or it is an opportunity to "trap" the error that is appearing now, by placing the tuning of the current note in the best compromise position relative to the notes from which it is being tuned. This may happen on smaller instruments with more acute inharmonicity changes.

Scale *plasticity* and interval *sensitivity*

Consider for example one note being tuned from two others, such that we are tempering both a slow and a fast interval, schematically illustrated in Fig. (9.18).

Fig. 9.18

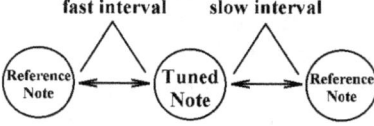

A specific example might be the notes G – B – E, in which the major third G to B is a fast interval whose beat rate is approximately 8 Hz, and the fourth B to E is a slow interval with a beat rate approximately 1 Hz (Fig. 9.19):

Fig. 9.19

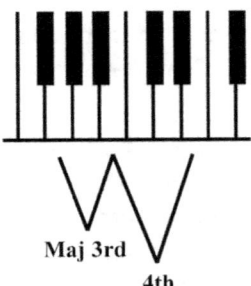

We will assume the fourth is wide and the third is wide also, as required by equal temperament, and that this remains the case throughout. In the "traditional" model a small alteration to the B by lowering it, will slow the beat rate in the third (G − B), and increase the beat rate in the fourth (B − E) by a "corresponding" amount. Similarly, raising the B a little will increase the third's beat rate while *decreasing* the fourth's beat rate by a "corresponding" amount.

In the contemporary model this "correspondence" is not straightforward. In effect, the beat rate or decay pattern in each interval has its own *sensitivity* to a change in the tuning position of the B, which will change that interval's beat rate or decay pattern. The degree of this change induced in each interval by altering the B, will depend on how far the B is moved (in absolute terms of a change in tension), *and* how sensitive that particular interval is to changes in the B.

Now in the "traditional" model the *sensitivity* of each interval is fixed – it depends purely on fixed arithmetic ratios. But in the contemporary model the *sensitivity* of intervals depends *on the individual interval*, and can be different at different positions in the scale. It also depends on *the degree of tempering* that is present in the interval. It is entirely unpredictable. In this example, the fourth's *sensitivity* to changes in the B, can itself change as we move the B.

The practical upshot of this for tuners is that we may find (1) a small but noticeable change to the beat rate of the third, is not *necessarily* going to produce a correspondingly noticeable change to the fourth, or (2) a small but noticeable change to the decay pattern in the fourth, is not *necessarily* going to produce a noticeable change to the beat rate in the third.

This effect is not confined to combinations of slow and fast intervals, but may also be encountered where both intervals are slow, Fig. (9.20):

Fig. 9.20

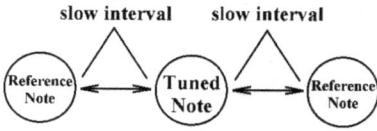

Some examples of this on the keyboard are shown in Fig. (9.21):

Fig. 9.21

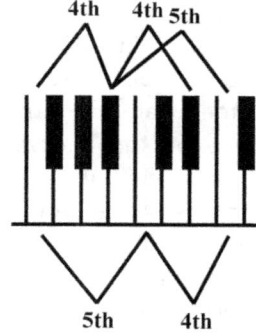

Here, we might be altering the B flat in relation to the F below it, and the E flat above it. From "traditional" theory we might expect the B flat to E flat fourth to be more "sensitive" to changes in the B flat, because in this interval the 4th partial of the B flat is involved, whereas in the fourth F to B flat, its 3rd partial is involved. In practice, the relative "sensitivities" of the intervals depend on each specific instance, with no such fixed rule.

The combination of the equally tempered fourth and the equally tempered fifth makes an octave. In "traditional" theory the fourth and the fifth must always have the same beat rate, if the octave is in tune. We must remember that the octave F to F here is itself a "slow" interval, even if its theoretical "beat rate" is zero. The fourth, fifth and octave are three connected "slow" intervals, which are related through "elasticity" in the scale. This can now have several different consequences. The fifth and the fourth will not *necessarily* have equal beat rates when the octave is "pure". The octave may not *necessarily* be within its *fine tuning region* when the fifth and fourth are both "pure" ("beatless"). Because all three intervals may have a *fine tuning region*, there may be more than one relative arrangement of note tuning "positions" that satisfies the criteria for "equal temperament".

The generalised practical rule at which we arrive is:

- It is possible for an interval's decay pattern or beat rate to be relatively *insensitive* to changes in one of its notes being altered, whilst at the same time another interval that is affected by the same note being altered, is very *sensitive* to changes in that note. This has a bearing on which interval we may choose to alter when improving part of the scale, or when error trapping.

Improving a scale – the advanced approach

When error trapping, there will be *two or more* intervals that need to be simultaneously considered. Having found an "error" or even just a part of the scale that could possibly be improved, first determine all tuning intervals that will be affected by altering either of the offending interval's notes. Always then re-tune on the interval that is going to be least amenable to the change you are about to try and make, and then check the other interval's responses. An expert tuner can keep track of all intervals and their responses, all locations of possible "error", and all locations of possible "give" from the scale's plasticity, and will be able to adjust accordingly, considering the scale *as a whole*, rather than just as intervals in isolation.

However, this level of proficiency, and this mental holding of the scale's state and behaviour in one's head, requires absolute familiarity with the scale's construction and behaviour, and its beat rate relationships. It takes experience and a long time to achieve. It is not an exercise for the less experienced. It is an advanced technique, but it comes naturally with sufficient experience.

The two "error" types

For good tuners with sufficient experience errors of judgement are not really an issue, because if they occur, they are discovered as part of the process, and simply corrected without fuss. What may appear to be "error" can also be a product of inharmonicity, and manifests as lack of progression and/or unsatisfactory fifths and fourths. Even if inharmonicity precludes very good progression, a good solution must nevertheless be found. Neither a good progression of beat rates in the thirds at the expense of the fifths, nor a good set of fifths at the expense of the thirds' beat rate progression, will constitute a good scale.

The overall "error" due to inharmonicity that would appear either in the slow intervals or in the fast - whichever set would need to be compromised for the sake of the other – must be "distributed" or "hidden" as much as possible throughout the scale. Just as in the "traditional" model we distribute the Pythagorean comma throughout the scale, so in the real world we *also* sometimes have to distribute any "distance" by which the actual physics of the scale deviates from the "traditional" model. The main cause of this is *rate of change* of inharmonicity over the scale, rather than inharmonicity itself. We achieve the distribution, in effect, by "error trapping".

The psychology of beat rates – judging progression

Most good tuners, students as well as professionals, can often tell you the beat rate of a fast beat pattern in the scale range, say, within around plus or minus half a beat, or maybe one beat per second. Some tuners, however, perhaps without the benefit of proper teaching, imagine that their beat rate judgement is infallible, and that they can tell you a definitely correct rate for any interval played in the scale.

The beauty of digital technology is how easily it can expose the facts. One can be sure that where this kind of belief is held, the tuner has not thoroughly put it to the test against digital verification. Such an attitude, in fact, gives away weaknesses in a tuner's knowledge of beat rate phenomena, the psychology of perception, and indeed self knowledge. All of these are essential ingredients for good tuning.

In tuning a scale, one thing we are aiming for, is a scale in which fast beat rates or beat rate *effects* on tone, progress evenly from the bottom to the top of the scale, and when tuning extends each side of the scale, they continue their even progression outside the scale area. This is very important. The *quality* of the fast intervals needs to change gradually and uniformly, throughout the scale, but without just sacrificing fifths and fourths to achieve this.

In aiming at this, there is one outstanding thing that should be remembered when tuning tempered intervals. If we are listening for a *fixed beat rate* with a *constant beat amplitude*, we will generally not find it. But we may *think* we have found it. *False expectations can lead to misjudgements* about what we are actually hearing, and even possible misjudgements of beat *rate*. A reduction of beat amplitude (depth or prominence) is easily mistaken for a reduction of beat rate, and in tuning tempered intervals both beat *rates* and beat *amplitudes* can vary.

The assessment of the rate for a *varying* beat rate (and most beat rates vary over the decay of the envelope), can depend on which part of the envelope happens to receive most conscious attention. There is no scientific *rule* which says you *should* listen to the first two second's worth, or that you should only listen after two seconds. The behaviour of intervals and the nature of piano tone is far too complex and varied to make up such simplistic rules.

One should therefore not underestimate the influence of the psychology of perception. Accordingly, as there is most definitely an unavoidable *psychological element* in beat rate judgement (as there is in the subjective judgement of *anything*), the progression of beat rates in the scale should *always* be judged *both* ascending and descending, as this helps to circumvent natural ambiguity arising from both the subjectivity of judgement, and the natural variability in beat properties.

The importance of interval quality

We cannot know in advance precisely what each instrument is going to require, in order to successfully accommodate its individual acoustical properties – those that are not addressed by "traditional" theory. However just knowing the size, age, condition and make of the instrument can tell us a great deal. Second to the actual beat rate, the most important guide in setting the scale is the *quality* of the intervals. Many tuners choose the major third C to E as the starting interval for the scale, for two good reasons.

Firstly the beat rate itself has a particular "quality" about it, that when it is correct is clearly recognisable without counting. Secondly, the *quality* of this interval at the correct beat rate is like a "litmus test" for the instrument as a whole. As a result of this one interval alone, it is often possible to make judgements about what is going to be required in the overall approach. In the case of a very small instrument the *quality* is usually sufficient to make a judgement about adjustments to beat rates to accommodate inharmonicity. Many expert tuners probably never even consciously think about this – they just tune the third until it is "right".

From experience, they just "know" it is going to be right for that instrument, and will enable a successful scale.

Very small pianos

The tuning of small pianos is dealt with primarily in the chapter 13, but some comments on their tuning can be included here. The size of a piano always has to be taken into account. The chapter on inharmonicity illustrates how the size of a piano and its scaling can affect scale tuning. However, it must be remembered that the effects illustrated in that chapter are for a *frequency definition* scale – a scale defined by the set of fundamental frequencies dictated by "traditional" theory. The artist tuner does not aim for these results. Rather, the aim is to get the "musical acoustic" qualities and properties of the scale correct.

This still involves ensuring uniform *quality* of the intervals and even progression of the beat rates or beat rate *effects*. A scale would hardly be considered good by tuners just because it has the "correct" frequencies at the fundamentals, according to a frequency meter, if the fast beat rates are disorderly or the fifths are unpleasantly fast. We would expect a musician to notice the consequences of this - the beat rates largely determine the quality of the scale intervals and will affect the larger intervals across the compass.

The results shown in the chapter on inharmonicity illustrate the tendencies in the nature of the scale, that the tuner now has to try to overcome in the case of a very small piano. We do this by working *with* these tendencies in order to work *around* the detrimental properties they would otherwise introduce into the scale. The aim of progressive "beat rates" in the major thirds is then more or less attainable without compromising the slow intervals. The "rules of thumb" to bear in mind are:

- The major thirds should still progress as evenly as possible, but the beat rates themselves may vary from the norm, and the *rate of change* of beat rate from one third to the next, may not be the same. Typically, the beat rates in the lower part of the scale are slower then the norm. If we begin with an "error trap" we can determine the likely situation. For example, we may begin with C40 (C4), and tune E44 (E4) and G35 (G3) from C. The fourth G – C cannot be so fast as to be unacceptable, even if much of the movement is due to falseness, and if it is, we may even need to adjust it by making the G higher than it would otherwise be. The third C – E, however must then be sufficiently fast or slow to yield a good beat rate in the sixth G – E. Generally on a *small* piano we will need to find intervals that "work against" each other in some way, and set a "box trap". The "error" inherent in the system needs to be suitably distributed between the three intervals of any "box trap", with a satisfactory compromise if necessary. No one of the three intervals should be wholly compromised for the sake of the other two. In the chapter on small piano syndrome, we look at linking the scale tuning to the necessary octave stretch. The overall approach is to "fit beat rates together" by effectively tuning multiple intervals simultaneously, rather than attempting to "put beat rates in" one or two intervals at a time, using idealised beat rates.

- Where slow falseness or some other cause of a modified beat rate is present in a major third - which is common on such instruments - we must not be misled by a *changing* beat rate that increases or decreases during the overall decay. The beat may for example begin at a rate that seems too slow, but later in the decay become much faster than the "traditional" rate. On very small pianos a perfectly regular beat rate in the major third may typically be expected to be set *slower* than usual.

- The fifths (as always) should be tuned listening to both the lower and upper beat rates, simultaneously. On the smallest instruments, it will be not be a question of specific beat rates, it will be a question of finding the slowest possible overall movement, taking into account the relative prominence (loudness) of the lower and higher beats. The slowest overall movement may well occur with the lower beat pure or beating *wide*, whilst the upper beat, beats *narrow*.

- The fourths should be allowed to be somewhat faster than their "traditional" rate, if necessary.

- The sixths also may be compromised a little without too much loss of *quality*.

- On many small instruments the rates and the nature of the progression *must* differ from that tuneable on a better instrument, but the overall effect should still be one of regularity of tone, equality of intonation, and reasonable progression of fast beat rates.

We might ask "If the major thirds are slower than normal and the fifths and fourths are faster than normal, can we not just correct this?" The answer is that attempting to make the major thirds faster, or closer to normal, and the slow intervals slower and therefore closer to normal, will produce lack of progression on any instrument whose size and scaling creates the "small piano syndrome". Setting intervals to normal beat rates actually then includes inharmonicity "errors" that could have been avoided, and these will accumulate as scale tuning proceeds. We have to counter these "errors" by putting deliberate changes in the beat rates, so that they differ from the "traditional" value, but in such a way that in the final result we can still achieve a *progression* of beat rates, giving the scale its *equality* of tempering in real acoustical terms. Without this strategy the resultant lack of progression can in some pianos be very severe.

There are some very small "baby" grand pianos for example, where trying to tune "traditional" rates can lead to the appearance of some thirds or sixths that are practically pure, together with some unacceptably fast fifths or fourths. The full price of such lack of progression will then be paid in the tuning of the overall compass, which will never be satisfactory in compound intervals over several octaves.

Scale sequence – limitations

A straightforward "Circle of Fifths" sequence is not practical for the professional tuner. Starting on C this might be, for example, C – G – D – A – E – B – F# - C# - G# - D# - A#(B flat) – F, using fifths and fourths, rising and falling, and sometimes octaves. In this sequence the first four slow intervals are tuned in isolation (without "checks"). If on tuning the E a conflict arises between the fifth A –E and the third C – E, there then is no logical way of isolating the "error" correctly. The sequence, in other words, does not allow "error trapping".

In any sequence, the second note of the scale can only be tuned from the information provided by *one* interval, the interval it makes with the starting "pitch note". We must be sure of the tempering in this interval, without reference to another. If necessary, in the case of very small pianos, we might even have to be prepared to return to it and alter it. The chapter on small piano syndrome looks at the alternative strategy of beginning with an octave. Many tuners begin with a favoured third, say, C to E, of F to A.

Every scale sequence within an octave must contain at least one "crossroads" where in theory we logically *may have to go back* and retune the previous note, depending on the piano's tuning characteristics. This is a point in the sequence where we have to tune (at least) *two* notes, from two different intervals, before we can trap any error using the third interval made between the two notes. At this point, we trap any tuning error, or distribute any difficulty due to inharmonicity. This will become clearer in the examples below.

Scale tuning sequence example number 1

This sequence[67] covers F33 – F45 (F3 – F4) starting on C40. (C4) It requires complete confidence with scale logic, and a good practical knowledge of beat rates. It is not really a "beginner's sequence", but if diligently applied by a more experienced tuner will reliably produce good results. The advantages are:

- The sequence maximises the opportunities for setting notes from at least two others simultaneously.

- It maximises the opportunities for each interval being tuned, to have an adjacent interval of the same kind available and already tuned, from which beat rate progression can be judged.

- It gives no particular precedence to slow or fast intervals, or to intervals of any given kind, an therefore allows better error control.

1. Tune C to the fork.
2. Tune E from C. We consider the *exact* beat rate if the beat is clear and perfectly regular. If not, we consider the apparent beat rate *effect* in the context of the interval *quality*, as already described above. Many tuners find this third has a particular, unique, distinctive character when its *quality* and beat rate, or beat rate *effect*, are correct. We must be absolutely satisfied with this first interval's correctness before proceeding. Because the slow intervals are subject to possible *irregular decay* patterns or other generic decay patterns, it is arguably better to choose a fast beating interval as the first one tuned.

---

[67] Appears also in Capleton B, 'Piano tuning techniques and the tuning characteristics of pianos considered as an effect of inharmonicity and mode coupling', *Journal of the Institute of Musical Instrument Technology*, 4, 3, 1991, pp. 91-126.

3. Tune G from C and E. The fourth G to C and the sixth G to E must be satisfied. If both intervals suggest the same - either that the G should be lowered, or raised – then we can follow this. If the intervals *disagree* in what they suggest then we must be sure the fourth is "on the right side" (wide). This may be misleading if falseness is present. We must then find the "best fit" position for the G in relation to *both* intervals, the fourth and the sixth. This may even involve having the fourth less wide than "traditional" if prominent falseness is "spoiling" the interval too much. The sixth must not be slowed too much as a consequence. If ever we "alter" a slow interval in this way, in order to prevent falseness "spoiling" the interval too much, we must remember we have done this, in terms of whether the note is "lower" or "higher" than it "should" be, in the scale. That way, the additional "error" we have just introduced into the scale can be accommodated in the rest of the scale, using the appropriate application of scale building logic.

4. Tune B from G and E. Again, both the third and the fourth must be satisfied following broadly similar principles.

5. The next two intervals constitute the "crossroads" in this scale sequence. We must tune one of two major thirds: A flat - C or B – D#. Whichever one we tune, it must progress properly from the adjacent third already tuned.

6. The second of the two "crossroads" thirds must now be tuned, *and the fifth made between A flat (G#) and E flat (D#) must be good*. Any "error" must be addressed by adjusting the two thirds, and the "check" on this is the progression of each third's speed from its adjacent third. The sequence ensures that the upper third (B – D#) is "progressed" down from the third above it (C – E), whilst the lower third (A flat – C) is "progressed" up from the third below it (G – B).

7. Now we must tune F# from D# and B. Both the fourth F# - B and the sixth F# - D# must be satisfied. Progressions can follow from the adjacent sixth G – E and the adjacent fourth G –C already tuned.

8. A# is tuned from F# and D#. Both the third F# - A# and the fourth A# - D# must be satisfied. The adjacent third G – B and the adjacent fourth B – E are already tuned and available for testing the progression of the beat rates.

9. Tune D from G and A#. Both the third A# - D and the fifth G – D must be satisfied. The third B – D# is already tuned and adjacent to the third A# - D, so the beat rate progression can be ensured. Similarly the fifth G# to D# is already tuned.

10. Tune bottom F from B flat, C and D. The fourth F – B flat, the fifth F – C and the sixth F – D must all be satisfied. Each one of these intervals has an adjacent one above it, already tuned, for progressing the beat rates.

11. Tune A from bottom F, D and E. The third F – A, the fourth A – D and the fifth A – E must all be satisfied. All have adjacent intervals already tuned, from which beat rates can be progressed.

12. Tune C# from F#, G#, and A. The fifth F# - C#, the fourth G# - C# and the third A – C# must all be satisfied. This is the last note. All other intervals are already tuned and available for progression testing.

Major thirds progression on a very small instrument:

The "error" due to changes in inharmonicity over the scale, will appear in the lowest fifth in the scale, F33 – C40 (F3 – C4), in which the position of C40 (C4) can be taken as error free (the long bridge position on the soundboard is most compromised at this position). This inharmonicity difference is why many tuners will find it becomes difficult to avoid too fast a beat (in the 2nd adjustable partial) in this bottom fifth of the scale (F–C). Because C40 (C4) is "fixed", the F33 (F3) position is the note responsible for the inharmonicity "error" here, which tends towards causing the F-C fifth to be too fast.

In an extremely inharmonic scale with non "traditional" beat rates in the major thirds, progression can usually be achieved by tuning the fifth F33 (F3) – C40 (C4) *first*, to ascertain what the greatest effect of inharmonicity is likely to be. We set this interval with the 2nd partial narrow but as fast as remains acceptable. Very high inharmonicity change over this interval may mean it is impossible to have both lower and upper beats on the "same side", *i.e.* both narrow. For more on this see the section of perfect fifth tuning, and the chapter on small piano syndrome.

Having satisfactorily tuned the F, we can set the C-E as normal, and tune the A as a descending fifth from E, so that as much as possible it "matches" the quality of the F-C fifth. How does this now leave the F-A third? It may be a little slower than normal, but if it is really too slow or almost pure then the tuning of the E must be revised, and the A re-tuned until a realistic progression is achieved between the F-A third and the C-E third, without completely spoiling the fifth A-E. If you find this impossible, it will be because the tuning of the fifth F-C was over-optimistically slow, and it will just have to be made a little faster.

Working with the notes F, A and E, the best possible tuning must be found that allows (a) similarity between the two fifths, (b) the rates of the F-A and C-E thirds to be in reasonable proportion, and (c) neither the F-C nor the A-E fifths to be too fast. It is then possible to revert to the normal sequence, and fit it around what has just been tuned.

Scale tuning sequence example number 2

A sequence following similar principles but starting from A37 (A3) is as follows:

1. Tune A37 (A3) to the fork.
2. Tune F33 (F3) from A.
3. Tune F45 (F4) from F33 (F3).
4. Tune C# from A37 (A3) and F45 (F4). Now the four notes already tuned divide the *diesis* between F33 (F3) and F45 (F4), so any "errors" should be retuned now. The beat rates of the three thirds, and the octave, should all be satisfied. The thirds are at the extremes and middle of the octave, so the inherent inharmonicity characteristics of the overall scale can be accommodated now. This stage is in effect, this sequence's "crossroads". The rest of the sequence allows each new note to be tuned between two others, forming one slow and one fast interval at each step.
5. Tune D from A and F33 (F3). Satisfy both the sixth and the fourth.
6. Tune A# from F33 (F3) and D. Satisfy both the fourth and the third. Progress the third A# - D from adjacent A – C# already tuned.
7. Tune F# from A# and C#. Progress the third F – A from adjacent F# - A# already tuned, but also satisfy the fifth F# - C#.
8. Tune D# from F# and A#. Progress the sixth F – D from adjacent F# - D# already tuned. Compare the fourth A – D with A# - D# already tuned.

9. Tune B from F# and D#. Progress the third B – D# from adjacent A# - D already tuned. Compare the fourth F# - B with the adjacent F – A# already tuned.

10. Tune G from B and D. Progress the third G – B from adjacent F# - A# already tuned. Compare the fifth G – D with adjacent F# - C# already tuned. We can now test the thirds series F – A; F# - A#; G – B for good progression.

11. Tune E from G, B and A. The sixth, fourth and fifth must all be satisfied. The sixth G – E can be progressed from adjacent F# - D# already tuned. The fourth B – E can be compared with adjacent A# - D#. There is now a sixth series progression F – D; F# - D#; G – E.

12. Tune C from E, F33 (F3), F45 (F4) and G. Progress the third C – E from adjacent C# to F45, and B – D#. All major thirds from A – C# upwards should now progress. Compare the fourth C – F with adjacent B – E. All fourths from A – D upwards can now be compared. Compare the fifth F – C with adjacent F# - C#. Test the fifth series F – C; F# - C#; G – D.

13. Tune G# from C, C#, D# and F45 (F4). Test complete scale.

Generalised scale building principles

Whatever sequence we use for setting a scale, the general principles to be remembered are:

- A complete scale "test" covers not just one octave of the compass, but an octave and a major sixth.

- The scale should *not* be thought of as an isolated "unit" of tuning. Proper beat rate relationships should flow seamlessly beyond the "scale area" so that it is not at all obvious precisely where the scale was tuned.

- There is an important inescapable *temperament* relationship between the major thirds and the fifths.

- The position of the scale area in the physical compass has consequences, according to design and scaling of the piano, the position of the bass break, and the change from trichords to bichords, or from plain steel to covered strings.

- The non-"traditional" features of the physics of the scale have important and unavoidable consequences in practice.
- The scale is "elastic"

Now we shall look at some of these in turn.

A complete scale covers an octave and a major sixth.

In theory, the whole "scale" is "defined" by the relative tuning positions of the twelve semitones from which it is made. But the effects of its tuning appear in all the intervals that can be produced from the scale once it has been extended in octaves. It is not the case that what might appear as good progression of intervals *within the scale area over an octave of the compass*, will *necessarily* produce continued good progression out this "scale octave", just by virtue of well tuned octaves. The appearance can be deceiving.

A good progression of "beat rates" for any given species of fast beating interval, nonetheless should continue outside the twelve note scale area. To test this we need to test over at least an octave and a major sixth of the compass. For example, for a scale between F33 and F45 (F3 to F4), the lowest major sixth in the scale is F33 (F3) – D42 (D4). The highest one, before we start repeating an octave above ones already considered, is E44 – C#53 (E4 – C5). C#53 (C5) is of course outside what we consider to be the scale area. The major sixth is the largest tuning interval, so the total section of the compass that we need to "test" all the intervals that a scale produces, is an octave and a sixth.

Relationship between the major thirds and the fifths.

No matter what tuning sequence we are using, if we are aiming at equal temperament the major thirds and the fifths will be related. This relationship arises in Temperament Theory itself. We can refer again to the Pythagorean Circle, or Great Circle of Fifths, Fig. (9.22).

Fig. 9.22

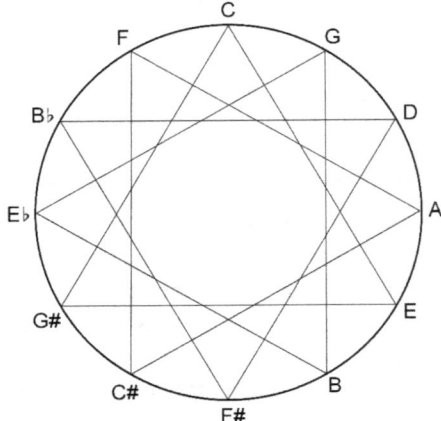

In this particular example the equally tempered fifths circulate clockwise to G# and anticlockwise to E flat. In equal temperament the "tuning sequence" makes no difference here – the note names are purely conventional and the interval G# to E flat, nominally a diminished minor sixth, still constitutes a tempered fifth as far as tuning is concerned. The straight lines connect notes that define equally tempered major thirds. Similarly, any of the straight lines across an arc containing the G# to E flat interval still constitute major thirds as far as tuning is concerned, even though the note names signify a diminished fourth.

In the practical application of temperament we dismiss the difference the Pythagorean and the Syntonic commas, and refer to them both simply as "the comma". By Temperament Theory, the sum of the amounts by which each of the fifths round the circle deviate from perfect ("pure"), must equal +1 comma (a positive value for fifths means *narrow* tempering). Also, the amount in commas by which any major third is consequently tempered, will be 1 minus the sum of the deviations or temperings in the four "fifths" across the arc for that third (a positive value for the major thirds means *wide* tempering). Thus, for example, the tempering in the major third F – A will be 1 minus the sum of the temperings in the fifths F – C, C – G, G – D and D – A. In Equal Temperament these four fifths are each tempered narrow by 1/12 comma, so all four constitute a sum of 1/3 comma. Therefore the tempering in the third F – A will be 1 minus 1/3, *i.e.* 2/3 comma. All Equally Tempered major thirds are thus wide by 2/3 comma.

There is a relationship between how wide the major thirds are, and how narrow the fifths are, when we deviate from Equal Temperament, by setting fifths or thirds consistently too fast or too slow. As far a this relationship is concerned Equal Temperament sits mid way between two extremes. At one extreme is *Pythagorean Tuning*, and at the other is *Quarter Comma Meantone Temperament*.

Pythagorean Tuning extreme

In Pythagorean Tuning as many fifths as possible are pure – so 11 fifths are pure and one is a "wolf fifth" that is far too narrow, by a whole comma. As a result 8 of the major thirds are far too wide, whilst 4 are pure. These 4 are the thirds whose arcs include the "wolf fifth".

Quarter Comma Meantone extreme

In *Quarter Comma Meantone Temperament* as many thirds as possible are pure, so 8 thirds are pure and 4 are far too wide. We cannot get more than 8 thirds pure, because to make *one* pure third we have to have the four fifths on its arc narrowed by ¼ comma (thus the tempering in the third equals 1 minus [¼ + ¼ + ¼ + ¼], which equals zero). We can only temper 11 fifths narrow by ¼ comma – the twelfth must make up the difference to make the whole circle back to 1 comma, so the twelfth fifth must be wide by 1 ¾ comma. There are then still four major thirds whose arcs must include this wide "wolf fifth".

In tuning Equal Temperament, we will attempt to equalise the size of the major thirds, and similarly the fifths, by attempting to achieve progressive beat rates. However, inaccuracy in judging the necessary speeds of major thirds or fifths will tend to shift the resulting temperament away from the middle of the range towards one of these extremes. We will not generally deviate as far as producing Pythagorean Tuning or Quarter Comma Meantone Temperament, but as we deviate one way or the other, our tuning will inherit some of the characteristics of the system towards which we are leaning.

The general relationship is shown in Fig. (9.23):

Fig. 9.23

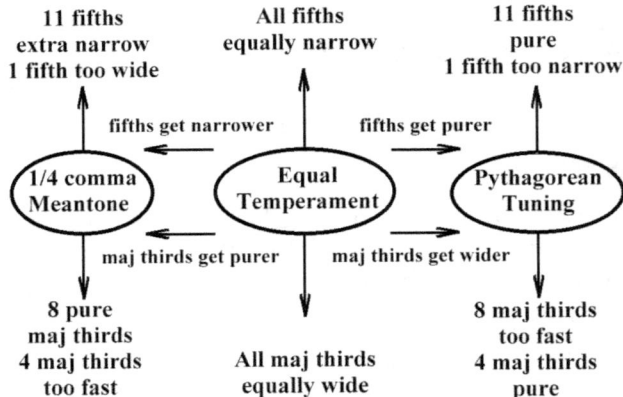

- If in setting the scale we tend to tune fifths (and their inversions, fourths) too "slow", then we head towards Pythagorean Tuning – one fifth (or fourth) will remain too fast, 8 of the major thirds will be too fast, and 4 of them will be too slow. The thirds will therefore fail to progress in their beat rates. The ones too slow will get slower, the purer the fifths are set, whilst the ones too fast, will get faster.

- If in setting the scale we tend to tune fifths (and their inversions, fourths) too "fast", then we head towards Quarter Comma Meantone Temperament. The result is that 8 of the major thirds will be too slow, whilst 4 will be too fast. Progression will again fail.

- If we are setting beat rates in the major thirds too slow, then we can expect up to 11 of the fifths or fourths to be too "fast", whilst one will be too "pure".

- If we are setting beat rates in the major thirds too fast, then we can expect up to 11 of the fifths or fourths to be too "pure", whilst one will be too "fast".

These rules are general, and will tend to be obscured by falseness. Nonetheless, if we consistently set fifths too slow or too fast then the progression in the major thirds will fail. Similarly if we set major thirds too fast or too slow then at least one fifth or fourth will be too fast or too "pure", and four thirds will remain out of progression.

The position of the scale area

If the scale area includes a part of the compass in which there is a rapid change of scaling factor, or a change of soundboard bridge, or a change from plain to covered strings, or any combination of these, then this will have consequences in the tuning of the scale. There may be rapid changes or jumps in inharmonicity from one part of the scale to the next. As in the general case for a small piano, good progression of "traditional" beat rates may be compromised, or impossible. In order to maintain control of the situation it is important to use a tuning sequence that incorporates "error trapping" so that excessive movement in any of the slow intervals can be avoided, and reasonable progression in the major thirds can be maintained.

The non-"traditional" features have consequences

The non-"traditional" features are inharmonicity and coupling effects, but we might also include the effects of falseness. Any interval may not behave in the simple, "traditional" way:

- The fast intervals may exhibit changing beat rates, or even dual beat rates.

- The slow intervals may exhibit *generic decay patterns* other than the Regular Beat.

- The relationship between beat rates in the first and second adjustable partials of fifths and fourths may not be fixed at the ratio 1:2.

- The beat *prominence* or amplitude in the *second* adjustable partial of a fifth may be sufficient to "spoil" the interval if its beat rate is not taken into account.

- A beat rate's *sensitivity* to change as the string's tension is altered, may be different in different intervals of the same kind, e.g. in two different thirds or two different fifths. The three tuning intervals formed between any three notes may have different *sensitivities* to change.

- The "traditional" beat rates may not be applicable in cases of large inharmonicity changes from one end of the scale to the other.

- A beat rate set in an interval between single strings using a muting felt or wedges, or between a unison group and a single string using a wedge, may change once the tuning of the unison group with the single string is completed. This applies even if the single string is completely stable.

- The scale is "elastic". Different intervals, although their theoretical beat rates may be logically related, may have different *sensitivities* to change. Consequently it may be possible to alter the tuning position of a note so that it affects one related interval more readily than another. This provides a degree of "elasticity" in the tuning positions of the scale notes.

Remember that as far as aural tuning is concerned, the criteria and considerations for an excellent scale are:

- Evenly progressing[68] beat rates in the thirds
- Evenly progressing beat rate in the sixths
- Minimum audible intrusion of movement due to beating partials in the fifths and fourths, including movement due to falseness and/or inharmonicity, consistent with progression in the thirds and sixths.
- The above will only be achieved when there *is* (the right amount of) movement in the fifths and fourths, and the beat rates are "correct" within certain limits. Movement in the fourths and fifths will be *unavoidable*, and there must be more movement in the fourths than in the fifths if the above criteria are to be met. We are not free to put whatever beat rates we like, wherever we like, if we are to fulfil the criteria. The beat rates are all interrelated in a complex network.
- Taking inharmonicity into account, where necessary, beat rates in the first adjustable partials of the fifths are usually necessarily slower than those specified by "traditional" theory, if the fifths are not to be obviously different from pure perfect fifths, or even unpleasant, to the sensitive musician's ear. This is also necessary to reduce beat rates in prominent second adjustable partials, and to utilise inharmonicity for the purpose of octave stretching.

---

[68] "Progression" of beat rates means that the beat rates increase steadily in even increments, with rising positions of the intervals in the compass.

407

- The characteristics of the scale, particularly the progression of beat rates in the major thirds and sixths, must extend without interruption into the intervals outside the scale area, also enabling good progression in the beat rates of the major tenths and major fifteenths.

- The limitations imposed on these criteria by inharmonicity and falseness must be handled intelligently and artistically for the best possible result.

"Error" diagnosis

Finally, let us reiterate the essence of *diagnosis*, as for students of tuning this is typically part of formal examination in tuning. How does one diagnose where the problem lies in a scale that is less than satisfactory? For example, we might find the fourth A-D too fast, but both fifths G-D and A-E are satisfactory. Perhaps the thirds A-C# and B-flat to D also sound correct. Where is the root of the problem?

One approach is to test through a fixed tuning sequence to try to find the root of the problem, and often this is easy enough. However, sometimes the mental "overlaying" of a tuning sequence does not reveal clear indications of the root of the problem.

Improvements to a scale can often be made without necessarily imposing a *sequence* on it. Any interval that is not satisfactory will by itself, in isolation from other intervals, *suggest* not one, but *three* possible actions to "correct" it, because every such interval has *two* notes, and correction can be made by altering (1) the lower note, or (2) the upper note, or (3) both notes. What one needs to do in diagnosis is to look for *two or more* unsatisfactory intervals that suggest the *same change* to the same one note, *i.e.* they both suggest raising it, or both suggest flattening it.

No change should be made on the suggestion of just one interval, but where two or more intervals make the *same suggestion* for changing a note, then generally, this suggestion should be followed. This is a good principle to start with. It is generally reliable, but not totally infallible. The greatest reliability lies in tracing intervals back from each note of the unsatisfactory interval to the "pitch" note (that can always be tested against the fork) by the *shortest possible routes* of tuning intervals, rather than through the route one would have taken in one's favoured tuning sequence.

Chapter 10 - Octave tuning

The 2:1 fallacy

Rudimentary theory says simply that the *frequency ratio* of an octave is 2:1. As we have already explored, a musical interval produced by an acoustic instrument generally requires far more than a single arithmetic ratio to describe the acoustical properties that affect its tonal and pitch qualities. As we have seen, the tone of the any piano note, and of the octave, is a complex recipe of many partial frequencies, partial fluctuations and decays. Individual *frequency components* group together into the *partials*, many of which are individually audible ingredients of the sound. Each partial has its own *envelope* shape when represented as a signal over time.

The shape of these envelopes in relation to tuning is a more complex matter than rudimentary theory supposes, and their shapes contribute as a collective whole to perceived tone quality. Practical octave tuning demands that the octave sounds at its best, in relation to its intonation and tone, and in relation to the intonation and tone of the instrument as a whole. The condition of the octave that fulfils these requirements, like *any* condition of an octave or of *any* interval, cannot possibly be fully and accurately described by a single, simple ratio. A graphical description for a single octave would in fact require a more complicated 3-D graph, or *waterfall* representation.

Initial issues

The tuning of an octave has three aspects. These are:

1) Its intonation
2) Its tone
3) The intonation and tone of other intervals, that share either of the octave's notes.

"Traditional" tuning theory assumes simply that all three demands will be satisfactorily met by tuning the octave to a so-called "beatless" condition. In practice, the concept of "beatless" applied to the octave turns out to be a rather "fuzzy" one. It is wise to appreciate from the outset that the term "beatless octave" is indicative only. It is a useful term of reference of aural perception, but it also it depends on the acuity of aural perception. In many instances, if we were to digitally analyse the soundscape of a "beatless" octave, we would find beating and fluctuations were actually present in the soundscape, as it is acoustically impossible for the soundscape of many octaves to be genuinely beatless, as a result of *any* tuning condition. The acoustical reasons are mainly an inharmonicity difference between the upper and lower note, and the natural false beat phenomenon.

The best intonation and best tone of an octave do not necessarily always conveniently coincide, in other words, the tuning condition giving the most desirable tone is not always the same as the one giving the most desirable intonation, especially on very small instruments. In practice, one must consider all three aspects of the octave tuning, and just as unison tuning is ultimately a question of tone, rather than just beating in some specific partial, so octave tuning is ultimately a question of tone and intonation, with due regard for the intonation and tone of other intervals that share one of the octave's notes. Even though the simple edict to tune the octave "beatless" turns out to be relatively crude or even sometimes meaningless, when analysed, this in no way provides an excuse for perfunctory octave tuning. Rather, it demands much more careful consideration to the whole question of octave tuning than is given in "traditional" theory.

Every octave is potentially subject to both natural false beat phenomena and inharmonicity. Falseness, as already discussed, is the natural presence of partial beating and fluctuations that are not fully *adjustable*. There will always be a difference of inharmonicity between the upper and lower notes of an octave. It is this *difference*, rather than inharmonicity *per se*, that makes octave tuning a more complicated affair.

The piano's scaling design determines the rate of change of string speaking lengths, and is often compromised towards the bass end of the long bridge, in order to keep the bridge position away from the stiffer edge of the soundboard. On large instruments there may be no acute problem arising from the change of inharmonicity over the octave, but on smaller instruments, including many ordinary sized upright pianos, a conflict can arise even in the first three adjustable partials of an octave in the central compass.

Digital analysis confirms the aural problem that is often encountered. Typically, the third adjustable partial (at the meeting of the 6th partial of the lower note with the 3rd of the upper note, pitched an octave plus a fifth above the pitch of the octave's upper note) can demand a tuning position incompatible with the octave's two lower adjustable partials' requirements. In other words, in order to tune the octave's 3rd adjustable partial "beatless", the tuning of the first two adjustable partials (pitched at the fundamental of the octave's upper note, and one octave above it) would have to be sacrificed. Some compromise is called for in these more extreme situations. This is *often* found in *central compass octaves* on smaller pianos.

No beating, or non-beating partial, is important just as a partial on its own. Partials are important for their contribution to *tone quality*. There is therefore no strict rule for how to tune an octave, or no ideal tuning solution, *just in terms of one individual partial*. The *fundamental* partial, especially, is by no means the only important partial. Octave tuning always requires attention to the *tone* of the octave, as a whole, in combination with attention to the other intervals that it affects.

The octaves are an important set of intervals in their own right, but they cannot be considered in complete isolation from the other compound intervals, or in particular, from the tuning of the scale. In the learning stages tuning the piano is often approached as a combination of three essential tuning processes, namely, the tuning of unisons, the tuning of octaves, and the tuning of the scale (the central chromatic scale of 12 or more adjacent notes, with tempered intervals between the notes). The procedure would typically consist of first tuning the scale, which incorporates unison tuning, followed by tuning outwards from the scale in octaves (and of course unisons).

This purely technical division can be misleading. From the musical point of view, the compass of the piano does not divide into two separate blocks, consisting of an isolated central scale abutted on each side by octaves. Musical intervals other than octaves are everywhere in the compass, so it is not the case that the importance of the tempering or tuning quality of intervals is something that only applies to the "scale area", a central octave's worth of the compass. Depending on the music being played, the central octave or so may not even be the most important part of the compass. Particularly important in determining the tone of the instrument as a whole, is the intonation of the compound intervals.

Tuning the scale in a central octave of the compass, does enable precision relationships between the intervals, in terms of their tempering and tuning qualities, to be made. Tuning outwards through the rest of the compass in octaves and multiple octaves then has *two* functions which are: (1) Proper tuning of the octaves and multiples octaves, and (2) To ensure that the relationships between the intervals in the scale, are properly translated throughout the rest of the compass.

Arguments about stretching

It has sometimes been said that octave stretching makes the instrument tonally "brighter", and this has often been argued as the justification for it. Conversely, it is sometimes said by some tuners that octave stretching *spoils* the tone of the instrument, *and that their clients agree*.

There are several different issues here. Both assertions, as they stand, tacitly imply, quite incorrectly, that the choice is simply one of either *octave stretching* or *no octave stretching*, on the rather shaky assumption that we all know what we mean by *octave stretching*. The assertions ignore the issue of exactly what "octave stretching" does mean, precisely *how* octave stretching is applied in each octave, and what precise characteristics and magnitudes it has over the compass. We will see below that in these very issues, lies the benefits or detriments in the way octaves are tuned.

Another glaring issue here is simply one of the psychology of persuasion. Anecdotal reports on the efficacy or otherwise, of octave stretching, do not really establish anything solid about it. Are those *for*, and those *against* something they are both calling "octave stretching", actually talking about the same thing? In "testing" octave stretching to see if it is good or bad, there will be a difference in a test carried out *ad hoc* or following one set of principles, and another test carried out *ad hoc* or following some other set of principles. It is hardly surprising that some tuners report positively, whilst others report negatively, about octave stretching.

Doing some arbitrary experimental octave tuning and asking a few colleagues or clients what they think, is of course no way to scientifically investigate either the acoustics of octave stretching or the psychology of it. The reason such approaches would be immediately rejected out of hand in rigorous scientific investigation, is that whilst they may seem anecdotally convincing, the conviction is usually based on what turns out to be little more than basic psychological principles of persuasion. Such "testing" can be highly misleading, and does not generally establish any useful principle or fact.

In practice, neither scientific investigation nor anecdotal "evidence" can tell us very much about *the art* of octave stretching. There is certainly some science behind the art of tuning, and science is useful for illustrating what is actually going on, in many instances. However, we don't of course *need* "scientific proof" for the *aesthetic* results of any aspect of the art of tuning. Nor should we place any value on hearsay reports of unscientific or pseudo-scientific "testing" of artistic practices like octave stretching, that purport to "establish" its validity or otherwise.

The argument for a "brighter" tone through octave stretching implies an *improvement* in the tone of the instrument is possible. The argument against this implies there is a *detriment* to the tone of the instrument that is possible. The salient question is therefore not "should we stretch octaves", but "what factors in octave tuning improve the tone, and what changes spoil it?" In other words, we should be asking "Is it possible to tune octaves wider than their minimum acceptable size, in such a way that there is improvement, but not detriment?"

The essential idea of octave stretching

From the length of this chapter alone, it should be clear that octave stretching is an important issue in tuning, but having said that, I should say immediately that octave stretching is not some set of values to be *imposed* on any instrument by the tuner. In octave stretching the tuner should absolutely take the lead from the individual instrument being tuned, and indeed from each individual octave, and not from any ideology, personal pitch preference, or theoretical edict. It is the ability to understand the principles, and appreciate the individual tuning characteristics of the instrument, that lies at the core of the matter.

So let us be clear straight away about what correct stretching implies, because it certainly does not imply tuning upper treble octaves that are fast-beating wide, in order to try to overcome *pitch flattening*.

- Firstly the soundscapes of aurally "beatless" piano octaves are not necessarily *acoustically beatless* throughout the soundscape. This is relatively easily revealed by more acute listening over a longer time (the beating is very slow), and/or by digital analysis.

- Secondly, it is not established that the most beautiful octave tone is one in which all low partials decay purely exponentially, rather than say, with a single null pattern and a prolonged *aftersound*, or even with a (sufficiently) slow beat.

- Thirdly, it is not the case that in the finest tuning, a change of tension that significantly improves intonation, will necessarily increase beat rates in the lower partials.

None of this should be misconstrued. Octave stretching is *not* a question of saying octaves beat anyway, so therefore let them beat. That would not of course be a formula for good tuning. Rather, we are saying that the octave does not, in any case, just "beat" or "not beat", as "traditional" theory suggests. Just like the unison, the octave's beat patterns are subject to decay, and therefore to the kind of behaviour that we have already seen in unisons. We shall see that what is technically called *mistuning* in unisons, has a counterpart in octave tuning. Just as *attenuation* plays an important part in unison tuning, so attenuation also plays an important part in octave tuning. In octave tuning, however, it is not necessarily *bridge coupling* that leads to the attenuation (although it can), but often, simply wave superpositioning.

The octave partial *recipe* is also different to that of the unison. Every partial in the unison soundscape is adjustable, all its decay patterns therefore being sensitive to tension changes. Only half of the partials in the octave's soundscape are adjustable, and the octave does not, and should, not in any case sound musically like a unison. In fine tuning the unison there is no relative *intonation* issue between the strings. In the case of the octave, there most definitely is a relative intonation issue between the lower note and the upper notes.

Tailored stretching

Octave stretching that arises in response to the behaviour of the individual instrument cannot be accurately determined in advance. It is not just a matter of inharmonicity, although inharmonicity is one of the most important factors that plays a part in octave stretching. We shall see below, that the art of octave stretching lies very much in the utilisation of *attenuation* in order to achieve intonation and tone. *Attenuation* we have met in earlier chapters, and is the reduction of fluctuation amplitude, as distinct from the reduction of beat rate.

Octave stretching is very much a matter of how the piano, *i.e.* the string-bridge-soundboard system as a whole, behaves. Stretching octaves in a properly controlled manner, led by the instrument itself, is *much* more demanding on skill and time, than simply tuning without paying any attention to it. It requires more acute hearing, more knowledge of tone, more effort, more concentration, more patience, more skill, and more time. It is, in short, strictly an advanced technique, but when mastered, it becomes second nature in the art of tuning, takes advantage of direct tone and intonation perception, and yields results that are well worth the effort.

The need to stretch octaves in fine tuning

Octave stretching occurs in fine tuning for two main reasons, other than treble stability issues. Firstly, it is to take account of *inharmonicity*, which we shall look at shortly. Secondly, it is to produce the best *intonation* over the wider compass, and the best tone over the compass as a whole.

Briefly, owing to inharmonicity, it may be necessary to stretch individual octaves in order to avoid triple octaves, and sometimes double octaves, beating narrow. In practical terms, narrow beating in double or triple octaves is essentially a *tone* and *intonation* issue.

The beating in the tone is often not very intrusive in the isolated triple octave, but together, such octaves accumulate to a less than polished tone and intonation overall, and will cause or exacerbate *pitch flattening* in the treble. This is the natural tendency for treble to sound a little flat compared to the bass or mid-compass. The difference in tone and intonation of the isolated double or triple octave, made between stretched or non-stretched octaves, may not seem important at first sight. This is rather like the difference between one fine tuning of a given unison, and another. However, over the whole compass, it is these differences that accumulate to "more than the sum of the parts". They determine the overall tone and intonation of the instrument as a whole. One of the least appreciated things about piano tuning and piano tone, is the way very small differences in tuning and tone, occurring in individual notes and octaves, accumulate to produce a large difference in the overall tone and tuning quality of the instrument.

Musical expectations and inharmonicity

There is a notable difference between the musical assumption about octaves, and acoustical reality. Three side-by-side octaves, each with proper intonation, do not necessarily produce a properly intoned triple octave between the two outside notes. The intonation between the lowest and highest notes, may even be different depending on whether they are played together, one is played after the other, or one is arrived at via an *arpeggio* between the two. This is not just a psycho-acoustic effect, but also has causes in objective acoustical features, in particular, in beating.

In terms of beats, inharmonicity often leads readily to a situation in which three side-by-side "beatless" octaves do not necessarily produce a "beatless" triple octave. These effects point towards two thing: (1) That there is a small *range* of possible sizes for an octave, and (2) That octaves may need to be tuned beyond the *minimum* size at which they have "good" tone and intonation.

Octave stretching in rough tuning

If we start widening octaves relatively crudely, for example for stabilising reasons, or to counter settlement during a large pitch raise, then we are dealing with an entirely different class of "octave stretching" to what we have just been talking about. This may be acceptable where we are dealing with a pitch raise, or an instrument that has been restrung and is subject to string stretching. In any instance where the instrument is likely to fall progressively flatter towards the treble, we might sometimes stretch the octaves to counter this. For example, an instrument that has stood without proper tuning for too long, might typically fall flat unevenly over the compass. It would not be unusual to find the treble much flatter then the bass, the flattest region being the highest part of the compass.

It is often sensible in these circumstances to carry out a rough tuning followed by a fine tuning. In the rough tuning the treble will need to be raised more than the centre and bass of the compass, and will usually have a tendency to fall further as tuning takes place. Countering this requires raising the treble still further. With good judgement based on experience, it should be possible to raise the treble by the right amount, so that after falling it is at the correct pitch ready for fine tuning, rather than flat or sharp. This may require relatively coarse octave stretching in the first instance, although this diminishes as the pitch falls.

Such crude octave stretching, however will have a largely (or entirely) chance effect on the tone of multiple octaves. The same is true of octave stretching by applying any pre-determined rules for the amount of stretching, rather than stretching each octave in response to the unique characteristics of the instrument, at each octave. These practices have a place, but do not represent the true art of octave stretching in fine tuning, that we are about to describe.

Octave tuning and the so-called "checks"

In tuning octaves, piano tuners often refer to *checks* or *check intervals* that are to be used in the tuning process. These are intervals other than the octave itself, but that include one note of the octave. They are intervals that are listened to, and possibly adjusted, in order to confirm or improve the tuning of the octave. The use of the word "check" can unfortunately give the wrong impression about the tuning of these intervals.

If, for example, I am about to "check" the tuning of the octave C40 – C52 (C4 – C5), by listening to the major tenth G#36 – C52 (G#3 – C5), then the reality is that the tuning of *both* these intervals are important, *each in their own right*. There will be a certain beat rate required for the tenth, and a need to avoid beating in the octave. It would not be proper to say that the tenth is *merely* a "check", or that the octave tuning *will* be "correct" when the tuning of the tenth is deemed to be "correct". *Both* intervals are equally important.

In reality, in this example, the tenth and the octave are two intervals that share a mutual note, the C52 (C5), and each interval is making its own demand for the tuning of that one note. Setting up the tuning of the piano, so that *both* demands *can* be met, despite the effects of inharmonicity, would be an important part of the tuning strategy *from the outset*. If tuning prior to this note C52 (C5) has not been sufficiently carefully handled, with regard for what is likely to happen when tuning this octave, then I could find it impossible to satisfy the demands of both intervals. The fact that, say, I may "get away with" compromising the tenth more easily than compromising the octave, would be no reason for regarding the tenth as relatively unimportant.

Unresolved conflicts between so-called "check" intervals and octaves, are in general a result of improper or absent tuning strategy from the outset. This is not simply a question "tuning a good scale", but on many instruments implies taking into account inharmonicity and the scaling of the instrument, throughout the whole tuning process, including the scale tuning. This, of course, requires experience and understanding, and even the making of strategy judgements from clues seen in the tuning behaviour in the earliest stages of tuning, and even from the appearance of the instrument, before commencing tuning.

If one is always tuning the same make and model piano, then this judgement, for that piano, is not so difficult. Indeed, if the pianos one is tuning are all high quality concert instruments, then the approach is very straightforward and hardly ever varies. To be able to get the best out of *any* instrument, however, no matter what it is, whether concert grand, upright, or any other size grand, requires the ability to "diagnose" the required approach, based on the appearance of the instrument, sensitivity to its tuning characteristics from the outset, plus experience and understanding of the principles involved.

The octave's partials and false beating

Remember that all soundscapes, whether of single strings, notes, or intervals, consist of a recipe of partials. The soundscape of any *single string*, sounding on its own, will always contain *only non-adjustable* partials (except under certain conditions of unusual physical faults). These non-adjustable partials may beat, or not, as the case may be. However, they cannot have beating *introduced* or changed, by altering the string tension over the normal fine tuning range. Any beating found in these partials will be *false beating*, whose beat rate cannot be adjusted in fine tuning.

In contrast, soundscapes produced by *more than one* string sounding together, in other words, soundscapes of *unisons* and *intervals*, *always* contain at least some adjustable partials, whose beat rates *can* be adjusted. Remember that an *adjustable* partial is defined as one that has a *beat rate* that is adjustable by altering the string tension. It is the adjustability of the *beat rate* rather than any other feature, that qualifies a partial as "adjustable". If the beat rate is unadjustable, the partial is "non-adjustable", but we still may be able to change *other* features such as decay rate, through tuning.

In addition to the adjustable beating in an adjustable partial, there may also be additional *false beating* in that partial, that *cannot* have its beat rate adjusted, even though it is occurring in an otherwise *adjustable* partial. This is because the soundscape of any interval will tend to *inherit* any false beating from any string in the interval, that is false. False beating, which itself is always non adjustable, is indiscriminate, and can "infect" both the adjustable and non-adjustable partials of an interval. Where it occurs in adjustable partials, it can occur in addition to the adjustable beating.

So, a partial of an interval can contain false beating if it has *inherited* it from one or more false strings. However, once it has been *inherited* by an interval, it is part of the interval's own soundscape. It is no longer simply the offending false string that you are hearing. It is the *inherited* false beating *in the interval's own soundscape*, which is not simply the two soundscapes of the individual strings. Any false beating in the interval, therefore, may be different in character to the false beating in the individual false string (or strings). It is important to realise this, to later understand how it behaves in octaves, and how it can be addressed.

Which of the octave's partials are adjustable?

Which partials in the soundscape of the octave will be the adjustable ones? For *any* interval, each note *sounding on its own* would have a soundscape with partials arranged in an approximately harmonic series. When both notes sound together, the sound is a soundscape of the *interval*, in its own right. It is not *necessarily* the same as just overlaying or mixing the two soundscapes of the individual notes.

In any interval's soundscape, an adjustable partial appears in any frequency region where we would find a partial in *both* notes, rather than just one or the other. We can call these frequency regions *meetings*. In other words, wherever a partial from one note *meets* with a partial from other (*i.e.* it has a similar frequency), there will be an adjustable partial in the *interval soundscape*, when both notes sound together. This adjustable partial of the interval, in effect, *replaces* the two partials of the individual notes, that meet.

A spectrum or soundscape for an octave interval is formed when two notes an octave apart are played together. Only the *even numbered* partials of the lower note meet with partials of the upper note, but *every* partial of the upper note will meet with one partial from the lower note. For example, the even partials numbered 2, 4, 6, 8 and 10 of the lower note will meet with the partials numbered 1, 2, 3, 4 and 5 of the upper note, consecutively. This means that *half* the partials in the soundscape of the octave are adjustable partials, generated by the strings of both notes. The other half are generated by the strings of the lower note only. They are created by the odd numbered partials of the lower note.

The overall spectrum for the octave interval contains, *just like the unison*, one set of partials arranged in an approximately harmonic series. This series of partials has the lower note's fundamental as the first in the series. But unlike the unison, in tuning the octave only the *even numbered* partials in this series are adjustable through altering the string tension. Half the partials in the octave remain *unadjustable*. By rough analogy, instead of painting a picture in which we can control the whole pallet (like the unison), we are now painting one in which we can vary only half the colours. For those learning to tune, this can sometimes contribute to more difficulty in fine tuning octaves, than in fine tuning unisons – differences in tuning are not so obvious.

When we are tuning tempered intervals, there are generally only one or two adjustable partials whose decay curves we are conscious of adjusting. This constitutes one extreme tuning situation. At the other extreme is the unison, in which we are adjusting every partial in the spectrum. Between these two extremes are the octaves.

The octave's dependence on unison tuning

A common error made by students of tuning, is attempting to fine tune an octave, when the unison being tuned from, is itself not properly tuned. Adjustable beating in this unison, that could have been reduced, is *inherited* by the octave, but cannot have its beat rate altered in the octave tuning. In this respect, once inherited by the octave, it is rather like false beating. However, unlike false beating, all we have to do is to go back and tune the unison properly, to reduce or "eliminate" its beat rate. The tone of the octave will always be *limited* by the tuning of the unisons, so in the first instance, the unison tone *must* be good. In the piano tuning as a whole, *everything* rests on the quality of the unison tuning. Unisons are the basic building blocks of the tuning. Without beautiful unisons, the rest of the tuning will never be beautiful.

Tone and intonation

Tuning the octave affects its tone in a way noticeably different to tuning the unison. Partial fluctuations in the octave soundscape do not have precisely the same psycho-acoustic consequence as they do in the case of the unison. This is not so surprising given that the octave's spectrum is quite different to that of the unison. Firstly, there are typically many more prominent partials visible in the spectrum of an octave in a given part of the compass, due to the fact that two different physical notes, and hence more strings, create its spectrum. The octave's soundscape is actually a richer tone recipe.

Secondly, every odd numbered partial of the octave is *unadjustable*, so these remain relatively unaffected by tuning (their frequencies of course change with tuning). This means, roughly speaking, 50% of the soundscape of the octave is determined by fixed factors essentially unaffected by tuning, whilst only 50% of the soundscape is affected by tuning. This does not mean the octave tone is insensitive to tuning, but it does mean its response is different to that of the unison. The tone of the unison is more critically affected by small tuning changes. However, in the case of the octave the *intonation* between the upper and lower notes is a critical issue. This can be very sensitive to its precise tuning condition.

Pitch flattening

A natural feature of piano tone found on many instruments, particularly if smaller than concert size, is that the treble compass can tend to sound flat when the instrument has had the octaves tuned "beatless". On some instruments the bass may also sound correspondingly sharp. In both cases the effect is worse, the further away we are from the centre of the compass. The effect is linked with both inharmonicity and the psycho-acoustics of pitch perception when listening to piano tone.

It is perhaps most easily discerned by playing upward *arpeggios* into the treble, or downwards into the bass. The pitches of the notes appear to get progressively flatter than they apparently "ought to be", the higher in the treble they fall, or progressively sharper, the lower in the bass they fall. The effect varies widely from instrument to instrument, but is also highly dependent on how the instrument has been fine tuned. On most pianos, if octaves have been tuned only as single octaves, with no regard for compound octaves and other compound intervals, then the effect will tend to be even more noticeable. On smaller instruments, the corresponding bass sharpening will occur.

In the treble, this effect is sometimes called *pitch flattening*. Remember that *pitch* is our *subjective sensory response* to the incoming acoustical information. It is *not* a property of the sound itself. Pitch flattening itself, is therefore a psycho-acoustic effect that can occur despite the fact that the notes in the treble are not actually tuned flat, in the sense that the octaves are not tuned beating narrow.

One may well ask how higher notes can sound flat compared to notes a number of octaves below, if each octave in between sounds like a good octave in its own right. In asking the question, we must see that we are just *assuming* that if the intervening octaves sound right, then the top and bottom notes must also sound right in relation to each other. We only think this in the first instance because of prior conceptual conditioning about musical intervals and pitch. Surely three good single octaves side by side, *must* make a good triple octave between the outside notes?

Well, actually, no, not necessarily. The *fact* is that on the piano this simplistically assumed pitch relationship is *not* always true. When we find the facts do not fit our ideas about how things work, we should question our ideas more closely. In this respect the piano tuner, as an artist, should take the same attitude as the scientist confronted with evidence disproving a previously unchallenged theory. The fact is that the acoustical and psycho-acoustical reality of musical tones does not always conform to the relative simplicity of *music theory*.

A "cause" of pitch flattening often cited is inharmonicity, but again we must remember that pitch is a psycho-acoustic *perception*, not an objective measurable quantity. So inharmonicity itself, which *is* an objective measurable quantity, can at best only play a part in the pitch flattening effect. Inharmonicity, even without considerations of pitch intonation, as we shall shortly see, does often itself require octave "stretching" (*i.e.* making octaves wider) in order to avoid unwanted beating in multiple octaves.

The top octave mistake

Pitch flattening is felt most acutely in the upper treble, which sounds flat relative to the centre compass. This can present the temptation to stretch octaves in the attempt to raise the pitch of the upper notes in this part of the compass, by introducing rapid beat rates in the octaves. Stretching the uppermost octaves so that they are fast beating simply because this is where the need of stretching is felt most, is the worst possible way of addressing pitch flattening. Not only does it produce the worst results in terms of detrimental effect on tone, but it is also happens to be the least effective way of tackling pitch flattening overall.

Firstly, octave stretching should not necessarily imply obviously beating octaves, but this is what will inevitably happen if the *top octaves* are stretched by even a *small amount*. Secondly, the most important part of the compass for octave stretching, as we shall discuss shortly, is well below the top of the treble. Once successful octave stretching has been achieved in the lower and mid-treble, the upper treble tends to fall relatively easily into place without the need to tune coarsely wide top octaves.

Pitch versus beating

Stretching to overcome pitch flattening, and stretching to overcome the tendency of inharmonicity to produce narrow beating triple octaves, are different principles but they are connected. Correct stretching to avoid narrow beating triple octaves, in other words, just to take into account inharmonicity, may be sufficient to overcome pitch flattening, and will in any case greatly improve the tone and intonation of the instrument. They are, however, subject to two different mechanisms. Let's begin by looking at the theory behind this.

A change of musical *pitch* is usually associated with a change of frequency *ratio* (one frequency divided by another). Although pitch depends on other factors also, it is the frequency *ratio* between two fundamentals, for example, that correlates to the perceived pitch interval between the two notes. Beat rates, on the other hand, are dependent on frequency *differences* (one frequency minus the other). Let us just assume for a moment that for some reason we *are* widening an octave by deliberately introducing beat rates. A consequence will be that raising the upper note of a wide octave by a fixed pitch change corresponding to say, n cents, will require an increase in *beat rate* in the octave, that depends on where the octave is in the compass.

In fact, the beat rate *doubles* each time we go an octave higher in the compass, for a given pitch change. Suppose we wanted to widen an octave at the top end of the compass by raising the upper note a microinterval of n cents. The beat rate we will need to introduce here will be *eight times greater* in this high octave, than it would be in an octave three octaves lower in the compass, if widened by the same amount in cents.

In other words, *if we wanted to avoid or minimise beating in octaves*, but we wanted the upper compass to be raised by n cents in relation to the bass or mid-compass, it would be far more effective to stretch octaves in the lower parts of the compass, rather than in the highest. Any octave that is stretched, will of course raise all the octaves tuned above it. Rather than raising a top note of a triple octave by n cents, we could "divide" the n cents over the lower two octaves. In practice, of course, we can utilise *attenuation* to stretch octaves without necessarily having to leave octaves with clear beat rates, but the basic principle still applies. Most of the stretching in individual octaves, carried out for the sake of triple octaves in the treble, needs to be applied to the lower two octaves of any triple octave.

To avoid problems manifesting between notes three octaves above the scale, and notes in the scale, and to avoid excessive stretching in the highest of the three octaves, the intervening octaves should therefore be stretched, as soon as possible as tuning proceeds upwards from the scale. However, we cannot stretch in an uncontrolled way, or spoil the tones of the octaves. What actually happens is that the instrument allows the effects of stretching in potentially producing fluctuations or beats, to be reduced or hidden, with judicious tuning, through the effects of *attenuation*, which we will describe shortly. As it turns out, the acoustical mechanisms in the piano that allow us to do this, are not the same throughout the compass. In general, the largest attenuation is available in the *middle* of treble, rather than in the centre of the compass or the higher treble.

The fine tuning region

When learning to tune octaves we may have acquired a mental picture, a mental model that perhaps may only be held subconsciously, in which we believe there is one string tension and one tension only, at which the octave is "beatless", and therefore "in tune". This residual, rudimentary mental image of the dynamics of octave tuning, must be overcome if the true potential of octave stretching is to understood. As in all tuning, we must begin to appreciate partial fluctuations *as a tonal feature* in the context of the tone as a whole, rather than as abstract phenomena that we happen to be able to hear.

For many octaves there is not necessarily any one tuning position at which fluctuating movement in *all* the partials is simultaneously eliminated, but it is usually possible to find a position at which movement is effectively minimised. We can represent this situation graphically in a similar way to the unison or tempered interval. Fig. (10.1) represents the possible tuning for an octave from a position where it is has clear regular beating too narrow, to a position at which it is has a clear regular beating too wide. In the small *fine tuning region* in between, the overall state of the octave interval's spectrum is very "still" – the lack of movement indicates an "in tune" condition. The fine tuning region is exaggerated in

Fig. (10.1) for clarity. In practice the region is very small, but nevertheless highly significant, and definitely manageable by the expert tuner.

The region should *not* be conceptualised as a region of acceptable "error" either side of a point of "perfection". This is not the case. The "best" tuning, even in terms of the tone qualities of that octave in isolation, is not necessarily in the middle of the region. As already said, the "best" tuning is not necessarily the one with statistically minimum partial fluctuation, but may be, for example, a tuning in which some significant partial has a single null decay, rather than a plain exponential decay. Additionally, the tuning of an octave is something that must fit into a much larger picture than the single octave itself. Octaves are not like "standard bricks" from which the final tuning is built. Octaves and the overall tone and intonation of the instrument are connected in a much more complex way.

The fine tuning region's presence becomes especially clear when attention is paid to the beat rates in the so-called "check intervals". It is sometimes detectable as a region in which a change of pitch intonation or tone quality without a significant or detrimental change of "beat rate", is possible. At other times, it may be a region in which a small amount of fluctuation can even be introduced. It is not equally present in every octave, and tends to be less present in the central compass close to the scale area, than in the mid-treble.

Fig. 10.1

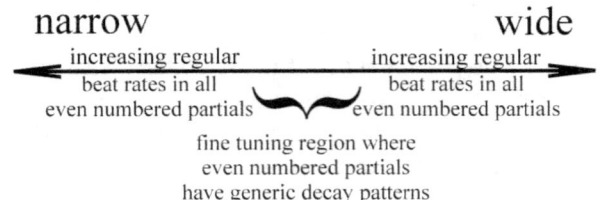

If the octave is tuned a certain amount narrow, or a certain amount wide, then the adjustable partials will in general have a *regular beat* decay pattern (envelope modulation pattern) which is detrimental to tone. In between, in the fine tuning region, at smaller "mistunings", the soundscape may appear "beatless" or may have other fluctuations slow enough to be comparable to the decay itself.

The general situation in octave tuning, in which movement cannot necessarily be entirely eliminated, is particularly obvious aurally, in the bass octaves. There may be a suitable tuning position at which the octave is clearly "best" by aural judgement, but close aural examination of the spectrum easily reveals that at any such point there are beating decay patterns in many higher partials. For bass octaves, and for many treble octaves, there may never be *any single position* at which all partials cease to beat or fluctuate. This is also easily demonstrable using digital analysis of bass octaves. Whilst it is more obvious in bass octaves, the same general principle and causes still applies to octaves in the treble.

Tuning octaves by intonation

The perceived *pitch intonation* of one note relative to the other must be attended to in octave tuning. One should simultaneously listen to relative pitch intonation, and tone (containing fluctuating partials) when tuning octaves.

Let us suppose we are tuning a rising octave, and we have arrived at an apparently "still" soundscape, indicating a "pure" octave, in normal piano tuning parlance. Associated with this position is also the perceived *pitch relationship* of the upper note relative to the lower note. The perception of pitch of course involves significant subjective, psychological factors, of which one must be aware, and about which one must have good theoretical knowledge and self-knowledge. With sufficient experience it is possible to tune the octave *by intonation*, keeping a check on its tone, in terms of fluctuation in the soundscape. However, there are a number of caveats.

The faculty of "perfect pitch" (or "absolute pitch") is of course a red herring here, in relation to fine tuning, because we are talking about *intonation* relative to the lower note. Pitch sense, either relative or "absolute", certainly cannot be used as a sole technique for fine tuning, because fine tuning is as much about tone as it is about intonation. Tone does not simply follow from pitch judgements. The psychological perception of pitch is in any case not always stable – it can change with exposure to stimulus and other factors, physical, psychological and physiological. In piano tuning, the extended exposure to relatively loud high frequencies in the treble can, for example, cause the ear-brain system to perceive pitches as progressively flatter as exposure continues, therefore demanding progressively higher frequencies to sound as a given pitch.

"Absolute pitch" sense is also not immune from such influences. Students I teach who have "absolute pitch", invariably suffer changes to their pitch perception when they are unwell, or under stress. However, experienced tuners are involuntarily, constantly "training" their sense of pitch or intonation under the relatively "objective" guidance of beats and other soundscape properties. An experienced tuner's sense of intonation in piano tones can therefore become more reliable, not least because it always takes into account a simultaneous awareness of tone, as beats and other fluctuations.

Advanced tuners use this more developed sense of *pitch* to modify the tuning condition of the octave by an exceedingly small amount, using pitch sensitivity, in a very controlled way. Suppose we are tuning a note up from flat, to form an octave above a lower reference note. Having arrived at the apparently "still" soundscape for the octave, somewhere inside the fine tuning region, it is often then possible to raise the upper note still more by a very small degree, using *pitch* sense, remaining inside the fine tuning region, without detriment to the overall soundscape, and often improving it.

Despite how we might mentally *conceive* it, pitch is not really a single, simple thing. It is a complex, psychoacoustic effect. The idea of a single, "correct" pitch, is actually a false paradigm. Within the micro-fine tuning range, there is not just one "correct" pitch to be perceived, so there is no purpose in searching for such a thing. Nevertheless, for the advanced, *experienced* tuner, the sense of *relative pitch* or *intonation* can be as quick a guide to what is necessary in octave stretching, as anything.

However, pitch intonation awareness must always be accompanied by awareness of the tone quality of the soundscape, including any beating. Below, we will look at this process in much more detail, because there is far more to it than simply making intonation judgements in the octave. It requires perception beyond that of the musician who is not also an expert tuner. A master tuner might tune an octave simply so that the "stretched" intonation is correct both in itself, and in its relationship with the tone of the octave, but this description is not much help to someone who has not yet mastered the art. Hence, there is a great deal more detail to follow.

Tone versus intonation of an octave

It is sometimes the case that "optimally" tuned octaves in terms of beating, are perceived by the musician as having "out of tune" intonation. It is also a possibility for an octave to be perceived as "correct" or "musically in tune", *i.e.*, with good intonation, even when regular beating (albeit slow) is present somewhere in the spectrum to some degree. Furthermore, depending on the partial recipe, an octave may be beating *wide*, whilst the perceived relative pitches of the notes suggests a *narrow* octave. In other words, an octave or multiple octave may be wide or pure "by beats", whilst at the same time being perceivable as narrow in its intonation. This is particularly true if a high degree of inharmonicity change over the interval is present.

Basically, then, intonation sense is something that can be, and ultimately *should be* used in octave stretching, but it must be used in conjunction with tone control, by listening to fluctuations in the soundscape, and certain technical procedures which we shall discuss shortly. The "stretching" of octaves by relying only on pitch sense *alone* is bound to be unreliable and improperly controlled. Also, "stretching" by a fixed number of beats in the octave would be an error, giving arbitrary results in terms of tone and intonation over the compass as a whole.

Inharmonicity - relation of single and multiple octaves

Let us first reiterate the situation for a single rising octave. Fig. (10.2) represents this as a schematic diagram, with a fixed lower note on the left, and the upper note being tuned, on the right.

Fig. 10.2

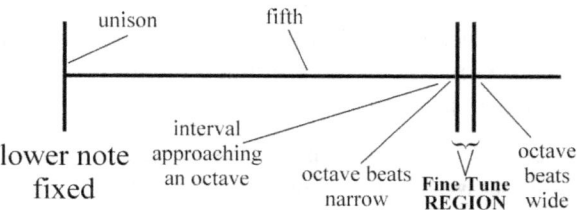

Fig. 10.2. From left to right represents perceived pitch, schematically. Vertical lines represent positions in the range of perceived pitch, from one note on the left, to a note an octave above it.

Here, there is not a single point position for a line on the right hand side to represent an octave, but rather, a small fine tuning *region*. Anywhere inside this region will constitute a "good" octave, in isolation from all other intervals. For visual clarity, Fig. (10.2) is schematic, and not to scale. The *fine tuning region* is in practice *exceedingly small* compared to the "tuning distance" between the lower and upper notes (the "distance" one would have to raise the lower note to make it the same pitch as the upper).

Let us now consider tuning multiple octaves. Fig. (10.3) represents the tuning of a triple octave. The grey bars represent the size of the octaves, double octave or triple interval, when they are tuned "beatless" (the *fine tuning region* now disappears into the thickness of the horizontal black lines). We can imagine these octaves are tuned "one on top of the other", bringing the upper note in each case, up from flat. Again, the proportions in the diagram are for visual clarity rather than to any scale.

Fig. 10.3

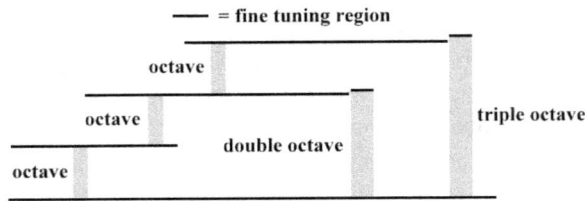

Fig. 10.3. Representation of the tuning of a triple octave in the treble. The vertical grey bars represent the sizes of the "beatless" intervals. Three "beatless" octaves may constitute a smaller interval than the "beatless" triple octave itself.

From the diagram we can see that there is a discrepancy between the sizes of the single octaves added together, and the sizes of the double and triple octaves, when the octaves are tuned "beatless". This is primarily due to differences in inharmonicity.

We can describe the reasons for the basic mismatch between single and multiple octaves, fairly simply, if we ignore further complicating factors like false beating and possible bridge coupling. Basically, inharmonicity raises the frequencies of the partials of any note, from the true harmonic positions, but the higher the partial, the more severe this raising of frequency becomes. Tuning the fundamental of a note to the 2nd partial of a note *one* octave below, requires raising the tuned note's fundamental by the same amount the lower note's 2nd partial is raised, due to inharmonicity. Because the raising in frequency due to inharmonicity, gets more, the higher the partial number, a given note's 2nd partial is not raised nearly so much as, say, its 8th partial.

If we had been tuning to a note *three* octaves below, we would have been tuning our fundamental to the 8th partial of the lower note. As a result, it is possible to tune, say, three octaves upwards, in single octaves, by matching each fundamental with the 2nd partial of the note an octave below, but then find that the fundamental of the top note does not match with the 8th partial of the lowest note. The actual result will depend on how the inharmonicity varies over that portion of the compass in which the three octaves fall, in other words, it depends on the scaling of the piano. Generally speaking, however, the result is similar to that shown in Fig. (10.3), and there will often be greater mismatches on smaller instruments.

Fig. (10.3) shows how we might tune up three consecutive octaves, each octave being good, only to find the resulting triple octave is narrow. The black lines at the top of the grey bars indicate the tuning positions of the upper notes for these intervals, tuned from the fixed lower note (at the base of the diagram). Anywhere inside a grey bar represents a beating interval. Here, the tuning positions of the notes for the double and triple octaves do not coincide with those for the individual octaves.

The notes tuned to form the individual "beatless" octaves collectively add up to intervals less than the beatless double and triple octaves. In other words, all three single octaves may be tuned "beatless", but if they are, the double and triple octaves made between them may be exhibit beats – they will be narrower than they need to be in order to be "beatless". Raising the frequencies of the notes in order that they fall within the fine tuning regions for the double or triple octaves, may result in them falling *outside* the fine tuning regions for the single octaves.

These scenarios are not just hypothetical. The principle is present to some degree on all pianos, but can be very marked on some smaller instruments. It is not just the *size* of the piano that determines the severity, but the *scaling* of the instrument, because it is how inharmonicity *changes* over the compass, that will have the greatest effect.

One solution is obviously to "stretch" the intervening octaves so that together they create beatless multiple octaves, rather than narrow ones. If the octaves are tuned without regard for the need to stretch them in order to match single octaves with multiple octaves, some musicians may not notice the resulting narrow mistuning in any given multiple octave. This does not, however, alleviate the need to address this feature of the piano's tuning characteristics. The improvement in the way the instrument musically works as a whole, and the improvement in the overall tonal and musical quality of the instrument that comes about from correctly dealing with this characteristic, is usually still noticeable to the musician in terms of the perceived overall tone and intonation of the piano.

Octave stretching, however, should not only address the mismatching of octaves and multiples octaves due to inharmonicity, but should also address treble pitch flattening (which may also be due to the effect of inharmonicity on perceived pitch), and the tone of the instrument as a whole. It should be thought of as a natural feature inherent in the instrument's requirements, not as a technique being imposed on the instrument. Its effect on any individual interval may be large or small, but the collective effect on the instrument as a whole is in any case large.

Stretching the octaves by arbitrary amounts will not do. The latter will be no more likely to find the optimum inherent in the instrument, than not responding at all to this natural, acoustical feature of the piano. As we shall see, uncontrolled stretching can cause over-stretching which can have undesirable effects, not only failing to realise the potential tonal beauty of the instrument, but also (counter-intuitively) often actually *worsening* the pitch flattening effect, rather than improving it.

The octave as a building block

The tone of the octave is a critical building block for the tone of the instrument as a whole. The better these building blocks fit together, the better the tone of the instrument as a whole. Other building blocks include the scale, and the unisons, the unisons being the most fundamental and critical building block. The following is an advanced technique that demands excellent scale tuning, and excellent unison tuning, in order for it to work.

The final tuned state of the piano is of course one in which all trichords are sounding. Tuning intervals by muting the outer strings of the trichord and "filling in" the unisons later, will not succeed nearly so well, because this process ignores the very mechanisms that octave stretching exploits. These mechanisms arise from the fact that the spectrum (or tone recipe) of an interval - be it a unison, tempered interval, octave or multiple octave - depends on the number of strings present and sounding, and not just on the interval itself.

Exercise : Finding the fine tuning region

Less experienced tuners should experiment with this on a number of different instruments. Start with a good scale, with properly progressing major thirds, and excellent unisons. It is in fact *imperative* that the thirds progression is good, and that the unisons are excellent, and stable. It is not that the fifths and fourths are not important in the scale itself, but the thirds will be the most important feature when it comes to tuning the octaves.

Take a major third in the scale, usually a third above G-B is more likely to work. Tune one string of the note an octave above the upper note of the chosen major third, so that the octave is good. Now test the beat rate in the tenth to ensure it is the same as in the third. The exercise is to move the upper note's position again by a very small amount, without destroying the octave, whilst changing the beat rate in the tenth between the lowest and highest notes. The object is to change the beat rate in the tenth by a very small amount – typically around 0.5 Hz. This is equivalent to giving the tenth not the same beat rate as the major third, as "traditional" theory dictates, but about the same beat rate as the *next higher* major third in the scale, assuming good beat rate progression in the scale.

If the beat rate in the tenth was the same as that in third, it should be possible to *increase* it significantly without destroying the octave. Some instruments will not allow any change to the octave that does not deteriorate it, in this part of the compass – this corresponds to a very narrow *fine tuning region*. Where this is the case try tuning a 17th (an octave-tenth) instead. In this case you must first tune the tenth so that its rate is the same as the third. You should not be judging the octave purely on any specific partial beating. You should be aware of any beating, false or adjustable, but should be focussed on the tone and pitch intonation of the octave, *as a musician would hear it*.

Having discovered the amount by which the tenth or octave-tenth can be changed without spoiling the octave (in the case of the seventeenth, this is the upper octave), we now know how large the fine tuning region is, and what level of fine tuning we are dealing with. If there seems to be no fine tuning region, it may be necessary to experiment with another instrument, or slightly higher in the compass, but it should be remembered that the process does require exceedingly good tuning lever and pin control. Without this, even a wide fine tuning region may not be discovered.

The upper unison can then be completed. Listen again to the octave when two strings of the upper unison are tuned, and again with all three. See what effect two strings has, over just one, and then how three strings differ.

Having become used to the level of precision with which we need to work, the next stage is to *increase* the stretching we can carry out, utilising further tuning characteristics of the piano. This involves deliberately slightly *spoiling* the octave when tuning just the first string of the upper trichord. This may seem strange, but all will be soon be revealed.

Exercise : The naturally beating octave

Let us now consider the case when the first string of the upper trichord sounding on its own, has a slow, shallow false beat in its fundamental. The beat rate may be only one every two or three seconds, and this will not necessarily be noticed. However, *this can have a significant influence on the tuning behaviour, nonetheless*. Typically, when tuning up the first string of the upper note of an octave from flat, we can reach a suitable "optimum" mistuning at which the false beat in the octave is minimised in prominence, possibly before the zero mistuning position is reached. This, just as in the case of the unison, will have been due to beat attenuation, as much as beat rate reduction. In practice, the false beat may not even have been noticed – working fast, we would simply be tuning the octave to its minimum beating state.

From the tuning point of view, it may well seem that this position *is* the position of the "in tune" octave. In reality, the upper string may still be flat, in terms of the *required* position. Whilst this may not seem a problem in isolation, continual repetition of this procedure can produce octaves that in combination, form compound octave intervals that most definitely will be flat, and probably even obviously beating narrow, *i.e.* they fail to address the problem of matching single, double and triple octaves due to inharmonicity.

The required technique is that in tuning the octave *on the first string of the upper trichord*, any false fluctuations should generally be *allowed* to be fully inherited by the octave, rather than attenuated. Any position of mistuning that produces an apparently optimum octave when tuning up from flat, must generally be passed through, and the tension raised more, at least until the false beat is definitely fully inherited in the octave. Whether the string tension is finally raised or lowered into position will depend on the process of setting the pin. As you do this, also listen to the *intonation* of the octave, and the beat rate in the tenth and/or seventeenth. Never move the octave tuning without listening to its intonation with the lower note, simultaneously. If this doesn't seem to have much meaning at present, it will eventually.

This deliberately inherited, rather than attenuated, false beat, is not a problem because we will *hide* it in the octave spectrum when the second and third strings of the upper trichord are tuned. The approach to the unison tuning remains the same as normal. The only difference is that any slow falseness in the upper trichord is now consciously *noticed* in the tuning of the octave, and is deliberately *not hidden* in the octave tuning with only one string of the upper trichord sounding, even when it could otherwise have been. One must consciously think about doing so, because hiding any false beat otherwise naturally tends to happen simply as a result of tuning the octave to its most "beatless" tone, even if one is not aware of attenuation taking place.

The octave is tuned by *raising* the upper string until the false beat is obvious. The danger is that there is an "invisible line" that must not be crossed in doing this. There is often a *range* of tuning over which the beat does not change very much, but at the upper end of the range, "beyond the line", so to speak, adjustable beating is also prominent and it will not be possible to subsequently attenuate the beat very much when the unison is completed. At the lower part of the range, better attenuation will be possible.

The position of the complete final upper trichord *must* be checked using a tenth or octave-tenth check, and if this is found to be too slow, the procedure must be repeated, setting the first string higher, even if this now involves setting it outside the octave *fine tuning region* and introducing a small amount of additional beating. Remember that the beat rate in the tenth (or octave-tenth) when just the first string of the upper trichord is sounding, *will not necessarily be the same* as the beat rate in the tenth with the whole upper trichord sounding.

Once all three strings are sounding, the tenth or octave-tenth may be slower (the rates in any case are not always stable, and can change as the interval decays). The rate must be *faster* than the corresponding third on the bottom note of the tenth (or octave-tenth), but should not be too fast. The tenth (or seventeenth) should shimmer brightly with a faster beat rate than the third, but should not sound coarse or "ripping".

In actual practice, we would be comparing this rate to the ones below, to ensure progression. However, even this "progression" is really more a matter of *tone quality* than quantified rate. Recognition of fast beat rates and their tonal effect, and the ability to make them similar and progressing, becomes an important part of the art. In terms of beat rates, the default arrangement is to have the tenth's beat rate equal the rate of the third that about two semitones higher in the compass than the normal third at the bottom of the tenth. This will be illustrated in the example below. Larger stretches can also be carried out, but always with due regard for tone.

If after completion of the upper trichord, the octave is found to have been compromised, with detrimental false beating or detrimental beating from wideness, then retuning will be necessary. If you are going to octave stretch, and you are going to do it conscientiously, then you must *expect* to have to re-tune occasionally. If *time-taken* is still your biggest problem in tuning, or your main concern, then you should perhaps not be undertaking advanced octave stretching of this kind at the current time. You can still tune octaves, say, a little wide, but you must not expect the effect to be the same.

The beat rates in the tenths may typically "lead" over the beat rates in the thirds by *at least* 0.5 Hz. For example, suppose the third C40 – E44 (C4 – E4) has a beat rate of 10.4 Hz. "Traditional" theory says the beat rate in the tenth C40 – E56 (C4 – E5) should then also be 10.4 Hz, if the octave E44 – E56 (C4 – E5) is "beatless". We are now saying the tenth might have a beat rate of, say, 11 Hz, whilst the octave remains good. Neither the third nor the tenth generally have truly constant beat rates, however. Typically, the beat rate will change after the initial part of the decay, but both increases and decreases in beat rate are possible. Often, the beat rate cycles faster and slower, or *vice versa*.

In the actual practice, beat rates in the seventeenths will not generally be the same as in the corresponding tenths. They will typically be faster. The "lead" (increase in the beat rate) of the seventeenth or tenth, over the thirds, will not be the same in all parts of the treble compass. Also the point in the compass at which "lead" begins to happen, will generally be different for the seventeenths, than for the tenths. The lead in beat rates in the tenths and seventeenths becomes greatest for notes in the mid-treble. We will look at this in more detail shortly.

435

Example analysis

As an example, let us take a look at some partial envelopes in this process, for a double-octave tuning on the Yamaha G3 grand piano. Fig. (10.4) shows the enveloped for the first adjustable partial of the double-octave C40 – C64 (C4 – C6), when only the first string of the upper note, C64 (C6), is sounding. The envelope shape is shown at five progressive tuning positions of the C64 (C6) string, starting at (A) with a narrow double-octave, and ending at (E) with a double-octave wide by about two beats per second.

Fig. 10.4

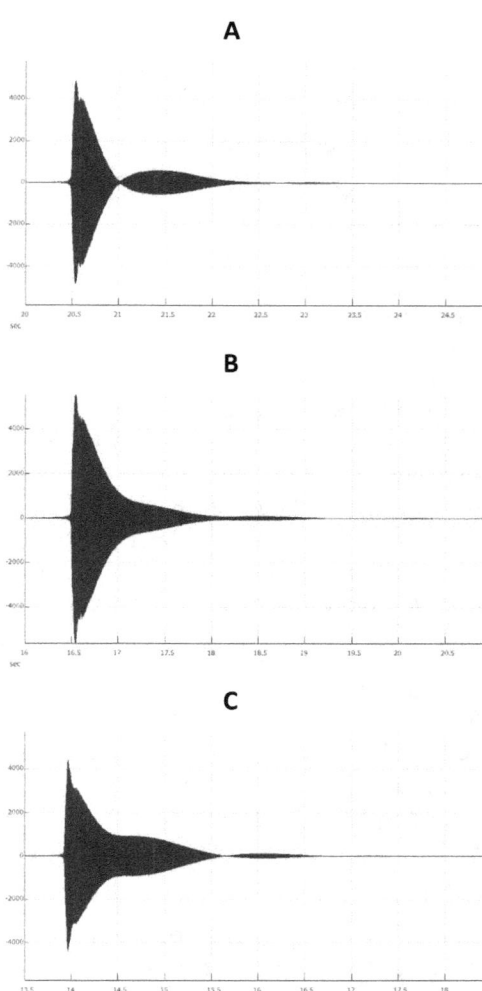

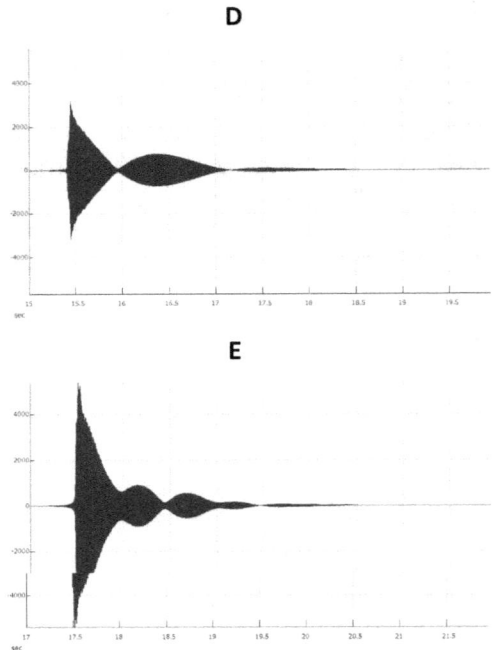

Fig. 10.4. The lowest adjustable partial of the double-octave C40 – C64 (C4 – C6) on a Yamaha G3 grand piano (at the pitch of C64). The top C64 of the interval is raised in stages from Fig. 3A to Fig. 3E, and checked by the major seventeenth A-A-flat 36 (Ab3) – C64 (C6) (two octaves plus a major third), which is a wide interval. As the C64 (C6) is raised, the beat rate in the wide seventeenth increases. At (A), the double-octave is narrow, but the seventeenth is, as normal, wide. At (B) the double-octave is "pure" aurally – the decay rate is not constant, but no beat is aurally detectable. It has a definite wide beat at (D), and by (E) is 2 beats per second wide. The beat rates (beats per second) in the seventeenth are: A 9.0, B 9.4, C 9.7, D 9.9, E 11.3.

Looking at Fig. (10.4), the adjustable partial at the pitch of C64 is shown for five stages of tuning, progressively raising the C64. The major seventeenth A-flat 36 (Ab3) – C64 (C6) is checked at each stage for its beat rate. At (A) the double-octave is narrow, and exhibits a beat of around 1 Hz (a *decaying regular beat*). From (A) to (D) the C64 (C6) is raised by almost 1 Hz according to the change in the seventeenth, which increases its beat rate by 1 Hz. The double-octave has also becomes wide after this change, with a beat rate of about 1 Hz.

In order to pass from one beat narrow to one beat wide in the double octave, we would by "traditional" theory expect to increase the beat rate in the seventeenth, which is always wide, by *two* beats, not just one. (On the double octave we pass from one beat narrow to one beat wide – that is a total change of two beats in the seventeenth).

In fact, we are not really passing from one beat narrow to one beat wide in the double octave, in the "traditional" theory's sense. From Fig. (10.4A) to (10.4D), as we raise the tension, we are in fact *attenuating* a false beat in the fundamental of the upper note, and then unattenuating it. It is best attenuated at Fig. (10.4B). The culprit false beat that is inherited by the double octave, originates in the fundamental of C64 (C6).

Fig. (10.5A) shows the 4th partial of the C40 (C4), and Fig. (10.5B) shows the fundamental of C64 (C6). The 4th partial of the C40 (C4) is "pure" but the fundamental of C64 (C6) has a false beat rate of around one beat per second.

Fig. 10.5

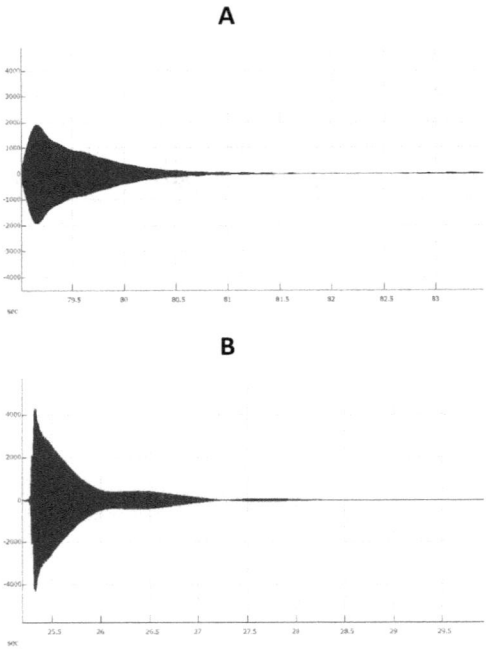

Fig. 10.5. Decays from the double-octave tuning on the Yamaha grand piano. A: The 4th partial of C40, which does not display a false beat. B: The fundamental of C64, which does display a false beat of about 1 beat per second. (The fast fluctuations on the edges of the decay at A, may be due to effects from adjacent, damped strings).[69]

We would normally *expect* such false beat to be inherited in the double-octave, and indeed it is. It is fully inherited at Fig. (10.4A) and (10.4D), and most attenuated at (10.4B).

In octave stretching, rather than tuning the first string of the double octave's upper note C64 (C6) so that the beat is best attenuated, as it is at Fig. (10.4B), we deliberately leave the double-octave inheriting it, as at Fig. (10.4D). We then proceed to attenuate the beat just by tuning the other two strings of the C64 unison.

---

[69] See *Beating frequency and amplitude modulation of the piano tone due to coupling of tones*, Cartling, Bo, *JASA*, April 2005, 117, 4, pp. 2259-2267.

Fig. (10.6) shows side by side at the same scale, (A) the double-octave's lowest adjustable partial when only the single (false) string is sounding in the double-octave, together with (B) the partial when the whole C64 (C6) trichord, tuned to an excellent unison, is sounding in the double-octave. At Fig. (10.6A), we see the false beat we have deliberately allowed to be inherited by the octave.

Fig. 10.6

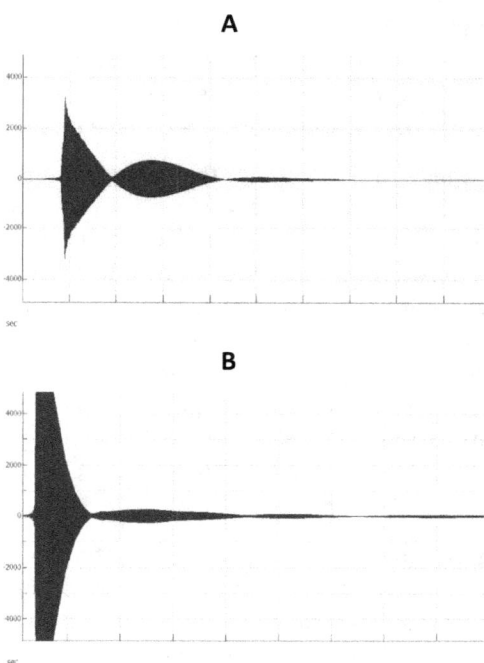

Fig. 10.6. The first adjustable partial of the double-octave C40 – C64, when (A) one string only of C64 is sounding, and (B) when the full trichord of C64 is sounding (tuned as an excellent unison). At (B), despite the partial starting with greater amplitude, the beat amplitude is substantially reduced.

When the whole upper trichord is sounding, however, shown at Fig. (10.6B), the partial begins louder but the beat amplitude is substantially reduced, compared to its amplitude when only the first string of C64 (C6) is sounding, Fig. (10.6A). In this way, the beat, which is at least partly due to the false beat in the C64 (C6) string, is "hidden" or reduced in prominence once the double-octave tuning is completed by completing the tuning of the upper trichord.

To reiterate then, the double-octave is initially tuned on the one string of the upper note, in this case, C64 (C6). This string exhibits a slow false beat, which will tend to be inherited in the double-octave, but can be "hidden" at this stage, simply by tuning the double-octave to its optimum available purity. However, this is not what is done. The double-octave can be "stretched" so that a beat rate equal to the false beat rate appears in the double-octave. This seems undesirable at this stage, but the double-octave behaviour is about to change, when the rest of the upper unison is tuned. Completing the unison to an excellent unison tuning, then "re-hides" (through natural *attenuation*) the beat again, so that the double-octave is tonally good, *and* at the same time the double-octave is "stretched" wider than it would otherwise have been.

Referring again to Fig. (10.4), the change in the beat rate of the seventeenth between stage (B) and stage (D), *i.e.* from "purity" tuned on the single top string to the preferred stretched position, is 0.5 Hz. The double-octave has thus been stretched by 0.5 Hz without introducing any more beating into it than is already present due to the shallow false beat. Stretching in this order means that the seventeenth beat rate is about one step higher in the progression of beat rates than "traditional" theory dictates. Larger stretching is often possible on many instruments. We should remember that that the amount of stretching available in this way will typically vary from one part of the compass to another, on any given piano.

There are of course some additional artistic decisions to be made in this process, concerning the *quality* of the double-octave. It is not necessarily only the first adjustable partial that determines the quality of the double-octave. The kind of behaviour just described may possibly apply to other adjustable partials, and there are additional factors other than the beat behaviour of a single adjustable partial, that contribute to the double-octave quality. The final tuning decision is an artistic one, but the principle of stretching noticeably on the single string tuning, and then reducing the detrimental effect of this stretching on completion of the trichord remains one of the basic techniques for double-octave stretching in fine tuning, where the upper string has a slow false beat. *Slow* false beats in the fundamentals of treble strings are *very common*.

The case of no falseness – larger stretching

Even when an octave is stretched so that *adjustable* beats are actually introduced into the octave on the tuning of the first string of the upper trichord, attenuation of the beating in the octave can take place on completion of the trichord tuning. Attenuation can either turn a regular beat into a single null, or render a beat sufficiently shallow that it does not spoil the tone.

Let's take a look at an example of stretching with beating from the Steinway model M, in which this time, the first string of the upper trichord is *not* false in its fundamental. Fig. (10.7A) shows a beat of approximately 1.6 Hz in the first adjustable partial of the stretched octave A37 (A3) – A49 (A4), when just the first string of the upper trichord A49 (A4) is sounding in the octave. This is not an inherited false beat, it is an adjustable beat rate and the octave is widened. Fig. (10.7B) shows the same octave tuning when all three strings of the upper trichord are added and sounding. The vertical scale is identical in both instances.

With just the first string of the upper trichord sounding the beat is more prominent (both visually and aurally) than when the rest of the upper trichord is added. It is should be emphasised that it is not a question of the original beat (with the single string) being masked by additional sound when all three strings are sounding. The beat amplitude, and hence its prominence, *decreases absolutely*, when all three strings are sounding. However, the upper unison tuning must be good to achieve this, and the effect is dependent on the stretching not being too great. Beyond a certain degree of stretching, the effect is lost.

The stretching in this case is indicated by (C) and (D) showing the beat rate in the third F33 (F3) – A37 (A3) and the rate in the tenth F33 (F3) – A49 (A4). The rate in the tenth is faster (8 - 9 Hz) than the rate in the third (approx. 7 Hz). The octave is therefore stretched by around 1 – 2 Hz on the tenth, by the reckoning of "traditional" theory.

Fig. 10.7.

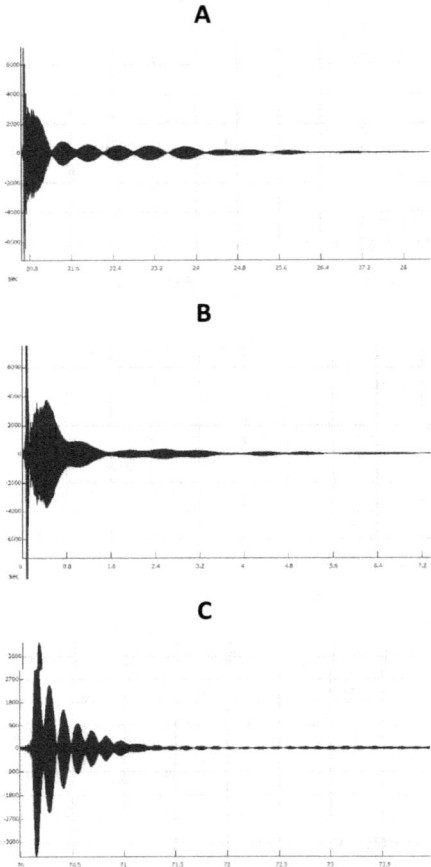

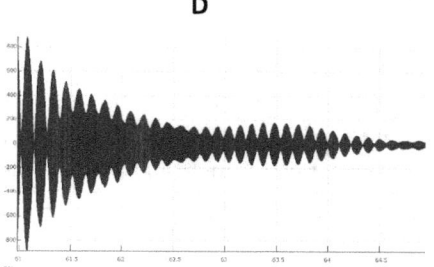

D

Fig. 10.7. The first adjustable partial of the single octave A37 – A49 on the Steinway model M, (A) when only the first string of the upper trichord is sounding, and (B) when the whole trichord tuned to an excellent unison is sounding. (C) is the beat in the major third F33 – A37, and (D) is the beat in the major tenth F33 – A49 (first adjustable partials in each case).

Fig. (10.7D) also shows another slow beating modulation of the partial envelope, together with corresponding variation in the beat amplitude. Such behaviour is common in the beat patterns of the tenths, and can affect the tone quality of the interval.

If stretching by introducing adjustable beats is necessary for purposes of tone and overcoming pitch flattening, then it is imperative that the beating is attenuated through the full trichord tuning. Attenuation will be lost beyond a certain small degree of stretching, and this point should never normally be crossed. The difference between enhancing the tone and spoiling it, must be respected at all times, so octave tuning should *never* be carried out in a mechanical way without constantly being conscious of tonal nuances.

We are currently talking about stretching where falseness does not seem to be present, but this is never out of the context of false fluctuations in the soundscape. Beat movement in the soundscape of the octave does not *per se* necessarily spoil beauty of tone, precisely because false beat movement is a natural part of the tone. False beat movement in an octave is therefore not necessarily to be eschewed. Any attenuated *adjustable beating* should normally be within limits set by the degree of false fluctuation naturally present in the tone, overall (although it may not be present in the particular note being tuned). There is a fine dividing line between fluctuation that is part of the beauty of tone, and beating that diminishes the beauty of tone. It is *very easy* to over-stretch. Where falseness *is* present, even slightly stretching beyond the limit of attenuation, will invariably spoil the tone.

The so-called "check" intervals are also to be evaluated in terms of tone. Treating *beat rate* as a kind of measured quantity with no regard for its relevance to the *tone* of the interval, is a very crude approach, that rather misses the point. The beating in the "check" intervals is often not constant in rate, and not constant in amplitude, as Fig. (10.7D) illustrates. One *must* take into account the overall *effect* of the beating on the tone, which is determined by intonation, tone quality, beat rate and beat amplitude. Above the mid-compass, the beating, both in terms of its rate and amplitude, provides an acoustical "shimmer" that enhances the piano tone provided there are no sudden changes in it from one interval to the next adjacent one, and provided it is not so fast and prominent (high amplitude) that the tone becomes coarse.

Beat patterns are seldom regular beats of constant rate and beat amplitude. *They are what they are*: envelope shapes that can be complex, causing beating to fade and grow, speed up and slow down. Beats can even be formed in such a way that beats are alternately large and small in amplitude, creating a psych-acoustic "double beat rate", in which the faster perceivable rate is twice the slower one.

In all cases, the overriding considerations are consistency (which is what is often called *progression*), beauty of tone, and utilisation of the relationship between the instrument's tonal and intonation characteristics.

The case of no falseness - small stretching

Smaller stretching can have significant effects for the overall tuning strategy, and in particular on the raising of the upper treble. It can still be carried out where there is no significant falseness. The principle is much the same as for the case where slow falseness is present in the tone of the upper note. Smaller stretching is typically needed in the octaves close to the centre of the compass, because here, on most pianos, the degree of stretching that can still be attenuated in the trichord tuning, is usually smaller.

The first string of the upper note can sometimes be raised from an apparent position of minimal beating, often sufficiently for an actual micro-change of pitch to be heard, but *without leaving a position of apparent minimal beating*.

This is a baseline strategy, but must be carried out in conjunction with tuning the "check" intervals. Then, if the need of stretching is greater, the string can be raised very slightly *beyond* this point, on the understanding that when the upper trichord is complete, the resulting disturbance in the octave will be attenuated.

The degree of disturbance, if any, that can be introduced into the octave soundscape on tuning the first upper string, and the amount of attenuation available on completing the upper trichord, can only be ascertained through aural acuity, experience, and good judgement. Completion of the upper trichord always carries with it the possibility that the octave may need to be retuned, if the relationship between the attenuation found to be available, and the requirements for the tuning of the first upper string, do not match. Experience of piano tone and its complex tuning characteristics, provides sufficient skill to avoid this on most occasions, but the possible need of retuning an octave *always* still remains.

To emphasise again, there is a fine dividing line between what will result in an enhancement of the instrument's tone qualities, and what will result in crudity. Always the tuning of the octave must never be carried out in isolation, but must be considered in conjunction with the tuning of the other intervals, with regard for their tone quality. These will include both the tempered "check" intervals and multiple octaves.

Beat amplitudes and rates – tone quality

Now let's repeat some important points. When tuning an octave, the beat rate in the third or tenth is not necessarily constant, just as the rates are not constant in the example shown. Also, the beat rate in the tenth when only one string of the top trichord sounds, may be different to when all three strings of the trichord are sounding, with an excellently tuned unison. If the beat rate in the tenth when the full upper unison is tuned, is too great, the tuning of the tenth on the first string alone may need to be slowed, even if it sounds incorrect on the first string only. Similarly, a beat rate in the tenth that seems too fast on tuning the first string, may not necessarily result in a rate too fast when the full trichord is tuned. These effects are not merely subjective – digital analysis shows that such differences can actually be present. This means it imperative to re-check the quality of the tenth or seventeenth on completion of the upper trichord, and *re-tune* the trichord if necessary, which may involve setting the first string to beat rate apparently too slow or too fast, to counter the effect of completing the trichord.

The beat rates in the tenths or seventeenths are not *merely* beat rates. They contribute to the tonal quality of the interval. There can be a fine dividing line between an interval with a fast beat rate that gives it a shimmering tone quality, and a just slightly faster beat rate that turns shimmering into tonal coarseness.

A beat in a tenth or octave-tenth (seventeenth) can speed up or slow down, in addition to the beat amplitude varying. The way in which a beat rate *varies over the decay time* may depend on the precise fine tuning condition itself. The overall consideration in tuning must take into account at least several second's worth of decay time, and how the tone of the interval is affected by these features, compared to adjacent intervals of the same kind. It is easier to produce a beautiful tone with faster beating if the beat amplitude is not too great. If the beat amplitude is high, that is, the beat is very prominent, then the same beat rate may equate to coarseness of tone.

The constant endeavour should be *equality of tone quality* between adjacent intervals. In terms of beat rates, this implies *progression of beat rates*. There can, however, be exceptions for the sake of tone. If, for example, one interval in a series of adjacent ones, produces very shallow beating only, it may be effective to allow the beat rate to be higher, in order to match the quality of adjacent intervals, provided the other intervals, especially the octaves, allow this.

How to address pitch flattening – inharmonicity

Pitch flattening usually seems worse in the top octave or so, of the compass. We have seen that addressing inharmonicity's effect over an interval of several octaves requires adjustment to the individual octaves making up the multiple octave. The effect of pitch flattening that manifests in the top octave or so, of the compass, is similarly best addressed through the octaves *below* the top of the compass. Stretching is most effective in the mid treble, and it is here that the maximum effect of attenuation occurs.

The first consideration even in overcoming pitch flattening, is still inharmonicity. This will tend to cause three "beatless" octaves to form a narrow beating triple octave. By stretching the intervening octaves, narrow beating triple octaves can be avoided. This also works towards overcoming pitch flattening, and by "matching" the intonation with the inharmonicity, greatly improves tone. However, it is often possible to stretch beyond this *without* introducing wide beating triple octaves, and without spoiling the tone of the single octaves, in the manner already discussed. There are two limits – (1) The tone of the individual octaves must not be spoilt, and (2) The triple octave must not beat wide. With skilful tuning, utilising attenuation in octave stretching, the single, double, and triple octaves will fit together without any one of them suffering a spoiling of tone through beating.

What about the quadruple octaves or wider compound octaves? In most cases, *correct* attention to single, double and triple octaves will solve any potential issues with larger compound octaves. It is not, after all, inharmonicity as an isolated factor that causes the need to stretch octaves. Rather, it is the complex connection of inharmonicity together with other factors such as the frequency region of the partials it may cause to beat, the ear's *dominance region*, and other aspects of the whole psycho-acoustic picture.

This, together with the difference in attenuation between the treble and the bass, means that correct octave stretching in the bass is a very different affair from what it is in the treble, which we will discuss later. Stretching the treble fully but strictly within the limits discussed - the limits that are inherent in the individual instrument, and inherent in each individual octave of that instrument - will provide the best possible compound octaves of larger size on most instruments. Except in very occasional cases, the musical tone of quadruple octaves or larger is not so critically affected simply by adjustable beating, not least because of where the beating occurs in the spectrum.

In the quadruple octave the first adjustable partial occurs at the meeting of the upper note's fundamental and the 16th partial of the lower note. Bass string spectra from the mid-bass on a reasonably large instrument can typically show more than 40 partials above the noise floor, so the 16th partial in these cases is still of significant relative magnitude. However, in this part of the spectrum the pitch separation between partials is only a around a semitone or less, so the spectrum here consists of a dense semitone cluster of partial pitches, many of which beat in any case due to the false beat phenomenon. The 8th partial at the triple octave is, conversely, only at the lower end of pitch clustering in the spectrum.

In unusual, extreme cases - and such cases do exist, usually on very small pianos - we may have to stretch octaves to a degree at which they actually beat wide, in order to reduce the rate at which a multiple octave is beating narrow, if the beating is very prominent. This can be experienced in some drop action uprights and in some miniature grands. It may even not be possible to eliminate a beat rate entirely in a triple octave, without introducing a wholly unacceptable beat rate in at least one of the intervening octaves.

The only option in this case is to compromise. When we are dealing with octaves that may actually be beating by necessity, we are facing another "hiding" situation similar to that when we are dealing with falseness. In this case, we might find preferable ranges of octaves or multiple octaves in which beats can be better "hidden" because they occur in less prominent partials. In smaller instruments where this is a problem, there is usually at least one multiple octave that requires this kind of "tempering" of its internal single octaves, in order to reduce its own tempered beat rate. Merely doing nothing to address the inharmonicity issue, will result in a very flat sounding treble, and triple octaves that are fast beating narrow.

In all of these cases, we are "stretching" octaves for no other reason than to *eliminate* or *minimise* beating in octaves and multiple octaves. This is a long way from the idea of deliberately introducing beats into the octaves or multiple octaves for the sake of some other ideal, such as overcoming an apparent severe flattening of pitch in the treble. Furthermore, the amount of "stretching" introduced arises entirely from the individual instrument and the individual octave. The overall octave stretching "picture" for any instrument is very much something "inherent" in the instrument itself, and "utilised" by the art of the tuner. The tuner must "read" the instrument as it is tuned.

Octave stretching that does not soil the tone of the octaves is *imperative* for a good tuning. Without it, we would be ignoring important aspects of the physics of the instrument that have tuning consequences, and our tuning would not be fully controlled. After *controlled* stretching the net result should be that treble pitch flattening effects are very effectively reduced, the treble sounds more alive, cleaner, more "inviting" of the higher register, "sings" better, and matches the bass better. Tonally, without any controlled stretching, the treble will sound relatively dull, "flat" in a tonal sense, and the instrument as a whole will be tonally more mediocre and uninteresting.

The wide beating "pitch correction" mistake

It is a conceptual error to assume that top treble octaves widened by introducing noticeable beats will make the treble sound "less flat" than octaves tuned "beatless". In fact, the psychology of pitch perception is more complex than this assumes. One of the important factors contributing to the perception of an octave or multiple octave as "in tune", is the absence of musically obvious "beating".

Because beating partials have *beat amplitude* and *decay* properties in addition to the more often cited beat *rate*, it is not only the beat *rate* that determines whether or not beating will be "significant" to the ear. For instance, to take an extreme example, even a beat whose rate is in the range to which we are most sensitive, will actually be inaudible if the beat amplitude is small enough. Beating that is obvious will be beating that is "significant" because of its rate *and* its amplitude, *and* its decay rate.

Once beating is "significant" to the ear, it is a psychological signal that the octave is not "perfectly tuned". The presence of "significant" beating cannot, however, *by itself* indicate specifically that the interval is wide or narrow – only that it is incorrect or imperfect one way or the other. We have to first add to this the other factors determining the *perceived pitch* of the upper note relative to the lower, in order to psychologically "decide" whether the octave (or multiple octave) is musically wide or narrow. Any suggestion of an *intonation* "error" will be reinforced when we hear aurally "significant" beating, but it is not the beating that tells us which way the intonation "error" seems to occur, flat or sharp. It is the *perceived intonation* that tells us which way the intonation appears to be in "error". In certain circumstances it is perfectly possible for the intonation of the top note of an octave to sound flat, even though the octave is slightly beating wide. The acoustical factors affecting pitch perception are not precisely the same as those that create beating.

Unless we have widened the octaves sufficiently to convince the ear through intonation alone, that an octave or multiple octave is *definitely* not narrow, or that it is clearly now *wide*, then any beating we introduce by widening the interval may actually *accentuate* the existing psychological assessment that it is *narrow*, even if it is beating wide. If there is already a tendency for it to sound narrow, according to the *pitch perception* factors, then it may still sound narrow if these factors are still present, even when it is widened by introducing beating. The wide beating plus the psycho-acoustics pitch flattening effect makes the octave sound even more narrow.

In other words, widening octaves that tend to sound narrow pitchwise, by introducing beats into them, can make the octaves sound *more narrow*, pitchwise, not less, until we reach a certain very crude point of widening. This is a serious problem in uninformed octave stretching in the top octaves, and it will introduce tonal coarseness through the beating. On the other hand, removing beating from a multiple octave that is actually beating narrow, will improve matters enormously – even if sometimes in the unusual extreme cases, this *does* involve *slowly* beating individual octaves.

In summary, any sense of "flatness" of the treble is just as likely to be made more acute by introducing indiscriminately widened top octaves through stretching of the *second kind* – the widening of octaves with disregard for spectrum matching. The introduction of widening beats into the octaves can make the treble sound flatter still, or not really improve matters, unless the widening is so severe that tuning integrity is completely destroyed.

Sometimes one finds an instrument with over-stretched treble octaves. Rapidly and *clearly* beating wide upper octaves can often be made more beautiful by bringing them *down* into a purer condition, without introducing "flatness" in the treble, provided matching between bass and treble is properly controlled. Often, the treble will actually sound less flat as a result.

Inharmonicity and instrument size

Octave stretching is not something confined to small, inharmonic instruments. All instruments are subject to some inharmonicity, even the largest. It is the *rate of change* of inharmonicity with position in the piano's compass that is important in producing matches or mismatches between spectra, or parts of spectra. Mismatches in the spectrum, or between spectra, are not the fault of inharmonicity itself, but rather, result from a mismatch of inharmonicity. In other words, it is the fact that different strings have different inharmonicity that causes the main complications we have to deal with.

A large instrument may be subject to less inharmonicity overall, but its *rate of change* of inharmonicity with position in the piano's compass may nonetheless be significant enough to have an effect on tuning. Also, a large, excellent instrument, will often tend to produce very clear spectra with clear and distinct partials in the upper spectrum. This may mean that what in the spectrum is physically subdued or lost to the ear on an inferior instrument, on a good instrument may be "loud and clear".

Beat rate prominence – maj 21st and dominant sevenths

Beats due to tempering are not generally noticed as such by musicians. What musicians may notice in tempered intervals is the tonal *quality*, coarseness or pleasantness, and the *intonation* of the intervals. This is especially true of larger intervals over several octaves. Whether or not consciously noted, a beat rate in a "fast beating" interval can sometimes make its presence felt more when it is *slower*, rather than faster. If we were to have some thirds in the scale too fast, we would not make it more likely that the actual *beats* in these thirds would be heard. We would however make it more likely that the thirds would be perceived as coarse in tone or unpleasant in intonation.

It must be remembered that *all* the intervals except the unisons and octaves and their multiples, are effectively tempered, not just those we actively temper. The tempering in these other intervals is a direct *result* of the tempering we carry out. For example, consider the minor 21st interval. This is the interval of two octaves and a minor seventh, say, C28 (C3) to B-flat-62 (Bb5). It constitutes the outside two notes of a root position dominant seventh chord. It is a tempered interval with a beat rate, even though not one that is "traditionally" tempered – and it will occur in every root position dominant seventh chord with the bass doubled two octaves below. The first meeting in the interval's spectrum is the 7th partial of the lower note with the fundamental of the upper.

Without inharmonicity, the interval C28 (C3) to B-flat-62 (Bb5) would beat wide with a beat rate of 17 beats per second. If the prominence of the 7th partial of C28 (C3) is not too great, *i.e.* if it blends into the rest of the spectrum or is not too distinct and clear, then the 17 Hz beat rate in the combined spectrum may not be too prominent. But what if the seventh partials on the instrument are *prominent* and clear?

If this is the case, then the 17 Hz beat rate in our interval may well be objectionable, or it may at least give the interval an objectionable quality. Let us now also suppose that there is also a tendency for the treble to sound progressively flat, so that open root position dominant seventh chords do sound flat in *pitch* at the seventh. This is then a case where octave stretching may be used firstly to raise the apparent *pitch width* of a set of tempered intervals, and secondly to *increase* beat rates in order to make them *less obvious*. Increasing beat rates in dominant seventh chords such as those that include the 21st interval, and in intervals other than the standard tuning intervals, may be possible without necessarily introducing coarseness.

The root position dominant seventh chord with the bass doubled two octaves below, is in fact a useful "check" or test for the need for octave stretching to overcome pitch flattening.

Procedure for controlled stretching - beginning

The compass can be viewed in terms of its physical break sections. There will usually be a region of the central break below the scale area, and tuning the octaves in this region first, can be advantageous. Between the scale and the bass break on small and medium sized instruments can be a region of poor inharmonicity matching, resulting sometimes in octaves that have *no obviously beatless* tuning position. The octaves here can however be tuned together with the descending fifths, fourths, sixths and thirds. Remember that inharmonicity mismatching can affect the fifths almost as much as the octaves. Poor fifths cannot be just ignored.

If the scale is tuned within an octave, a tenth "check" will not be possible until the fourth note below the scale, but the minor third / major sixth test can be carried out from the first note. This is a most reliable test and will usually ensure a good octave where there is any uncertainty. Exceptions certainly do exist, but only in a minority of cases.

The minor third / major sixth check

A descending octave can be "checked" with the minor third / major sixth test. This involves the two notes of the octave, plus a note a minor third from the bottom note of the octave, which is also a major sixth from the octave's top note. The musical pitch of the partial to which we listen here can be quickly found in the following way.

Take the minor third whose lower note is the note we are tuning, and whose beat rate we are checking. Now think of completing a minor triad on this minor third – by adding a note a perfect fifth above the lower note being tuned. The pitch we want is two octaves above that added note. The reason I suggest thinking in terms of a minor triad is that it makes musical sense of the pitch we are listening for, in relation to the minor third which makes it easy to remember and find. The adjustable partial is at the pitch of the 6th partial of the lower note being tuned – there are many other partials audible which may be beating but are *not the right ones to listen to*, so by making musical sense of the correct partial's position in relation to the minor third, it becomes easy to identify quickly.

The pitch of the partial for the corresponding major sixth is precisely the same (it makes a first inversion of the same minor triad). By "traditional" theory, the rates should be the same in the minor third and the major sixth. Any conflict between this test and the major tenths progression should be investigated.

The minor third / major sixth relationship is sufficiently accurate that a good bass octave can frequently be first tuned by this alone. It does not, of course, take precedence over the octave itself, but if you are not sure about the latter, the "check" is generally very reliable.

First check the beat rate in the major sixth (whose lower note is a minor third above the note we are about to tune). Now tune the note at the bottom of the octave (an octave below the upper note of the sixth) against the note a minor third above it (the bottom note of the sixth) until the beat rate in the minor third is precisely the same as the sixth. Remember of course that the minor third is *narrow*, so unlike tuning tenths and octaves tenths, lowering the note will make its beat rate *slower* (assuming it is already "on the right side").

Successful octave stretching, as we will discuss shortly, does not usually require deliberate stretching to the bass, beyond what will just happen naturally as a consequence of inharmonicity in each individual octave. Stretching in the bass is not subject to the same advantages of *attenuation* as the treble. If you do want to stretch the octave, then in the minor third / major sixth test the beat rate in the minor third must be slightly *slower* than that in the major sixth (the minor third is a narrow interval). If the rates are to be allowed to differ, then it is important not to exceed a "good" octave, and to ensure that progression of tenths or octave tenths is not obviously compromised.

The function of the minor third / major sixth relation

In tuning bass octaves it is not generally sufficient just to tune so that the first adjustable partial (at the pitch of the upper note's fundamental) is beatless. Good tone and intonation generally demands at least a beatless second adjustable partial. This is at the same pitch as the partial heard in the minor third / major sixth test. When the beat rates in the test are the same, the 2nd adjustable partial of the octave itself, is generally beatless.

The *function* of the minor third / major sixth test is not to attain minor third and major sixth relationships, but to produce a beatless second adjustable partial in the octave. If one just tunes this second partial beatless when tuning the octave, it invariably has precisely the same effect as tuning by the minor third / major sixth test. This, however, does not mean that one should generally tune bass octaves by listening to this specific partial (or any other) rather than listening to the whole. What one should be listening to, especially in the lower bass octaves, is discussed more, below.

Some bass octaves on some pianos will *not* be at their best when the 2nd adjustable partial is "beatless". One cannot simply create a "rule" that says bass octave tuning is *defined* by the 2nd adjustable partial being "beatless". The tone recipe of the octave is far more complex than that, and so is the psycho-acoustics of octave tone and intonation perception. Some octaves will require, possibly, a slight beat in the 2nd adjustable partial in order for the soundscape *as a whole* to be at its best. In such octaves, the minor third / major sixth test can be expected to fail.

Continuing down to the break

In this region there may be more rapid inharmonicity changes due to the compromised or reverse curvature at the end of the long bridge, changes from plain steel to covered strings, from trichords to bichords, or even a change of bridge (see also the chapter on small piano syndrome). Such physical features can manifest in a conflict between the attempt to give progression to the falling major third, and the demands of the minor third in the minor third / major sixth test. Conflicts of demand between other interval can also occur. Conflicts require first ascertaining that the conflict does not arise from correctable error in previous tuning. Then they require artistic compromise taking into account the relative demands of the conflicting intervals. As soon as the fourth note is tuned, there is both a falling tenth "check" plus the minor third/major sixth test available simultaneously for each note.

In this region it is not usually necessary to make any particular effort to stretch the octave for the sake of the octave. However, falling fifths may demand it because of a tendency to be too fast (again caused by the rate of change of inharmonicity, typically due to bridge compromise). Each note's tuning should take into account the demands of all the tuning intervals in which it falls.

Continuing down to the bass

Tuning the bass before the treble, has advantages. Tuning the bass first provides the necessary notes from which to test beat rates in tenths and octaves tenths as one tuned upwards from the scale area. The difference in tuning characteristics between the area just below the scale, and the area just above it, with respect to *attenuation*, on most pianos, is marked. However, on medium to large grands, where trichords continue below the scale area, significant attenuation here may still be possible, and hence stretching may be accomplished.

Remember that in aural tuning "stretching" does not simply refer to the side-effect of tuning with inharmonic partials. It is a much more deliberate technique of working with inharmonicity, attenuation, tone and intonation. Bass strings are inharmonic, but on most pianos, in practice it is not necessary to deliberately stretch octaves in the bass in order to match the treble well with the bass, provided properly controlled stretching is expertly carried out in the treble, especially the mid-treble.

"Stretching" will be occurring when bass octaves are tuned without any conscious intention to stretch, due to the presence of inharmonicity. But this is not something the tuner has to make any effort to do. It is just a side-effect of inharmonicity. In terms of applied practical art and technique, utilising attenuation, the bass octaves do not generally *need* to be "stretched", except on some very small instruments, and even then, it is still questionable. The idea that stretching bass octaves will make the treble seem relatively sharper, is a somewhat crude conclusion based on rather simplistic reasoning. One *can* stretch bass octaves, but the effect is not simply to improve the relative intonation in the treble.

In the treble one can take advantage of *attenuation*, but in the bass, the soundscape is very different; the higher partials can fall within the *dominance region* of the hearing, and *attenuation* effects are, over most of the bass, negligible in their impact. As a general rule, bass stretching beyond the natural results of inharmonicity, is simply not necessary, provided the stretching in the treble is properly handled.

Tuning upwards from the scale

The first few octaves tuned upwards from the scale area into the treble must be tuned together with the fourth, the fifth, the third, and the sixth. Remember that the octave tone is complex mixture of ingredients. On the one hand, the octave tuning cannot be "sacrificed" for the sake of these other intervals, but on the other hand, the octave tone is *not* a single, simple entity that demands just one tuning position for "perfection", that the other intervals are merely used to "check" on. The idea that everything fits together like rigid bricks in building a real version of the "theoretical model" is as simplistic as the "traditional" theory the idea comes from.

An octave in this region can often be stretched using the perfect fifth as a gauge in addition to the third or tenth. An initial apparently "perfect" octave tuning may result in an *obviously* beating narrow fifth, and if the beat rates in it are as fast as "traditional" theory purports they should be, then the beating will indeed be very obvious. But often (though not always) if the fifth is re-tuned slower or even almost pure by raising the upper note, the resulting octave *once its trichord is complete* will *still* be perfectly good. Octaves can exhibit the same kind of "plasticity" that we met in the scale intervals. Unlike the picture painted by "traditional" theory, it does not follow that moving the upper note by x beats when it is sounding in some other interval, will introduce the same beating in the *completed* octave itself.

Tuning down to the break in the bass before embarking on the treble, gives tenth "checks" for the first octaves upwards into the treble. Tuning the whole bass first, provides the maximum support for tuning upwards from the scale.

Let us assume for example that the scale is completed between F33 (F3) to F45 (F4), which is not a rule, but is quite common practice. We now need to tune the notes F#46 (F#4), G47 (G4) and G#48 (G#4). Each note must be set as far up in the octave *fine tuning region* as possible, with consideration for the fourth, fifth, third and sixth, of which it makes the upper note. Most instruments, however, will not allow much or any stretching here, because available attenuation in this region is often too small. The opportunity to attenuate usually begins to become more significant half an octave or so above the scale, but there is no fixed rule for all pianos.

The beat rates in the thirds and sixths must continue to progress well, even outside of the scale region. Progression is still necessary even though the tenths and seventeenths may "lead" over the thirds.

Inharmonicity determines that the fifths will generally sound good because they will be slower than their "traditional" rates, whilst the fourths *may* be faster than their "traditional" rates. (Think about what happens to the rates of an adjoining fifth and fourth, when the octave between the two outside notes is slightly stretched. The fifth will become slower and fourth faster). But we must be careful that the fourths are not *too* fast, especially if their beat is *prominent*.

The thirds and sixths must continue their beat rate progression outside the scale area, but the rate of change of these and other "fast" intervals may increase above the rate of change in the scale, so that the rates may be faster than their "traditional" rates. The beats in thirds and compound thirds contribute to the tone of the interval. We have to be careful that the beat rates in the thirds and sixths are "shimmering", but not causing coarseness in tone. All adjacent tenths and seventeenths must be well matched, so that no one stands out as different.

On most instruments each note in this part of the compass can be raised as far as the width of the fine tuning region for the octave allows. As already stressed, often in this mid range of the compass the width of the *fine tuning region* may be practically zero – only an initial trial and error will reveal its nature for the instrument in question. The width of the fine tuning region will likely then increase as we rise up the compass, as far as the mid treble. On very few instruments the region's width is quite pronounced even in the mid compass adjoining the scale area. However, the following should be noted before "stretching" octaves immediately outside the scale area:

Caution for the first few octaves

If maximum stretching is begun immediately as tuning outside the scale area progresses into the treble, this *can* create a problem. Unless similar "stretching" is carried out into the immediate bass side region of the scale, which is not usually possible without tonal detriment, a sudden jump in tenths beat rates will occur outside the scale in the bass direction progression.

Usually the notes on the bass side just outside the scale area are not as amenable to lowering without spoiling the octave, as they would need to be to prevent this jump. Therefore, it is advisable to begin stretching on the treble side *progressively* - full stretching should only be reached in the octaves above notes in about the middle of the scale (around middle C), rather than from the bottom. This is another reason why it may be argued that overall matching of the bass and treble is best achieved by tuning the bass, or part of the bass below the scale area, first. Good practice from the point of view of *stability* is also to tune down at least to the bass break, first.

Continuation of stretching

Tuning A49 (A4), a major tenth is made with F33. (F3) "Traditional" theory dictates that the beat rates in a major third, tenth and octave tenth all on the same lower note, should have the same beat rate. However, we must aim for the first string of the A49 (A4) trichord to give us a faster rate in the tenth with F33 (F3), than the third F33 to A37 (F3 – A3). The difference in beat rate will depend on the instrument, and on the variation that the octave's fine tuning region allows.

As a "rule of thumb" for a reasonable sized instrument, the beat rate in the tenth may be say, the same as the beat rate of the major third *not* on the *same* bass note, but rather, on the note one tone above. Thus say, the tenth F33 (F3) to A49 (A4) may beat with the same rate as the third G35 (G3) to B39 (B3), if the width of the octave fine tuning region allows – but note the caution above regarding the progression of beat rates into the bass area. Also, one should check that the fourth up to the note being tuned, and the sixth, are not unduly compromised (the fifth is not generally at risk).

The increment in the tenth's beat rate over the third may certainly be more or less than this, but this is a reasonable default to aim for, initially. One should learn to almost instantly judge very fine differences of beat rate in the context of tone, which is something that comes only with experience, and is an essential part of the art. The allowable increment in beat rate of course depends on the *tonal properties* of the tenth – for any tenth there will be a maximum beat rate above which the tenth will be tonally too coarse.

Thus we consider both the beat rate and the tonal *quality* of the third, tenth or octave tenth. It is *not* a question of putting a number on the beat rate. Every piano is different, and the tone of the tenth will depend as much on the beat *amplitude* as on the rate. Once the ability to small differences in fast beat rates reliably, is developed, it is necessary to be able to translate this acuity of hearing back into an assessment of *tone*. Thus, we are listening for *shimmering* or *coarseness* at the same time as quantifiable beat rates.

In any case, we are aiming for the beat rates in the tenths and octave-tenths to "lead" over the rates in the corresponding thirds, by at least 0.5 Hz. In tests I carry out on students each year using digitally recorded intervals, those students with several months experience invariably can hear a difference of this magnitude, reliably.

Remember that the amount of stretching that will be possible, increases as we rise in the compass above the scale, until its maximum in the mid treble. We will are aiming to reduce stretching again in the higher compass, so that the top octave or so, of the compass, contains octaves that are *not* deliberately stretched. All this is because *attenuation* is best in the mid treble compass, and because stretching by a given microinterval produces worsening results in terms of beating, the higher in the compass we apply it. We do *not* want to stretch the top octaves to the point where they are beating wide.

After tuning the second string of the octave's upper unison, the effect on the other intervals must be checked. Sometimes, the tenth will have slowed a little. It may sometimes even be necessary to stretch the octave on the first string of the unison apparently *outside* its fine tuning region. As a single string, the octave will now appear too wide. But by the time all three strings of the unison are tuned, the octave between the two trichords may still be within its fine tuning region – not because the strings are unstable, but because the unison as a whole will behave differently to the single string. *Attenuation* will come into play.

A couple of instances by trial and error should be enough to ascertain the pattern for the whole mid-treble range. The degree of stretching that is likely overall can be ascertained by listening to the instrument first, both in terms of relative pitch throughout the compass, and beat rate relationships. Familiarity with the individual instrument, or the size and make, is good. It's always a good idea to check some multiple octaves and their attendant beat rates in the associated tenth and octave–tenth, before starting to tune.

We are saying that the beat rates in the tenths should progress, but may remain always slightly faster (or in some instruments, considerably faster) than their "traditional" rates, sometimes more when setting the first string of each trichord. To reiterate: how fast a tenth can be, is, as in the case of a third, a question of the individual instrument's characteristic tone and the acoustic environment. The beat rate of a third or tenth, or octave-tenth, is in itself an absolute quantity (although it may vary during the decay), but its *effect* on the *quality* of the interval depends on *other* factors, like the beat amplitude, and the recipe as a whole. The beat rate "sits" in the tone of the interval, affecting it. Some pianos can take faster fast beating intervals than others.

Always, avoiding the need of excessively widened single octaves in the upper treble, is achieved by widening most in the mid range, even though there, at the time of tuning, it may not seem strictly necessary. This "pushes up" the upper octaves relative to the notes below the mid-treble, and allows the upper octaves to "spectrum match" with the central compass and bass, and not to sound "pitch flat". The overall stretching in the compass is sufficient to make the treble sound "pitch bright" and at the same time the upper single octaves have a natural silvery brightness, and sustained "singing" quality, rather than coarseness.

This comes from the fact that they themselves are not particularly "stretched", but are well "spectrum matched" with the lower notes. The undamped upper notes then act as very finely tuned aliquot strings, affecting the tone of the whole piano. Remember that the degree of widening possible will always be limited by the width of the individual octave *fine tuning region*, which varies over the compass, and on some instruments may be practically zero just above the scale. Be prepared for this, and to find the point in the compass where stretching can "take off".

There is therefore more than one scenario for octave tuning or stretching, which we can represent approximately in the following diagram, Fig. (10.8).

Fig. 10.8

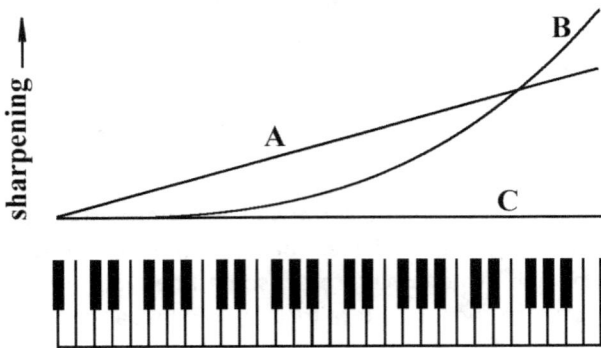

Fig. 10.8. Approximate representation of different types of "tuning curve" or "octave stretching curve".

The section of the keyboard compass represented here is from A#38 (A#3) to C88 (C8). The bass is not shown. Tuning by single octaves alone, without even checking multiple octaves, would tend to result in no appreciable sharpening of the notes, represented more or less by the line **C**. This is a line of "no stretch", but will typically result in a duller sounding instrument, that is both tonally "flat" and with appreciable pitch flattening in the treble. If we were to start correcting multiple octaves that are found to beat flat, by widening the upper octaves, inharmonicity alone will dictate a curve somewhat like **B**.

The degree of correction necessary will also probably result in the upper most octaves beating considerably wide, in order to reduce the beating in the multiple octaves. This always spoils the tone of the upper compass. If, however, we adopt controlled octave stretching by tuning to the sharp most outside of the *fine tuning regions* without introducing detrimental beating, and carefully controlling and progressing the other intervals such as octave-tenths, then a curve somewhat like **A** is more likely to result. This is the kind of "tuning curve" (a term now used by makers of digital pianos and tuning meters) that minimises any tendency for *pitch flattening*, but does not introduce detrimental beating into the octaves.

Note that the height of the line *does not mean that the upper individual octaves are stretched more than the lower ones*. The height of the line represents the *overall total sharpening* above the position on the line **C**, which will at any point on the line, include the sharpening of notes on lower octaves. Hence the sharpening *accumulates*, and the line gets higher, from left to right. Actually stretching upper octaves more than lower ones, results in a curve more like **B**.

The most important point is that each octave must be stretched according to its own behaviour. In other words, each individual octave will *allow* its own amount of stretching. Stretching should not be *imposed* on it. On many pianos, the resulting "tuning curve" will be "smooth", but may not necessarily be like **A**, **B** or **C**. On other instruments, the curve is not necessarily entirely smooth, but may contain "jumps", even though good progression of beat rates in the tenths and octave tenths, etc., is maintained.

From lower to higher treble – a summary

The principle to remember is that usually, acute spectrum mismatching is most likely to first start showing up in the triple octaves, so we need to prepare for this in the tuning of the *intervening* octaves, *before* we actually reach the triple octave test. This is done by always checking the double octaves, and graduating the octave tenth rates slightly faster than their corresponding thirds, whilst not exceeding the *fine tuning regions* of the octaves. Thus we minimise the possibility of having to stretch the highest (third) octave in a triple octave beyond its *fine tuning region*, in order to match it with the lowest note in the triple octave. It is necessary to always keep a check on the progression and quality of the tenths and seventeenths, together with the double and triple octaves.

Remember that as much as possible of the raising of the "pitch" of the top note of a triple octave, should be achieved on the lower two octaves. Raising by a given number of cents requires a smaller "beat rate" increase, the lower in the compass it occurs. The tenth and octave-tenth checks are most important *before* reaching triple octaves, e.g. before reaching the note three octaves above the lowest note of the scale. Controlled stretching in the lower and middle treble compass, will make the task of octave and triple octave matching much easier. If the first two-octave range of the compass above the top of the scale is judiciously stretched, then above this point, correct stretching for the third octave can often be achieved just by tuning the triple octave itself. After this, the highest octaves in the compass generally fall into place without much effort.

On a good large instrument with clear spectra, proper octave stretching can certainly produce superior results on the instrument overall. When the multiple octaves are in proper tune, the natural brightness of the instrument due to its own inharmonicity is brought out, without the need to tune any fast beating and overly coarse octaves. Properly tuned multiple octaves with particular attention to the mid-treble, is usually all that is required to avoid "pitch flattening" in the upper treble.

Tuning major tenths, seventeenths, and triple octaves

The are two separate issue with the relationships between major tenths, seventeenths, and the octaves, in the treble tuning. These are:

(1) The major tenths should progress in beat rates, but be a little faster than the major third at the bottom (sharing the same lower note). The same is true of the major seventeenths (octave-tenths). However, the rate in a seventeenth *will not necessarily be the same as the rate in a tenth*, on the same bottom note. This is at variance, of course, with "traditional" theory. Typically, with correct stretching, inharmonicity dictates that the rate in the seventeenth will be faster than the rate in the tenth. It must not be allowed to be coarse.

(2) When tuning a treble note, it can be checked with the note a major tenth below, *and* with the note an octave below that, *i.e.* a major seventeenth below the note being tuned. As already discussed, the

seventeenths should be faster than the major third at their base (sharing the same lower note). Care must be taken that in making them faster, the interval to a major tenth below the note being tuned, is not made too fast or coarse.

Tuning *any* treble note as an octave, one should also tune with the major tenth below, the note a major seventeenth below, and the notes two octaves below, and three octaves below. This should be routine. In the higher treble, the intonation over the triple octave must be good, and not allow the upper note to sound flat. Tuning by *intonation* especially in the upper compass, becomes important as an advanced technique, once full control of intervening stretching, intonation and tone, has been mastered. Tuning by intonation is *not* a good idea until one intuitively understands both the objective and subjective aspects of intonation, and has mastered the other techniques so that the sense of intonation is not isolated from the sense of tonal nuance.

Bass octaves – first principles

I stressed earlier that what are referred to as "test" or "check" intervals are not *merely* "test" or "check" intervals at all. They are all intervals in their own right that need to be considered and tuned. Experienced tuners will often find they can tune the lower bass octaves without the need of "checks". With experience comes the ability to tune effectively by the octave alone, especially on excellent instruments. This invariably leaves the so-called "check" intervals correct in any case. Unless the tuner has this degree of experience and certainty, however, time consuming error correction at a later stage will be avoided by tuning all "check" intervals as tuning proceeds.

The tuning dynamics of the bass are very different to the mid-compass and treble. Soundboard coupling effects, for example, as experienced in the mid-compass and treble, are not really significant in the bass. This is possibly due to the much smaller partial decay rates in the bass, as dissipative coupling effects are more pronounced, the greater the (single string) decay rate. The spectra encountered in bass tuning differ significantly from those in the mid range and treble in a number of respects. The differences include the number of prominent partials present, the relative distribution of prominence, the decay rates, and, in particular, the frequency ranges.

The latter is important because the human ear has a *dominance region* of frequencies to which it is more sensitive. The position of the ear's *dominance region* in relation to the spectrum will be very different in the bass, to the mid range and treble. In the bass octaves, partials with high numbers fall readily within hearing range and their frequencies coincide with those found in the treble of the piano's compass. They may also contain more energy than those of higher notes. The higher parts of spectra of the bass unisons and octaves (and indeed other intervals) are therefore very audible – they are very clear and generally very rich in audible partials right up into the upper spectrum.

Falseness in the bass

Bass strings are always likely to exhibit audible falseness in their soundscape. As a broad "rule of thumb", the lower in a string's spectrum falseness occurs, the more detrimental it will be to the resultant tone. Higher partials that are both false and exceptionally prominent can also be detrimental to tone. Bass strings that are false in the fundamental or lower partials will also reduce the quality of all the intervals of which they form a part. It is not usually possible to successfully "hide" falseness in bass strings in quite the same way as it might be in the treble strings. This is theoretically due to the smaller partial decay rates and consequent lack of attenuation effects available in the bass.

Pitch and tone in the bass

The fast beating intervals such as the major thirds, tenths, and octave-tenths (major seventeenths), must progress steadily and evenly in beat rate if no individual interval is to stand out as a tonal anomaly in the context of the whole. Ensuring beat rate progression and beat rate relationships in the bass are good, is of course a pre-requisite. Some irregularity can sometimes be expected, however, where there is a change of stringing or change of bridge.

The status of fast beating interval bass "checks", particularly the more obscure ones (other than major tenths, major seventeenths, and the minor third/major sixth test), is often misunderstood. There is a multiplicity of "checks" that can be found for bass tuning, arising from the fact that the bass tone is so rich in prominent partials, particularly in the lower bass, that fall within the ear's *dominance region*.

There is little merit, however, in tuning the lower bass by a series of obscure "checks" unless this actually produces optimal tuning in the octaves and multiple octaves themselves. The complexity of the acoustics in the bass is such that, even if one is sure of listening to the correct adjustable partial in the pool of all the rest, and even if one is sure one is taking into account false beating (which is very prevalent in bass string partials), actual optimal octaves will still not necessarily follow. Therefore, tuning the lower bass octaves ultimately demands *tuning the octaves*, and not tuning them by tuning some other substitute interval. The "checks" are basically a support system in instances where there is some ambiguity, insufficient experience, insufficient aural perception, or insufficient knowledge of what the octave itself should sound like.

Bass octaves are not some kind of acoustical engineering problem that must be solved using acoustical "tricks" of beat rate relationships. They are *musical intervals*. They must *sound right* in their own right, in the musical context, which means *intonation* and *tone*.

Intonation versus tone in the bass

Natural beating is a tonal feature, and *beyond a certain degree* its presence is associated with the "out of tune" condition. Contrary to the common musical presupposition, *tone* and *intonation*, with respect to what sounds "in tune", do not always agree or coincide. In the lower bass, the relationship between the relative pitches of the notes and the tone of an octave is not always comfortable. You might expect, on the basis of musical "intuition" or supposed musical "logic", that in tuning, the best intonation always coincides with the best tone. Unfortunately, nature, as the acoustics and psycho-acoustics behind the perceived sound, does not always oblige us by supporting our musical preconceptions.

On smaller instruments (at least smaller than concert size) the optimal tone does not necessarily coincide with the optimal pitch relationship of the notes, and *vice versa*. One should not drastically sacrifice the tone for the sake of pitch, but there is also a limit to just how unsatisfactory pitch intonation can be allowed to be.

It is also good practice to wean ourselves off the idea that intonation is an *objective property* of the sound. One can even learn how to subjectively induce purely psychological changes of pitch perception, in order to ascertain which pitch relationships "could" be perceived as right. In particular, it is most important to listen to both intonation and tone (which includes any beating) *simultaneously*. Just as in tuning treble octaves, one must be able to find the changes of intonation that can be made without spoiling the tone of the octave, but in the case of bass octaves, it is not *attenuation* that allows this.

A lower bass octave may have a number of possible tuning conditions within a small range, that seem to be "satisfactory", yet not wholly satisfying. In this case, the solution does not lie in some "test" interval providing a supposed "theoretically correct" answer. Tuning in this way abdicates from responsibility for the octave itself. Part of the problem lies in the subjectivity of both tone and intonation (pitch), and the tendency is to want some "objective" test to replace this subjectivity. The true solution lies in learning how to listen to the whole, with acuity and depth of perception, but *without analysis*.

One should *not* be focussing on *individual* beat rates in the partials, in fine tuning the bass octave. There is always a very large number of beat rates, many of them are false, and the relationship between them and the tone is very variable and complex. They will, however, play their part in the tone, because a part of tone quality is movement in the soundscape. The tuner must listen to the *tone*, and to the *effect* of beating rather than the beating in specific partials themselves, and to the *intonation* of the interval, which are basically the qualities that the musician hears. This is the art. In teaching, it is teaching by demonstration, imitated by the student, *without analysis*, that is perhaps the best way.

Pitch ambiguity

It is in the lower bass octaves that the effects of *pitch ambiguity* are most noticed. Pitch ambiguity is a well known effect in psycho-acoustics, and it arises where the acoustical components that help to determine the perceived pitch, themselves deviate from the kind of arrangement they have in musical tones that are perceived to be more clearly defined. Pitch ambiguity can be deliberately created in electronically generated tones through fairly simple strategies, the well known *Shepard's scale* being one example.

In Shepard's scale a rising scale is heard, and when the octave is reached, it appears, apparently quite normally, that a note one octave above the starting note has been reached. In fact, this note is precisely the same as the staring note, so an apparently endlessly rising scale, but that never really rises, can be heard. It is a kind of auditory equivalent of the endlessly rising staircase in Escher's well known picture 'Ascending and Descending'.

Escher's picture is of course a deliberate optical illusion, but then pitch is also, in many instances, at least in part an illusion. Shorter bass strings in pianos often have relatively weak or even absent fundamentals, but this does not result in us hearing a musical pitch at the pitch of the second partial. The ear-brain system seems to recognise the presence of the approximate harmonic series in the partials that *are* present, and generally interprets the pitch as being at that of the "missing" fundamental.

The effect is widely present in all kinds of instances from telephones to tubular bells. The frequency that would normally correspond to the fundamental pitch of a person's voice that one seems to hear in a telephone conversation, is generally not present, for technological reasons. It doesn't need to be. We still hear the person's voice at the normal pitch, but with a perceived alteration *tone* or *timbre*. Similarly, orchestral tubular bells not produce frequencies directly corresponding to their perceived pitches. Their partials do not fall in an harmonic series, or anything much like it, but a few of the partials are related in a way that seems like a small *part* of an harmonic series. This seems to be sufficient for the ear-brain system to "fill in" the missing fundamental from the series in a kind of involuntary psycho-acoustic assumption, so we hear a corresponding musical pitch.

In the bass of the piano, the partials of shorter, thicker bass strings, can be very inharmonic, and relatively weak in the lower part of the partial series. Also, whilst the lower partials fall in a relatively insensitive part of the hearing range, many of the higher partials may fall within the ear's *dominance region*, where we are very sensitive to them. Considerable pitch ambiguity can result, even to the degree that sometimes, because changes of string tension can alter the tone recipe, *raising* the tension can appear to *lower* the pitch, and *vice versa*.

Bass octave reinforcement – the simple difference tone

On good, larger instruments there is often a definite point in the tuning of a single bass octave where the tone seems richer and "reinforced", with the lower and upper notes seemingly in "perfect agreement", even though some beating in the partials inevitably remains present in the spectrum. This "definite" point is possibly not, however, due to the agreement between partial frequencies in the individual string spectra, and "beatlessness" in certain critical partial meetings. It is true that beatlessness in the 2nd, 4th and 8th partials may be particularly important, since these partials are relatively low in the spectrum and are all the same note (name) as the fundamental. However, the sense of tone and intonation agreement or reinforcement in the well tuned lower bass octave, comes primarily from listening to the lower note of the octave, and may be reached when some higher adjustable partials are not perfectly beatless.

Some say the effect may be due to the *difference tone*. For the bass octave, the *simple difference tone* may be significant. This tone is produced by the difference in frequencies of the fundamentals of two notes (theoretically it may also occur from other partials), so in the octave it always occurs at the pitch of the lower note's fundamental when the octave is in tune. This may possibly be responsible for the sense of strengthening of the octave's bass note that sometimes occurs on a good instrument, when the octave is in good fine tune. If the octave is narrow, with respect to the difference tone, the frequency of the difference tone will be below that of the lower notes' fundamental. If the octave is narrow, it will be higher in frequency.

A diversion

I would not recommend this as a substitute for tuning an octave with excellent tone and intonation, but if one is "academically" interested in so-called "checks" for the tuning of the very lowest notes, then the following one just happens to be an interesting case on many larger instruments. I include it as little more than a curiosity.

The lowest bass notes on a quality grand are rich in distinct partials, offering a number of "tests" for the octaves, depending on the instrument. Often the seventh partial and those around its higher octaves are clear, and can be utilised. The "check" will generally only work on the best, large instruments, and even then there is no guarantee. Only if the beat relationships described below are already clearly present after tuning the octaves by more conventional means, should this control test be employed. The test involves the compound intervals of a minor seventh and major sixth, each plus two or three octaves, depending on what is clearly audible. Fig. (10.9) is a schematic diagram of the test:

Fig. 10.9

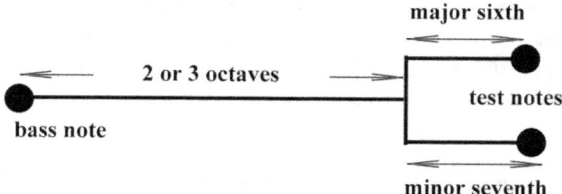

Fig. 10.9 schematic diagram of the lower bass equal beating test. It is only generally of help on high quality instruments.

This test is empirical, rather than derived from "traditional" theory. The test note for the sixth is one semitone below the test note for seventh. The minor seventh may be considered *narrow* (contrary to "traditional" theory that ignores inharmonicity), whilst the major sixth is wide. We are going to be comparing adjacent partials (13th and 14th) in the spectrum of the bass note with the corresponding "test" notes.

When the tuning of the bass octave is sufficiently close, both the sixth and the seventh will produce a beat. It may be necessary to fine-adjust the tuning before both beats can be heard. When they are both present they are generally clear and distinct. Raising the bass note will generally *increase* the beat rate in the seventh, but *decrease* it in the sixth. Conversely, lowering the bass note will *decrease* the beat rate in the seventh, but *increase* it in the sixth. There will be one tuning position for the bass note at which both beat rates are either the same, or in the same proportion as in adjacent intervals. The notes "tested" should be an *adjacent* group. The "test", if it is audible on the given instrument, will usually work well, right down to, and including the very bottom string.

Tuning the bass first – especially on mini pianos

We have already seen that by tuning the bass first, or at least the first octave below the scale area, there is an advantage in that the tuning of the first part of the treble can be matched against bass notes below the scale. If spectrum matching of bass and treble is going to be an issue, then somewhere in the tuning process some compromises are going to be made. On a very small instrument with severe inharmonicity problems it is particularly beneficial to tune the bass first. This is one case where deliberate bass stretching may *sometimes* be employed. Mutually exclusive adjustable beating generally remains unavoidable and the amount of stretching over the compass necessary just to match the relative pith intonation of the bass and treble reasonably well, may be considerable. The opportunity for attenuation is generally absent, but the business of producing consistent intonation and tone qualities in the bass octaves remains the main task.

Unusual trebles – high inharmonicity inequality

The octaves progressing into the treble close to the scale area, may on some poorer or smaller instruments behave in such a way that the normal elimination of movement in the octave is impossible, not because of falseness, but due to excessively large inharmonicity differences between the notes. When this happens, it may be that to eliminate the beat in the second adjustable partial of the octave, the first adjustable partial must be left beating relatively fast. Alternatively, the second adjustable partial must be left beating fast in order to eliminate beating in the first.

This kind of disagreement can happen between any two prominent adjustable partials, not necessarily just the first and second. Moreover, it is possible for the octave to be beating very *narrow* at one adjustable partial, whilst at the same time beating very *wide* at another adjustable partial. This means that within the *fine tuning region* we find that as one beat rate reduces, the other increases. Outside the region, *both* beat rates are too fast. Without accurate perception of beat rate behaviour, the difficulty in tuning such instruments is often blamed on supposed "falseness", when falseness is not at all the root of the problem.

The terms "wide" or "narrow", as always, must only be applied *specifically to an individual adjustable partial* rather than to the interval as a whole. "Wide" or "narrow" is determined by whether we have to increase or decrease the tension in the tuned string in order to slow down or speed up the beat rate for that particular adjustable partial. An octave is never, in fact, entirely "stretched", unless *all* its partials are definitely beating wide.

However severe the case, the solution, as always, still lies in *tone* and *intonation* and the relationship between them. There is no "beat rate" rule to apply, that will work as well as knowledge and control of *tone and intonation*. Ignoring tone and intonation and attempting to use beat rates as a "tool", has the worst consequences in these cases. Tuning where there are very high differences of inharmonicity between upper and lower notes, is, as ever, very much an art, and it develops with experience.

When it is right, tone, intonation, and even beat rate progression all follow.

The top treble octaves – additional considerations

It is the extreme bass and treble of an instrument that usually suffers most from inexperienced tuning, but of these two extremes it is the top treble that invariably requires the greatest experience to master. Unlike the bass, there are fewer easy tests by beat rate available at the top treble, should we feel the need of them, and we are also here dealing with frequency ranges outside the ear's dominance region, to which the ear is generally less sensitive. It takes some years for a tuner to reach mastery of the top treble octaves and unisons, and to be able to tune them "cleanly" with complete control, with reliable stability and a view to maximising the tone of the instrument.

It is not rare to find that the upper treble of an instrument has been tuned in an altogether uncontrolled way, often with a large degree of questionable octave stretching. As we have said, it *is* important that the *pitch relation* of the notes in the upper treble is suitably matched to the mid range and bass, but we should not aim to be stretching the octaves in only say, the top octave and a half, in order to achieve this. Controlled and expert octave stretching from the outset of tuning will usually avert this need.

Judgement of the instrument's tuning condition from a *musical point of view*, involving the sort of "tests" a musician client might carry out, is not sensibly left until after tuning is complete. How the instrument responds to tuning, in terms of tone and pitch, is something we need to monitor as we proceed with the tuning, so we can modify our approach if necessary. This means that in the treble we might want to test various intervals *as complete chords and arpeggios*, and also with the use of the sustaining pedal. The main issues in the top treble are (a) the quality of the *unisons* in affecting the tone and "singing" of the instrument in this range, and (b), the tuning of the octaves in relation to the tuning of the multiple octaves and compound intervals.

The *pitch intonation* of multiple octaves is most obviously apparent when the notes are sounded sequentially rather than simultaneously, and particularly with arpeggios "filled in" between the lower and upper notes. Thus tuners may frequently "check" the tuning of the upper notes with the use of rising arpeggios or chords from an octave below the tuned note, or from the mid range of the compass, or even across the whole compass from the bass. From this, it might seem to the observer that the tuner is tuning by pitch judgement alone. But a skilled tuner will always be using tone and beat phenomena to control the tuning, even when testing for *pitch relations* of the notes. For the artist tuner, even when tuning by interval *tone quality* and *intonation* the judgement is subtly informed by the tuner's aural sensitivity to beat patterns.

Often, the most complained about "hazard" in the treble of most instruments is *falseness*. To the less experienced tuner the difference between false beating, and beating due to avoidable excess "mistuning" is not always immediately apparent. Fine tuning of the top treble octaves involves the ability to hear the *fine tuning region* clearly despite any falseness, and to be able to distinguish constantly between decay movement that is controllable and false movement that is not, despite the fact that both are ingredients in the overall tone.

Falseness that is judiciously "hidden" in a unison may reappear in an octave of which the unison is a member. As we have seen, often, the condition of an octave with a false string in a trichord, is improved once the whole unison is tuned, but conversely, the octave condition can sometimes be deteriorated once the unison is tuned, if judgement fails. It may be necessary to reset the tuning of an octave once knowledge of a unison's characteristics has been gained.

Moving the head - position of listening

Finally, the octaves depend absolutely on the unisons. The tone of the high unison is, like all tone, dependent on *psycho-acoustic* influences. The incoming sound information affecting the psycho-acoustic assessment of tone may itself be affected by the room acoustics and *position of listening*. The position of the head may also affect phase relationships between the ears, and other factors. Therefore, if unsure, in the case of tuning where falseness is present, *listen from different positions*. Ultimately, what counts is *tone*, not some principle in terms of beating. All kinds of "false" fluctuation patterns happen in the fundamentals of the upper notes. There is typically a *range* of what could be called tunings of *minimum beating* in any unison, over which the *tone* varies.

One must not adhere to the idea propagated by "traditional" theory, that these unisons *should* be "beatless". If you have been conditioned by this notion, you will nearly always be disappointed, and you will miss the difference between beauty of tone *with fluctuations*, and less than the most beautiful tone possible, *with fluctuations*. False beat fluctuations are always to be "minimised", but this is a fuzzy concept. A top treble unison is not necessarily unbeautiful because its tone contains some fluctuation. The difference between a beautiful tone and a less than most beautiful tone of the unison, is one of those things that circumvents the mind of the rationalist: you cannot necessarily define it, *but you know it when you hear it*.

The musical picture

Testing for progression of the fast beat rates right through a central scale octave and out beyond it on both sides, is part of assuring the underlying technical structure of the tuning. One also looks, of course, for equality of tempering in the fifths and fourths, and in particular the absence of unpleasantness of tone in these intervals due to too much movement in the soundscape. This depends, of course, not just on beat *rate*, but on beat *amplitude* and the *tone* of the interval. The progression of beat rates in the major tenths and octave-tenths (major seventeenths) over the compass is important because these beat rates and their relationships determine the smooth progression and equality of tone and intonation over the compass. They are critically related to octave stretching and hence the intonation and tone of the instrument as a whole.

In the final analysis, the musician is not usually aware of technical structure, but *is* aware of tone and intonation. Not only is the musician aware of poor tone and intonation, but discerning musicians are generally aware of *improved* tone and intonation on a given instrument when it is presented to them. The improvements in musical tone and intonation, achievable through judicious and artistic octave stretching that empathises with the piano's characteristics, are well worth the extra effort they demand in tuning.

It is, on the other hand, perfectly possible to end up with satisfactory technical structure, but unrealised tonal potential, or even unsatisfactory intonation over the wider compass. Tone and intonation over the compass as a whole is not determinable simply by listening to beat rates! It is easily exposed in a suitable *musical* test. Tuners have their own favoured "test pieces" to assess the overall tuning, but we do need a musical test that exposes all *individual notes* in the *musical* context, over the width of the compass. For the treble stretching, one such assessment, somewhat simplified, is as follows:

Here, the upper notes in the right hand are repeated to assess their tone and intonation. They can be repeated more than once, the score produced here being only indicative. In fact, playing the note three or four times gives a good indication of tonal beauty or lack of it. Firstly, the unison tone can be assessed. Secondly, by quick arpeggiation of the chords, and through the way they expand position over the compass, the intonation between bass and treble can be appreciated. It is more "musical" to cover all the notes through a cycle of fifths, rather than chromatically, so each "test" here modulates into the next through the use of the first inversion.

Again, the illustrated score only indicates the *kind* of approach that can be taken. Many other possibilities are of course possible. The point is to expose *individual notes*, for tone and intonation. Playing just any piece of music, however impressive, will not necessarily do this.

In the method outlined here, after the whole cycle of fifths has been passed through, *every treble note* above the scale would have been *individually exposed* in a tone and intonation context. The whole cycle takes less then three minutes with careful listening, when there is no interruption for any re-tuning. Of course, if the need for re-tuning for intonation or tone is exposed, re-tuning must be done.

Tuning two or more pianos together

Tuning two pianos together is a specific task set at the higher levels of study and training in piano tuning. The need to tune two pianos together is not uncommon, as many teachers use instruments together. Of course, there is some repertoire in which tuning *more* than two pianos together is called for, however, such situations are generally *less* of a problem because typically, identical pianos are provided. The problems arise when the pianos are different.

If the two pianos are the same make, model and age, then there is generally no special problem in the task. Often, however, two pianos being used together will be very different is size, sometimes one being an upright and the other being a grand. In this case, there may be significant inharmonicity differences between the instruments. The general rule is to tune the best instrument first. If a large grand is being tuned with a smaller one or an upright, then this must be tuned first.

Discrepancies between individual notes of the pianos can arise even in the scale area, so when the scale is tuned on the inferior instrument, it must be checked note for note against the superior one. In these situations some discrepancy must be expected, and up to a point, musicians regard small discrepancies as part of the normal sound of two pianos being played together. The larger or more reverberant the acoustics, the easier it will be to tolerate differences. Where intolerable discrepancies are found, always alter the inferior instrument, in preference to the superior one.

Clients will not necessarily understand the technical issues. The superior piano should display *excellent tuning*, and in general any shortcomings in the inferior one will then be explainable by the inferiority of the instrument, and the task of "matching it" to the superior one. In the final result, the tuning of the inferior instrument must be compromised in order to improve the match of individual notes and octave intonation over the whole compass, with the superior piano. Where a tempered interval on the inferior instrument must be compromised, the compromise should not necessarily fall wholly in the interval, but may need to be shared between compromising the interval, and compromising the matching between the notes on the two pianos.

When a suitable scale on the inferior instrument has been produced, the octaves must be tuned. Some "test" notes at the top of the compass should be tuned first, by tuning three abutting octaves from the scale up to the test note. and compared the test note to the same notes on the superior piano. Typically, C, A and F are a good choice. There *must* be a good match here between the test notes on the two pianos. The test notes will indicate the degree of stretching necessary on the inferior piano.

The intervening octaves must then be tuned, also matching individual notes frequently between the pianos. Any wholly unacceptable tuning between a scale note and the octave or compound octave note on the inferior piano itself, cannot be allowed, and if necessary, the match between the two pianos must be compromised to avoid this.

Summary of common "checks" for octaves in the treble

<u>Double and multiple octaves:</u> Beat at pitch of upper note.

<u>Tenths (tempered wide):</u> Beat pitch one octave above upper note.

<u>Major octave-tenths (major seventeenths) (tempered wide):</u> Beat at pitch of upper note.

Summary of common "checks" for octaves in the bass

<u>Double and multiple octaves:</u> Beat at pitch of upper note.

<u>Major tenths (tempered wide):</u> Beat pitch one octave above upper note.

<u>Major octave tenths (major seventeenths) (tempered wide):</u> Beat at pitch of upper note.

The minor third (narrow) / major sixth (wide) check for bass octaves:

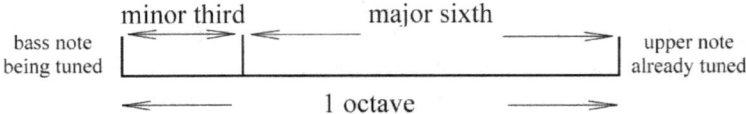

Beat pitch two octaves plus a fifth above bass note being tuned.

Chapter 11 - Setting the Pin

To the layperson, the term "setting the pin" could possibly be the most ambiguous term in the piano tuner's vocabulary. It is of course the *string* that is being tuned, and it is the string-pin system as a whole that is being "set" in a particular condition.

Even today, one sometimes still hears the meaning of the phrase "setting the pin" described as "taking the twist out of the pin". This is a very ambiguous description that suggests a lack of understanding of the basic physical system. If "twist" in the pin refers to elastic, torsional distortion of the pin when force from the lever is applied, then one cannot say the twist would need to be "taken out" as a deliberate act, after finishing turning the pin one way or the other. Any such "twist" that is introduced as the lever is used, will naturally be lost as soon as the turning force is removed. As soon as the force from the lever is removed the system will tend to equilibrium. If this equilibrium is an unstable one, which it may well be, it will not be due to *torque in the pin* that can be "taken out". As we shall see, it is not movement of the pin itself, either turning or twisting torsionally, that is the primary cause of tuning instability on a good instrument.

The fact is that whilst elastic, torsional distortion of the pin does play a part in the dynamics of setting the pin, it is only one of the physical factors that are involved, and "setting the pin" is not a process of removing it, as such. Setting the pin may be "intuitively" achieved quickly and easily by the experienced, expert tuner, but what the tuner is doing is not something that can be described simply as "taking the twist out of the pin". What is being "taken out", properly speaking, is the tendency for the tension in the speaking length of the *string* to change easily, especially when the note is struck hard.

Another idea often encountered is that setting the pin is just raising the tension a little above a desired "target" and then dropping it back down, or alternatively lowering below, and then bringing back up.

Sometimes a proper pin setting action may happen to be just this kind of action, but it in no way represents a *general definition* of what pin setting involves. Often such an approach would result in *instability*, as we shall soon see.

Let us first be quite clear about what we mean by tuning stability. On a piano in good condition, after a note is tuned, it should be possible to strike the note harder than any pianist is likely to strike it, several times in succession, without any resulting change to the tuning. This means striking the note forcefully, loudly, even with more than one finger at a time. Assuming the piano is in good physical condition, without loose pins, then if such striking easily "knocks out the tuning", the pin was not set. Quite simply, tuning without proper "pin setting" results in tuning that changes rapidly as the instrument is played. Even if the instrument is not played hard, then the change still takes place, but not quite as rapidly as when notes are struck hard. This is distinct from the falling out of tune that may also take place due to changes of temperature and humidity, or moving the instrument.

If the "pins are not set", instability of tuning on a good piano *will tend to arise quite naturally* as a result of *the proper design of the instrument*. It arises primarily from the fact that the string is divided into a number of sections, *deliberately* separated by friction points, with resulting tension differences between the sections. It is primarily the handling of these tension differences that pin setting is all about. Of course, the tuner does not tune by measuring or feeling the tension differences as tensions alone, but rather, by being aware of how the soundscape changes and responds to the "feel" in the lever.

The discussion set out here on setting the pin may seem so detailed as to be daunting at first sight. The apparent complexity arises only because the dynamics of pin setting are described here in unusual technical detail and depth. A practical and intuitive understanding of these dynamics are usually learnt from many years of piano tuning experience. The essence of what is described will therefore already be familiar to expert tuners at a practical and intuitive level, at the very least as subconscious knowledge gleaned from practical experience.

Does one need to understand the technical details? If you are a theorist you will probably want to. If you are a practical tuner you can skip over them if you wish, and still learn to set the pin from guided experience. However, if you quite literally want to "know what you are doing", then naturally, looking at the technical details can be enlightening.

To begin with then, here is an "intuitive" description, without too many technical details, that will serve as an introduction in any case:

A practical overview

When turning force (torque) is applied to the pin through the lever, the pin may or may not turn in the wood, because of the friction between the pin and the wood. Either way, the top stringing tension *changes*, because the pin may still torsionally twist or bend.

Now if we are to be pedantic, we should mention that the wood close to the pin may itself distort, and that certain extremely small movements of the pin relative to the wood may take place, that are indiscernible by the tuner. We are not, however, going to get into this kind of pedantry, because we are primarily interested in how the basic mechanics affects the practical task.

For our purposes, therefore, we shall consider that at any instance the pin either rotates in the wood or not, as the case may be, and look at the basic behaviour under those conditions. Tuning involves very fine precision changes to the soundscape, but the means of making these changes, nonetheless, are relatively crude. There is a pin inserted into a block of wood, and it must be turned manually, against high friction and high tension, using a simple lever. Sometimes the pin discernibly turns, whilst other times it discernibly twists slightly, torsionally, without appearing to turn. One of the first things a tuner must learn to recognise, is the difference between a turn and a twist of the pin.

The string may pass under a pressure bar and over a top bridge on the frame, or just through an agraffe, or if it is a bass string, it may pass a side draft pin on the frame. On a grand piano it may pass under the *capo d'astro* bar. All of these points are high-friction metal against metal points. The string will only slip over any of these points when the tensions in the parts of the string on either side of the point are at a certain minimum ratio or more. In other words, you can only raise the tension in the speaking length, when the tension in the top stringing is a certain percentage *above* the existing tension in the speaking length. (Of course, there may actually be more than one tension in the top stringing because it is in different sections, but the general idea will suffice for now without this complication). Similarly, you cannot lower the tension in the speaking length until the top stringing tension is a certain percentage *below* the existing tension in the speaking length.

Once the string is slipping over the friction points, the friction reduces, and it is then easier to keep it slipping. One way to get the string slipping when it is close to doing so, is to cause the string to have large vibrations, *i.e.* to repeatedly strike the note hard.

There is thus a kind of "disconnection" between the pin movement and the speaking length tension change, which is called *hysteresis*. Setting the pin requires understanding and working *with* this hysteresis, and not fighting it.

Now when you turn the tuning lever, the pin may just *twist* torsionally, or may actually also turn in the wood, depending on how far you turn, and how much friction there is between the pin and the wood, *etc.* When you let go of the lever, any torsional *twist* in the pin, will naturally turn back, or *untwist*. As it does this, the tension in the top stringing will drop a little after trying to raise the tension, or may rise a little after trying to lower the tension. This may have negligible effect, or may be highly significant, depending on how much the pin twisted, and what the percentage difference between the tensions in the top stringing and speaking length, need to be, for slippage across the bridge to occur.

Basically, it is necessary to know when the string is slipping over the bridge or other friction points, and *when it is about to do so*. One must not leave the string in a condition where it is about to slip either way, and could do so following playing the note hard, or after further small changes in tension anywhere in the string, brought about by temperature or humidity changes, or small shocks.

Consider a trichord. If the speaking length tension changes a little due to the crown on the soundboard changing a small amount, the effect on the tuning will be far worse if, say, one of the trichord's strings is already about to slip across the bridge. The change in crown can "tip" the tension ratio over the critical point, and slippage may occur. Without this problem, the "pitch" of the piano may change very slightly, but without necessarily "upsetting the tuning" so much. Consider also that string in the event of a hard strike from the hammer. Slippage will be encouraged by the strike, so hard playing will tend to cause the note to fall rapidly out of tune.

The top stringing acts as a kind of "buffer" between the tuning pin and the speaking length, and this buffer must be correctly handled. To be in complete control, one must especially know when the tuning pin is *twisting* without turning in the wood, or when it is actually turning in the wood. Both actions can affect the "buffer" of the top stringing.

Changes in the audible soundscape do not, in general, therefore follow turns or twisting of the tuning pin in a straightforward relationship. When the pin is turned or twisted, the only *immediate* effect is a change in the *top stringing tension* just beyond the pin. From here, the string passes over the friction points. Only a change to the actual speaking length tension produces a change to the soundscape. In order for the speaking length tension to change, the string must slip over friction points, but it will only do this under certain conditions, already outlined.

As you tune, there will be a "target" tuning that is "discovered by doing". You hear where you want to go. The nature of the soundscape and the way it behaves close to a good tuning condition indicates what kind of result should be possible on that particular string or note. Merely finding the tuning "target" and leaving the tuning there, *is not sufficient*. One has to *get to the target* through a route of movements or changes in the soundscape such that when the target is finally tuned and then left, the tuning is stable there. The string may not be slipping across the bridge (or under the *capo d'astro*, etc.) now, but it may do if the string vibrates enough. Striking the string hard has the same effect as reducing the necessary percentage difference between the string tensions for slippage to occur. If you leave the string with a tension difference close to that percentage, striking the note hard can make the string slip, with the result that it goes "out of tune".

If the tensions in the parts of the string between the friction points are left close to the condition that enables slippage, then slippage can easily subsequently occur with changes of temperature, humidity, or just from the physical shocks of the hammer striking the string in normal playing. When the "pin is set", hard striking of the note, or small forces applied to the tuning lever in either direction, will not upset the tuning.

The sequence of lever movements necessary to arrive at the "target" with the tuning stable, are those that enable proper control of the changes in speaking length tension during tuning, and also leave the various parts of the string between the friction points, with their tensions set up in such a way that further slippage of the string across the friction points is least likely to happen.

Basically, the practical effect is that at any time, the soundscape either appears to *resist* or *co-operate* with the attempt to change it in the direction it is being tuned, or it follows some condition between these two extremes. It mechanically "wants" to move towards, and settle down naturally at, some tuning condition, which the tuner needs to make equal to the chosen target condition. In the final tuning movement, the soundscape and the piano tuner should, so to speak, be "co-operating" with each other, towards a final tuning condition that is the desired one. In general, this does not just happen, but has to be "engineered" or "set up" in the route taken through a reducing range of tuning movements that "close in" on the target.

As the tuning pin is turned to raise or lower the tension, some of the "effort" applied to move the tension in the speaking length and change the soundscape, can be "taken up" in the top stringing before the change occurs in the speaking length. There may even be a palpable "delay" as the pin is turned or twisted, before any change to the speaking length occurs. The amount that the top stringing intervenes in the process of trying to alter the speaking length tension, is not always the same, but depends on a range of physical conditions that the stringing may be in at any time.

At one end of the range, there is practically no intervention if the tuner attempts to *raise* the tension, but at the same time there will be maximum intervention if a *lowering* is attempted. At the other end of the range, there is practically no intervention if a lowering of tension is attempted, whilst at the same time a raising of tension is met with maximum intervention of the top stringing. In the middle of the range, there will be equal intervention either way. The maximum intervention may be anywhere between negligible and large, depending on the piano and the pin.

This maximum amount of intervention by the top stringing is technically called the "hysteresis width". Where the string lies in this range at any time, depends on whether the tension has just been raised or lowered, *and* on the hysteresis width, *and* on how tight the pin is in the wood.

The tuner "feels" and hears this intervention from the top stringing and friction points. One knows from the feel in the lever when the pin is turning in the wood, and when it is moving elastically. One can also *hear* when the string is slipping through or over the end of the speaking length, from the change in the soundscape. Importantly, it is possible to hear the *rate* at which the string is slipping over the end of the speaking length, compared to have much the lever is moving. It is possible to hear the difference between movement in the soundscape due to the mistuning, and movement due to *changing* mistuning.

This may all seem rather complicated, but really it is no more so than many other tasks we perform, many of which take some time to learn. For example, consider the way many vehicles behave in steering. A large boat or plane will not necessarily just turn instantly with a tight response when the rudder or engine power is altered, but this does not make it impossible to steer or control. Expert pilots can still put the largest vessel exactly where they want it.

This is crudely analogous to the large hysteresis width. Many small cars, however, will appear to respond instantly to the steering wheel and engine, which is analogous to the small or negligible hysteresis width. This is generally easier to control. Just as different tuning pins can behave in different ways even on the same piano, so the car in the example can behave differently on different surfaces such as wet or ice, at different times. This still does not stop the expert driver controlling the vehicle.

Steering vehicles requires seeing exactly where the vehicle is at any time, and where it is going, in relation to how the steering is responding at the time. Tuning requires hearing exactly where the soundscape is, and where it is going, in relation to how it is responding to the lever movements at any time.

The tuning condition can usefully be *likened* to the classical model of mechanical equilibrium, although it is *not* physically the same thing, as it can be subject to much more complex behaviour. Fig. (11.1) shows representations of the classical three different equilibrium states.

Fig.11.1

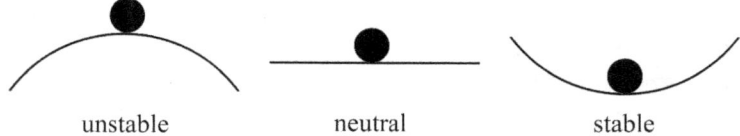

unstable neutral stable

Fig. 11.1. The three "classical" different equilibrium states. The ball rests on a surface. In unstable equilibrium the ball may not be moving, but a little disturbance will cause it to roll a long way. In neutral equilibrium a little disturbance will move the ball, but not much further than the disturbance itself moves it. In stable equilibrium, the ball will tend to return to its original position despite any disturbance.

In terms of a *very small* turning force subsequently applied to the tuning lever after tuning the string, the soundscape for a well set pin will show no tendency to change more one way than the other. For a small force nevertheless large enough to *cause* a change to the soundscape, subsequently applied, the situation for the well set pin should resemble the neutral equilibrium model. For a pin that is not properly set, the situation will resemble more the unstable equilibrium. A small force on the pin in one of the two possible directions, will then tend to *easily* cause a change to the soundscape. As far as the ball representing the soundscape is concerned, a situation like *stable equilibrium* cannot generally be relied on to occur. If it did, the note would have a tendency to retune itself after any detuning. Not in fact, theoretically totally impossible, but a rather demanding expectation!

On starting to tune a string, there may well be a certain small range of pin movement that can be made *without* causing any change to the soundscape. In other words, the soundscape only begins to change after a certain minimum pin movement is made. This is due to the *hysteresis* already mentioned, and which we will deal with in more detail below. In general, though, as far as pin movements that *do* cause a soundscape change are concerned, then if one is about to succeed in setting the pin, the final fine tuning movements should *feel* like the situation shown for neutral equilibrium, and definitely not as though one is trying to "balance" the desired tuning on top of a convex surface, as in the unstable model.

However, just *outside* this small, final fine tuning range, as one is first bringing the string into tune, the situation should *feel* as though one is in something like the *stable equilibrium* scenario, rather than the neutral, with the ball slightly off-centre, where the ball or tuning actively co-operates with the attempt to move it "in" towards the "target" tuning at the centre of the surface. The initial tuning movements need to set up this state of affairs, first. At the very centre of the curve in the model, however, *for very small movements*, that very small portion of the "curve" in stable equilibrium, is in any case is very similar to a flat surface, and resembles neutral equilibrium.

Suppose one is lowering towards the target tuning. If the soundscape has a tendency to fall easily below the desired target, a larger pre-setting movement upwards may be necessary above the target before lowering again towards it. Having then reached the target, there may still be a tendency for the soundscape to rise or fall, *unless* the *feel* is that tuning is now on the "neutral plane". If it is, the pin is probably set, if the soundscape shows no tendency to change with hard striking. If the tendency is to rise, then pushing below the target may be necessary before rising back towards it. The same "neutral plane feel" then applies.

If the tendency is for the soundscape is still to fall, then another movement above the target must be made first. One may end up lowering or raising the soundscape towards the target, deliberately passing through the target each time. In general, the route into the target may be a decreasing series of movements in both directions, "homing in" on the target.

This does not mean a pin can never be set in just one movement. It sometimes can. But there is difference between just making one movement, and making one movement with the knowledge that the pin is being set.

Any string at any time will have a natural tension towards which the speaking length will tend to move, through slippage over the friction points, representing maximum equilibrium or stability at the time. This tension can be above or below the current speaking length tension. The slippage is assisted every time the note is played, and harder striking assists more. Setting the pin is about getting this tension to coincide with the desired soundscape. To do this one must always be aware of how co-operatively or otherwise, the soundscape is changing *as* it is being tuned. The final bringing of it into tune *must be* while the string is co-operating, and if necessary, the top stringing must be retensioned up or down, and/or the pin repositioned in the wood, in order to bring this about.

Some tuners believe they are setting the pin when in fact they have no real practical understanding of it. or at least are not doing so by virtue of any real technique they are using. The "classic" misconception is to regard pin setting as simply raising the note a little above the desired target, and then dropping it in. In many cases, this simply will not work. Many Steinway upright pianos are examples of such cases, as are small grands with large bearing angles in the top stringing. Below, we shall see in more detail how such an approach may leave the tuning unstable, because of the way *hysteresis* operates. It may sometimes work up to a point, but finer tuning demands a much more developed technique, allowing complete freedom to enter the final tuning position from above or below, as each individual situation dictates.

If you are learning to tune, the above overview of setting the pin may suffice. You may, however, still find the sections below useful. If you are of a mathematical turn of mind, the technical description may be of interest. If not, the mathematical parts can simply be skipped.

If you want to study in more detail, but only in terms of the practical approach, then from here you can safely skip to the section *From theory to practice: setting the pin*.

If the following theoretical description of the process seems rather involved and complex, this faithfully reflects the nature of the art of setting the pin, when it is explained in technically detail. The art, as it is practiced by expert tuners, simply cannot be described with any real fidelity, in only a couple of lines of explanation. The human body and mind is of course able to perform enormously complex tasks without necessarily considering the technical details conceptually. A violinist, for example, may play the violin beautifully without considering graphs or equations for the actions of the strings and fingers. But playing the violin beautifully is nonetheless an enormously complex task, and if one were to describe it in the same technical way as I am about to describe some aspects of pin setting, even greater complexity would be found.

Theory in detail

Tuners who set the pin with expertise and have paid attention for many years to the behaviour of the pin and string and soundscape in the tuning process, will have noticed that there is not one, fixed response dynamic for every pin and string. For a given string, just how the audible soundscape alters when a given force is applied to the tuning lever, depends on a number of factors:

- It depends on whether we are raising or lowering the tension.
- It may depend when we strike the note and how hard we strike it.
- It generally depends on whether we have just raised or lowered the tension, and by how much.
- It depends on the individual pin and string, and whether the pin is already "set".

Students of tuning often find the response of the pin and string is not at all what was intuitively expected, and may find that a piano's response to tuning attempts is counterintuitive and confusing on many instruments. This is because the physical part of tuning is not simply about "turning the pin to alter the tension". The system is just not that simple. It is absolutely essential in successful tuning that not only do we arrive at the right audible soundscape, but also that *how we arrive at it* is also right. The expert tuner achieves this by being acutely aware of *how the soundscape is changing* in relation to the "feel of the pin" is changing, which is actually the feel of the string as well as the pin. Sometimes a tuner may be heard taking a note repeatedly up and down in pitch by small amounts. This is not in order to merely hear the sound, but it is to detect how the soundscape changes in relation to the "feel" of the pin.

The "feel of the pin" is not constant, but constantly changes according to where one is in the process of setting the pin. Whilst this may seem to indicate something inscrutable is going on, a very large part of the process *can* be explained in objective terms, which is precisely what we shall now do. However, the tuner's final decision on *when* the pin is properly set, and *where* in the range of soundscape possibilities it is set, remains without doubt one of the factors that ultimately defines piano tuning as a practical *art*. Piano tuning without any regard for setting the pin, is not proper piano tuning at all. It only becomes piano tuning proper, in the process that takes control of the relationship between how the soundscape changes, and "feel of the pin". In piano tuning, the end result never justifies the means of arriving at it. Tuning carried out in that way will be unstable, and worse, will actually upset the instrument. On the contrary, the *means of arriving at the end result*, which is called *setting the pin*, is what creates the right end result.

The master tuner can always readily tell if an instrument has been previously tuned by a tuner who sets the pin properly. Tuning without sufficient pin setting skills leaves an instrument inherently unstable, and requiring much more skilled work to bring it into better stability, and hence finer tuning.

Basic practical observations

The student of piano tuning will be able to observe some basic features of the piano's behaviour. These will vary from piano to piano, which is why it is important to practise on as many different instruments as possible. These basic features are as follows:

- Turning the pin by a small amount in one direction or another, does not necessarily cause any immediate change in the soundscape.

- The soundscape can easily continue changing after the pin has been turned.

- Sometimes the soundscape changes very easily with only a little force on the lever. Sometimes it seems to be highly resistant to change. This varies according whether the pitch has just been raised or lowered.

- Striking the note hard can cause the string tension and soundscape to change.

- A string whose tension has just been altered may seem just as likely to spontaneously rise in pitch, as to fall, when no attempt to tune it is being made.

- The response of the string and soundscape may seem to display a "delayed reaction" to the attempt to change it.

Whilst these characteristics may seem like "faults" to the learner, they are a natural part of the mechanics with which we have to deal. We have to understand these features and become the master of them.

The characteristics can be superficially explained by the following. When the lever is turned, the pin may twist torsionally without turning in the wood, and it may bend a little. This will cause a change to the tension in the *top stringing*, above the bridge (or agraffe). The pin may also, at some point, actually turn in the wood, which will also cause a change to the tension in the top stringing, above the bridge (or agraffe).

However, a change to the top stringing tension does not necessarily result in a change to the speaking length tension (and hence the sound), because of friction at the bridge or agraffe. Consequent changes to the speaking length tension can only take place if the string actually slips over the bridge, or through the agraffe. It will only do so, if the *ratio of the tensions* in the top stringing and speaking length is great enough. The tension in the top stringing may even be reduced, without necessarily affecting the tension in the speaking length. Often, it may be possible to cause the string to start to slip over the bridge or through the agraffe, just through the action of striking the note, and sometimes it may continue to slip in other circumstances.

One of the major features of the natural mechanics is often a perceived "gap" or "disconnection" between a given small turn of the pin, and resultant changes to the speaking length tension, and hence changes to the soundscape. This "gap" in response, which is sometimes experienced as a "delay", is a mixture of two effects which we will deal with in detail. One is *mechanical hysteresis* and the other is *latency*.

Setting the pin requires a technique that takes into account definite, natural physical features of the system, and at the same time uses them to advantage, in order to attain tuning stability. The primary three physical features are:

- The elasticity and recoil in the tuning pin, top stringing and tuning lever.
- The friction between the tuning pin and the pin block.
- The friction between the string and friction points such as the top bridge or agraffe.

As far as the practical technique of tuning is concerned, these three physical features translate into four specific *parameters* that lie behind the practical strategy for setting a particular pin. These parameters are:

- *Hysteresis width* – a measure of how much the tension in the top stringing may be increased or decreased without affecting the speaking length tension (and hence the soundscape);
- *Pin torsion* – how far the pin must turn under *torsional distortion only*, before the pin starts to actually *turn in the wood*;
- *Spring* – how far the lever must be moved before the pin actually turns in the wood.
- *Latency* – the delay in slippage of the string over the bridge (or through the agraffe) causing changes *after* turning the pin.

These will all be dealt with in detail.

All these features are variable from piano to piano, and often even from pin to pin. There is therefore no one, fixed prescription for how to set a pin, simply in terms of one particular set of tuning lever movements. In addition, a pin subjected to too much movement in tuning, may heat up because of friction in the plank, which can alter these features on a given pin.

Understanding friction – the main characteristics

The string passes from the tuning pin to the top bridge, or to the *capo d'astro* bar, or to the agraffe. Depending on the piano, and where the string falls in the compass, it may pass under a pressure bar before passing over the top bridge, or it may also be drafted sideways past a bridge pin on the iron frame. On a grand piano the string may pass under the *capo d'astro* bar. At the other end of the speaking length, the string passes over the long bridge (soundboard bridge) or bass bridge, where it is drafted past two bridge pins, before passing across the duplex scale bridge (on the iron frame) if one is present, and finally terminating at the hitch pin.

Looking at the situation the other way around, tracing the string path from the hitch pin to the tuning pin, we can say the system is comprised of a high tension string that passes across a number of *friction* points, before being coiled around the tuning pin. From the hitch pin to the soundboard bridge is the *backstringing* length. From the soundboard bridge (long bride or bass bridge) to the top bridge, agraffe or *capo d'astro* bar, is the *speaking length*, whose tension determines the audible soundscape. From the top bridge, *capo d'astro* or agraffe, to the tuning pin, is the *top stringing* length.

When the tuner turns the tuning pin, this only gives *direct* control over the length of top stringing. The tension in the other string sections cannot be altered without the intervention of at least one friction point. It is the friction points, the elasticity in the string and pin, and the friction between the tuning pin and the pin block, that primarily creates the need to "set the pin", but also makes stable tuning possible.

There are two kinds of friction at these friction points: *kinetic friction* and *static friction*. Kinetic friction operates between two contacting surfaces when the they are in relative motion. Static friction operates when two contacting surfaces are not in motion relative to each other. For a given pair of surfaces pressed together under the same forces, static friction will usually be greater than kinetic friction. Static friction is proportional to the normal force (the force at right angles to the surface), whilst kinetic friction may also be a function of the velocity between the two surfaces, and typically, both will depend on temperature. In the case of kinetic friction, temperature will usually increase as motion takes place, and this may have the effect of *decreasing* the kinetic friction.

Friction will also depend on the specific materials from which the surfaces are made, and the condition of the surfaces. It does not necessarily follow that a smooth or shiny surface will have low friction.

Friction is most dramatically reduced not by smoothness, but by lubrication.

In the case of a piano string passing across the top bridge or through an agraffe, there will always be friction at these points. When the tuning pin is turned, the immediate effect is that the tension in the top stringing will increase or decrease. The tension in the speaking length will not necessarily be affected immediately, because the friction at the bridge or agraffe may prevent the string slipping over the bridge or through the agraffe. This lack of immediate "connection" between the tension in the speaking length and the tuning pin movement, is part of the effect known as *hysteresis*.

Elasticity – the main characteristics

Elasticity plays an important part in setting the pin. For small movements in any simple elastic system, the displacement due to elasticity is generally proportional to the force applied. This means that roughly speaking, in turning the tuning lever in order to overcome elastic forces such as the top stringing stretching without moving across the top bridge or agraffe, or the pin twisting without turning in the wood, or even just the tuning lever bending a little, the force the piano tuner must apply up to the point the pin turn in the wood, will be approximately proportional to how far the tuning lever is moved. It follows that for relatively small lever movements, relatively little elastic force is involved. For larger tuning lever movements, greater elastic forces will be involved.

Mechanical hysteresis in the string-bridge system – technical description

The friction points across which the string passes on its way out from the tuning pin towards the speaking length, may include the top bridge or agraffe or *capo d'astro* bar, the bearing plate and the pressure bar.

The tension state of a string passing over a friction point is described by the *capstan equation*

$$T_o = T_i e^{\mu\theta}$$

in which T_o is the "outgoing" tension in the string past the friction point, T_i is the "incoming" tension in the string approaching the friction point, μ is the coefficient of friction and θ is the angle through which the string passes. Fig. 11.2 shows an example:

Fig. 11.2

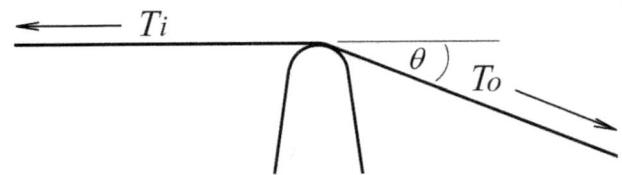

From the capstan equation it can be seen that the ratio of the tensions on each side of the friction point is *independent* of the radius of the friction surface. For μ as the coefficient of static friction, the ratio T_o/T_i is the maximum tension ratio in which the string will remain in equilibrium – attempting to increase T_o further when the ratio is already equal to $e^{\mu\theta}$, will result in the string slipping over the friction surface.

Consider now the string passing through an agraffe. As the tension in the top stringing (between the agraffe and the tuning pin) is *increased* by turning the tuning pin, the tension in the speaking length will *not increase* until the ratio of the tension in the top stringing, to the tension in the speaking length, exceeds $e^{\mu\theta}$. The tension in the top stringing will, of course, change *rapidly* as the pin is turned, because the length of top stringing is relatively short. If the pin is now turned back so as to lower the tension, the tension in the speaking length will now *not decrease* until the ratio of the tension in the speaking length to the tension in the top stringing, exceeds $e^{\mu\theta}$. To lower the speaking length tension, the tension in the top stringing will hence need to be lowered *beneath* the current tension in the speaking length, before the speaking length tension will start to reduce.

In practice, there may be more than one friction surface in close proximity, such as the top bridge and pressure bar on an upright piano. Where multiple friction points are considered, $e^{\mu\theta}$ is simply replaced by $\exp(\mu_1\theta_1 + \mu_2\theta_2 + \ldots \mu_n\theta_n)$ for *n* points and angles.

Let's now look more closely at the case of a single agraffe. The situation can be represented with a *hysteresis graph*, Fig. (11.3).

Fig. 11.3

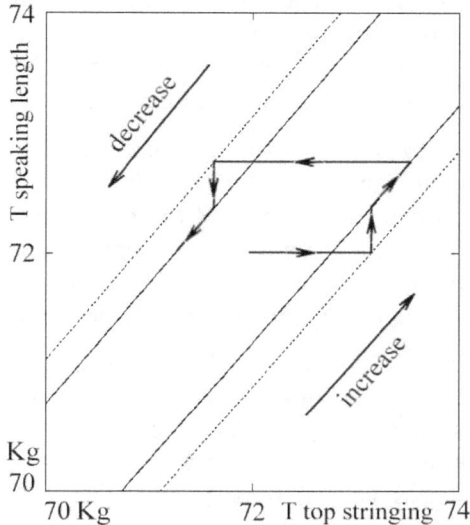

Fig. 11.3. Hysteresis loop for a piano string in tuning. The "increase" pair of lines apply when tension is being raised, whilst the "decrease" lines apply when tension is being lowered.

The graph has two pairs of lines, which in the small portion of the range shown, are effectively parallel. The outer (dotted) lines are for *static friction* - they show $\exp(\mu_{static}\theta)$, the maximum tension ratio between the top stringing and the speaking length, when the string is *not moving* through the agraffe, using the coefficient of static friction. The inner (solid) lines show $\exp(\mu_{kinetic}\theta)$, and are for *kinetic friction*, when the string is *sliding through* the agraffe. The vertical axis gives the tension in the speaking length, and the horizontal axis gives the tension in the top stringing length. The parameters illustrated for the fiction coefficients and angle are not unrealistic, but are chosen here for illustration clarity: $\theta = 3°$, $\mu_{static} = 0.3$, $\mu_{kinetic} = 0.2$.

The graph is actually *two* pairs of graph lines, for both static and kinetic friction. On the right are the dotted and solid lines (dotted for static friction and solid for kinetic friction) that apply when the tension in the top stringing is being *increased*, whilst on the left are the lines for when the tension in the top stringing is being *decreased*. This dual situation is typical of hysteresis graphs. Each pair of lines represents the tension values corresponding to the *maximum tension ratio* that can exist between one tension and the other. Neither the top stringing tension nor the speaking length tension can appear outside the outermost graph lines.

The heavier arrowed lines near the middle of the graph represent the tensions in both sections of the string under the following sequence of events. To begin our understanding of the dynamics, we will assume for the time being, that the tuning pin turns perfectly smoothly in the pin block - there is no tuning pin recoil due to friction between the tuning pin and the pin block, and there is no elasticity in the tuning pin to consider.

We begin, for convenience, in the centre of the graph with equal tension (72 Kg) in the speaking length and top stringing. Now follow the arrowed line. As the tuning pin is turned to increase the tension, the top stringing tension (horizontal axis values) increases but the speaking length tension (vertical axis values) remains unaffected. This is because the friction at the agraffe is preventing the string from being pulled through it. The pin turns but there is no change to the soundscape.

As soon as the tension in the top stringing reaches the dotted line for the static friction *limiting ratio*, the ratio between the top stringing tension and speaking length tension reaches its maximum, and the tension in the speaking length *jumps* up to the inner, kinetic friction line, as the string begins to slide. The soundscape starts to change. As the pin is turned further, the ratio of tension in the speaking length to that in the top stringing, remains fixed by the kinetic friction and follows up the solid *increase* line. At the top right hand corner of the arrowed path, the pin turn in stopped when the tension in the top stringing is about 73.5 Kg, and the tension in the speaking length is about 72.75 Kg.

Now we begin to turn the tuning pin back so that tension decreases. We are now entering the left pointing horizontal arrowed line at the top of the circuit. As top stringing tension decreases, the tension in the speaking length stays the same, so the soundscape stays the same. The ratio between the tensions decreases until both tensions are the same, at about 72.75 Kg, in the centre of the graph width. The soundscape still will not have changed. As the pin is turned down further, the tension in the top stringing now falls below the tension in the speaking length, until the static friction line (dotted) for *decreasing* tensions is reached, at about 71.5 Kg. The tension ratio is now great enough for the static friction to be overcome, and the string starts to slide through the agraffe, towards the speaking length. The soundscape then starts to change. At this point the ratio for kinetic friction takes over, and the tension in the speaking length suddenly jumps down to about 72.5 Kg on the kinetic friction (solid) line. The pitch of the string would suddenly drop. As we continue to turn the tension down, the ratio between the string portions follows down the *decreasing* kinetic friction line, until we stop turning the pin when the tension in the top stringing is around 71 Kg, and the tension in the speaking length is 72 Kg.

What we have seen in this particular example, is that when the tuner turns the pin to raise the string's pitch, there is at first no apparent response in the string's sound. Then at a particular point the pitch suddenly jumps and then increases smoothly as the pin is turned. As the tuner attempts to lower the pitch, at first there is no apparent response from the string. The pitch then suddenly falls and then continues to fall smoothly as the pin is turned.

This sudden jumping in pitch which can happen even when the tuning pin itself turns smoothly, would be experienced by the tuner as a "sticking", "clicking" or "pinging" string. It will be only be obvious when the distance between the solid and dotted lines of each pair is large. The "delayed action" effect in tuning, when the pin turns but the soundscape does not, is however distinct from the "pinging", and it is only this that can be called *hysteresis*. This "delay" is also distinct from the *latency* that can also occur, where the slippage of the string over the bridge continues after the tuner ceases turning the pin.

How noticeable hysteresis is, depends on the piano, and on the individual string in the piano. It depends not only on the friction points themselves, but also on length of the top stringing compared to the speaking length. The hysteresis width will be less noticeable if the top stringing length is relatively short, than if it is relatively long. As the wrest (tuning) pin is turned, the tension in the top stringing will increase or decrease more rapidly in a short top string length than in a long one, so the ratio of top stringing and speaking length tensions more rapidly reaches the *limiting ratio* necessary to make the speaking length tension change. The effects of hysteresis can be anywhere between severe, and unnoticeable.

Now let's look at how the parameters affect the response. Clearly, the general width of the graph between the "increase" lines and the "decrease" lines, indicates how severe the hysteresis is, for a given top stringing length. The horizontal distance between the two pairs of lines, is the *hysteresis width*. It is the range through which the tension in the top stringing can change, with no response coming from the speaking length. This overall width of the graph increases with increasing angle θ through which the string passes. If the string passes through several angles and friction points, all these angles must be added together. The greater the total angle, the wider the hysteresis width. A piano whose bearing angles are collectively large, will usually have more noticeable hysteresis than one whose bearing angles are small, especially if it also has a longer top stringing length.

The distance between the two solid lines on the inside, increases as the coefficient of *kinetic* friction increases, and similarly, the distance between the two dotted lines on the outside, increases as the coefficient of *static* friction increases. If the friction points produce a large *difference in the coefficients of static and kinetic friction*, then the distance between the solid and dotted lines on each side of the graph, will be large, resulting in string "pinging", or "sticking". On the other hand, if the coefficients of static and kinetic friction are closer, the dotted and solid lines will be closer, and there will be no appreciable sudden jumps or "pinging" in the pitch of the string - but there could of course still be a large overall *hysteresis width* resulting in a large "delayed action" between turning the pin and changes to the soundscape. A large difference between kinetic and static friction could be caused by rust or other imperfections. Pinging or sticking is then more likely to occur.

Summary of hysteresis

When the tuning lever is turned clockwise, the tension in the top stringing increases. When it is turned anticlockwise, the top stringing tension decreases. However, the tension in the speaking length does not necessarily change, because friction at the top bridge or agraffe and other bearing surfaces, may prevent the string from slipping over the friction surface. Slippage *will* occur, when the tension in the top stringing and the speaking length are in a particular *ratio*. If this ratio is large, *i.e.* there needs to be a large difference between the tensions in the top string and the speaking length before the string will start to slip, then we say the hysteresis *width* is large. Then, a large change to the tuning pin may well be necessary before the tension in the speaking length, and hence the soundscape, changes. The pin *may* have to be moved a certain minimum amount before any change to the soundscape occurs. With a small hysteresis width, the soundscape *always* changes more quickly and readily, in response to a change at the tuning pin.

However, even where the hysteresis width is large, it is still possible for only a small change to the pin to produce an *immediate* change to the soundscape. This occurs if the tension ratio between the speaking length and top stringing is *already* sitting close to the critical ratio for slippage to occur. Then, just a small change to the pin and tension in the top stringing, can tip the ratio easily up to that necessary for slippage, and the soundscape changes almost immediately.

If, in the case of a large hysteresis width, the string and pin is left in a state where such a small change could produce an immediate effect on the speaking length and soundscape, the tuning could be unstable. Any slight change in the top stringing tension could produce slippage, and the speaking length would then go out of tune.

The hysteresis generally means that we must pull up the top stringing tension a certain amount before the speaking length tension rises, and also that we must push down the top stringing tension before the speaking length tension falls. The "distance" between the two points at which slippage occurs up or down, is the practical hysteresis width. Where the pin and top stringing is left "sitting", in relation to this hysteresis width, determines how far we might have to move the pin and top stringing tension, before slippage occurs.

Hysteresis in just *one* of the factors affecting setting the pin. We must consider it in conjunction with the other factors for a fuller understanding.

The hysteresis width, and hence the amount the lever may need to be turned before a change to the soundscape is effected, increases as:

- The total of the bearing angles increases
- The ratio between the top stringing length and the speaking length decreases

With experience, one can therefore see in advance, just by looking at the design features of the piano, what kind of hysteresis one may be dealing with when one starts tuning. Pinging or string sticking and clicking, increases as the difference between kinetic friction and static friction increases. The factors causing this, are generally less obvious, visually.

Assessing hysteresis

Hysteresis width must be assessed not just by one lever movement, but by a minimum of two. It is possible for a soundscape change to be made by a very small lever movement, even though the hysteresis width is large, if the tension ratio is already close to the critical ratio. The hysteresis width is felt between the position of the lever at the point where one stops changing the soundscape (by removing the force on the lever), and the point where the soundscape starts to change again after moving the lever back in the opposite direction.

Recoil

We must now add tuning pin and string *recoil* into the picture. The tuning pin fits tightly into the wrest plank (pin block) and there is therefore considerable friction between the pin and the plank (pin block). It is possible for the front of the pin, furthest from the plane of the pin block, to rotate or twist slightly under torsional elasticity, whilst the back (threaded part) of the pin inside the pin block does not move. Also, because the tuning lever head fits only onto the *front* part of the tuning pin, above or beyond the coils, the tuning pin itself acts as a cantilever. This cannot be avoided.

The position of the tuning lever head on the pin, which is at a distance from the plank surface, ensures that there will *always* be component of the *turning force* applied by the tuner to the tuning lever handle, that acts *parallel to the pin block*. This will cause a *bending moment* transverse to the pin's length. Again, this *cannot* be avoided by any special hand position, positions of the digits, or physical technique. This will be present even when correct lever technique is used, and no incorrect "pin bending" is being done. This is a consequence of the physics, of the geometry of forces, and is present whether we like it or not. Which direction the bending moment acts in, will depend on the direction of turn, and the position of the lever on the pin. *Both* the torsional *and* bending elastic components can cause *tuning pin recoil* after the tuner stops applying the turning force to the tuning lever. The effect is that the tension in the top stringing may then drop a little, or increase a little.

The *torsional* twisting of the pin will always tend towards a change in top stringing tension after force is removed from the tuning lever, *opposite* to the force that was being applied. In other words, if we are *raising* the tension, then when we let go of the lever, torsional recoil in the pin *always* tends towards *lowering* the tension in the top stringing, whether or not a lowering actually results. But the *bending moment* to the pin may act *either* to effect a decrease, *or* an increase in the tension, depending on the situation. It is the combined effect of torsional twisting *and* bending moment that produces different results in different situations, that we are about to examine.

Only if the bending moment is at right angles to the string (and parallel to the plank) will tension be least unaffected by it. For this reason, tuners attempt to position the lever in line with the string, but this is not always possible. Using a "star" shape tuning head there are still only eight different positions for the lever to fit on the pin, which is itself square in cross-section and may be at any rotational angle in the plank. Also, the casework or frame often prevents the ideal lever position. The lever is in actuality seldom exactly in line with the string, with the result that some bending moment not at right angles to the string, is generally present.

The bending moment will actually cause the pin to bend against its own elasticity, if only very slightly. Any bend in the pin other than perfectly at right angles to the string and parallel to the plank, will contain a component in line with the string. If the pin is consequently bent towards the speaking length, *bending recoil* will tend towards increasing the top stringing tension. If the pin is bent away from the speaking length, recoil will tend towards a fall in the top stringing tension.

On a grand piano, if the lever is positioned on the right of the string, the bending moment and torsional twisting act together in the same direction, increasing recoil. If the lever is on the left, the two tend to counter each other, which can reduce the overall recoil. On an upright piano, the sides are reversed. On the upright, torsional and bending moment act together when the lever is on the left of the string, and against each other when the lever is on the right. In other words, on the upright piano, the lever must be on the right to minimise the combined effects of torsional and bending recoil. This is contrary to the "old school" adage that it should be on the left, and that upright piano tuning should therefore be left handed, in order to *lift against* string tension and protect the plank and pin. In fact, proper, skilled lever technique, will not damage the plank (pin block) or permanently distort the pin, whichever side the lever is used, and there are distinct advantages in the pin setting process when the lever is on the right, due to the two elastic distortions countering each other.

The net result from both forces will depend on the situation, which varies from one instrument to another, the age and condition of the instrument, and even from one pin to another. Ideally, from the point of view of reducing bending recoil to a minimum, a lever position parallel to the string is best, but from the point of view of muscle efficiency and sensitivity, this is not necessarily always the best position. It also does not necessarily follow that smaller recoil is always an advantage over larger recoil. However, *always* tuning such that the bending moment is maximised in the same direction as the string tension acting on the pin, is clearly to be avoided.

The pin should *never* have bending force *deliberately applied* by pulling or pushing on the lever in a direction at right angles to the plane of the pin block. The resultant bending moment is much larger than that due to turning the pin, and will rapidly result in damage to the pin or plank. Such actions cause easily changes to the soundscape, because they circumvent the normal mechanism of tuning pin recoil and friction in the plank. For this reason poor tuners often rely on this highly damaging method in tuning. However, tuning this way will also circumvent the piano's natural mechanism for setting the pin, and will leave the tuning very unstable indeed. Damage to the instrument, both physical, and in terms of tuning stability, can be severe.

An unstable tuning due to relatively large tuning pin recoil that a tuner fails to take into account, is illustrated in Fig. (11.4). Now, here we illustrate a situation in which the difference between kinetic and static friction at the agraffe, is hypothetically negligible, so there is no string "ping", and there is in effect only one increase line and one decrease line. There is, however, a noticeable hysteresis width. The parameters here, are again chosen primarily for simplicity and clarity of illustration.

Fig. 11.4

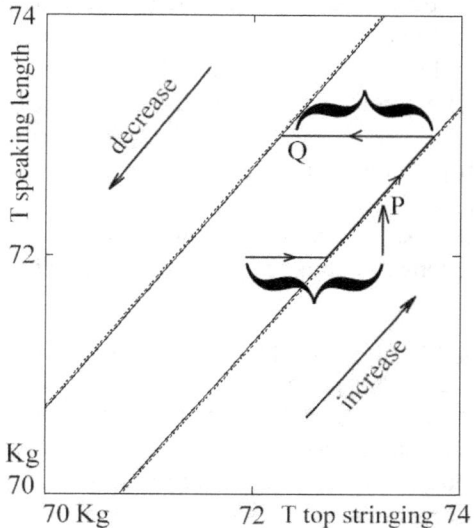

Fig. 11.4. The tensions in both parts of the string begin at 72 Kg, in the centre of the graph. The tuner starts to apply turning force to the tuning pin, which increases the tension in the top stringing. However, the pin is tight in the plank, and to begin with, the pin twists without actually turning in the plank. The tension change whilst the pin is twisting rather than turning, is represented with the lower horizontal curly bracket. The pin starts to turn in the plank when the tensions in the two parts of the string are at point P. When the tension in the top stringing reaches about 73.75 Kg (the upper right hand corner of the arrowed path), the tension in the speaking length is about 73 Kg, and the tuner stops turning the pin.

The upper horizontal curly bracket shows how far the tension in the top stringing now reduces, as the pin and top stringing *recoils* when the tuner releases the turning force on the lever. The tension in the speaking length, during this recoil, remains the same, because the reducing tension in the top stringing remains within the area between the graph lines. The top stringing tension comes to rest at point Q, at just over 72 Kg, just before reaching the decrease line.

The distance of Q from the *increase* graph line is relatively large. A subsequent change to the system that tended to *increase* the tension in the top stringing, would have no effect on the tension in the speaking length, because for increases in top stringing tension the speaking length tension will not start to increase until the top stringing tension is back at 73.75 Kg (the upper right hand corner of the arrowed path). However, the distance of Q from the *decrease* graph line is very small. If the top stringing tension *decreases* for any reason such as the pin bending or "settling" a little, the decrease line will be soon reached, and the tension in the speaking length will start to fall along the line. Only *a little* disturbance to the top stringing of a suitable kind, therefore, will easily cause a drop to the pitch of the string. In other words, by leaving the tension in the top stringing at Q, the system is left in *unstable* condition. The tuning will not be stable, and the tension in the speaking length is likely to fall quite soon. Stability of tuning is not, therefore, simply a question of allowing the pin to recoil until the desired string pitch is reached.

The recoil may of course be *greater* than the hysteresis graph width, or it may sometimes be negligible. Either way, stability demands that for any given desired tension (or pitch) in the speaking length, the tension in the top stringing is placed *between* the lines of the hysteresis graph, in such a position that disturbances to the top stringing tension will not disturb the speaking length tension. If there is as much tendency for the top stringing tension to *increase* after tuning, as there is for it to decrease, then the position should be *central* between the lines.

Alternatively, somewhere rather closer to the *increase* line than to the *decrease* line, may produce the most stable results where the top stringing tension is likely to decrease after tuning, which is more common. Only rarely, do we find that the tension in the top stringing has a tendency to *increase* after tuning, but this can happen. If the tension in the top stringing is left at a position *on* a graph line, then the slightest change to the top stringing tension can cause an almost immediate change to the string's pitch, depending on which line the value falls on.

In such a case, stability depends on the distance between the lines for kinetic and static friction. Although in the example we made this distance negligible, in practice there will always be some difference between kinetic and static friction. In practice the reason that no string "pinging" may be noticeable when the difference between friction types is small, depends also on the relationship between elasticity and the *hammer strike effect*, which is dealt will below.

Summary of essential points so far

Where the hysteresis width is large or noticeable, the final tension in the top stringing should generally be left close to the middle of the hysteresis width. This mean there is approximately an equal distance for the top stringing tension to increase or decrease, before changes to the speaking length and soundscape would take place. Another way of looking at this, is that there is a "dead" distance for turning the pin over which no soundscape change takes place.

With the pin set, the position of the pin is about half way across the "dead" distance. The exception to this happens where *latency*, which we will come to shortly, is causing changes to the tuning that may be countered by leaving the tuning close to one end of the hysteresis width. The normal "falling in pitch", over time, of a piano, is one latent effect, contributed to by factors beyond the piano, such as humidity. However, other factors may cause latent alterations sooner than this, even during tuning, and latency can also cause a *rising* in pitch.

The tuning pin recoil will affect the position of the top stringing tension within the hysteresis width. This must be taken into account, through the feel of the pin, and through knowing the difference between the feel of the elastic responses, and the feel of the pin actually turning in the wood.

A typical scenario may be that illustrated by Fig. (11.4). After raising the tension, the recoil leaves the top stringing tension close to, or on the decrease line. In this case, the soundscape must be first raised above the desired position. Then easing the tension down a little will be easy, due to the top stringing tension already being close to or on the decrease line. When the speaking length tension is as desired, the recoil will then bring the top stringing tension further back inside the hysteresis width. The recoil on decrease will typically be less then that on increase. The relative amounts of raising the tension first above the final desired value, and then easing it down will depend on the relative hysteresis width and recoil parameters.

It should be remembered that this is only *one scenario*. Raising the tension a little above and then easing down, is not a universal solution to setting the pin. The final lever movement depend on where the top stringing tension is within the hysteresis prior to the final movement. There are unfortunately further complications affecting this, which we must now examine.

What we have just described is a schematic *and still simplified* representation of part of the dynamics of setting the pin. It helps to understand the basic principles and effects of hysteresis before refining the description. Before moving on, however, it is worth stressing now that we are not talking here about *longer term* changes and instability due primarily to humidity and temperature changes affecting the wood and metal of the instrument. Such changes are as much likely to affect the speaking length directly as the top stringing, and cannot be avoided. Rather, we are talking about *tuning instability* which is potential in most instruments when they are inexpertly tuned. Tuning instability will, however, also cause changes due to humidity or temperature to have greater effect, sooner. It is up to the piano tuner to understand the instrument being tuned, and to *provide* tuning stability for it.

In effect, each string and pin needs to be "intuitively" assessed by the tuner for

- Elasticity and Recoil - related to how "tight" the pin is in the wood. It is also related to the top stringing features and bridge frictions. One must be able to accurately sense when the pin is moving under elastic forces, and when it is actually turning in the wood.
- Hysteresis - related to the top stringing features and bridge frictions. One must be aware in practice, of the hysteresis width.
- String "ping" - technically a "fault", but not rare.
- Where the tensions lie in relation to the hysteresis.

The tuner does not do this as a mental exercise, any more than the concert pianist has to think hard about which finger to put where, when playing Mozart or Beethoven. Expert tuning lever technique and an acute ear, enables the tuner to rapidly make this assessment through "feel" and hearing, as the tuning is being carried out.

Elasticity and recoil includes contributions from the pin, the top string length, and the tuning lever itself, which will bend slightly along its length. The difference between the pin turning in the wood and the combined elastic components, can be felt by the tuner through the lever. The pin turning in the wood is not the only valid motion for setting the pin at the finest tuning. Final fine adjustments through elastic torsion are valid, but the position of the pin in the wood must first be correct, as there may be more than one position of the pin in the wood that allows the same final fine adjustment to be made through elasticity. This does not mean the tuner needs to "bend the pin", but rather, generally refers to very small torsional movements.

505

The effect of the hammer strike

Having seen what hysteresis is, and how it may be used to advantage, we now come to the effect caused by striking the string with the action's hammer, *i.e.* by playing the note hard. This complicates the picture of hysteresis illustrated above, but again is *used* by the tuner to advantage. The tuner in effect uses three "tools" to achieve stability of tuning: the tuning lever, the ears, and the piano key struck by the other hand.

When the string is struck reasonably forcefully, this has a similar effect to temporarily reducing the friction slightly at the bridge or agraffe. In the hysteresis graph, striking the string is equivalent to moving the lines *closer together* and towards the centre of the graph. Look now back at Fig. (11.4). If the graph lines were moved closer, the point Q could be actually *on* the decrease line or even *outside* it. Fig. 11.5) shows the same point Q but with the graph lines moved inwards, representing their temporary displacement due to a forceful hammer strike on the string.

Fig. 11.5

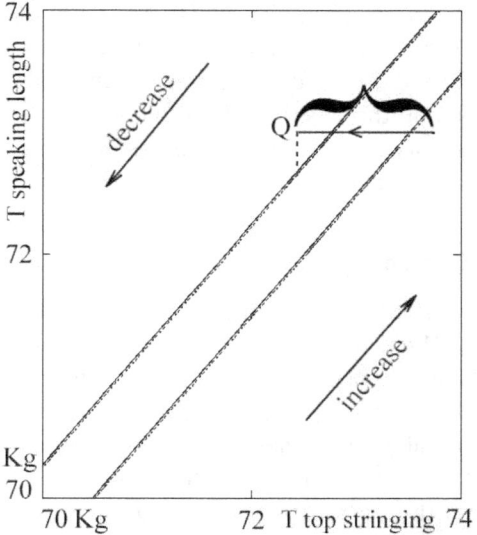

Fig. 11.5. The point Q represents the tension in the top stringing after recoil, which is a value along the horizontal axis of just over 72 Kg. No values can exist outside the graph lines, so the value of the speaking length tension on the vertical axis at this point will now be about 72.5 Kg on the new decrease line – in other words by striking the string and moving the graph lines inwards, the tension in the speaking length *drops* from around 73 Kg to 72 Kg, and hence the

frequency of the string drops. If Q had been in a different position, close to the *increase* line, and after moving the lines inwards it had ended up *outside* the increase line, then the tension in the speaking length, and hence the frequency of the string, would *rise* on striking the string.

If, however, Q had been placed more or less centrally between the graph lines, and remained between the graph lines when they moved closer together, then on striking the string there would no tendency for the frequency of the speaking length to either rise or fall. In other words, the tuning would be more *stable*.

Large hysteresis compared to the elastic range

The hysteresis width and the elasticity are parameters that can vary greatly, and their relative size determines the practical actions involved in setting the pin. In Figs. (11.4) and (11.5) we can think of the width of the curly brackets as the *elastic range* of the *pin and top stringing*. This is the *maximum* range through which the tension in the top stringing can be moved by elastic movement alone, without the pin actually turning in the wood. Where the hysteresis width is relatively large, and the pin is not too tight in the wood, the elastic range may become relatively unimportant in setting the pin, and the position of the pin in the wood will suffice. In this case, it is necessary for the tuner to be aware of the hysteresis width in terms of the pin position, and to set the pin in the middle of the width.

Small hysteresis compared to the elastic range

Many of the finest instruments exhibit only relatively small hysteresis in their tuning. There will nevertheless be some hysteresis width because there will always be some friction at the agraffe or bridge *etc.* A narrow hysteresis width does not in itself mean that tuning is likely to be more unstable – instability only arises in this context where there is likely to be a subsequent change to the top stringing tension, represented by a distance across the hysteresis width *greater* than the distance from the tension's current position to the edge of the graph. On a fine instrument, unless it has recently been restrung or has loose pins or other isolated problems, any changes taking place soon after tuning will be due to less than perfect pin setting only, which is up to the tuner to eliminate as far as possible.

The final tuning movements of the lever and pin are often (but not always, depending on the acoustical *fine tuning region* of the soundscape) exceedingly fine. In the case of small hysteresis width compared to the elastic range, the final movement of the pin (with good pin tightness in the plank) is be made using the elastic movement alone. The pin's ability to allow elastic turning whist not turning in the wood is in fact then essential. This ability is lost where the pin is either be unable to support the tension without turning back, or is left of the verge of being able to turn back in the wood (*i.e.* it would be a *loose pin*). As already stressed, in fine tuning it is very important for the tuner to tell the difference between the pin turning in the wood, and elastic movement, and when to use which movement.

The final pin movements now depends on the relative magnitude of the hysteresis width and the elastic range. The larger the elastic range, the less predictable will be the final tension in the top stringing after the force on the tuning lever is removed. A little consideration should show that whatever the elastic range, it would be best in this case for the final movement of the lever not to involve the full elastic range, but to be *as small a movement as possible*, involving as little recoil as possible. In other words, if raising the tension, the clockwise elastic movement made, and the following recoil, should both be kept small. This will be conducive to fine acoustical adjustments in the fine tuning region of the soundscape.

As outlined above, elastic movement works in such a way that the displacement is proportional to the force applied. Similarly, if we are utilising small displacements, then the elastic force employed will also be relatively small. Therefore, in the example of hysteresis width that is not too large, the final tuning movements of the lever *should not involve very much physical effort*. Rather, in this case of small hysteresis width compared to the elastic range, they involve *small, easy movements of the lever*. Setting the pin in this case does however demand first that the pin's rotational position in the wood is suitable, in order to allow this final tuning, and to achieve this, may first require large, more forceful movements. This scenario is most typical of the best instruments. Where fine tuning is carried out on an already good tuning condition, no initial large movements to turn the pin in the wood may be necessary.

From theory to practice: setting the pin

We can now put the above theory into more practical tuning terms. The following describes the basic *mechanics* of the practical task of setting the pin. The finer details of the underlying physical mechanics of the system are inevitably more complex, but this description is adequate in the first instance.

Starting from rest, the distance the *tuning lever itself* has to be moved before the tuning pin will turn in the wood, is the *spring*. The "feel" of this movement, and the degree of it, includes elastic contributions from the lever itself, the pin and the top stringing. We do not of course always have to use the full *spring* in the lever, to turn the pin in the wood, in order to actually change the tension in the speaking length – merely torsionally *twisting* the pin *a little*, can often do this. In other words, the *spring* alone, or part of it, can sometimes change the soundscape. The *spring* or a part of it will always cause a change to the top stringing tension, but as we have seen, because of hysteresis (*i.e.* because of the friction points), changing the top stringing tension will not always cause a change to the speaking length tension.

Whether or not the pin moves in the wood, there is, because of the hysteresis, a minimum movement that must be made by the lever in order to bring about an actual change in the *speaking length* tension, and hence a change to the soundscape. This is the *travel* of the lever. It is the minimum amount, at any time, that the piano tuner has to move the lever, whether or not the pin turns in the wood, before causing an actual change to the speaking length tension and hence the soundscape. The movement may consist of just some of the *spring*, or of the full *spring* plus a turning of the pin. It may be anywhere from zero, to a distance that follows from the full hysteresis width, depending on where the tensions lie in the hysteresis loop.

If the total *spring* is larger than the *travel*, then how much *spring* movement is required to change the soundscape, depends on the rotational position of the pin in the wood. Setting the pin then involves positioning the pin such that the final required adjustment to the soundscape uses minimum spring and recoil. If the *travel* is larger than the *spring* (a very wide hysteresis width) then setting the pin is more a case of just leaving the pin positioned in the middle of the hysteresis width.

The *spring* and the *travel* must both be discernible by the tuner. This means the tuner must know how much *spring* and hence *recoil* the "pin" has, and also precisely when the pin turns in the wood. This enables positioning of the pin ready for the finest tuning. The *spring* and the *travel* may be about equal, or either one may be greater than the other. The difference is not just a feature of the piano or pin, it is also a feature of how the pin is currently set, *i.e.*, how the tensions in the top stringing and speaking length are currently set up. The *setting* can vary from pin to pin, and even from time to time, on the same pin. Furthermore, the *spring* and the *travel*, will each have different values depending on whether the tension is being increased or decreased, because they will depend on the tension in the top stringing and where the tensions lie in relation to the hysteresis width. It may even differ according to the position the lever is placed on the pin, owing to the bending moment.

It should be clear then, that whilst setting the pin has one objective, the actual action or technique of setting the pin is in no way merely reducible to *one action* for which there can be a single set of specific, prescribed lever movements, for all pins and all pianos. Nevertheless, the technique can be accurately described, providing the physical dynamics of setting the pin are understood.

The pin is normally properly set when the *travel* is at about an equal maximum for both lowering and raising the speaking length tension, for that particular pin and string Bear in mind that the tuner does not need to keep thinking conceptually "this is spring", or "this is travel", any more than he might think "I am now listening to the soundscape". The process involves *feeling* through the lever when the soundscape is being changed through *spring*, and when the pin is actually turning in the wood. It is knowing how far one will have to turn to bring about a required soundscape change in either direction, and how much spring that will involve. This is coupled with awareness of *how easily or otherwise* the soundscape changes or will change in either direction, in order to leave the potential *travel* at a maximum at precisely the desired soundscape.

Pianos in good condition do not fall out of tune because the pins "unwind" in the wood. They fall out of tune because the tension in the speaking length changes. There are numerous ways *other* than an "unwinding" of the pin, by which this can happen. Typically, the soundboard crown may change slightly. Such "wholesale" changes due to unstable temperature or humidity can cause a "pitch change" to an instrument without necessarily being disastrous for the interval tuning, unisons, and octaves. Any pin and string with poor *setting*, however, can react differently to small changes in speaking length tension or top stringing tension, in the first instance, spoiling the unison of which it is a member. One way, in practice, to know whether a pin and string is likely to react this way (*i.e.* the best way to know its *setting*), is to be aware of the potential *travel* on that pin, for either lowering and raising of the speaking length tension.

The following are the *basic* scenarios. Which of the these scenarios applies in a given case, may depend on the individual pin, whether the tension is being raised or lowered, and even what rotational position the tuning lever is placed in, on the pin.

It should be stressed that these scenarios are for a motion beginning *from rest*, and that *the scenario for a given pin may actually be different according to how the pin has been previously moved, or how well it was previously set*. Thus, these scenarios change from one situation to another, as tuning takes place. This dependency also means that an instrument's tuning characteristics depend very much on how well it has previously been tuned – not in terms of its acoustical state of tuning, but in terms of pin setting. An instrument tuned by a tuner who fails to properly set the pins, will present extra difficulties to the next tuner. Persistent good pin setting will improve an instrument's tuning characteristics over time. It will make a piano more and more stable over time, and easier to tune, at the next tuning. This is a most important truth about the good maintenance of pianos that is not very widely appreciated.

Scenario 1: *Spring* smaller than the *travel*

If the *spring* is smaller than the *travel*, then the pin must turn in the wood in order for there to be a change to the soundscape. The scenario is encountered as a prevalent condition on a piano particularly where the pins are tight (a large *spring*), the hysteresis width is large, and there is a large *travel* when the pin is set. Some Steinway uprights and very small grand pianos with tight pins and large bearing angles have this feature.

Remember that whether or not this scenario applies in any given direction, depends on the current pin *setting*, *i.e.* the relative tensions in the top stringing and speaking length. Even on a piano with very tight pins and a very large hysteresis width, the tensions can still be set up when the pin is not properly set, so that the *travel* in one direction is very small, and thus smaller than the *spring*. In such a case, the tuning will generally be unstable.

Scenario 2: *Spring* greater than the *travel*

If the *spring* is greater than the *travel*, then a change to the speaking length tension and soundscape will take place *before* the pin begins to turn in the wood. When this is the prevalent condition on the piano the pin must first be carefully positioned in the wood, often by first making larger changes to the soundscape, before fine tuning takes place using the elastic torsional movement. Even where this scenario seems to be the main feature on the piano, it is still often possible for the *travel* to be become larger than the *spring* in one direction, at a point where the pin is not properly set. In other words, it can be that a small turn on the lever without the pin turning in the wood, is sufficient to cause a tuning change.

Scenario 3: *Spring* equal to the *travel*

If the *spring* is approximately equal to the *travel*, then the pin will turn in the wood more or less as the change to the speaking length tension and soundscape, takes effect. The practical effect is that the pin always seems to turn, or it may be difficult to discern when it is turning and when it is not. This is in general allows the easiest tuning and pin setting, especially for inexperienced tuners.

Tuning slowly and smoothly

It is not always possible to turn the *pin in the wood* very slowly and smoothly, because of the difference between the static and kinetic friction of the pin against the wood. If this difference is relatively large, the pin will tend to turn suddenly, after a certain amount of torque has already been applied to the pin. This can happen without string "clicking" or "pinging", but in extreme cases the pin can turn with a sudden crack or bang. It is the *difference* between the kinetic and static friction that determines this, rather than the magnitude of friction alone, so it is generally possible for both tight and not-so-tight pins to turn smoothly.

Many pins will not turn perfectly smoothly, but this is not necessarily a sign of poor quality. In practice, the turning characteristics of pins often varies with the rotational position of the pin in the wood. A very tight pin may turn smoothly, but after a few movements the high friction will generate sufficient heat to dramatically increase the difference between the static and kinetic friction. The pin itself will then stick and click, often with a large cracking sound at each movement.

Even if the pin is capable of turning smoothly in the wood, it is still not always possible to tune by smooth, continuous motion of the lever. The turn of the lever may result in no change to the speaking length tension, followed by a sudden, uncontrolled change. This is the hysteresis in action. Smooth tuning, is, however, possible on many pianos.

Experienced tuners can often exert control through smooth turning, even where there is noticeable hysteresis, because they are aware, from the sound and the lever "feel", of the *rate* the string is slipping over the friction points, compared to the spring in the lever. This allows the tuner to both control the rate, and leave the top stringing in the middle of the hysteresis width, at the desired soundscape. This technique *looks* easy but it is far from easily acquired. It takes years of experience to do this *and* always reliably set the pin.

Inexperienced tuners should not just rely on trying to emulate this. Many students on beginning to learn piano tuning, turn the lever exceedingly slowly, listening intently to the soundscape, for the smallest change, and taking a very long time indeed in attempting to tune each string. Apart from being extremely time consuming, this is not correct technique, will seldom allow the finest tuning, and will not generally allow setting of the pin.

The expert tuner makes the final very fine adjustments to the soundscape after getting into the *fine tuning region* of the soundscape very quickly, in one or two movements, much faster than could be achieved by such painstaking slow, smooth pin turning. This error made by students of tuning arises from the assumption that the soundscape must alter in relation to the pin position (whether twisted or turned) rather like the volume or tone alters when turning the volume or tone control on an amplifier or radio. As we have already seen, the behaviour of the piano is nowhere near as simple as this.

Coaxing

Proper pin setting technique requires lever movements that provide constant feedback on the pin position in the wood, the elastic movement, and thus where the tension in the top stringing falls *in relation to the hysteresis width*. This is experienced by the tuner as feedback heard and felt through the pin and lever. As a tuner, one must *always* be aware of where the soundscape is, in relation to the hysteresis width and the spring. This is experienced as awareness of where the soundscape is, in relation to its stability, or tendency not to change, in relation to forces the tuner applies to the lever.

Good tuning also requires that the total distance the moves through, over the total tuning time, is reduced to a minimum. One must tell from the "feel" of the lever when the pin is turning in the wood, and when elastic movement is taking place. Where tuning by smooth pin turning will not allow proper pin setting, the proper *coaxing technique* can be used. *Coaxing* has been misunderstood and dismissed in some texts, and it must be linked to proper understanding of the mechanics of pin setting. As in all tuning, in coaxing one must be aware of where the soundscape is, and which direction it is moving in. It is perfectly possible in some instances for the soundscape to be "rising" (getting "sharper") whilst the pin is being turned *anticlockwise*, and *vice versa*. It is only great experience that teaches the tuner to recognise precisely where the soundscape is in relation to the desired soundscape.

Correct coaxing consists of a series of quick, small forces applied in one direction on the lever, resulting in a series of quick, small movements mostly within the *spring*, between the force applied, and its recoil. Each actual turn of the pin in the wood is achieved as a very small turn just outside the maximum spring.

If the pin is very tight, the movements are not through the whole range of the spring, but from just inside its maximum. Each time the pin turns a small amount, the lever will be just back inside the extent of *spring* again, and is then the process is repeated.

Thus, on analysis, even though the pin turns in the wood, for much of the overall lever movement in coaxing, the pin is not actually turning in the wood. The force applied to the lever in coaxing must be torsional – no unnecessary bending moment or "pin bending" must be introduced. The tuner may often begin by lowering the tension with a small, quick movement, but may also begin by raising the tension, depending on the situation. This is to ascertain where the tuning currently lies in the hysteresis width.

Inexperienced tuners typically alter the tension too much. Experienced tuners will always keep changes to a minimum. However, the size of the initial change will depend how far away from the desired soundscape we are when we start. If a string is say, 6 beats flat, it is generally no use just raising the soundscape slightly above what is desired, and then fine tuning. We would have to initially raise by an amount more than this is we wanted to stabilise the string.

Let's assume the tension has dropped and is now to be raised. Within the first, small turn on the lever (which happens in a fraction of a second), the tuner can feel first the elastic movement, and can distinctly feel through the lever, the extremity of the elastic movement and the point at which the pin turns. This happens very quickly. From this the tuner can gauge the *spring*. Often the spring will be comparable to the hysteresis width, but somewhat smaller. This means that the speaking length tension (and soundscape) will not change until the pin actually *turns* in the wood.

The moment it turns, the tuner can feel it, and hear the change in the soundscape. There must a perceived connection between "feel" in the lever and changes in the soundscape.

It may now be that the soundscape indicates the tension requires further increase. If the increase required is still quite large, the tuner may continue to turn the pin in the wood in one movement. More typically, the further tension raise required is very small, but may still require an exceedingly small pin turn, rather than just fine tuning through the spring. This is achieved now by another coax. Both 'coaxes' of the pin in the tension increase direction, happen quickly, perhaps in half a second. In effect, the lever oscillates back and forth on the torsional elasticity of the pin, initially applying somewhere between zero torque and the torque necessary to cause the pin to *just turn* in the wood.

Repeated motions of this kind cause the pin to turn in very small stages in the same direction until the soundscape is improved. To continue moving the pin in the wood, after a first larger turn of the pin, the torque range used for coaxing is about the same, and never much more than the *spring*. The pin is "nudged" round using a series of very small force increases from just below the necessary force to just above it.

Coaxing can go both up and down in tension. The coaxing motion explores the elasticity and enables the precise magnitude of the *spring* to be clearly perceived, at the same time as raising or lowering the speaking length tension and adjusting the soundscape. At each point of turning, the point at which the pin will turn has already been sensed through the lever, so that *when it turns*, how far it turns in relation to what the soundscape is doing, is under fine control through the lever movement.

If the spring is larger than the hysteresis width, then coaxing can still be used to set the position of the pin where a relatively large initial soundscape change is required, but the fine tuning will be within the spring. If the hysteresis width is larger than the spring, then the final soundscape change will happen on a coaxing movement, but the spring may still be used to place the top stringing tension in the middle of the hysteresis width.

If the tuner can hear sufficiently, but is simply unable to effect lever movements that keep the speaking length tension within the *fine tuning region*, then fine tuning and pin setting will be impossible. Tuning expertise always requires excellent "connection" between the sensitivity of the ear and the lever skills.

Many pins will not turn "infinitely" smoothly, *i.e.* they will not respond as if there were no difference between the static and kinetic frictions of pin against wood. Rather, the pin will turn quite suddenly when sufficient torque is applied. The sudden movement is not necessarily large. This does not create such a difficulty if coaxing is mastered, because the *coaxing* technique will deal equally effectively with smooth turning pins, and the most sticky of "clicking" pins.

Coaxing allows in most cases (except where the hysteresis is much larger then the spring) the rotational position of the pin in the wood to be "set", so that final fine tuning can be made smoothly on elastic movement, *i.e.* within the spring.

The audible soundscape

Having looked at the basic *mechanics* behind in setting the pin, we should now consider in a little more detail the *soundscape* to which the tuner listens. For convenience, we shall assume for the time being that we are dealing with the tuning of a unison. It is necessary to have read the previous chapters on tuning unisons to appreciate why we must speak of a soundscape rather than say, "the beat".

It is imperative that the movement of the tuning lever is guided by what the ear hears in the unison soundscape. In setting the pin, it is not only what the soundscape is like at any particular time, that is important, but also *how the soundscape is changing* as the tuning lever moves. *How the soundscape is changing* in relation to the movement and force on the lever, is how the tuner knows where the tensions are, in relation to the hysteresis. In practice, this comes down to being aware of where the soundscape is in relation to the soundscape's tendency to change.

The soundscape contains sonic movement. This is true even though the aim might be to reduce movement as beat patterns to a minimum, or to eliminate them. The very fact that the soundscape consists of decaying partials, decaying at different rates, creates movement and change in the soundscape. The pattern and form of movement would be the same every time the note was struck, if the tuning and soundscape were perfectly stable. However, this pattern and form can itself change, and it is this overall kind of change that indicates tuning instability. It is possible to hear the difference between the kind of "movement" that can take place in a stable soundscape that will be the same each time the note is struck, and "movement" that indicates the form of the whole soundscape is changing slightly.

A soundscape whose condition is changing overall, may be doing so as the lever movement causes the change, or due to instability, indicating the pin is not set. It is important for the tuner to hear how readily the soundscape changes in relation to the force applied by the lever. In other words, the tuner must perceive how easily the soundscape can be changed, as force on the lever is applied. In the final lever motions, the soundscape usually changes more easily in one direction than in the other.

One must be able to tell from the soundscape, whether it would show a tendency to "rise" or "fall" after striking the note, if left to itself, without any input from the lever, but this has to be done *during* input from the lever, by noting how easily the soundscape can be changed in relation to the "feel" of the lever. If the soundscape has a tendency to move away from the desired soundscape, *i.e.* it resists the lever's attempts to make it reach that soundscape, then more pin setting, of the preliminary coarser kind that may reposition the pin in the wood, is necessary. The final lever movement should ideally concur with the rising or falling tendency of the soundscape, and hence serve to "take out" this remaining tendency by depleting it.

To be able to develop sufficient aural sensitivity to soundscapes and knowledge of their behaviour, takes at least many months, or more usually many years, of continuous practise, depending on the aptitude of the individual. As described in the chapters on tuning, this is *not* merely a question of "listening to beats" or "eliminating beats". Beats are a guide, not a definition. Every soundscape of every excellent unison will in any case, in general contain "beats" or fluctuations in the decay curves of certain partials, somewhere in the spectrum. We are looking to produce the best tone within the range of possibilities that constitute a minimised movement in the soundscape.

Pin position - avoiding recoil

In setting the pin we really need to avoid recoil as much as possible. When the pin recoils, we are not really in *control* of the tension change and of course we need to maintain maximum control over the tuning at all times. We cannot alter the *spring* at any point, but we can minimise recoil, because the recoil itself depends on how much of the *spring* we are using to make the soundscape change. It depends on how far we are turning the lever. The smaller the turn, the smaller the recoil will be.

For this reason, the final tuning movements must be as small as possible. In order for this to happen, the pin must be positioned so that the desired soundscape can be reached from the smallest torsional movements. At the same time, in this position, the tensions must be set up in the middle of the hysteresis width at the point of the desired soundscape.

We may often need to raise or lower the soundscape beyond the desired point, in order to reverse back into the middle of the hysteresis width. To begin with, we may need to shift the pin position back and forth until the desired soundscape is within reach without needing to turn the pin in the wood. If the recoil is noticeable, then we may need to tune beyond the desired soundscape and let recoil move it back towards the desired sound. However, we do not then know where the tensions are in relation to the hysteresis. We should never really rely on recoil to do the work of tuning. Rather, after the recoil, we should need to also aim to apply an additional small reverse force to the lever to bring the soundscape fully back to the desired point, whose motion is sufficiently small that the following recoil from *this subsequent motion* has no effect.

Striking and tuning

Striking the note relatively hard effectively reduces the hysteresis width and enables smaller elastic movements to effect a change in speaking length tension and soundscape. This in turn enables the pin to be set more quickly, but will also ensure that the final tuning is made under conditions similar to those when the piano is being played by a powerful pianist. Always tuning by striking the notes with considerably less force than they will subsequently be subjected to by a powerful pianist, can easily lead to tuning instability.

Continuous *fortissimo* striking by the tuner is not necessary, however. The difference, for purposes of setting the pin, between a very light strike and a medium-heavy strike, may be critical, but the difference between medium and very heavy strikes do not generally provide the same advantages. The soundscape is different at different strike volumes, and it is not only loud notes that are important.

Generally speaking, the ideal tuning position for a loud note will be the same as for a quiet note, but in piano tuning, the micro-changes to the soundscape to which we should be listening, are not necessarily more obvious at a loud level. They are usually fully exposed at a medium level. Final fine tuning adjustments within the *fine tuning region* do not therefore generally need to be made *fortissimo*, but only at a medium level. The tone of a unison at *piano* or *pianissimo* level should also always be checked continually in the process of tuning, as it *can* differ, depending on the condition of the hammer and action.

Good tuning will therefore include loud strikes followed immediately by soft strikes, to both set the pin, and achieve the correct soundscape for both loud and soft playing.

Catching the movement

In fine tuning a unison, the soundscape *moves*, before it is made *still*. The movement is caused by fluctuations in the partials, but we need not isolate these consciously. We just listen to *tone*. As the tuning is finalised, the movement in the soundscape is being reduced. For example, the soundscape may be falling, and we are *assisting it* to do so, as we approach the desired sound, as part of setting the pin. How much of the audible movement taking place is due to remaining unwanted mistuning, and how much is due to the fact we are in the process of *changing* the tension?

The answer is that the two should initially *coincide*. One should feel the rate at which the soundscape is changing due to unwanted mistuning, and match the rate of change due to the tuning action, to it. If you think carefully about this, you will see that the movement rapidly decreases initially, leaving very slow or small changes taking place last of all. Imagine the movement in the soundscape represented simplistically as a beat. The time taken to move the lever to the desired soundscape should be about the same as the time it seems it would have taken the beat to complete. In practice, it may not be a "beat" as such, but may, for example, be one limit of a changing vowel sound moving to the other limit. If we move the lever and the tuning more slowly than this, we are both wasting time, and failing to control the tuning.

Bridge coupling and hysteresis

There is a very crude similarity between the effects of bridge coupling in unison tuning and the effects of hysteresis, that can cause the two to be confused. At very small mistuning, the effects of strong dissipative bridge coupling may be mistaken for the effects of hysteresis, if the necessary acuity of ear is not sufficiently developed. In fact, the two are quite different. In order to effect a change to the speaking length tension and hence to the soundscape, the adjustment *must* be made on one of the hysteresis lines. In other words, during the change itself, the hysteresis width plays no part - the action of the lever is "direct" on the soundscape without the hysteresis intervening. Only during the *change* from clockwise to anticlockwise force on the pin, or *vice versa*, does hysteresis intervene. Any change in the lever position that produces *no change* to the soundscape, is generally due to hysteresis.

Strong dissipative bridge coupling can also cause no apparent change for small lever movements. However, changes to the lever position *inside* the fine tuning range, *and* on a hysteresis line, do produce *fine changes* to the tone. If those fine changes are not heard, particularly by a tuner listening only for beats, it may seem as though one is inside the hysteresis width when one is not. The solution to this confusion is provided by better aural sensitivity to the exact condition of the soundscape at any time, and to be aware when it has changed, and when it has not changed.

Differences between bass, mid compass, and treble.

The wrest pins (tuning pins) are usually the same in the bass, mid compass and treble. The strings are, however, of very different lengths, and the decay rates, and hence coupling effects, are very different, the bass decay rates being much smaller than decay rates in the treble. Bridge coupling effects are for tuning purposes much more significant in the treble than in the bass. In general, a significant *coupling region* is not so obvious in the bass, but plays an important part in the mid-range and treble. Increase coupling effects in the higher treble can help to overcome the otherwise increased tuning difficulties of this part of the compass. A wider *fine tuning region* will make setting the pin easier, and in principle, pin setting is easier in the high treble.

However, the difficulty for learners is that lever control and knowledge of high compass soundscapes is most difficult to attain. High compass aural tuning skills usually take years of professional practice to become developed. Inexpert tuning of the higher treble generally owes its failure as much to never having produced the best soundscape, as to failure in setting the pin. The use of an electronic meter can therefore produce relatively good results in the high treble compared to the work of an aural tuner who cannot hear and control the soundscapes sufficiently. Nevertheless, the pins *must be* set in the higher treble, so the work of the expert aural tuner cannot be fully emulated by electronic means.

Multiple friction points

There are in practice, multiple friction points in the top stringing. The greater the number of friction points, and the greater the angles, then generally the greater will be the effects of hysteresis. Even just the appearance of the piano design gives a good indication of what sort of behaviour one is likely to encounter, once one begins tuning. For example, very acute bearing plate angles and large bearing angles of the string through the agraffe on a grand piano, suggests large hysteresis width, and suggest that fine tuning is going to involve relatively large lever movements. It suggests that one may be finally setting the pin with a lever movement that does not produce a change to the soundscape, but does ensure the top stringing tension is well within the hysteresis lines. Multiple friction points will mean that there may be several ratios of tension within the top stringing itself. Whilst this complicates a description of what is going on during tuning, the effect in practice remains similar to what has already been described.

Backstringing and "re-setting" the string tension

The tension on the string is applied from the tuning pin end of its overall length. Possible "slippage" at the top stringing friction points is therefore considerably greater than at the friction points at the long bridge or bass bridge. Nevertheless, occasionally a quick, relatively large decrease and then re-increase in tension may be necessary before attempting to fine tune some strings. This may be in instances where previous tuning without regard for pin setting has "upset" the overall condition of the string and the tensions in its various parts.

It may also be necessary where there is a bridge side-draft pin fault, such as the pin being somewhat loose in the wood, so its position alters slightly as the string tension changes. The large movement prior to fine tuning "re-sets" the string tension. Large movements like this should only be made where appropriate, for example, where there is unusual instability and resistance of the string-pin system to being set stable, usually associated with previous lack of expert tuning. Where large changes in tension are not required, carrying them out will only serve to increase instability, rather than to improve it.

Latency

One important kind of tuning instability arises from the fact that once the string does start to move across the top bridge or through the agraffe, it may take some time before it finishes its possible movement and reaches equilibrium again, even if the tuning pin is not moved further. After turning the pin, the tension in the speaking length may continue to change, even if the note is not repeatedly struck hard. This means that in some severe cases (which certainly do exist) one may actually be turning the pin anticlockwise to lower the tension, whilst the tension in the speaking length in still *increasing*, or *vice versa*.

One must be clear from the changing soundscape whether the tension in the speaking length is increasing or decreasing, and at the same time one knows from the feel of the lever and pin, where the tension in the top stringing is, in relation to the elastic recoil and hysteresis width. If, for example, a string has a tendency for its soundscape to carry on rising *after* turning the pin clockwise, it may be necessary to turn the pin clockwise first, and then turn in back a little whilst the soundscape is still rising. One is still aiming for the final position to be one in which the top stringing tension is roughly in the middle of the hysteresis width, and not to one side of it, or on a hysteresis line.

The string may continue to slip over the bridge immediately *after* the force on the lever is released, even if the note is not struck. Hard striking will obviously encourage this to happen, and is therefore used as a technique in this respect. However, striking the note hard in order to cause the tension to lower without using the lever, is a common trap into which students of tuning often fall. The final fine tuning should always be *controlled* directly with the lever, in order to ensure that the pin is actually set.

From the fine behaviour of the soundscape one should hear whether the tension is lowering or rising. The final lever movement needs to be in the same direction as the natural tendency of tension change, so as not to leave the amount of this change to chance, but to actually exhaust it, so that it has fully taken place. For example, after raising the tension, if the pin is set, there will be no tendency for the tension to fall again if a very small force encouraging it to do so, is applied to the lever.

There should similarly be no tendency for the tension to rise. If, however, the tension is slipping fractionally *after* release of the lever force (assuming no loose pin), it will be very easily lowered by a very small lever force in the same direction. Fine tuning the soundscape downwards is then very easy, requiring little force, because the lever and the natural tendency are working together. One needs, for example, to set the tension through the pin position in the wood, above the desired position, and work *with* the latency tendency in lowering slightly, on elastic motion, in order to arrive at the final desired soundscape.

This is just like dealing with the mechanical hysteresis, except that the latter requires hard strikes to cause any change to the tension after the lever is released. In latency, tension change will take place *after* the lever is released, as revealed by the quietest striking (the pin being tight in the plank).

Latency in "settle"

In some instruments changes to the tuning may take place due to the individual string-pin-bridge system "settling" quite soon after tuning, or due to the whole or a substantial part of the whole string-bridge system "moving". It has been suggested by Fenner,[70] for example, that as tuning proceeds, the long bridge may "tilt" under the effects of increasing tension and the downbearing angle. Some instruments, part of the way through the tuning process exhibit destabilising *changes* in the tuning, which causes broad sections of the compass to rise sharp or fall flat, even if individual pins are properly "set" and the unisons remain stable.

Occasionally, an individual string may show instability, upsetting a unison, an interval or octave, even though pin was apparently set. These effects are relatively rare, but can be experienced, and when they do, the tuner needs to deal with this behaviour effectively. Such movement is certainly a tuning stability issue, but *appears* not to be a pin setting issue, as the individual unison tuning may not be affected. The tuner must obviously take steps to counter this tendency in the tuning (or retuning) of each *individual* note, even if the effect involves a whole section of the compass tending to rise or fall.

During a large pitch raise, or when tuning new strings, this kind of "settle" and instability is to be expected, and standard strategies are employed to overcome it, such as tuning to a higher fork frequency. In some cases the movement is connected with excessively large hysteresis,

---

[70] Fenner, K, *Causes of variable tuning characteristics of pianos*, Europiano Publications, 3, European Union of Piano Craft Guilds, Das Musikinstrument, Frankfurt, 1977.

and the measures taken to overcome the physical movement or settle may involve deliberate alteration to the way the pin is set.

Once one knows how far an individual note is likely to move where such a problem exists, the pin can be deliberately set in a position where the top stringing tension is left close to one side of the hysteresis width, appropriate to counter the direction of subsequent movement. This technique is quite distinct from the standard techniques, for example of actually setting strings sharp in order to counter flat wise movement, due to new strings stretching, a pitch raise, or an unstable instrument without a full iron frame.

Practical guidelines summary

1. Take the time to understand the nature of hysteresis.

2. Learn to recognise by ear when the soundscape is falling or rising, even by the minutest degree. Then, the precise effect of a small change in lever force can be heard, and which way the string is slipping over the top bridge or through the agraffe, can be heard.

3. When the pin is being properly set, the final movements of the lever are easy and small, not forceful or large, unless (more rarely) the hysteresis width is very large compared to the *spring*.

4. First, but only if necessary, re-position the pin in the wood, so that the above applies. This movement is coarser than the movements that follow. Then use finer (usually) elastic turning to micro-adjust the soundscape. The "target" soundscape should be arrived at with very little lever force, but the soundscape should show no tendency to change too easily, or to resist change, in relation to the lever force. Its relation to the lever force should feel neutral.

5. If, however, the hysteresis width is large compared to the *spring*, then the soundscape should be insensitive to such small elastic pin turns, when the pin is set. Final lever movements are then generally a little coarser than otherwise, and the aim is to leave the pin position so that the top stringing tension is in the middle of the hysteresis width.

6. Finally, if after tuning a string, you know that a few hard strikes of the note, or a small force on the lever one way or the other, will cause the soundscape to change, then in effect, you know that the pin is not set! In this case, more time and effort is required. Time and effort spent in experimenting to find the stable situation, is time well spent.

Chapter 12 - Setting the pitch

"Pitch", as we have seen, is a subjective perception, dependent on psychological and physiological factors. *Frequency*, on the other hand, is a specific number of cycles per second, and is an objectively measurable property of the sound. The phrase "setting the pitch" really means setting a *frequency* of the first scale note, which is going to be used as the frequency axiom for the entire instrument.

The standard fork frequencies are 523.3 Hz for the C fork, and 440 Hz for the A fork. The C fork is used to tune middle C, but its frequency is that of the fundamental of C52 (C5), an octave higher. Middle C's fundamental would require a fork at 261.7 Hz, but such a fork is not generally used. Tuning forks for frequencies lower than 440 Hz can be made, but tuning forks are a "technology" that only works well for practical tuning usage, within a relatively small range of frequencies. Forks for lower frequencies tend to "misbehave" when the base is placed on a sounding surface for amplification, producing strong higher partials, and only a weak fundamental.

As the absolute reference for the whole tuning process, the setting of the first note is critical – one must be able to rely perfectly on its accuracy and stability. A tuning fork's accuracy cannot necessarily be taken for granted. It is not uncommon when testing together forks of different ages and conditions, for discrepancies to be revealed. The forks will often beat together. An older or damaged fork should be tested against a new one or suitable electronic equipment. (Never assume, however, that any electronic equipment - whether a tuning meter or a keyboard instrument - is automatically accurate merely because it is electronic. Many electronic audio circuits are *not* frequency stable to a high degree – stabilised circuitry does exist but is specially designed and more expensive).

The fork will normally have been tuned to be correct at 20 degrees Celsius, which is the SI standard room temperature, so provided it has not just been removed from freezing conditions or exposure to severe heat, it will be at the ambient temperature and will be sufficiently accurate under normal conditions. The best "scientific" forks are made from *invar*, an alloy of iron and nickel, which has a negligible coefficient of thermal expansion, and hence is very frequency-stable despite changes of temperature.

The fork should never be struck on a hard surface, not only because of the possibility of damage to the fork, but because this will tend to produce distortion in the fork's tone, and the presence of at least one prominent, higher, inharmonic partial. The sound from a tuning fork is close to being a pure tone, a tone that has no partials, but even when the fork is properly struck, partials can still be produced, and other partials will tend to be produced in the sound when it enters any further system that distorts it, even slightly. Such systems can include the human ear. When we listen to the fork by holding it close to the ear, we should normally hear a pure tone. Placing the fork base on a wooden surface such as part of the casework or soundboard can sometimes produce distortion.

The purpose of placing the fork base on such a surface is to use the surface as a soundboard for the fork – being a larger surface area the wood amplifies the fork's sound and drains energy rapidly away from the fork. This rapid draining of energy causes the fork's sound to decay quickly, which means it soon has to be restruck. Some tuners use the technique of placing the fork stem between the teeth. This leaves both hands free and has the advantage that the sound is heard loud, because it travels directly through the bone to the ear. Whilst the sound is heard as loud, the decay is slow because the rate of energy loss from the fork is small. The head does not act as an amplifier and no one else listening close by, would normally be able to hear the fork.

Placing the fork between the teeth has one complication however – it tends to produce distortion in the fork's tone which can lead to the perception of one or more higher partials. These partials will invariably be considerably more inharmonic than the corresponding string partials, so it will beat against the string partial, while the fundamental of the fork is not beating with its corresponding string partial. Attempting to tune the string by the higher partial will lead to the fundamental being mistuned. The fork's true tone can be tested by holding it free in the air, close to the ear. It is possible for each ear to perceive a slightly different *pitch* for the fork, a condition known as *binaural diplacusis*, but the actual beating heard in tuning arises from *frequency* differences and will be more reliable.

Tuning the first string to the fork usually justifies a little extra time precisely because of the fork's absence of higher partials. In tuning trichord unisons, the higher adjustable partials are far more sensitive indicators of "pitch" condition than the fundamental. In tuning to the fork, only one partial of the string meets with the one fundamental partial of the fork.

It must be remembered that if tuning from middle C using the standard C fork, the fundamental frequency of the fork matches the *second* partial of the C40 (C4) string. Using the C fork to set a *frequency for middle C* above or below the fork, then requires that the lowest beat rate heard between the fork and the string is mentally *halved* in assessing the frequency change at middle C itself. Thus, for example, if we were tuning *middle C itself*, 3Hz sharp (so that its fundamental frequency is 264.625 Hz) we would require *6 beats* between the fork and the string, not just 3 (we ignore string inharmonicity here).

Only if the required "pitch" is quoted as a frequency for C52 (C5) can we use the frequency difference as the required beat rate itself. Thus, if we are required to tune to "C 528 Hz", this obviously refers to the pitch of C52 (C5) and not C40 (C4), so now we may tune the middle C sharp by 4.7 beats per second (the beat still appears at the second partial of the string). We could alternatively tune the C52 (C5) to the same beat rate.

Inharmonicity has consequences in "setting the pitch". The C fork frequency of 523.3 Hz is based on the A 440 Hz standard. The figure to two decimal places is 523.25 Hz, and is calculated simply as

$$440 \times 2^{3/12}$$

given that the C is three semitones higher then the A 440 Hz. The calculations ignores inharmonicity. Inharmonicity has the effect that if we tune the C52 (C5) fundamental "beatless" to the fork and then tune down the octave to middle C, the middle C's fundamental may then be flat. Even when we tune middle C's second partial "beatless" to the fork (which is more usual), its fundamental may still be flat. For a medium sized instrument the flattening is likely to be around 0.1 – 0.2 Hz, and the beat rate necessary to counter this by tuning sharp, is therefore roughly 0.2 – 0.4 Hz.

Correct tuning (where there is no expected subsequent change) therefore requires a slight beat movement between the fork and the string, the string being slightly "sharp of the fork". Usually, a higher beating partial is also prominent, involving the 4th partial of the string. The beating at this partial (which is more inharmonic) under the same tuning condition can be as much as 2 – 3 Hz, if the fork stem is placed on a sounding surface or held in the teeth.

Mostly, however, "pitch changes" or "pitches" (which of course are really *frequency* changes or *frequencies*) are specified for the A49 (A4) note. In the case of tuning to the A fork, the fork frequency is that of the fundamental of the A49 (A4) string. We may, however, tune A37 (A3) to the A fork, in which case it is the second partial of the string that is coinciding with the fork's fundamental. The required beat rate will be *the same* whether we are tuning A37 (A3) or A49 (A4). For example, the International Standard Concert Pitch usually quoted as *A440 Hz*, sets the frequency for A49 (A4) at 440 Hz, but some orchestras now use a higher pitch. A frequency standard of A 444 Hz can be tuned by setting either A37 (A3) or A49 (A4) four beats sharp of the fork, provided the *pitch of the beat* being used is always that of A49 (A4).

If tuning from the C fork but using an A49 (A4) frequency reference, the rule is that the required beat rate between middle C and the C fork should be 1.2 beats for every cycle per second the A49 (A4) is intended to differ from 440 Hz. Thus the higher orchestral A 444 Hz pitch requires C40 (C4) to be 5 beats sharp (to the nearest whole beat) of the fork (and not 4 beats). The pitch note (C or A) must be checked again against the fork after the trichord is completed because it is possible for the completed unison to behave differently to the initial single string.

Some tuners check the pitch of say, middle C, by comparing the beat rate of the major third G#36 (G#3) to C40 (C4), with the beat rate between G#36 (G#3) and the fork. The idea is that the beat rate should be the same if the C string "matches the fork". This *may* work reliably, but certainly cannot be guaranteed. One should understand the mechanism on which this method relies.

On playing the actual keyboard notes, the 5th partial of G#36 meets with the 4th partial of C40. This meeting is at an *octave above* the pitch of the 523.25 Hz fork. The fork, whose fundamental is an octave above the C40 string, must produce a significant *2nd partial*, an octave above its fundamental, in order for a real meeting to occur with the 5th partial of G#36. The fork's inharmonicity in this case cannot in general be expected to be the same as that of the 4th partial of the actual C40 string on the piano, so the beat rates of the major third played on the keyboard, and played between the G# and the fork, cannot necessarily be expected to be the same. Similar unreliability occurs if the beat heard is not caused by such a meeting of partials external to the ear, but is psycho-acoustically formed from distortion in the ear.

Chapter 13 - Small piano syndrome

Small piano syndrome is a set of (generally detrimental) effects inherent in the design of small pianos, but the causes can be regarded as being present in some degree in most pianos. Its presence, when very noticeable, is not necessarily due to design "fault", but due to difficulties in achieving the desired tonal and tuning characteristics when building a piano to limited size specifications. The larger the instrument, the less noticeable the effects, and even on just a reasonably large but well designed instrument the effects may be negligible. It is on small and miniature pianos that we may see a set of very obvious characteristics that collectively constitute small piano syndrome.

Compromises in piano design parameters arises in a number of areas in the construction of small pianos. These include:

1. The length available for the bass strings. The bass strings need to be long for the best tone. Shorter, and therefore thicker bass strings, produce an inferior dull, "tubby" or "woolly" tone, and a tone in which the musical pitch is ambiguous or less well defined. The overall length of the grand piano or height of the upright piano limits the maximum possible length of the bass strings. Another result from very short bass strings and the consequent high inharmonicity, is "pitch sharpening" in the bass. This means that when bass octaves are tuned to an optimum condition according to tone and beat rate phenomena, the perceived pitch relationship of the octaves may not seem to be correct (The perception of pitch depends on the relationship of the partials, which are very inharmonic). On some strings, the fundamental may be weak or even absent, leading to the perception of a "psychological" fundamental only, which can have a large effect on pitch judgement.

2. The long bridge position and corresponding string speaking lengths, as the bridge approaches the bass break, is

compromised. In order to prevent the bass end of the long bridge becoming too close to the edge of the soundboard or the bass bridge, its curve in this area must deviate from that employed higher in the compass. There is some compromise on most pianos, but on smaller pianos the compromise is greater. In general, the bridge starts to curve back in the opposite direction, reducing the rate of increase of speaking lengths. In this area, generally on the bass side of the centre of the compass, the scaling factor or speaking length ratios between strings, changes rapidly approaching the end of the bridge. The rate of change of inharmonicity from note to note, then also changes relatively rapidly. This causes a corresponding rapid tone changes, and can severely compromise the possibility of regular beat rate progression in tuning.

3. The bass break occurs relatively high in the compass, often within the scale area itself, causing a sudden change of inharmonicity, and compromising the possibility of well progressing beat rates in the scale. Often part of the scale may have covered strings, and part plain steel. Sudden tone changes are also caused.

4. At the bass break, the lack of space in which to fit the overstringing crossover plus the underdampers, will in a short upright piano often necessitate compromised damper lengths, with the result that the break notes may be damped less efficiently than the other dampers.

5. The smaller soundboard (together with shorter, thicker strings and reduced tension) produces an inferior tone and a less powerful combination of sustaining power plus volume.

6. The string speaking length ratios over the octaves and multiple octaves (which are closer to unity than on a larger piano) result in larger changes of inharmonicity over the octave, even on parts of the bridge where the scaling factor may be constant. This can led to difficulty in obtaining octaves, double-octaves and triple-octaves, that are all simultaneously beatless. This effect, whilst specifically part of small piano syndrome, is not exclusively due to small size of the piano. It is the rate of change of inharmonicity across the compass that is the primary problem, and it is also possible to have an unsuitable rate of change on a larger piano. On the smaller instrument, it is more difficult to avoid this.

Piano manufacturers can employ design features to counter these effects in some degree, but small piano syndrome cannot simply be avoided altogether. It is a set of potentially adverse consequences, inherent in the principles of piano design.

Here we are principally interested in how small piano syndrome affects the immediate task of the piano tuner. For this we are primarily concerned with what happens to scale tuning as a result of the bridge-curve compromise at the bass end of the long bridge, the effects of small scaling on the tuning of octaves, and its effects on balancing tuning across the compass as a whole.

In the tenor and treble parts of the compass, the speaking lengths of the strings increase from treble to bass according to a *scaling factor*, or a *scaling function*. A scaling factor is the number by which we would have to multiply the speaking length of a string, to get the speaking length of a string one octave below in the compass, assuming a section of speaking lengths over which this holds true. A scaling function would be the function (of note number or position on bridge) that gives the speaking length, where the speaking lengths increase in some other way.

It is important to remember that not only is the change in speaking lengths over the compass important, but also the *rate* of change. As long as the progression of speaking lengths is steady, *i.e.* follows a "smooth function", then from the scaling factor it is relatively easy to find a ratio for the string speaking lengths between the lower and upper notes for any interval in that section of the compass. If the scaling factor remains constant over a couple of octaves, the speaking length ratio (SLR) for any interval in that region can be given by

$$\mathrm{SLR} = S^{\frac{n}{12}}$$

where S is the scaling factor and n is the number of semitones in the interval. The scaling factor will generally be less than 2, but it should be noted that the actual curve and position of the bridge on the soundboard is dictated by a number of practical building considerations, not just an idealised scaling factor.

The beat rates specified by "traditional" theory for the tuning of the scale, only apply with precision where there is no inharmonicity present. Inharmonicity is present in every piano string, and depends on the string's tension, its speaking length, its diameter, and the partial number. The inharmonicity raises the frequency of a partial from its harmonic value.

Ignoring potential bridge coupling effects, a general guide to the effect of piano scaling on beat rates can be obtained by considering the inharmonicity. Knowledge of the precise effects on a given instrument would require precise knowledge of the speaking length and diameter of every string in the region of the compass being considered. This is not a difficult task for computer analysis, given the necessary information. There are, however, general principles that will apply.

A more general guide to the principles can be obtained by looking at the continuous functional relationships between speaking length, diameter and beat rates, where inharmonicity is taken into account. In this approach, rather than considering a specific set of speaking lengths and diameters at discrete positions along the bridge, we can look at the more general way in which the beat rates change when inharmonicity changes smoothly from one end of the scale to the other. Using this approach, general tuning characteristics can be found for the case where the position of the bridge over the scale area is compromised. Either it fails to increase the speaking lengths at the same rate of increase as it does higher in the compass, or it curves back in the opposite direction to the curve that would otherwise have been dictated by a constant scaling factor. The beat rate for a given interval will be determined by the actual speaking lengths of the strings for the two notes of the interval, determined by the actual curve of the bridge.

In the following description of effects on beat rates in the scale area, a "smaller" piano means one with a shorter speaking length for the longest string in the scale. The "bridge end compromise" means the amount by which the curve of the long bridge deviates from what it would have been if the string lengths had obeyed a constant scaling factor.

We then find that for notes of an interval whose fundamental frequencies are in the "traditional" theory's equally tempered ratio, the beat rates will be affected as follows (*n.b.* this is *not* the tuning strategy):

<u>For the tempered fifths</u>

The smaller the piano, the *more* the fluctuation (overall beating).

The more the bridge end compromise, the *faster* the beat rate.

<u>For the tempered fourths</u>

The smaller the piano, the *slower* the beat rate.

The more the bridge end compromise, the *slower* the beat rate.

<u>For the major thirds</u>

The smaller the piano, the *slower* the beat rate.

The more the bridge end compromise, the *slower* the beat rate.

<u>For the major sixths</u>

The smaller the piano, the *slower* the beat rate.

The more the bridge end compromise, the *slower* the beat rate.

These indicate *not the tuning strategy itself*, but the tendencies in the *tuning characteristics* of the piano, that is, the change to beat rate for a given fundamental frequency ratio for each interval. Small pianos, with compromised scaling, which will be most compromised at the bass end of the long bridge, roughly where the scale falls, require these characteristics to be addressed.

Consider tuning the scale beginning with the sequence C up to E, down to G, then up to B. With small piano syndrome the beat rate in the third C – E will be slower than usual if the interval's fundamental frequency ratio is "correct" according to "traditional" theory. We may of course not want a slower beat rate, and therefore tune the usual beat rate. This makes the E "higher than the piano requires". In tuning the fourth C down to G, we might similarly tune the G "lower than the instrument requires". This does not necessarily mean the resulting sixth G – E will be too fast, because the inharmonicity may result in this beat rate being slower than normal. The actual result depends on the inharmonicity and bridge compromise on the individual instrument, and hence on the beat rate relationships on the particular piano being tuned. On tuning the B, we might still find little sign of inherent problems for essentially the same reasons. However, if we continue tuning in the "usual" beat rates, the scale will inevitably not work out at some point. If we do not tune slower thirds, then somewhere in the scale one or more fifths may need to be *wide*, to allow the thirds to have the "normal" beat rates we gave them. Also, because these thirds, when they are at the "traditional" beat rate, are wider than the instrument requires for an equal distribution of the actual acoustical comma, other thirds will necessarily become narrower, destroying progression.

Where inharmonicity *changes* over the scale, as it does on a small, scale compromised piano, the traditional theory of temperament itself fails, because the size of the comma is not constant for all interval relationships. The task facing the tuner is not the equal distribution of one comma, but achieving *progression* of beat rates. The best progression of beat rates will be achieved when the principles outlined above are remembered. If tuning by intervals other than those listed, the principles can be remembered simply as:

Wide intervals may need to be slower;

Narrow intervals may need to be faster.

Finding the actual beat rates that are necessary depends on the instrument, and requires the "trapping" techniques described in the chapter on scale tuning, but working around a *tuning sequence* specifically designed for smaller pianos may be especially useful. We will look at this on a moment.

General approach to tuning very small pianos

(1) Study to this chapter. (2) Refer to the chapter on scale tuning. The principles discussed there will be present, but more acute. Be familiar with scale logic, and try to get beyond dependence on a fixed tuning sequence. In tuning the scale always use box trapping to find the optimum beat rate relationships. Work fast, and not slowly just because of the perceived additional difficulty. The "traditional" beat rates remain the starting guide, but be prepared to change your mind on the tuning of each particular interval as you proceed, and concentrate on *progression* rather than preconceived absolute beat rates. Also remain aware of the *pitch intonation* and *tone* of the fast intervals, in relation to their beat rates. With practise and experience, you will be amazed how good a guide this can be to actual the required beat rate.

Be prepared to work with the piano's natural tuning behaviour, rather than the "traditional" model, and stay with it mentally, understanding it as you proceed. Especially, remember that the fifths will typically have a prominent upper adjustable partial, and that they will easily reach the tuning state in which they may be *wide* at the lower adjustable partial whilst still *too narrow* at the upper one. One *must* be aware of this, and understand it, not confusing it with falseness, in order to control the tuning of fifths most effectively. Also, listen to the *tone* of the interval and get the "bigger picture" at the same time as controlling the beat rates.

Successful small piano scale tuning is much more about proper understanding of the tuning behaviour encountered, and the ability to take intelligent control of it, than it is about any "magic" sequence or rule of thumb. The aim, as always, is *progression*.

Tune the bass first (if you do not already do this anyway). There will be an immediate and pressing issue with pitch matching the treble with the bass. This means octave stretching will be imperative. Stretching the bass "downwards" is a tempting but deceiving thought. The scaling on most very small pianos means that the tone of the octaves in the bass is inevitably very compromised, and is not going to benefit from further compromising due to widespread stretching.

Often, over most of the bass compass no really satisfying tuning for an octave and its multiple octaves can be found. The tuning must primarily satisfy the demands for *octave tone*, then *multiple octave* tones, in terms of beating, and octave intonation. Further interference in the attempt to pitch match with the treble is not desirable, and is in any case not necessary if correct stretching in the treble is carried out.

In tuning the treble, introduce octave stretching *immediately* from the scale, using the bass also to tune from, and ensure optimum progression of the compound major thirds (tenths and seventeenths), making their beat rates *lead over* (faster than) the corresponding thirds in the bass. Stretch to the maximum, but do not destroy the octave tone. A little compromise may be allowable in accordance with the indications of the compound thirds, which should *lead over* the thirds, and progress. One should not *aim* to stretch the top octaves, but rather, aim to complete overall stretching as much as possible *before* reaching the top part of the compass. In this way, almost all the severe conflicts usually met between octaves and multiple octaves, due to inharmonicity, will be resolved, and the pitch matching of treble and bass will be optimised without ruining the tone. There may still be occasional anomalous octaves.

Seeing inharmonicity changes

Inharmonicity is the raising of partial frequencies due to string stiffness. It is very dependent on the string's speaking length, the shorter the string of any given diameter, the higher the inharmonicity. The main problems in tuning arise when strings have *different* inharmonicities to each other. Wherever the following changes in the stringing are observed across the intervals being tuned, one must expect to make greater effort to work around the inharmonicity change, in order to get progression of beat rates in the thirds and sixths, and to avoid bad fifths or fourths:

- A change from plain steel to covered strings
- A change from one bridge to another (a *break*)
- A change to trichords to bichords
- A rapid change in speaking length

The smaller the instrument, the more of these features are likely to appear within the scale area. The smallest instruments typically have all four features. There is no solution in simply avoiding the central compass and attempting to tune a scale elsewhere. The central compass must still be well tuned, and the best way to tune it is to address it directly.

The scale puzzle and the trick of solving it

The real "trick" of scale tuning on small pianos with severely compromised scaling features, lies not in a specific "magic" tuning sequence, but in the mental ability to solve the puzzle of how the scale beat rates are behaving on the particular piano being tuned. This means two main things, that are dealt with in the chapter on scale tuning. Firstly, it means recognising that fifths over a large change of inharmonicity, may typically need to be simultaneously *wide* at the lowest adjustable partial, whilst *narrow* at the second adjustable partial. Secondly, it means mentally grasping the beat rate relationship between at least four notes at a time, and preferably of all thirteen notes in the octave, and being able to see what needs to be done in order to attain the desired result. Admittedly, this is difficult for a beginner, but continued practise, especially in scale *diagnosis*, leads eventually to this ability.

Tuning the scale – an example

There are numerous approaches one could take to tuning a scale where high inharmonicity changes are present. The principle must be to discover what the inharmonicity problem is, in each case, and to accommodate it accordingly. Applying pre-decided beat rates will not work as a universal approach. Rather, we must find out what tuning is required for each interval, *as we proceed*. The aim is still progression of the fast beat rates, and similarity of the slow ones.

The following is one approach I sometimes favour. It is not a sequence of tricks to be followed for a magic "solution". Rather, it is *one approach* that illustrates how the *general technique* can be flexibly used to *lead* to the solution. I emphasise again, the solution for any piano is dependent on the tuner's skill at solving the puzzle dynamically, as tuning proceeds, rather than on any universally applicable "trick". This particular approach can be adapted for any starting note, following the same broad principles, but here the sequence starts on C40. This is an advanced discussion, so it is advisable to use an actual keyboard to follow it.

Normally (in the absence of small piano syndrome) one would expect that in a perfect fifth and perfect fourth put together side by side to make an octave, the beat rate in the fifth would always equal the beat rate in the fourth, provided that the octave was "beatless". Thus, for example, the beat rate in F33 – C40 would equal the beat rate in C40 – F45, provided the octave F33 – F45 was "beatless". Where there is large change of inharmonicity between the lower and upper notes of the octave, however, it is a different story.

On a small piano, where there is such a change, the bottom half of the scale generally needs to be "spread downwards" more or "lowered" in order to avoid beating that is too fast in the lowest fifths. High inharmonicity leads the natural progression of beat rates in the fifths to be reversed from its "traditional" prediction, because of the reverse curve on the end of the bridge. In other words, beating in the lower fifths will be faster than the beating in the higher ones, not slower, unless a strategy is employed specifically to avoid this, as much as possible. Where the break itself falls within the scale, this tendency *may* be suddenly reversed back again below the break, but may also abruptly worsen the effect. Also, a change of stringing from plain steel to covered strings, may behave in the same way. As long as we "discover" each interval's behaviour as we proceed, we can still "build" the best possible scale.

A large change in inharmonicity between F33 (F3) and C40 (C4), for example, will often typically cause the fifth F33 – C40 (F3 – C4) to behave with "classic" small piano syndrome. In such fifths, the inharmonicity is such that even with the lower adjustable partial "beatless", the higher adjustable partial still beats unpleasantly fast. The problem is not really so much one of tempering the fifth, as of simply finding a tuning at which the interval does not contain unpleasant beating. No "beatless" state is possible for the fifth as a whole, because always when one adjustable partial is "beatless", the other is beating too fast. And yet we only ever really want *slow* beating in a fifth, for an acceptable tone.

The only solution is to have the lower adjustable partial slightly *wide*. Both partials then beat, but the overall amount of beating can be more acceptable for the tone of the fifth as a whole. The lower partial beats *slowly* wide, whilst the upper one beats *slowly* narrow. Alternatively, the higher adjustable partial may be acceptable when the lower adjustable partial is "beatless". Either way, as far as the lower adjustable partial is concerned, the fifth is considerably wider than in its "traditional" tuning. The choice of tuning, however, depends on the *prominence* of each partial. There is no point in worrying about a very fast beating higher adjustable partial, if it is not very prominent, and is not in fact spoiling the tone of the octave.

Starting with C40, one can tune the fifth F33 – C40 (F3 – C4), when it is impossible for it to be "pure" or "beatless" anyway, so that its beating is acceptable. This involves, in effect, a very slightly *wide* "stretch" on the fifth, on its lower adjustable partial, in order that the 2nd adjustable partial is not too fast. In doing this we are responding to the inharmonicity characteristics of that individual instrument, which is essential if we are to find a scale solution. Next, we tune the F45 (F4), both for the octave F33 – F45 (F3 - F4), and the fourth C40 – F45 (C4 – F4).

Once again, the "traditional" rules simply do not apply, because of the inharmonicity change over the scale's width. On a sufficiently "scale challenged" piano, even for the *octave* F33 – F45 (F3 – F4), the lack of any available tuning position at which the octave could really be regarded as "beatless", is strikingly obvious to the astute ear. When the first adjustable partial is beatless the second may be fast. But when the second is beatless, the first is beating wide. The octave must *not* be narrow, in fact, it must be stretched, but must not have its tone completely spoilt by the stretch. The fourth *must* be wide (at its lowest adjustable partial), faster than the "traditional" rate perhaps. Satisfying these principles for both the fourth and the octave, will determine the tuning of the F45 (F4).

We now have the starting "anchor" for the scale. The same principles are then applied in tuning the B-flat38 (Bb4). Both the fourth F33 (F3) – B-flat, and the fifth B-flat – F45 (F4) must contain suitable "tempering".

The next step is to tune the E-flat43 (Eb4) plus the E-flat31 (Eb3) an octave below it. The fourth B-flat38 – E-flat31 (Bb3 – Eb3) must be wide, and must relate as properly as possible to the fourths C40 – F45 (C4 – F4) and F33 – B-flat38 (F3 – Bb3) already tuned. It should preferably not be faster than C40 – F45 (C4 – F4). The octave E-flat to E-flat must of course be good, and can be stretched, but the fifth from the lower E-flat up to the B-flat must also be acceptable. It can be even be wide at the lower adjustable partial, providing the octave is good and the fourth is wide. These factors will all usually work against each other, which is good, because it directs us to the necessary point of compromise determined by the inharmonicity. We are seeking to use the tuning of these intervals to find the correct position for the upper E-flat 43 (Eb4).

Now at last we are in a position to find out how the inharmonicity is affecting the beat rates in the major thirds. The E-flat43 (Eb4) is now fixed, according to the inharmonicity in the scaling from E-flat31 to E-flat43 (Eb3 –Eb4). We now need to tune the A-flat36 (Ab3) (G-sharp) from the E-flat43 (Eb4) above it. This will give us the beat rate for the third A-flat36 to C40 (Ab3 – C4). This fifth *must* relate properly to the other fifths already tuned. The deviation of the third from the "expected" rate will indicate which way the fifth needs to be "compromised", if any. If there appears to be no conflict, all well and good. Otherwise, the antagonism between the third and the fifth must be resolved mostly by compromising the third, not the fifth. Then we have the first indication of how the beat rates in the thirds are going be. Invariably, we will be looking at a slower than "expected" beat rate in the third.

At the same time, we now have a *minor third / major sixth* "test" that we can perform, in F33 – A-flat36 – F45 (F3 – Ab3 – F4), but we are going to use this "test" in a different way from normal, because we are not actually tuning the octave F33 – F45 (F3 – F4). The position of the A-flat that we are tuning actually doesn't matter as far as this "test" is concerned. What we are looking for is the (probably unequal) relationship between the minor third and major sixth beat rates. Recall that normally, one would expect them to have equal beat rates, no matter where the A-flat is positioned, providing the octave is good. It is, after all, a "test" for a good *octave*, that is not dependent on the size of the sixth. A large difference in inharmonicity or a large stretching between lower and upper notes of the octave will cause this principle to fail. If the beat rates *are* the same, then the aim to now to set the beat rate in the third B39 – D#(E-flat)43 (B3 – D#4) to about the same beat rate. If the beat rates minor third / major sixth differ, then we can take an average.

We then have in the scale area F33 – F45 (F3 – F4), the following notes tuned (in rising order): F33 (F3), A-flat, B-flat, B, C, E-flat and F45 (F4), all based as much on the effects of inharmonicity as on the attempt to temper the intervals. "Filling in" the rest of the intervals for progression, using "traps" if necessary, is then straightforward, basing the approach on whatever is your favourite sequence.

For example, if starting on C40 (C4), I would normally then tune the E44 (E4). In this instance, in the approach for a very small piano, the C – E can be modified from its normal tuning, based on the adjacent B – D# already tuned, from which it should progress. The G can be tuned as a major third below the B already tuned, based on progression from the adjacent A-flat to C.

The secret of success will still now lie in a thorough understanding of scale logic, tuning specifically to discover the extent of the effects of inharmonicity, deliberately tuning by "traps", and being prepared to modify previous tuning. However, this initial "sequence" sets up critical intervals that tend to provide a foundation for quick success.

Tuning pianos together

When two pianos are to be tuned for use together, small piano syndrome can create specific problems that need to be addressed, in matching the tuning of one piano to the other. Whilst there will usually be difficulties in matching the scale on one instrument with the scale on the other, even greater mismatches are likely to occur in the rest of the compass, in the octave tuning. For this reason, tuning pianos together is dealt with in detail in the chapter on octave tuning, Chapter 10.

Chapter 14 - Hearing

The ability to detect fine relative pitch differences falls into two categories:
1) Notes that are heard sequentially
2) Notes that are heard simultaneously

For fine tuning purposes we often need to address as a matter of course, intervals of one cent or less – less than one hundredth of a semitone. Often we are dealing with a fraction of a cycle per second, in a partial's frequency. These changes are significant because of their effect on pitch intonation or tone or both, depending on the situation.

An important limiting factors on the perception of pitch occurs in the basilar membrane in the ear, which responds to sound and then stimulates sets of hair cells that communicate with the brain. Roughly speaking, which part of the basilar membrane will be excited by an incoming sound, will depend upon the frequency. This mechanism of discrimination is subject to what is known as *critical band* limitation. Two pure tones (or indeed partials) of different frequencies can be said to differ by more than one *critical band* if they fire two largely different sets of hair cells on the basilar membrane.[71] On the other hand, if the frequency difference falls within one *critical band* then the two tones will be difficult or impossible to distinguish because they fire the same set of hair cells.[72]

When we two hear pure tones simultaneously, we can adjust precise differences between them by altering the resulting beat rate with precision. This change of beat rate is a tone property change, and can be

---

[71] See, for example, Campbell and Greated, *The musician's guide to acoustics*, London, 1987, p. 59.

[72] Campbell and Greated, *op. cit.*, p. 83.

utilised for both high and low frequencies. However, the ability to distinguish the *pitches* of pure tones heard separately, diminishes significantly for lower frequencies. At pitches equivalent to the bass end of the piano compass, pure tones a semitone apart can be difficult to distinguish. At middle C we would normally expect to be able to differentiate pure tones that are 20 cents apart,[73] but would find pure tones only 5 or 6 cents apart (a difference of around 1 Hz at middle C) difficult or impossible to distinguish.[74] Actual notes on the piano are of course rich in partials, and this improves the sensitivity by involving a greater part of the basilar membrane.

This is an important principle to remember. The facts about *critical bandwidth* serve to emphasise the dependence of pitch perception on *sets of partials* rather than just on one frequency, *especially* for low notes. Sets of partials also constitute the ingredients for the perception of tone quality, so tone and pitch are inevitably related. It is sometimes possible to change the perceived pitch of a tone, essentially by altering its tone quality, rather than by altering the fundamental frequency by an amount larger than the critical bandwidth. This is something that can certainly occur in piano tuning, particularly in octave tuning (specifically, in octave stretching).

The dependence of pitch perception on *sets* of partials is particularly relevant to the perceived pitch of bass strings, especially on smaller instruments, because for these strings the part of the spectrum within the hearing's *dominance region* where it is most sensitive, is often significantly inharmonic. The dependence of pitch sensation on the partial recipe can contribute to the well known conflict between the perceived pitch of a bass note on a small piano and the suitable placement of the note with regard to beating.

Plomp and Nelson[75] have shown that the minimum detectable frequency separation of two *pure* tones heard *simultaneously*, and heard *as a pitch difference*, is about 20 Hz, which occurs in the most sensitive region around 200 Hz. Conversely, heard separately, tones only 2 Hz apart can be distinguished at all audible frequencies below about 500 Hz. Listening to two simultaneous *pure tones* separated by much less than 20

---

[73] From the graphs, Campbell and Greated, *op. cit.*, p. 95.

[74] From unpublished research by the author.

[75] See Campbell and Greated, *op. cit.*, p. 59.

Hz, even in the sensitive region around 200 Hz (around G below middle C), we tend to hear not two distinct pitches, but a single tone with a beat modulated amplitude - a *beat rate*.

Notes in this region of the piano are of course not *pure tones*, but *complex tones* rich in partials, so on the piano we can hear distinct pitches much closer together. However, in the lower register, the psychological dimension of pitch perception can become much more significant in tuning, and is sometimes experienced acutely when an increase of tension beyond the normal fine tuning range on a lower monochord bass string in a piano of lesser quality, appears to *decrease* the pitch, or *vice versa*.

The physiology of the ear is not the only biological factor affecting pitch perception. The brain itself must be taken into account, and here there appears to be a different order of complexity that is relatively little understood at the current time. Sound information is processed in the primary and secondary auditory cortices on the superior temporal lobe of the brain, and the primary cortex has a *tonotropic* organisation, that is, it is organised into bands of characteristic frequency, in which neurons are "tuned" to specific frequencies.[76] It is now widely recognised that learning can involve biological changes in the brain, so it is perhaps not so surprising that acuity of pitch and tone perception is not necessarily fixed, but can improve with training and experience.

Fine tuning a piano is no ordinary, everyday task, except of course for the professional tuner. After, say, ten years experience, a properly trained and talented tuner can perhaps be expected to have become very proficient in piano tuning. At the beginning of my own professional career, 20 to 30 tuning per week was normal. Therefore, given that a full-time professional tuner might well have made in the order of 10,000 thousand fine piano tunings in ten years of practice, which is well over 2 million strings finely tuned, we might reasonably *expect* the tuner to be making use of *extraordinary* aural abilities by that time. This kind of experience is completely outside that to which laypersons and even professional musicians are exposed. We should not be so surprised, for example, to find that despite the findings of the research already mentioned, the tuner's capability in pitch judgement has become so sophisticated, that often, specific frequency changes can be implemented through pitch

---

[76] Bear, MF; Connors, BW; Paradiso, MA, *Neuroscience*, Baltimore; Philadelphia, 2001, p. 381.

judgement alone. This is possibly not uncommon in octave stretching by master tuners.

Although we cannot change the physiological nature of hearing, the potential capabilities of the trained ear-brain system extend well beyond what might be expected on the basis of the idea that the finest perceivable differences are in terms of *beat rates* alone. A "natural" expert tuner with fully trained hearing should be able to switch back and forth from hearing a tone as discrete partials to hearing a composite tone, or to focus on any mixture of the two, at will. When this kind of ability is also linked with pitch sensitivity and awareness, a powerful synthesis in musical tone and intonation perception can result.

Aural sensitivity is also something that can be consciously adjusted by the tuner, once the hearing is sufficiently honed (which may take years). It is not always appropriate to listen with the highest *acoustical analytical* sensitivity. Often, it is more important to listen *as a musician*, rather than *analysing* tone by ear. Most pianists will not have the artist tuner's capability of instantly penetrating into the soundscape at will, and of being able to perceive what aspects of the partial structure are contributing to, or detracting from the overall tonal quality. Nevertheless, aural perception that is able to instantly analyse the soundscape for individual partial properties, is obviously superior in at least one way, to an aural perception that cannot do this.

This level of aural perception is not something that just naturally occurs without concerted training and practice. It is part of the art of higher aural perception, part of the repertoire of techniques available in the art of tuning. In most cases, as long as the aural perception of the tuner is above that of the musicians who are the tuner's clients, this will suffice. But unless higher aural perception, at least to some degree, is in place, then the final artistic control of the potential qualities in the finest instruments, will not always be realised.

Piano tone is an "orchestra" or "ensemble" of partials. Classically trained professional musicians, and indeed conservatoire students, do not hear merely the "overall effect" of an orchestra or ensemble, like the inexperienced, untrained or uninitiated auditor. Rather, they hear both the "overall effect" (texture, *etc*.) *and* the individual parts, both as instrument sections of the orchestra or ensemble, and as separate contrapuntal components of the music. By analogy, for the piano tuner, the spectrum of partials in the soundscape is in effect the orchestra, whilst the partials are the instruments or instrumental sections. One can listen to the overall effect, as tone, or to the individual "instruments" or "sections", as partials.

For many learners, once the ability to hear individual beating partials has been gained, it can be difficult to redirect attention away from them, and onto the *tone* or *intonation* that results from the synthesis of the whole set of components. It is true that once one has learned to *reliably* perceive component partials in sound, and has practised this way of hearing for long enough, that "mechanism" of hearing indeed then appears to "rewire" the brain and hearing, so that as an aural perception it cannot then be effectively "switched" off, at will.

Accordingly, one can always *hear* the beating, *etc.*, but this does not mean one should always be *listening* to it. To deliberately *listen* to something and to merely *hear* it, are not one and the same thing. There are certain things in the components of tone that we should pay attention to when we hear them, and then there are things that we *will* notice, because of our "rewiring", but need not pay attention to. When we *do* pay attention to something and change it, like a high partial that is too prominent, or a beat that is too fast, it is because of its contributing effect on the whole, not for the sake of *it*, as an isolated thing. Knowing the difference is not a question of *numbers*, or *rules*, or *edicts*. It is about sonic beauty.

This all comes to the fore especially in the tuning of stretched treble octaves or "unstretched" lower bass octaves. The idea that an octave *should not* contain (or is indeed capable of not containing) *any* fluctuation, is a simplistic, learned *concept*, not a truth issued from some authoritative oracle. It is, in fact, an edict without authority. One of the worst errors of judgement in practice, is in approaching the tuning of a piano just by *listening* to fluctuations in partials, through, so to speak, the "rewired brain".

Chapter 15 - The Kirk Experiment

The Kirk experiment, 1959,[77] concluded that the majority of a set of musically trained and untrained subjects, preferred a deviation of 1 or 2 cents maximum difference between the frequencies of the three strings of piano unison groups. This much quoted (and sometimes misquoted) experiment confirmed previous reports of the preferences of musically trained subjects for "detuned" piano unisons.[78]

The Weinreich model, even as a model for motion of just two strings in only one plane, already puts us in a position to reappraise such results. To put this into perspective, we can note that at middle C (= 261.6 Hz) a mistuning of 1 or 2 cents at the fundamental would be a mistuning of 0.15 Hz or 0.3 Hz respectively. Thus the 'preferred' maximum mistunings indicate a typical 'critical mistuning' or *coupling region* for middle C. (Kirk tested for notes from E32 (E3) = 165 Hz to G 59 (G5) = 784 Hz, which gives mistuning ranging from 0.1 Hz to 0.45 Hz at the G 59). Kirk also reported 6 to 8 cents as the maximum 'acceptable' mistuning (as distinct from 'preferred') for musically trained subjects. This would correspond to mistunings at middle C of 0.9Hz to 1.2Hz, respectively.

In the light of the effects of bridge coupling, therefore, there is little in Kirk's experiment to strongly suggest an actual preference for the presence of audible regular beat rates in the fundamentals of unison groups. A difference of 1 or 2 cents between string fundamentals falls within a 'critical mistuning' region of a unison with strongly coupled behaviour, so does not necessarily even correspond to the presence of beats at all, in the fundamental.

---

[77] Kirk, RE, 'Tuning Preferences for Piano Unison Groups', *JASA*, 31, 1959, pp. 1644 – 1648,.

[78] Martin, DW and Ward, WD, *JASA*, 26, 1954, p. 932

There are other points in relation to the Kirk experiment that are worth drawing attention to. Firstly, Kirk remarked:

> According to piano tuners, unison groups which are tuned too closely lack interest and diminish too rapidly.[79]

Bridge coupling theory does not support the idea that a unison group, however closely tuned, will have an *entire spectrum* that will always decay "too rapidly", or that the tone will "lack interest". The theory indeed shows that *for a given partial*, a certain mistuning can increase the decay time of the aftersound. However, this is only for the single partial to which that mistuning is being applied. We cannot apply the principle to the fundamental and simply ignore the effect the mistuning has on the other partials!

The idea has gained some currency because Benade proposed a hypothesised model that suggested this,[80] and the Weinreich model for a *hypothetical* purely resistive bridge does show this behaviour, but neither model fully represents the actual situation, or is applicable to the whole spectrum simultaneously. Given that piano string vibrations are not periodic and are inharmonic, both models can only apply to only one partial at a time, and furthermore, both models ignore the fact that real piano strings *in situ* vibrate in two planes with transverse parametric motion.

A misunderstanding about unison tuning arises here, encouraged by Kirk's emphasis on mistuning *in cents*, rather than in *Hz*, which obscures the importance of what actually goes on in unison tuning, with regard to the fact that the tone is a relatively large *set* of partials. The reason is as follows:

Inharmonicity aside, the mistuning *in cents* remains *the same in all partials*, so it seems unnecessary when talking about mistuning in this way, to refer to a specific partial. However, the mistuning as a *frequency difference* is different in every partial, and it is this that determines *beat rates*, *decay patterns*, and *decay rates* in any adjustable partial, and these features are *critically* important in the tuning.

---

[79] Kirk, *op. cit.*, p. 1644.

[80] Benade, AH, *Fundamentals of musical acoustics*, NY, 1976, 1990, p.336 *ff*.

Ignoring inharmonicity, the mistuning in Hz at any partial is the mistuning at the first partial multiplied by the partial number. For example, the third partial will have a mistuning *in Hz*, 3 times that in the fundamental, and the eighth partial will have a mistuning *in Hz* 8 times the fundamental's. Inharmonicity will alter this relationship, but for now, this basic model will suffice to illustrate the principle. We cannot really talk about introducing a mistuning in the fundamental, in order to change the fundamental decay rate, and ignore the possible effects this has on the state of the other partials, whose mistunings *in Hz* are proportionally multiplied. We cannot in any case simply sacrifice tone, as affected by all these other partials, in order to adjust the decay rate of a single lower partial. The tone of the unison, and indeed its decay time overall, is dependent upon these other partials – it is dependent on the whole spectrum.

Nevertheless, the idea as reported by Kirk is colloquially, or *roughly* describing that there is a relationship between mistunings and the tuner's work in achieving the best tone of the unison, which may well include some decay time factors in some of the partials. We would never, however, sacrifice the first principle of reducing movement in the soundscape to a minimum.

Another major possible reason for the presence of mistunings is falseness. As described earlier, the endeavour to minimise fluctuations in adjustable partials that have inherited false beats, may often involve mistunings, as the least fluctuating pattern does not necessarily occur at zero mistuning.[81]

Secondly, Kirk stated:

> The criterion for the zero beat condition was the absence of amplitude fluctuations in the level recordings …. [footnote] : Several unison groups could not be tuned to meet these criteria and were not included in the investigation.[82]

Kirk was presumably recording the whole spectrum of each mid-range note of the piano, as no particular partial or filter was mentioned. No

---

[81] See also Capleton, B, 'False beats in coupled piano string unisons', *Journal of the Acoustical Society of America*, 115, 2, 2004, pp. 885-892.

[82] Kirk, *op. cit.*, p. 1645.

amplitude fluctuation in the overall signal envelope does not *necessarily* mean no amplitude fluctuations in all the individual partials, and this does not necessarily mean no mistuning throughout the spectrum. However, the reasons for this are somewhat involved. Kirk's excluded unisons are most likely to have been instances of *falseness*.

Finally, we must be clear about the subjects used by Kirk, and dispel any misunderstandings that they were all, for example, professional pianists or other musicians, or even piano tuners! All were said to have musical experience, but the actual data is as follows. Kirk used 123 subjects, 102 of whom were students in Introductory Psychology classes at the University of Kentucky Northern Center. The other 21 subjects were not necessarily all music students, but *included* music students from Cincinnati College Conservatory of Music. The remaining were Baldwin Piano Company engineering and research personnel.

Part 3 - Advanced Theory

> Everything should be made as simple as possible, but not simpler
>
> - Albert Einstein (attributed)

This section contains some of the theory behind the material presented in the first two parts of the book. It outlines the basic relevant theory for the "classical" model of a vibrating string, on which "traditional" piano tuning theory is based. It also deals with the Weinreich model for bridge coupled piano strings, and then considers largely speculative extensions of the model for string motion in two planes, and for trichords. Lastly, it examines some issues and ideas concerning falseness and inharmonicity.

The primary aim is to elucidate the parts of the theory that at present, are most closely related to the more salient issues encountered in the experience of practical tuning. The section is not intended as a general survey of all related theoretical acoustics. The latter is important in its own right, but many of its most interesting questions, from the point of view of acoustics, have little immediate bearing on what have always been the most pressing issues in art of tuning. Conversely, much of the acoustical theory that truly illuminates tuning practice and provides useful understanding for the art, may seem unimportant from the point of view of acoustical physics.

A leap forward was made when inharmonicity became more widely accepted in understanding piano tuning, and this no doubt contributed to a raising of practical standards. Since the mid-20th century, the theory that has borne the closest and most natural relationship to the issues most keenly experienced in tuning practice, is in my view, the Weinreich model. Accordingly, it is dealt with in detail, because whilst it is already recognised as seminal in the field of piano acoustics, it still deserves to be better known in the field of piano tuning itself.

Chapter 16 - The single piano string in one plane

When any musical instrument string vibrates, the oscillatory motion is due to the propagation of transverse waves along the length of the string. Longitudinal waves can also be propagated, and these would become significant in relation to high amplitude transverse waves, but here we are primarily concerned with small amplitude transverse waves.

The simplest model for a tensioned string is the hypothetical ideal string tensioned between rigid supports. Real piano strings are not ideal strings, although their behaviour is often considered as similar, but with additional complications. By definition, the ideal string is uniform in diameter and mass per unit length, and is perfectly flexible. Waves propagating along the ideal string will be reflected at the end boundaries if they are rigid, so general motion consists of waves travelling in both directions along the string's length. The summation of these waves produces a resultant standing wave which does not propagate, but causes the string to oscillate transversely.

The structural and behavioural features of the waves that the ideal string can carry are particularly elegant. The wave motions and wave forms are periodic, and so can be represented as a Fourier series. Every wave form that can be carried by the string is resolvable as the sum of an infinite series of simple sinusoidal partial waves whose wavelengths are integer fractions of the string's speaking length. These partial waves are the Fourier components or harmonics that superposed make up any complex wave form on the string. Because the wave velocity along the ideal string is the same for all wavelengths, the corresponding frequencies of the partial waves, or harmonics, are integer multiples of the fundamental frequency, which is the lowest allowed frequency, the frequency of the longest allowed wavelength. The fundamental frequency v_0 is fixed by the string tension T, mass per unit length μ, and speaking length *l* of the string, according to Mersenne's equation:

$$v_0 = \frac{1}{2l}\sqrt{\frac{T}{\mu}}$$

1

The harmonics - the partial waves with frequencies that are integer multiples of the fundamental frequency - can have their frequencies arranged in a series $v_0, 2v_0, 3v_0$ ... etc, which is called the harmonic series. The constant 1,2,3... etc in this series is the Harmonic Number for that harmonic (note the fundamental is harmonic number 1). The frequency of the nth harmonic is thus:

$$v_n = nv_0$$

2

The harmonic series can also be considered as a corresponding set of wavelengths, or as a set of perceived simple tones. If an ideal string existed, and we were able to hear the sound it radiated, the harmonics would be heard as a set of simple tones with recognisable musical interval relationships between the harmonics in the first part of the series. In this scenario, of course, we would no longer be dealing only with the behaviour of the isolated *ideal string*. The behaviour of the string would be to some extent altered by the presence of the air into which the string radiates energy.

Similarly, attaching the string to a bridge and soundboard would alter the expected behaviour of the string. In practice, real musical instrument strings are neither *ideal* nor are they isolated from the effects of air, bridges and soundboards. The effects of these are not, however, generally sufficient to completely destroy the notion of a harmonic series being associated with a real string. Partial waves still exist on real strings, and their frequency relationships still correlate quite well with those of true harmonics, but we should nevertheless refer to the set of audible pitches from real strings just as *partials*, rather than *harmonics*.

The ratios of frequencies between adjacent and near adjacent harmonics in the first part of the harmonic series, are 'harmonic ratios', which are whole number ratios associated with the basic musical intervals of Western music. Harmonic ratios are found between different speaking lengths of identical strings at identical tension, when the notes of the strings are separated by musical intervals. Two audible frequencies in the mid-range will also generally sound a musical interval if they are in an harmonic ratio. These ratios include (amongst others) 2:1 for an octave, 3:2 for a perfect fifth, 4:3 for a perfect fourth, 5:4 for a major third, and 6:5 for a minor third. As musical pitches, the lower harmonics in the harmonic series are thus separated in sequence, from the lowest upwards, by the musical intervals of an octave, a perfect fifth, perfect fourth, major third and minor third. Sounding simultaneously, the first six harmonics produce a major chord - the lower part of the so-called "chord of nature".

The ideal string - general proof of harmonic partials

The existence of harmonic partial waves in the motion of the ideal string, can be proved in the following generalised way. With the speaking length along the x axis, the displacement y to the string caused by *any* wave moving in the positive direction along the string, with phase velocity c, will be a function of both x and ct, where ct is the distance moved by the wave in time t. For a wave moving in the positive direction, the general equation can be written:

$$y(x,t) = f(x - ct)$$

3

Similarly for a wave moving in the negative direction we put:

$$y(x,t) = f(x + ct)$$

4

Because waves will be reflected at the boundaries, the resultant displacement due to the presence of both waves will be:

$$y(x,t) = f(x + ct) + f(x - ct)$$

5

which we can alternatively write as:

$$y(x,t) = f(ct + x) - f(ct - x)$$

6

The solution to this equation will depend on the nature of f, which is restricted in that it must satisfy the boundary conditions at each end of the string. For rigid boundaries on a string stretched from $x = 0$ to $x = l$, these are that $y = 0$ at $x = 0$ for all time, and similarly $y = 0$ at $x = l$ for all time. From 5, the condition $y = 0$ at $x = l$ requires that:

$$f(ct+l) = f(ct-l)$$

7

This states that f must be periodic in x with a period $2l$, such that

$$f(x+2l) = f(x)$$

8

Because $f(x)$ is a periodic function it can be expanded as a Fourier series:

$$f(x) = \sum_{n=-\infty}^{+\infty} f_n \exp\left(\frac{i\pi nx}{l}\right)$$

9

Thus Eqn. (6) expanded becomes:

$$y(x,t) = \sum_{n=-\infty}^{+\infty} f_n \left[\exp\left(\frac{in\pi(ct+x)}{l}\right) - \exp\left(\frac{in\pi(ct-x)}{l}\right)\right]$$

$$= \sum_{n=-\infty}^{+\infty} 2if_n \exp\left(\frac{in\pi ct}{l}\right)\sin\left(\frac{\pi nx}{l}\right)$$

$$= 2\operatorname{Re}\left[\sum_{n=1}^{+\infty} C_n \exp\left(\frac{in\pi ct}{l}\right)\sin\left(\frac{\pi nx}{l}\right)\right]$$

10

where $C_n = 2if_n$.

Eqn. (9) describes a series of simple sinusoidal standing waves of the form:

$$y_n(x,t) = \operatorname{Re}\left[C_n \exp\left(\frac{in\pi ct}{l}\right)\sin\left(\frac{\pi nx}{l}\right)\right] \quad n = 1, 2 \ldots \infty$$

11

Putting $\omega_0 = \dfrac{\pi c}{l}$, we can write:

$$y_n(x,t) = \text{Re}\left[C_n \exp(in\omega_0 t)\sin\left(\dfrac{\pi n x}{l}\right)\right]$$

12

which is a simple sine wave shape along the length of the string, in the form

$$\sin\left(\dfrac{\pi n x}{l}\right).$$

The wave shape is stationary in x, but its amplitude oscillates according to

$$\text{Re}[C_n \exp(in\omega_0 t)].$$

This is the partial standing wave number n, or nth harmonic, whose angular frequency is $n\omega_0$, and whose wavelength in x is $2l/n$.

What we have just shown is that any standing wave on the ideal string between rigid boundaries is equivalent to the sum of an infinite series of simple sinusoidal standing waves whose wavelengths are whole number fractions of 2l, twice the length of the string. There is a corresponding series of travelling waves that moving in both directions, and responsible for creating the standing waves. The important features here are that:

a) The boundaries are rigid, so the displacement of the string at each boundary must remain zero. As a consequence, *all* waves on the string are *periodic* along its length. This periodicity means that the function for the shape of *any* arbitrary wave on this string can be expanded as a Fourier series.

b) All waves on the string propagate with the same velocity so the frequencies of these Fourier component waves are related in a harmonic series.

The harmonic spectrum

Traditional simple tuning theory tacitly assumes that piano strings behave as though these features were present, and that the decay of the tone can be regarded merely as an added, but otherwise independent feature. From these assumptions also follows the notion of the harmonic partial spectrum – a partial spectrum of all the simple harmonic partial amplitudes (amplitudes of the harmonics), of the frequencies that together make up the more complex overall motion. Understanding what the piano tuner listens to, or how the tone of a unison changes with tuning, is to understand the structure and behaviour of piano string spectra. It is not actually the string motion in itself that is considered at all in the process of tuning, but rather, the audible partial spectra, mostly radiated from the soundboard.

When we consider the spectrum of a string, rather than the waves on the string, we transfer our attention from the complex motion of the string as function of time, to the internal structure of that motion as a sum of its simple harmonic partial components, as a function of frequency. We move from considering the system in terms of string displacement in the time domain, to its description in terms of amplitudes in the frequency domain. In the frequency domain representation of the string motion we show what component partial frequencies are present in the makeup of the overall motion, and by what magnitude or amplitude each frequency is present. This is equivalent to showing the values of the partial amplitudes fixed by the C_n in the Eqns. (11), against their corresponding frequencies $n\omega_0$. An example first part of a spectrum is illustrated in Fig. (16.1):

Fig. 16.1

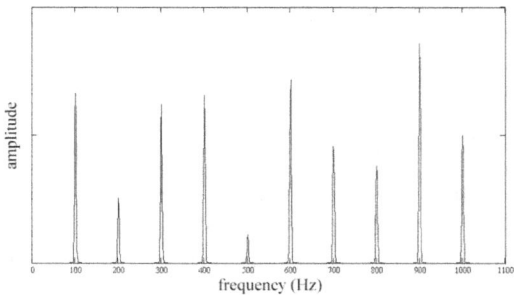

Fig. 16.1. Example spectrum of the first 10 harmonics of an ideal string, shown as harmonic amplitude versus frequency.

Fig. (16.1) illustrates how the first part of a spectrum for an ideal string might look, in its graphical representation. The various peak heights corresponding to partial amplitudes, are determined by the initial conditions of setting the string into motion, perhaps by striking or plucking. The appearance might not be quite what would be expected. The string motion $y(x,t)$ expressed as a Fourier series of harmonic partials as in Eqn (9), shows that frequencies in the spectrum only exist as exact, integer multiples of the fundamental – no other spectrum frequencies exist. The peaks in Fig. (12) however, show a small amount of frequency component either side of each exact harmonic frequency. Drawn with precision, the peaks in the spectrum Fig. (12), would be expected to be straight vertical lines at the exact positions of the harmonic frequencies. The width of the peaks that often appears in such plotted spectra arises from the fact that the spectrum graph is *not* generally produced by direct plotting from the mathematical analysis on paper, but is usually generated by the digital *fast Fourier transform* method.

This spectrum is two dimensional – amplitude is one dimension, frequency is the other. It is important to appreciate that this kind of two-dimensional spectrum cannot represent the whole perceivable structure of the actual piano string tone. This is not merely because the peaks for a real, stiff string might be in different positions on the frequency scale. More importantly, this spectrum of the ideal string represents steady state motion that does not decay, whereas the real piano string motion and sound does decay, and the magnitudes of the frequency components, and even the frequencies themselves, may not be constant over the decay time.

The Fourier transform of the ideal string spectrum

The relationship between more general one-dimensional bounded motion in the time domain, and the spectrum of amplitudes in the frequency domain, can be obtained through the Fourier transform. The Fourier transform of a time function $f(t)$, is a function of frequency $F(\omega)$ given by:

$$F(\omega) = \frac{1}{2\pi} \int_{-\infty}^{+\infty} f(t) e^{i\omega t} dt$$

13

where each frequency $\omega/2\pi$ in the spectrum has an amplitude determined by the magnitude $|F(\omega)|$ of the complex quantity $F(\omega)$.

We showed the equation of motion of the standing wave on the string was:

$$y(x,t) = 2\operatorname{Re}\left[\sum_{n=1}^{+\infty} C_n \exp\left(\frac{in\pi ct}{l}\right) \sin\left(\frac{n\pi x}{l}\right)\right]$$

14

The factor $\sin\left(\frac{n\pi x}{l}\right)$ in each term is purely a space function of x, so for any given value of x, Eqn. (14) becomes an equation of motion for the given point x on the string:

$$y_x(t) = 2\operatorname{Re}\left[\sum_{n=1}^{+\infty} (B_x)_n C_n \exp\left(\frac{in\pi ct}{l}\right)\right]$$

15

which is now motion as a function only of time. The real constant $(B_x)_n$ is here just $\sin\left(\frac{n\pi x}{l}\right)$ for the now constant x at each value of n. The motion at this point still consists of the component partial motions, and we can rewrite Eqn. (14) in the alternative form:

$$y_x(t) = \operatorname{Re}\left\{\sum_{n=1}^{\infty} A_n \exp[-i(n\omega_0 t - \phi_n)]\right\}$$

16

where now ϕ_n contains the phase information for each partial and A_n is real. It can be shown from standard theory that the Fourier transform of the complex Fourier series inside the parenthesis will be:

$$Y_x(\omega) = \sum_{n=1}^{\infty} A_n \delta(\omega - n\omega_0) \exp(i\phi_n) \qquad 17$$

where $\delta(\omega - n\omega_0)$ is the *Dirac delta function* of $(\omega - n\omega_0)$. The Dirac delta function is defined in such a way that $\delta(q)$ is zero if $q \neq 0$, and $\int_{-\infty}^{+\infty} \delta(q) \, dq = 1$, so at all values of $\omega = n\omega_0$ for any integer n, $\delta(\omega - n\omega_0)$ will be unity, and at all other values of ω it will be zero. Thus the spectrum $|Y(\omega)|$ plots as a series of equally spaced isolated peaks, each of amplitude A_n. As already mentioned, actual plots produced using digital technology will generally contain steep *bell curve* like peaks with a definite width, or over small parts of the frequency domain even angular spikes. This is due to the digital method of the fast Fourier transform (FFT) and the limitations imposed by the sampling rate in relation to the signal block length.

The partial spectrum of the real string in one plane

The motion and sound of the real piano string both decay. Whilst some decay is due to incidental frictional and radiation energy loss, it is not these decay mechanisms that interest us most. The main mechanism of piano string decay is "designed" into the system – energy is deliberately drained from the string through the bridge, to the soundboard, in order for the soundboard to act as an amplifier.

For a single piano note, the more the system is designed to amplify, the faster must be the decay, since there is no further energy input into the system after the note is played. The transmission of energy from string to soundboard requires that the soundboard bridge moves, so we no longer have the 'rigid' boundary conditions that dictate periodicity, and allow us to express motion as a Fourier series. Also, the considerable stiffness of high tensile steel strings results in mode frequency dispersion – waves on the string will not all propagate at the same velocity, but rather, the phase velocity will depend upon the wavelength.

The allowed wavelengths on the string are still mostly fixed by the (now more complicated) boundary conditions, but the dispersion of wave velocities results in partials whose frequencies are no longer related in a perfect harmonic series. In piano strings this effect is commonly known as inharmonicity. Inharmonicity causes the peaks in the spectrum to be placed at inharmonic positions on the frequency axis – the frequencies will not be integer multiples of the fundamental.

Owing to the decay of the piano string tone, its complete spectrum must be represented three dimensionally. A two dimensional relationship between amplitude and frequency can only ever be a representation of the amplitude-frequency relationship at a given instant in time, because the amplitude-frequency relationship is constantly changing as a function of time. The piano string spectrum to which the piano tuner listens, is not however merely a three dimensional version of the ideal string spectrum – it is *not* like an ideal string spectrum but extended into the time dimension. This is because there is another very important difference between the nature of the spectrum for the piano string *in situ* and that of the ideal string.

In the case of the ideal string between rigid supports, the two dimensional spectrum for steady state motion shows the Fourier amplitudes of the harmonics, which are the normal modes of the system. When we deal with the piano string *in situ* and are concerned with what we may hear, we have to address an audible spectrum whose discrete, audible partial components, as we shall see, are not necessarily all normal modes of the system.

We are interested in what we hear as the piano string motion decays, so we can use the motion at the soundboard bridge - which is the driving point of the soundboard motion – as a suitable indication for the source of sound radiation from the soundboard. We must not assume that bridge motion has only one degree of freedom, nor can we assume that only bridge motion in one dimension contributes to the supply of energy to the soundboard for sound amplification and radiation. The string itself has two lateral degrees of freedom (waves can occur in two orthogonal Cartesian planes), and there is evidence that sound energy is radiated from the soundboard through different radiating 'antennae'.[83] It is possible that in the general case, we are dealing with motion at the bridge that has two transverse degrees of freedom, and two possible orthogonal 'antennae' directions for the transmission of energy from the string.

To begin with, however, we can build a picture of the motion at the bridge by considering motion in just one transverse direction. How does this motion relate to the Fourier integral for motion on the string? Any bounded lateral motion in y at the bridge boundary of the string's speaking length, can have its displacement y described by the inverse Fourier transform:

---

[83] Weinreich, G, 'Coupled piano strings', *JASA*, 62, 6, pp. 1474 – 1484, 1977.

$$y(t) = \text{Re}\left[\int_{-\infty}^{+\infty} F(\omega) e^{-i\omega t} d\omega\right]$$

18

where $F(\omega)$ is the Fourier transform of $y(t)$. We can take a component motion $F(\omega)e^{-i\omega t}$ at the bridge where $y = f(t) \neq 0$ and consider whether a wave responsible for this motion can exist on the string. Waves of a given angular frequency ω and phase velocity c may propagate in both directions along the string, so this component string motion will be generally a combination of two simple harmonic waves. However, the principles of identical sinusoidal waves moving with the same phase velocity in opposite directions on a string, dictates that the frequency of the resultant standing wave is always the same as the frequency of either wave alone, so as long as we are interested only in frequency and not in relative phase angles, we need only consider one wave for the time being.

For any phase velocity c, the possible values for the wavelength λ of *any* wave on the string are limited by the length of the string and the boundary conditions, just as in the case of the ideal string. We take the top bridge or agraffe boundary to be rigid, and the soundboard bridge boundary to be non-rigid having a complex impedance $Z(\omega)$. If the boundary was rigid all wavelengths on the string would necessarily have to satisfy $\lambda = 2l/n$ for a string length l and any integer n, for $n>0$, in order that displacement remains zero at each boundary. Whilst $\lambda = 2l/n$ is real, the amplitude of any wave could nevertheless decay because of friction and air resistance, and such a decay factor could typically be expressed in the imaginary part of a complex ω in the equations of motion. We can however just as easily address λ as complex. The non-rigid boundary means that any value of λ must now satisfy $(2l/n) + \delta_n$ where the δ_n is the complex change to the wavelength as a result of Z's finite, complex value. The real part of δ_n which can be positive or negative, changes the real length of the wave, whilst the imaginary part introduces an exponential decay.

It may seem strange that the real wavelength can change whilst the string length is fixed, until we appreciate that the transverse motion of the boundary can allow a node position to fall close to the boundary but *either side* of the boundary in x, depending on whether $\text{Re}[\delta_n]$ is positive or negative. An actual node just outside the speaking length does not actually have to exist as a wave node in the next medium; the idea merely means that less than a complete number of wavelengths actually appears within the speaking length at any time, whilst a node just inside the speaking length means that a small length of wave continues beyond the node before reflection occurs. In other words the end behaviour of the string is somewhere between fixed and free, but closer to fixed.

The possible values of λ are therefore still a discrete, well separated series $\lambda_1, \lambda_2 \ldots \lambda_n$, provided $\text{Re}[\delta_n]$ is small compared to $(2l/n)$. The corresponding angular frequencies of these waves will be $2\pi k_n c_n$, where the wavenumber $k_n = 1/\lambda_n = n/(2l + n\delta_n)$ is complex, the imaginary part determining a decay rate, and c_n is the phase velocity for that wavelength. The quantity $n\delta_n$ is just a complex wavelength change associated with a particular value n and is more simply written Δ_n. Thus the only allowed values for ω on the string are given by:

$$\omega_n = n\left(\frac{2\pi c_n}{2l + \Delta_n}\right)$$

19

If $\text{Re}[\Delta_n]$ is always small compared to $2l$ and the dispersion between the c_ns is small compared to c_1 then clearly the real parts of the ω_ns must have a discrete, well separated series of values.

To reiterate, the general motion of any point on the string is a consequence of whatever (non periodic) wave shape is propagating on the string. However, the possibilities for the shape of this wave are limited by the string length, the limited dispersion factor, and the boundary conditions which are also limited. Because ω can only take the allowed values that fall in the discrete series fixed by the conditions in Eqn. (19), the only component waves of the kind suggested by the Fourier integral

$$y(t) = \text{Re}\left[\int_{-\infty}^{+\infty} F(\omega)e^{-i\omega t} d\omega\right]$$

20

that can be responsible for the non periodic motion at the bridge, are an infinite, *discrete* series of simple harmonic (sinusoidal) partial waves whose complex frequencies depend on $Z(\omega)$ and Δ_n, but have real parts that are close to a harmonic series.

With these restrictions to ω applied, the integral is thus replaced by the summation:

$$y(t) = \text{Re}\left[\sum_{n=1}^{\infty} C_n e^{-i\omega_n t}\right]$$

21

where both C_n and ω_n are complex, C_n containing amplitude and phase information. The spectrum for the bridge motion is thus similar to the ideal string spectrum in that the partial frequencies are discretely spaced at approximately harmonic positions on the frequency axis, but because ω_n is itself complex, each partial $C_n e^{-i\omega_n t}$ decays, its decay rate determined by the imaginary part of ω_n.

Eqn. (21) is the general motion at the bridge (in only one transverse dimension) expressed as a superposition of normal mode displacements at the bridge.

Chapter 17 - The Weinreich Model

Specific string behaviour or bridge-soundboard behaviour, where there are two transverse degrees of freedom, and possible coupling between the strings, is inherently very complicated, so we begin by examining the case for one transverse degree of freedom.

Tuning practice deals more with generalities of behaviour. For contemporary tuning theory, clearly we need a method of attack that enables us to describe *generic* string-bridge system behaviour patterns. At the same time however, we are particularly interested in a range of *different* behaviours that are of practical significance when encountered by piano tuners. We need to be able to examine generic, functional behaviour, in the absence of precise information, in considering the behaviour of a number of strings connected by the bridge as a single system. The approach we need is in some ways related more to that sometimes used in modern physics, than it is to classical mechanics, and it is from techniques perhaps more familiar in modern physics that we can benefit.

This approach to the behaviour of piano strings was taken by Weinreich, whose work is of great importance to the theory of tuning. Formerly, the edicts of "traditional" piano tuning theory were based on the idealised behaviour of single strings, and the treatment of multiple string behaviour merely as the sum of the these behaviours of isolated single strings. Weinreich's approach sufficiently emphasised for the first time the importance of the fact that piano strings in motion are members of a coupled, dissipative system, whose physics is quite different from the former, 'traditional' model, and even significantly different from that of coupled non dissipative systems, which are perhaps more familiar in engineering and mechanics.

According to the traditional theory, in tuning we are either dealing with heterodyne interference beats, or the absence of these beats, and the theory comprises the calculation of the required heterodyne beat frequencies for equal temperament. The new approach allows for the fact that in fine tuning practice one is equally dealing with the adjustment of dynamic spectral decay patterns, in which the movement and change that we adjust, is not a simple heterodyne beat interference phenomenon. It also provides a greater understanding of a whole range of phenomena that falls outside the scope of "traditional" theory.

We are going to be considering the forces on the soundboard bridge, at the boundary of the speaking length of one and more strings. We begin by considering a system that consists of a bichord (two adjacent strings of one unison), and consider the partial wave with one planar degree of freedom. We can say for convenience that the strings are horizontal as in a grand piano, and that they move only in the vertical plane. Also, we assume for now an isotropic (the same in all transverse directions) boundary admittance at the soundboard bridge (the top bridge or agraffe is considered perfectly rigid). When the boundary admittance is isotropic, the one 'vertical' planar degree of freedom can be taken as a co-ordinate in line with the hammer strike. It is immaterial if the hammer strike is not actually vertical.

At a first glance, there may seem to be three one-dimensional 'oscillators' involved - the two strings, plus the bridge which is itself an elastic oscillator. We would normally expect to look for *three* normal modes from a one-dimensional system comprising three simple oscillators. The system configuration is however of two oscillators (one for each string's partial), as would be a system of two masses coupled by a damped, massy spring mechanism, the third oscillator being driven.

The configuration at the piano bridge is not precisely the same as the latter, however. In the piano string case, the two strings of a unison group cannot be moving independently from the bridge at their point of their contact with it, because they are fixed to the bridge by the bridge pins. The displacement from equilibrium of both strings at the boundary position are therefore always equal, and equal to the displacement of the bridge. We must model this as two oscillators, coupled by both dissipative and reactive mechanisms.

Although each string at this position has the same displacement as the bridge, the force exerted on the bridge by each string may be different for each string. This is because the forces are due to incident and reflected waves in the strings at the boundary point, which superposed are the partial standing waves. The amplitudes, phases, and frequencies of these waves, can be different for each string. The force exerted on the bridge at any instant by each string, depends on the tension and incident angle of the string to the bridge, in the wave plane, which may be different for each string.

Two normal modes in one plane – Weinreich's model

The bridge, which is both elastic and dissipative, causes changes to the behaviour of each oscillator. The role of bridge movement is already well known - it causes perturbations to the phases and wavelengths of incident string waves. Thus in our system the bridge becomes a perturbation factor for each oscillator in the system. When the oscillators are identical in frequency, then for incident partial waves of a given partial number, the two possible normal modes of the system are:

(1) When the partial waves of both strings are out of phase with equal amplitudes, so that the net force on the bridge and hence motion of the bridge is zero, and

(2) When the partials waves are equal amplitude and in phase so that the force on the bridge and its motion is twice what it would be for just one of the strings.

In this condition mode (1) is the *symmetric* mode, and mode (2) is the *antisymmetric* mode. For the isolated antisymmetric mode where the waves are out phase the bridge does not move. The bridge in effect behaves for each wave as though it was rigid, reflecting all wave energy with no change to the frequency that would have been present if the boundary *had* been rigid. In this condition there is no frequency perturbation or decay introduced by the bridge, so the antisymmetric mode decay rate is zero (in practice, of course, decay would still occur for other minor reasons, including internal friction in the string, air drag, direct sound radiation from the string surface, etc).

Conversely for the symmetric mode where the waves are in phase, the bridge movement is twice what it would be for either wave alone. The function of the moving soundboard bridge is primarily to transmit energy to the soundboard, and for energy dissipation that is proportional to bridge velocity, the damping effect of the bridge on each string will now be twice what it would be if only one string was sounding. Also, the perturbation to the frequency for a rigid boundary will be greater than if one string alone was sounding. Thus the decay rate of the symmetric normal mode is twice the single string partial decay rate.

General motion, as a superposition of the normal modes, will have an overall decay rate depending on the admixture of the normal modes that make up the general motion. For identical strings and speaking lengths, the partial waves of both strings will initially be in perfectly symmetric motion if they have the same frequency and other initial conditions. In an idealised situation with a 'perfect' unison (equal frequency) tuned between the given partials of each string, and a 'perfect' hammer strike equal on both strings, we will excite only the symmetric normal mode and we would expect it to decay at twice the single string rate. In practice, a 'real' hammer strike is unlikely to create such 'mathematically perfect' initial conditions. The functional dependence is also such that even a very small presence of 'mistuning' (frequency difference between the partials if measured uncoupled), phase difference, or amplitude difference between the partials, will introduce a significant quantity of antisymmetric mode. This will substantially decrease the overall decay rate.

Effects of an 'imperfect' hammer strike – partial phase difference

Consider two partials tuned to equal frequency, having standing waves with equal amplitude, but that are not perfectly in phase, perhaps due to 'imperfection' in the hammer strike. Any phase relation between the waves other than perfectly *in phase* or perfectly *out of phase*, constitutes an admixture of the normal modes. The decay curve of general motion would then have two parts, a fast decaying early part in which the symmetric (in phase) mode is dominant, followed by a slowly decaying later part, in which the antisymmetric (out of phase) mode dominates after substantial decay of the symmetric motion has taken place. Thus, the system under these conditions displays a 'double decay curve' consisting of an initial *prompt sound* followed by an *aftersound* (Fig. 17.1).

Fig. 17.1

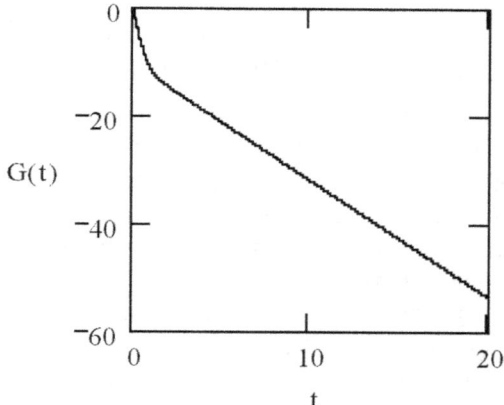

Fig. 17.1. Double decay curve for a perfect unison between two partials of a bichord, with one degree of freedom. Whilst the frequencies of the partials are equal, the partial standing waves differ in phase. General motion amplitude decay G(t) in dB (logarithmic scale) is the sum of the fast decay of the symmetric mode, and the slow decay of the antisymmetric mode. The summation of two ordinary exponential decay curves yields a double decay curve.

Fig. 17.2 shows the formation and the relationship of the double curve to its component parts.

Fig. 17.2

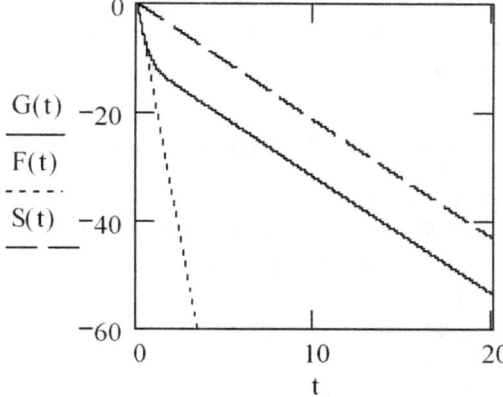

Fig. 17.2. The formation of a double decay curve G(t), which is the sum of the fast decay F(t) (dotted) and the slow decay S(t) (dashed). The decays themselves are exponential, and the vertical scale is logarithmic, in dB.

In mechanical terms, if the two string partials are slightly out of phase, the system condition changes as follows. For the small amplitudes in the system, the force exerted on the bridge by each string is proportional to the string's gradient at the bridge. For most of each cycle, the strings' gradients will be either both positive or both negative, because their phases differ only slightly. The rate of change of each string's gradient differ, because they are in different stages of a sinusoidal cycle, but during most of the cycle they will *both* be positive or negative, and the strings are thus in 'symmetric' *relative motion*. As it turns out, the gradients will only be *equal* twice every cycle, so the rest of the time the two forces on the bridge will not be equal, even when the gradients are both positive or negative.

The system also displays actual 'antisymmetric' *relative motion* between the strings. When one string reaches it maximum gradient, it then starts to decrease its gradient while the other string, lagging behind in phase, is still *increasing* its gradient. Now, during this time, the strings themselves are moving in opposite directions (but their *gradients* are nevertheless still *both* positive or negative). Their *relative motion* directions are opposite, as they would be in the antisymmetric mode. The second string then reaches *its* maximum, and then follows the first again, back in the same direction. As the overall decay progresses, the component of symmetric mode decays faster than the antisymmetric mode component, and the relative proportion of actual 'antisymmetric' relative motion between the strings in each cycle, increases, as the relative phases move apart, the phase approaching the 180 degrees *out of phase* condition of the pure antisymmetric normal mode.

Effects of an "imperfect" hammer strike – partial amplitudes

In the model, if an "imperfect" hammer strike results in the two string partials having *different amplitudes*, but otherwise equal phase and frequencies, some antisymmetric mode will be present, and the decay curve will again have two slopes. To begin with, the two strings at the bridge always move in the same direction *throughout* the cycle. Owing to the equal frequencies, the two string gradients at the bridge are always simultaneously *both* positive, negative or zero, throughout the cycle.

However this is not solely the *symmetric normal mode* itself, because the purely symmetric normal mode has *equal amplitudes* of each partial. The amplitude difference constitutes a degree of asymmetry. The mechanism by which this asymmetry persists is somewhat different to the case where it is due to phase difference.

As the symmetric normal mode decays, the string partial with the least amplitude approaches zero amplitude earlier than the second string. Because the second string is still sending incident waves to the bridge, and causing it to move, some of the energy that would have been reflected back along the second string to make its standing wave, is now passed into the first string, generating a new wave. The resultant new standing wave on the first string is 180 degrees out of phase with the standing wave on the second string. This is the normal behaviour for a driven oscillator in this situation. The two standing waves then asymptotically approach the same amplitude.

Frequency difference between the partials

In these scenarios, we have looked at the effects of phase and amplitude differences separately, while assuming the partials have equal frequency. In practice, slight differences of both phase and amplitude are likely to be present simultaneously, and perhaps even more likely will be the presence of 'mistuning' or frequency difference between the partials. When we introduce a difference in *frequencies* of the individual partials, the situation is more involved.

As it turns out, we will now not only affect the decay rate of general motion, but we can also alter the decay rate of each normal mode. Furthermore and the mode frequencies themselves, are affected. Because *frequency differences* are inevitable in the process of tuning, and because phase and amplitude differences are in any case a possibility, we need a full understanding of how the system behaves in relation to frequency differences, or 'mistuning' between the partials. This "mistuning" is what the piano tuner alters in the process of tuning.

The process of tuning involves changing the tension of one string at a time, and hence changing the frequency of any partials we may be listening to. We need a mathematical model that can display the system behaviour as a function of 'mistuning' between the isolated partials. Even when the individual, isolated partials have the same frequency, the *mode frequencies* of the coupled system can be different frequencies, which is typical of most coupled oscillator systems. Since general motion is the superposition of the mode frequencies, rather than the isolated partial frequencies, the response of the system to the action of tuning is quite different to what elementary tuning theory might predict, when the partials we are tuning are close in frequency.

Our unison system comprises two oscillators coupled by a common dissipative mechanism – the soundboard bridge. The analytical method first applied to this system by Weinreich is one of great utility (and is important in particle physics). We are by definition interested in partials that are close in frequency, and we are also dealing with a system whose normal mode frequencies will be close to the uncoupled component oscillator frequencies. We use a method based on perturbation theory.

Firstly, for one partial wave on each string we can view the string-bridge system as a system of coupled simple oscillators. To do this we define an *unperturbed* system in which the bridge is *rigid*, with one partial wave on each string, with partial number m, the partials on each string being perfectly tuned with equal frequency. There is no damping. For a string with its length along the x axis, the standing wave in y as a function of x and time t can be written in the form:

$$y(x,t) = \text{Re}\left[\sum_m C_m u(x)_m \exp(i\omega_m t - i\phi_m)\right]$$

1

The space function $u(x)_m$ is the mth eigenfunction of the single string system, with amplitude C_m. For a string uniform along its length, this eigenfunction is well known to be sinusoidal in x. This wave shape, with nodes and antinodes stationary on x, oscillates as a standing wave in time according to the time function $\exp(i\omega_m t - i\phi_m)$. The angular frequency of oscillation is ω_m and the phase angle is ϕ_m.

For a single partial wave on string k, we can rewrite Eqn (1) more simply as:

$$y(x,t)_k = \text{Re}[C_k u(x)_k q(t)_k]$$

2

Also, for the one partial we can set the phase angle to zero and put

$$q_k = A_k \exp(i\omega_k t)$$

3

where now the constant A_k can replace C_k so that we can write simply:

$$y_k(x,t) = u_k(x) q_k(t)$$

4

Now, rather than considering the displacement y(x,t) of the string, we can consider the behaviour of $u_k(x)$, which is fixed and known, as a function of time. We thus need only consider the time function $q_k(t)$ to analyse the system behaviour for one partial.

Our unison (which has two partial waves, so k = 1 or 2) can therefore be considered as a system of two simple oscillators $q_k(t)$, and we can set the unperturbed condition of the system by having identical eigenfunctions $u_k(x)$ on each string, and identical simple harmonic oscillations $q_1(t)$ and $q_2(t)$, with angular frequency ω_0. The unperturbed unison with two such oscillators is in fact a specific instance of a general system of an arbitrary number of uncoupled simple harmonic oscillators whose equations of motion are in the common form:

$$\ddot{q}_k + \omega_0^2 q_k = 0$$

5

and have the solutions:

$$q_k = A_k \exp(i\omega_0 t)$$

6

Now we must consider what happens when the bridge is not rigid, but acts as a coupling mechanism between the oscillators. The effect of the coupling mechanism in a system of oscillators is a perturbation to the otherwise simple harmonic motion of each component oscillator, due to the additional forces imposed on each oscillator by the coupling mechanism and the other oscillators. These forces are typically proportional to some series of displacements q for a 'spring like' reactive component, plus a series in $\dot{q}$ for a resistive, frictional or dissipative component, and $\ddot{q}$ for a 'mass like' reactive component.

The perturbation on oscillator k consists of replacing the zeros in Eqn. (5) with linear homogeneous functions of q_k, so that we have the set of equations:

$$\ddot{q}_k + \omega_0^2 q_k = f_k$$

7

which will have the normal mode solutions

$$q_k = \sum_n A_{kn} \exp(i\omega_n t)$$

8

The time dependency of q_k (which is now no longer simple harmonic) can be written as

$$q_k = \text{Re}(\psi_k)$$

9

It can be shown from the perturbation theory[84] that for such a system of coupled oscillators the equation of motion can be written:

$$\frac{d\Psi}{dt} = i\Omega\Psi$$

10

where Ψ is a vector of all the ψ_ks, and Ω is a *dynamical matrix* whose elements kk' are the frequency perturbation on oscillator k by the coupling mechanism and oscillator k'.

Eqn. (8) represents a system of coupled, linear equations of motion, one for each oscillator, and each one expressed in terms of the motion of each oscillator in the system.

Now to examine the generic behaviour, we do not need any more information about the ψ_ks because the solution to such a system of

---

[84] See Weinreich, *op. cit.*

equations depends on the eigenvalues of the matrix Ω. Furthermore, when the solutions of such a set are oscillations, the eigenvalues control the frequencies of oscillations of the system, which are the normal mode frequencies. To begin examining the system all we have to do is 'normalise' the matrix, setting the various sizes of its elements as proportions of unity (or in the case of imaginary parts, i) and we can see the dependency of the system upon various coupling or perturbation parameters. These parameters in our system are determined by:

(1) 'Mistuning', *i.e.* frequency difference between the isolated, uncoupled partials.
(2) The 'configuration' of the bridge-soundboard system as the coupling mechanism.

The 'configuration' of the bridge falls into two parts, a resistive (dissipative) part and a reactive part. The resistive part of the bridge coupling is the part of the bridge-soundboard system that drains energy from the strings (for amplification by the soundboard). There will also be a smaller part of this which is inefficient dissipation caused by friction *etc*. The reactive part falls into two areas – a 'mass like' or inertial component due to the effective mass of soundboard and bridge, and a 'spring like' component due to the elastic properties of the bridge-soundboard. When the incident partial waves on the string encounter the bridge, they encounter the admittance properties of the bridge - the bridge *can* move, but at the same time it *impedes* movement. This impedance at any moment depends not only on the physical bridge-soundboard properties, but also on the behaviour of the strings attached to it, which are exerting forces on it.

Any bridge movement causes a perturbation to the frequency ω_0 that would otherwise have been present had the bridge been rigid. Quantitatively, the perturbation caused by the bridge admittance for one partial is given by the well known relation:

$$\zeta = i\omega_0 Z_0 Y / \pi$$

11

where ω_0 is the unperturbed partial frequency associated with a rigid bridge, Z_0 is the wave impedance (characteristic impedance) of the string for that partial, and Y is the bridge admittance. The perturbation has both 'resistive' and 'reactive' parts, causing both real frequency change and decay. Since ζ is complex we can write it as $\xi + i\eta$. The real part of bridge admittance ranges from −1 to +1. A positive value would correspond to a 'mass like' or inertial bridge property, which acts to raise frequencies. A negative value would correspond to an elastic bridge, which acts to lower frequencies. A zero real part would indicate no reactive component of the bridge – this corresponds to a rigid bridge whose impedance (the reciprocal of the admittance) would approach infinity. In practice the bridge has both inertial and elastic properties, but overall the frequency shift from ω_0 must be either positive or negative corresponding to a dominantly 'mass like' or 'spring like' bridge condition.

The imaginary part of ζ can only be positive, ranging from 0 to i, and corresponds to the damping effect of the bridge, with no appreciable change to the frequency ω_0. We set the decay rates relative to the single string partial decay rate, designating this decay rate as $\eta = 1$.

For partials that are not perfectly in unison (the same frequency) we designate a relative (angular frequency) 'mistuning' of 2ε as the 'mistuning' perturbation to the uncoupled angular frequency ω_0 of the strings. String 2 is given the angular frequency ω_0 and string 1 the angular frequency $\omega_0 + 2\varepsilon$. (We follow Weinreich's use of a term 2ε rather than designating the whole 'mistuning' difference between the partials as ε. This is consistent with the approximation technique for producing Ω). For a rigid bridge condition with the mistuning as the only perturbation we would then have:

$$\Omega = \begin{pmatrix} 2\varepsilon & 0 \\ 0 & 0 \end{pmatrix}$$

12

Because the bridge is not rigid, the partials are coupled through it and we must add the corresponding perturbations. The strings are fixed side by side on the bridge so the perturbations due to the bridge motion are the same on all four elements of Ω. We therefore obtain:

$$\Omega = \begin{pmatrix} 2\varepsilon + \zeta & \zeta \\ \zeta & \zeta \end{pmatrix}$$

13

The behaviour of the eigenfrequencies in relation to mistuning now depends upon the value of ζ. In understanding how the system behaves, it is useful to first observe its behaviour under the extreme *idealised* conditions that the bridge coupling is either wholly resistive or wholly reactive. A real piano bridge-soundboard-string system is an elastic structure as already discussed, and of course it also has mass, so it will actually contribute a reactive component in addition to dissipating energy from the strings. Nevertheless, a good understanding of the actual behaviour is assisted by looking at the behaviour under these extreme conditions, because the behaviour under actual conditions will be somewhere between the two.

Purely resistive idealised bridge

Let us first consider what happens if the bridge is purely resistive. In this case $\xi = 0$ and $\eta = i$. Figs (17.3) and (17.4) shows the resultant behaviour of the real and imaginary parts of the eigenfrequencies of Ω, as a function of mistuning ε. The imaginary part of the frequencies corresponds to the decay rate.

Fig. 17.3

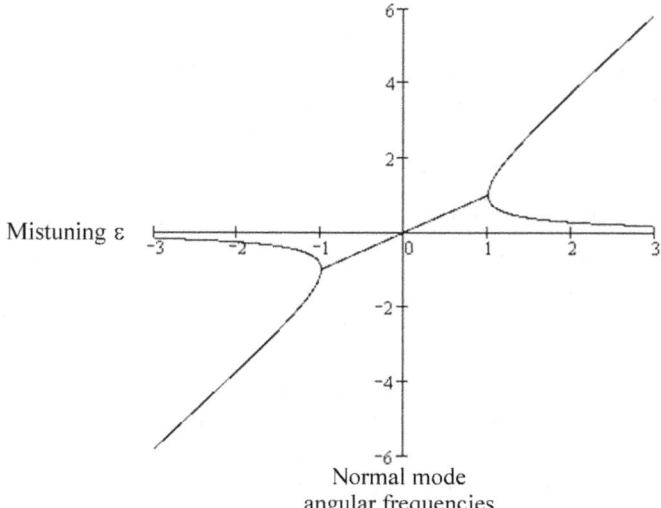

Normal mode angular frequencies

Fig. 17.3. The real parts of the eigenfrequencies of Ω, for a purely resistive bridge, when $\xi = 0$ and $\eta = 1$. The graph shows the variation of angular frequencies of the two normal modes of the partial unison, against angular mistuning ε (half the total mistuning). When ε is smaller than ± 1 (the unit being the single string partial decay rate in the absence of the other partial) the real parts of the eigenfrequencies are the same. General motion will then be a superposition of identical frequencies and there will be no beat pattern.

When both strings of the unison are sounded, it is the coupled system's normal modes that determine the bridge motion. We can see from Fig. (17.3) that when ε is between −1 and +1 the real mode frequencies coalesce and are the same. The total mistuning between the partials is 2ε and the graph unit is the single partial decay rate as it would be in the absence of the other partial. Thus, we see that if the total 'mistuning' (2ε) is equal to, or less than, twice the single partial decay rate, then the normal modes share the same frequency and there will be no beat pattern despite the remaining 'mistuning'. We do not have to have *perfect* unison (equal partial frequencies) to eliminate the beat pattern in the general motion decay curve.

The beat pattern that *will* occur in the unison when the 'mistuning' is greater than twice the single string partial decay rate, will not be a heterodyne beat as in simple 'traditional' tuning theory. It *is* similar to a heterodyne interference pattern, because like a heterodyne beat, it *is* a superposition of frequencies. Nevertheless it is actually a display of general motion of a coupled system, whose superposed components are the system's normal modes, and not the individual string's partials. Further outside the range in which the mode frequencies coalesce, the normal mode frequencies *do* approach the individual partial frequencies.

Fig. (17.4) shows the imaginary parts of the eigenfrequencies. These are their decay rates. Here, when ε is smaller than plus or minus one - the same points at which the real mode frequencies coalesce - the decay rates of the frequencies start to differ. *Inside* this region (the circle in Fig. 17.4) the general motion decay of the partial unison will be a superposition of *two* decay rates for any value of ε, and hence will exhibit the kind of beatless (because the real parts have coalesced) double decay curve we saw above. On the other hand, *outside* the circle only one decay rate exists. On superpositioning here, the frequencies outside the circle would have a exponential *overall* decay curve, but would exhibit a beat pattern due to the difference between the real parts of the frequencies.

When mistuning is exactly zero, the decay rate of one eigenfrequency is 2 units, or in other words twice the single partial decay rate. This is the decay of the symmetric mode proper. The other eigenfrequency has no decay, which corresponds to the antisymmetric mode. Only if the mistuning was exactly zero, *and* there were no initial amplitude or phase differences in the component partials, could such an exact normal mode occur. In practice, because in the piano unison system motion is initiated with a hammer strike, the initial amplitude of the antisymmetric mode would then be zero, and only the symmetric mode would occur. This would lead to an exponential decay curve, at twice the decay rate of the single partial.

Fig. 17.4

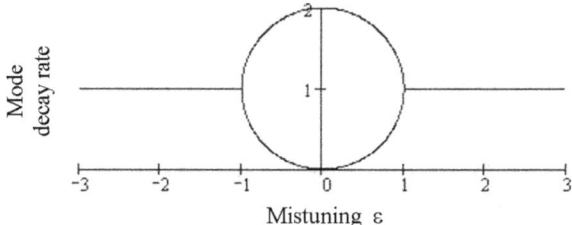

Fig. 17.4. The imaginary parts of the eigenfrequencies of Ω, for a purely resistive bridge, when $\xi = 0$ and $\eta = 1$. These constitute the decay rates applied to each real frequency. For mistunings larger than twice the twice the single partial decay rate, *i.e.* for ε greater than +1 or less than −1, there is one decay rate of the system, equal to the single partial decay rate, so the decay curve would be exponential as it would be for uncoupled strings. For mistuning less than this, there are two decay rates, so the system will generally exhibit a 'double decay curve'. At zero mistuning the decay curve when only the symmetric mode is initiated, would be exponential, at twice the single partial decay rate.

Purely reactive idealised bridge

At the other idealized extreme is the behaviour of the system when the bridge is wholly reactive, and does not dissipate energy. In this case,

$\xi = 1$ and $\eta = 0$. Fig. (17.5) shows the corresponding behaviour of the eigenfrequencies which are now real only, and do not decay.

Fig. 17.5

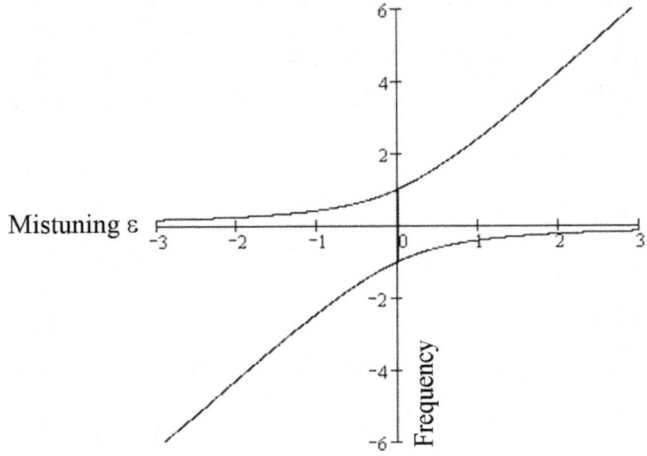

Fig. 17.5. The eigenfrequencies of Ω (with ξ subtracted, to centre the graph on the origin) when the bridge coupling is purely reactive, with $\xi = 1$ and $\eta = 0$. The units are the single partial decay rate. Now the matrix Ω and its eigenfrequencies are real only, so there is no decay. However the difference between the eigenfrequencies cannot be eliminated, even when mistuning is zero.

At this extreme of a purely reactive bridge coupling with no decay, there are *always* two different eigenfrequencies, even when the mistuning is zero. The general motion of the system will therefore always exhibit a beat pattern in the superposition of the its mode frequencies.

Realistic bridge – both resistive and reactive

A more realistic model must have a bridge coupling that is both resistive and reactive. As we might expect, the behaviour of the system is then somewhere between the two idealized extremes. Figs (17.6) and (17.7) show the behaviour of the eigenfrequencies for various combinations of reactive and resistive components in the bridge coupling.

Again, any reactive component, even when very small, means that there will always be some difference between the two eigenfrequencies.

Fig. 17.6

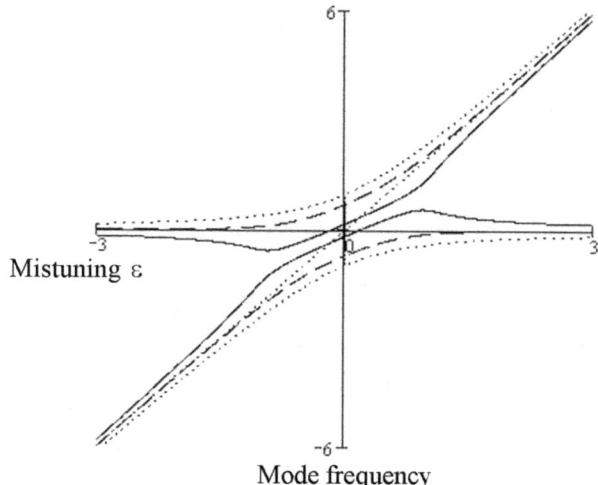

Fig. 17.6. Behaviour of the two mode frequencies with mistuning ε, for various combinations of resistive and reactive bridge properties: Outer dotted curves: $\xi = 0.99$ $\eta = 0.17$; Dashed: $\xi = 0.71$ $\eta = 0.71$; Solid: $\xi = 0.17$ $\eta = 0.99$; Straight dotted through origin: $\xi = 0$ $\eta = 0$. This is the behaviour with no coupling or a rigid bridge. The second eigenfrequency in this case follows the horizontal axis.

Fig. 17.7

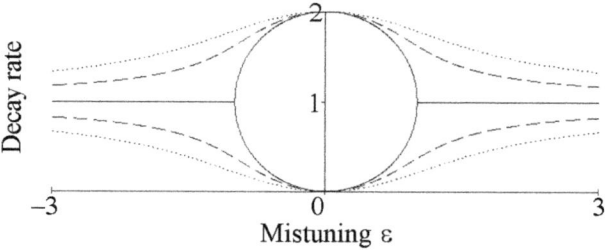

Fig. 17.7. Decay rates of the eigenfrequencies for various combinations of resistive and reactive bridge properties: Dotted: $\xi = 1$ $\eta = 1$; Dashed: $\xi = 0.5$ $\eta = 1$; Solid: $\xi = 0$ $\eta = 1$

We need to interpret the significance of these results in relation to tuning in practice. We saw that for a purely resistive bridge, as mistuning decreases, the real mode frequencies completely coalesce at $\varepsilon = \pm 1$ and two separate decay rates suddenly appear at the same degree of mistuning. When the bridge is reactive and resistive, the changes are not so abrupt, but the deviation from uncoupled behaviour does still significantly change by the time $\varepsilon = \pm 1$. For the time being we can generally regard ± 1 as a mathematical *critical mistuning* value for ε.

The 'unit' is the single partial decay rate, so the actual frequency difference between the isolated partials at this critical point, will vary from partial to partial. The decay rate of a note in the mid-range of the piano will obviously depend on the individual instrument and to some degree on the individual string. We can take a conventional *decay time* as the time taken for the amplitude to drop 60 dB, that is, to one millionth of its initial value. If we take the overall note decay rate as a limit to the possible decay time for any partial, then for a reasonably large grand piano we might be dealing with an overall decay time of around 20 seconds, but it could be more or less than this. Any individual partial in the spectrum may be faster decaying than the overall note decay rate. Fig. (17.8) shows the total *critical mistuning* in Hz ($= \varepsilon / \pi$) in relation to decay times in seconds.

Fig. 17.8

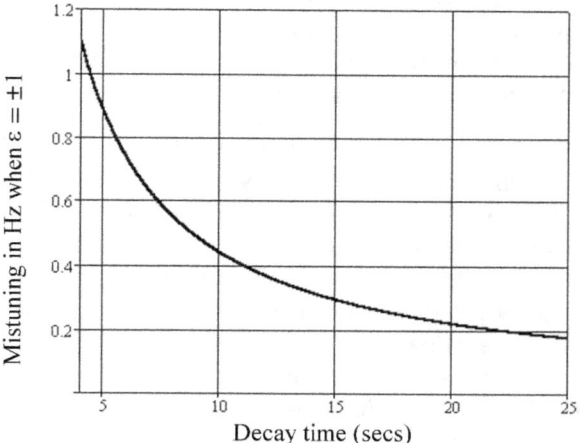

Fig. 17.8. Total *critical mistuning* in Hz (the frequency difference when $\varepsilon = \pm 1$) relative to decay time in seconds (time for amplitude to drop to initial amplitude X 10^{-6}). For a total mistuning M and a decay time T, the relationship is:

$$M = \text{Log}_e 10^6 / \pi T$$

For partial decay times typical of the mid-range the critical mistuning is small. For example a decay time of 20 seconds gives a decay rate of 0.691 and a *critical mistuning* frequency difference of 0.22Hz. We are dealing here with unison tuning, so this is not insignificant. In the case of the superposed spectrum of a unison there is a partial meeting at every partial number. If the mistuning at the fundamental is say, 0.2Hz, then the mistuning at say, the sixth partial, will be 1.2Hz (or more with inharmonicity).

If we are aiming at removing beat movement from the unison then clearly the tuning tolerance of the fundamental will be inside the *critical mistuning* range. We must now remember that (1) the decay time of the partial may be much shorter than 20 seconds, giving a wider critical mistuning range, and (2) the *critical mistuning* of $\varepsilon = \pm 1$ is merely an obvious *mathematically* significant point, and in practice coupling may have other effects significant to tuning, well outside the critical mistuning range. Later, we will deal in detail with what actually happens in fine unison tuning, and we shall see how mode coupling is important in ways that are not yet apparent. It is also worth mentioning here that if we were dealing with *tempered intervals*, the strings involved would rest at *different* points on the bridge, so the configuration of matrix Ω will be different. We would also be dealing with minimum 'mistunings' in the order of 0.5Hz (for the tempered fifths). We will deal with this later.

We can show the functional dependence of the mode frequencies on 'mistuning' in a way that relates directly to the practical task of tuning beats. In the following graphs Figs (17.9), (17.10) and (17.11), the vertical axis shows the frequency difference between the normal modes, versus the 'mistuning' in Hz on the horizontal axis. The 'mistuning' axis is effectively a scale representing a range of tensions over which one string is adjusted close to unison. The 'mistuning' value at any point is the frequency difference we would find between the partials of each string (of one given partial number), if they were sounded separately.

Negative values would correspond to the tuned string being 'flat' to the fixed one, and positive values, sharp. Any value will be proportional to the square root of the string tension (ignoring inharmonicity). The vertical value corresponds to a possible 'beat frequency' so we can see how this 'beat rate' behaves as one string is tuned, altering the 'mistuning'. This is a generalised indication, however, because as we shall see, the idea of a heterodyne-like 'beat rate' is not always an appropriate description of how the system behaves in response to a frequency difference between the modes.

Fig. 17.9

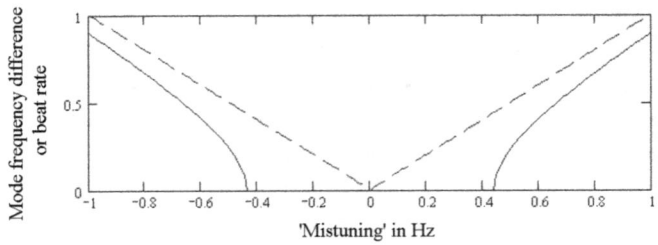

Fig. 17.9. Behaviour of the system for an idealised (unrealistic) bridge that is purely resistive, when $\xi = 0$ and $\eta = 1$, for partials with isolated, individual decay times of 20 seconds. The beat rate drops to zero at a 'mistuning' of

around 0.45 Hz, where the mode frequencies coalesce. The dashed line shows the beat rate that would occur without the effects of bridge coupling.

Fig. 17.10

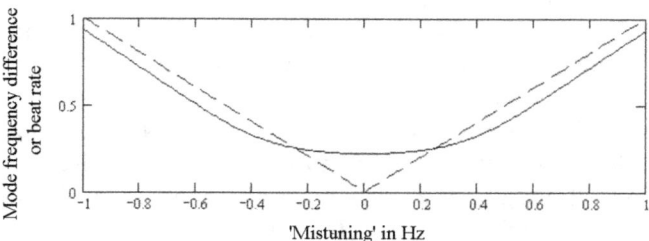

Fig. 17.10. Behaviour for a more realistic bridge, both resistive and reactive, when $\xi = 0.5$ and $\eta = 1$, for a single partial decay time of 20 seconds. For 'mistunings' of less than around 0.3Hz the mode frequencies are approximately the same. The dashed line shows the behaviour without the effects of bridge coupling.

Fig. 17.11

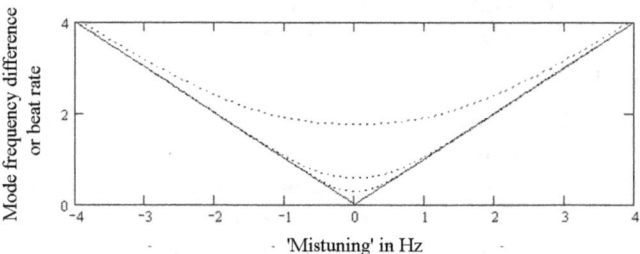

Fig. 17.11. Behaviour for a bridge with equal resistive and reactive parts. The solid line is the behaviour without the effects of coupling. Upper dotted line is for a decay time of 5 seconds, middle dotted for a decay time of 15 seconds, and lower for a decay time of 30 seconds.

The existence of a difference between the mode frequencies does not necessarily indicate a regular *beat rate* because the shape of the decay curve depends also on the mode decay rates, which for a bridge with a reactive component may differ at the same time as the mode frequencies.

Decay curves

The most relevant thing to the tuner is the activity at the various partial meetings – the beats, or the shape of the decay curves in time. When tuning a unison between two strings we are in general listening to the combined spectra of both strings, in which there will be a great deal going on simultaneously. When we begin with the unison 'out of tune' and beating regularly, the overall complexity of beating will be a superposition of all the beat rates occurring at all the adjustable partials in the combined spectrum. There will be many beat rates simultaneously sounding. The tuner can of course listen to one partial (one beat rate), or more than one simultaneously. Ultimately the tuner must listen to the *whole spectrum* of the unison because it is the whole spectrum that constitutes the note heard by the musician, and it is the quality of the whole that counts.

Nevertheless actual beat patterns (or other irregular patterns) occur in individual partials, so we need to examine the behaviour of the individual partial. The first thing to consider is the formal nature of beat patterns. Weinreich's model, even though restricted to waves in only one plane, immediately throws different light on the notion of 'beat phenomena' to that indicated by traditional theory. Rather than there simply being either regular equal beats, or no beats to consider, there is a larger range of characteristic decay patterns that are recognisable.

Fig. (17.12) shows a beat of 1 per second as we might expect it from traditional theory, assuming no effects of bridge coupling, and a beat caused by heterodyne interference. There is an overall exponential decay, and the only significant property of the beats with which the traditional theory deals is the beat frequency.

Fig. 17.12

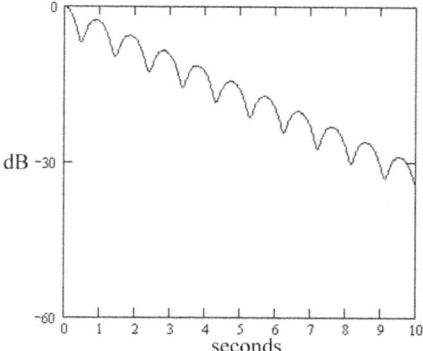

Fig. 17.12. Example of a beating decay pattern with a beat rate of 1 per second. All the beating decay curves of traditional theory represented in this way are either exponential (straight lines where there is no beat) or similar to this beat pattern, but with perhaps a different beat rate and overall decay rate.

As the mistuning approaches zero in this simple traditional model, the beat humps simply get longer in time (wider on the horizontal axis) as they get slower, until a straight line is reached at zero mistuning. In the Weinreich model taking into account bridge coupling, the behaviour will be quite different. The first difference to draw attention to is that regular beats have *two* significant properties, not just one. There is the beat *frequency*, with which we are familiar, but there is also the beat *amplitude* (the vertical 'height' of the beat). In the above example Fig. (17.12) the beat amplitude is more or less constant throughout the five seconds of decay shown. Fig. (17.13) on the other hand shows a typical beating decay curve over 10 seconds, when bridge coupling is taken into account.

Fig. 17.13

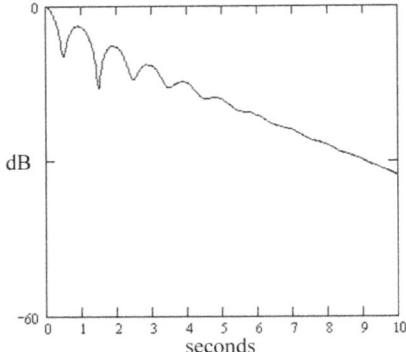

Fig. 17.13. Example of a beat pattern with a beat rate of 1 per second, when bridge coupling is taken into account. Here, the bridge admittance has equal resistive and reactive parts. The decay rate of the beat is different to the overall unison decay rate.

Because the bridge coupling affects the individual mode decay rates, the beat amplitude may decay *relative* to the overall unison decay rate. In this example the beat amplitude grows before it decays, and then decays faster then the overall unison decay rate. After about 5 seconds the overall unison decay is more or less beatless.

In tuning practice, the first three seconds decay is more than adequate to establish a beat rate, so in this example the predictions of traditional theory are not unduly disturbed. However, the middle section of the decay curve here, say from 4 to 7 seconds, is important because it demonstrates that there is such a thing as a 'shallow' beat (a beat with relatively small amplitude) compared to the beginning of the decay curve. This is enough to show that beat *amplitude* is significant, as well as beat frequency, not least because a normally significant beat frequency may be present, but rendered insignificant or inaudible because of its diminished amplitude. This is important, although we will not encounter in the Weinreich model a decay with such a diminished beat amplitude throughout the overall unison decay curve.

Decay curves for an idealised, purely resistive bridge

We can now compare the unison decay curves at various degrees of mistuning, for various bridge admittance properties. Firstly, the idealised case of a purely resistive bridge with no reactive component, will give an insight into the effects of the resistive element of bridge coupling, by showing the behaviour under conditions at this extreme end of the range of all possible bridge properties. Fig. (17.14) shows computer generated decay curves of general motion for a decreasing series of 'mistunings' between the isolated partials of the unison. In this and the following illustrations the partial amplitudes are equal.

Fig. 17.14

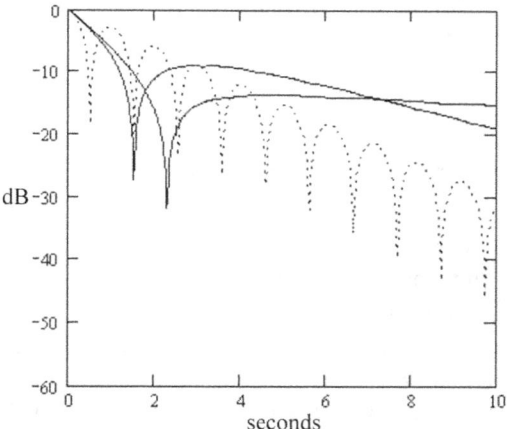

Fig. 17.14. Decay curves for a unison at one adjustable when the bridge is purely reactive. The overall decay time of each isolated partial is 20 seconds (time to decay 60 dB). The dotted curve is for a 'mistuning' of 1Hz, and shows a regular beat pattern. The curve with a single null at about 1.5 seconds and the steeper of the two aftersound decays, is for a mistuning reduced to 0.2Hz. An even smaller mistuning of 0.1Hz yields the later, single null just after two seconds, and a very slowly decaying aftersound.

At a 'mistuning' of 1 Hz there is a regular beat pattern, whose amplitude is constant, but the beat rate is slightly less than 1 beat per second – the regular beat rate will always be slightly less than that predicted by 'traditional' theory without the effects of bridge coupling. The overall decay rate of a regular beat (that will always occur when $|\varepsilon| > \eta$) is the same as that for a single partial, and is unaffected by bridge coupling. As the beat rate in the partial unison is reduced however, a point in fine tuning is reached (when $\varepsilon = \pm \eta$) at which a regular beat pattern is replaced by single 'beat' consisting of a prompt initial decay dropping into a single null, followed by the rising of a distinct aftersound, which has a much slower decay rate. As the mistuning gets smaller, the initial level of the aftersound gets less, and the single null 'beat' gets later. As 'mistuning' tends to zero, the decay rate of the aftersound tends to zero, but its amplitude also tends to zero. At zero 'mistuning' the decay curve is a straight line on the logarithmic scale, with a decay rate twice that of the single partial. The initial decay rate at $t = 0$ is always twice the single partial decay rate due to the phase coherence of the partial waves on the two strings, initially emulating the symmetric mode. This effect has also been discussed by Benade.[85]

Decay curves for a realistic bridge, both resistive and reactive

As we have said, the actual piano bridge boundary will be both resistive and reactive. The introduction of a reactive component to the bridge admittance has two important effects. Firstly, the beat amplitude of a regular beat is no longer constant. The beat amplitude will in general decay as the overall unison decays. Secondly, the degree of 'mistuning' at which the regular beat pattern begins to transform to an irregular decay, is increased. We can see from the graphs showing the real and imaginary parts of the matrix eigenfrequencies that this is likely to happen. For a purely resistive bridge there is are clear 'critical tuning' points where $\varepsilon = \pm \eta$ at which the real frequencies coalesce, and the imaginary ones split into two separate decay rates. Where there is a reactive component however, such a point of 'critical tuning' is 'blurred'. Fig. (17.15) shows three decay curves for progressively smaller 'mistunings', from 0.5Hz to 0.1 Hz, when the bridge is equally resistive and reactive.

---

[85] Benade, AH, *Fundamentals of Musical Acoustics*, New York, 1976, 1990, p. 336.

Fig.17.15

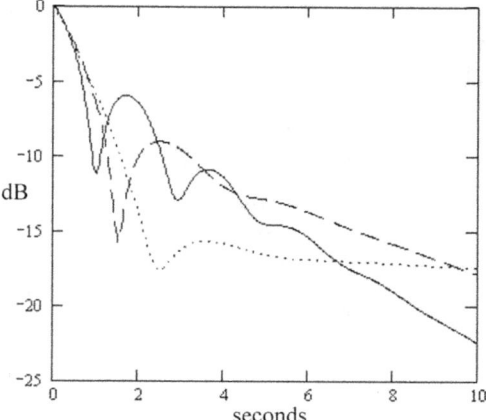

Fig. 17.15. Three decay curves for a unison between partials when the bridge is equally resistive and reactive, for 'mistunings' progressively closer to zero. Solid = 0.5 Hz, dashed = 0.3 Hz, dotted = 0.1 Hz. The single partial decay time is (time to decay 60 dB) is 20 seconds.

The precise form of an audible decay curve will depend on the partial decay rates and the configuration of the system's conversion of bridge motion to transmitted sound energy. Nevertheless there are distinct generic behavioural patterns to be noted. The following sequence of figures illustrates some variations in the decay representation with different system configurations.

Fig.17.16

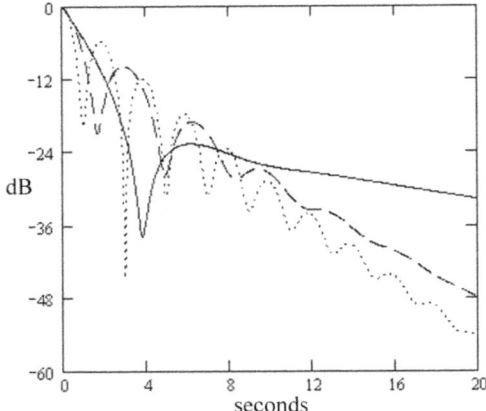

Fig. 17.16. Decay curves of a theoretical rate of energy transmission proportional to $|\psi_1 + \psi_2|^2$ to an equally resistive and reactive bridge, for an increased decay time of 40 seconds, and mistunings dotted = 0.5 Hz, dashed = 0.3 Hz, solid = 0.1 Hz.

Fig.17.17

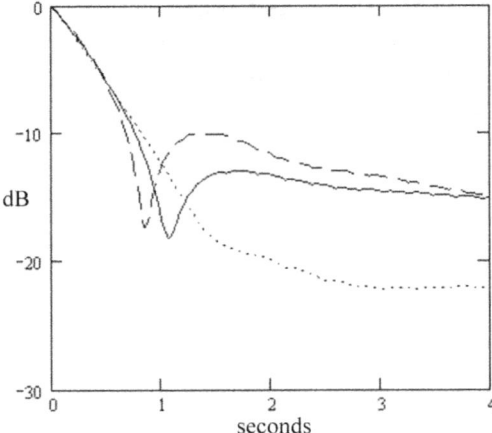

Fig. 17.17. Decay curves of general motion amplitude of an equally resistive and reactive bridge when the decay time is reduced to 10 seconds, for

mistunings dashed = 0.5 Hz, solid = 0.3 Hz, dotted = 0.1 Hz.

Fig.17.18

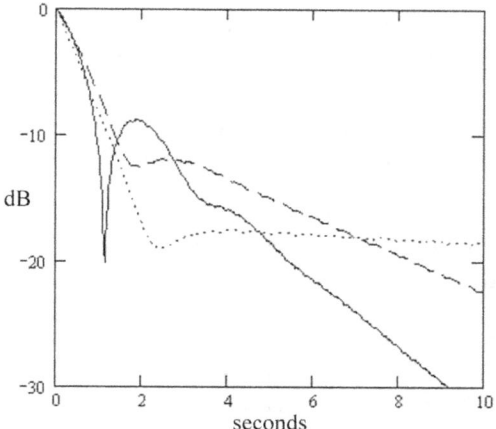

Fig. 17.18. Decay curves of general motion amplitude for a bridge with a reactive part half the magnitude of the resistive part, when $\xi = 0.5$ and $\eta = 1$, for mistunings solid = 0.5 Hz, dashed = 0.3 Hz, dotted = 0.1 Hz. The decay time is now 14 seconds, giving a decay rate of approximately $k = 1$ in the decay rate $\exp(-kt)$, or aprox. - 4.3 dB/s.

Generic decay patterns for two string partial unisons

It is important to grasp the generic features of the Weinreich model behaviour, because they persist as basic patterns in more complicated models. These generic features can be summarised as follows.

At larger mistunings outside the 'critical tuning' region, only a bridge that has no reactive component yields beats whose amplitude remains constant throughout the overall decay. Such a bridge is of course a theoretical ideal. A reactive component as little as a twentieth of the resistive component results in a clear beat amplitude decay over 10 seconds. Weinreich suggested that the reactive component is probably closer to being equal to the resistive component. A real bridge that is both resistive and reactive gives rise to beats that decay in amplitude over the overall decay time, the effects of this being stronger the smaller the mistuning.

In general there will be a somewhat imprecise 'critical' mistuning magnitude, typically somewhere less than half a beat per second for a mid-compass decay rate, that roughly separates regular beat behaviour from strongly coupled behaviour. When we reach the latter in the process of 'closing' the unison, there is typically one prominent beat followed either by a steady decay, or perhaps by another much shallower, less prominent beat before a steady decay.

There is always a dual overall decay rate in this region, even where there is no appreciable beat, except in the absolute case of zero mistuning, *and equal initial conditions* for each partial. This latter condition would very easily violated by any 'hammer irregularity' or other perturbing factor in the physical setup that introduces a small amount of antisymmetric mode even when mistuning is zero. Figs (17.19) and (17.20) show an effect of such a 'hammer irregularity'.

Fig.17.19

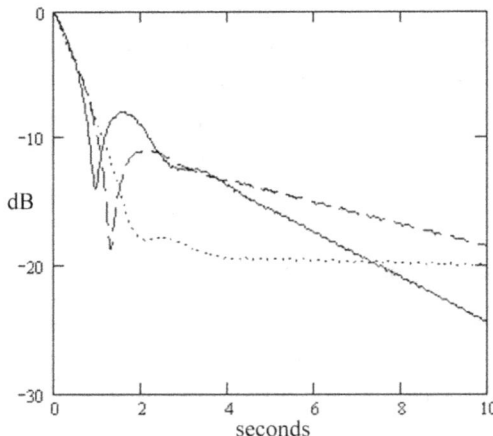

Fig. 17.19. Partial unison decay curves for equally resistive and reactive bridge, with a partial decay time of 14 seconds. Mistuning values are solid = 0.5 Hz, dashed = 0.3 Hz, dotted = 0.1 Hz.

Fig.17.20

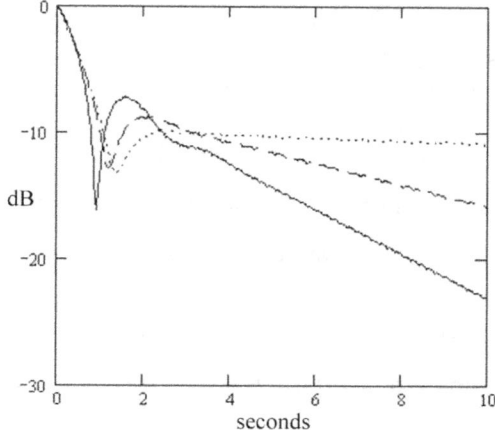

Fig. 17.20. The same unison conditions as above but with a 'hammer irregularity' of 0.1 introduced.

With or without "hammer irregularity" the "aftersound" part of the decay curve is a strong function of mistuning when the mistuning is small enough for the system to exhibit strongly coupled behaviour.

Inside the "critical mistuning" region it is not only the aftersound decay rate, but other factors such as the initial 'single beat' amplitude, the position of the beat in time, and any residual beat amplitude decay rate, that are the characteristics strongly dependent on mistuning. It is tempting to think of unison fine tuning in terms of the mistuning as it approaches zero in the model we have been examining, but it must be remembered that the piano string unison itself is not a single partial unison, but a very large series of such unisons, each with its own system configuration in terms of partial decay rate and bridge admittance. The bridge admittance is in general a function of frequency so we can expect it to differ for each partial.

Tuning in practice

The unison as a whole is a *set of partials*, not a single partial. Different partials will have different initial amplitudes, and the initial *individual partial amplitudes* will not necessarily all be the same. In the spectrum of a two string unison, there is energy from both strings contributing at every partial. In tuning other intervals, energy from only one string creates each peak in the spectrum where there is a non-adjustable partial. We might therefore generally expect the adjustable partials to be more prominent in the spectrum than the non-adjustable partials where there is no meeting. This is not of course an inviolable rule, but a generality. The audible *prominence* of an adjustable partial in the unison when mistuning is small, will at any time depend upon four inter-related factors:

1) The overall amplitude of the partial relative to other partials.
2) The rapidity of a beat rate or non-exponential movement in the decay curve.
3) The "beat" amplitude in the decay curve.
4) The position of the partial in the spectrum.
    1. The amplitude of the decay curve relative to the other partials, at any given time, is dependent on the relative overall decay rates and changing decay curve shapes of the partials.
    2. Within the bounds of fine tuning, the more rapid the beat or non exponential movement, the more prominent will be the partial.
    3. The greater the "beat" amplitude, the more prominent the "beat" will be.
    4. The position of the partial in the spectrum is important because of possible "masking". This is the psychological effect of a sound being "hidden" by another sound, and with regard to hearing partials, has been noted by Helmholtz and others. A partial frequency may especially be masked by another sharing (approximately) the same musical note name, or another that bears a musical relationship of a fifth or compound fifth to the first.

The 'mistuning' at any partial will be approximately the mistuning at the fundamental multiplied by the partial Number, so mistuning gets proportionally greater as we move up the spectrum to higher partials. We cannot just satisfy the demands of one adjustable partial and simply ignore the others. Whilst, say, the fundamental may be well inside the "critical mistuning" area and beatless, higher partials whose mistunings are multiples of the fundamental's, may still be beating, and their relative amplitudes may be more prominent.

The artist tuner's job in unison fine tuning is to adjust the quality and tone of the unison, and like any complex tone the quality of the piano unison is very much dependent on the high frequency elements of its spectrum. Since the partial decay rates do not steadily diminish with increasing partial number (roughly speaking, if there is any pattern at all the decay rates overall tend to increase, though not steadily), and since the greatest residual mistuning will always occur in the highest partials, prominent high partials can be very important components of the unison, that most definitely require the attention of the tuner.

Chapter 18 - Two strings, two planes

Real piano strings vibrate in two transverse planes even though struck in only one plane. Even a single string sounding alone will behave in this way, suggesting the presence of two orthogonal transverse normal modes. If modes in only one plane were present for each partial, then following a strike no orthogonal transverse motion could occur. Spectrum analysis of piano tone often reveals more than two frequency peaks for a single partial, but here we will consider the case where a partial may have just two transverse normal modes contributing to it. We can conjecture that in the actual situation there may be more complex mechanisms at work.

Consider two strings with two normal transverse planes in which waves can propagate. Considered as an extension of the Weinreich one-plane model, the situation at the bridge is now one in which we are dealing with four coupled oscillators, and a system with four normal modes. The configuration at the boundary of each individual string is as for the single string with two planes of motion. If both isolated strings have Cartesian mode axes that share the same alignment, or if only one string has Cartesian mode axes, then we can consider the 'vertical' oscillators and 'horizontal' oscillators as two normal pairs, since there will be no effect of 'vertical' on 'horizontal' and vice versa.

The resultant output may, however, be a superposition of both, if the "output" from the bridge through the soundboard system does not share the same alignment. We would thus expect two pairs of normal modes similar to those in the Weinreich model, except that one pair will be 'horizontal'. If both strings have Cartesian mode axes, but the axes alignment at the boundary of the strings differ, then relative to one set of axes the other string's modes will be superposed. We will examine in detail the case when only one string has Cartesian mode axes for the one partial number. This represents a situation where the partial of only one string is false. We would normally expect the strike axis and "antennae" axes (the axes for radiation through the soundboard) to be approximately or exactly coincident (and the same for both strings), because the hammer strikes perpendicular to the soundboard surface.

We define the unperturbed coupled system as having isotropic reactive boundary conditions for both strings. The resistive part is however anisotropic, the 'horizontal' being less than the 'vertical' (again because of the soundboard design and structure). Empirical results suggest that the 'horizontal' and 'vertical' bridge admittances are normally about the same in their reactive component, but for the resistive component a 'horizontal' value about a quarter of the 'vertical' would be reasonable.[86] This is only a rough relationship, and for simplicity we will take the resistive components when measured in line with the individual string mode co-ordinate axes, to be in this same ratio, even though the mode axes may in general be inclined to any resistive "antennae" axes that may exist.

We allow for two isolated string modes per string, for the one partial number we are representing, which have their mode co-ordinates in Cartesian co-ordinate directions. We number these such that 1 = 'horizontal' mode of string 1, 2 = 'vertical' of string 1, 3 = 'vertical' of string 2, and 4 = 'horizontal' of string 2. (These 'vertical' and 'horizontal' directions are aligned with the Cartesian normal mode directions for the isolated strings – in this case there cannot be more than one set of directions, so either only one string is false, or both have the same alignment of normal mode axes). As before, element mn of the matrix is the perturbation on oscillator m by the bridge and oscillator n. If the bridge and individual boundaries were all rigid and the only perturbation was a mistuning 2ε between the strings, we would have for the complete matrix:

$$\Omega = \begin{pmatrix} 2\varepsilon & 0 & 0 & 0 \\ 0 & 2\varepsilon & 0 & 0 \\ 0 & 0 & 0 & 0 \\ 0 & 0 & 0 & 0 \end{pmatrix}$$

---

[86] Weinreich, *op. cit.*

When we take into account the bridge's overall 'horizontal' and 'vertical' admittance, allowing for different values in the 'horizontal' and 'vertical' directions, the matrix now becomes:

$$\Omega = \begin{pmatrix} B_h + 2\varepsilon & 0 & 0 & B_h \\ 0 & \zeta + 2\varepsilon & \zeta & 0 \\ 0 & \zeta & \zeta & 0 \\ B_h & 0 & 0 & B_h \end{pmatrix}$$

2

where ζ is the same complex 'vertical' admittance perturbation as in the one plane Weinreich model, and B_h is the corresponding value in the 'horizontal' direction.

Next, we allow for localised perturbations to the 'horizontal' motions at each individual string boundary, which can occur as a physical condition of the side-draft arrangement on the bridge. We designate these H_1 for string 1 and H_2 for string 2. Similar perturbations local to the vertical motion of the individual string may seem less likely, but are still theoretically possible. These, however, would have the same relative effect on the matrix as altering the value of the mistuning ε, together with the values of H_1 and H_2, so we need not actually include them in order to examine the general behaviour of the system. With the localised horizontal perturbations included the matrix becomes:

$$\Omega = \begin{pmatrix} B_h + 2\varepsilon + H_1 & 0 & 0 & B_h \\ 0 & \zeta + 2\varepsilon & \zeta & 0 \\ 0 & \zeta & \zeta & 0 \\ B_h & 0 & 0 & B_h + H_2 \end{pmatrix}$$

3

The position of the zeros in this configuration mean that there is no effect of the horizontal string modes on the vertical ones, and *vice versa*, - in other words our vertical and horizontal axes are aligned with the individual Cartesian normal mode co-ordinates of the string(s). As we have said, either both strings are false (*i.e.* they have anisotropic reactive boundary conditions) and have their individual normal mode axes aligned together, or only one string is false, depending on whether H_1 and H_2 are non zero.

The eigenfrequencies as a function of ε can be found relatively easily by computer. As we already suggested, we set the vertical bridge admittance as equally resistive and reactive, and set the horizontal bridge admittance with the same size reactive component, but only a quarter of the vertical resistive component. This is generally in line with the empirical results found by Weinreich. Thus $\zeta = \xi + i\eta$, $\xi = 1$ and $\eta = 1$, but $B_h = \xi + 0.25\eta$. We can then test the response of the system for various perturbations H_1 and H_2 to the horizontal oscillator frequencies, representing localised anisotropy in the bridge admittance.

As was the case in the Weinreich model, the system's normal modes are symmetric and antisymmetric, but in this case they are both horizontal and vertical. Thus there are two symmetric modes, one vertical and one horizontal, and similarly, two antisymmetric modes. In the hypothetical case of a purely resistive bridge the real and imaginary parts of the eigenfrequencies appear in two corresponding pairs, recognisable from the one-plane model:

Fig.18.1

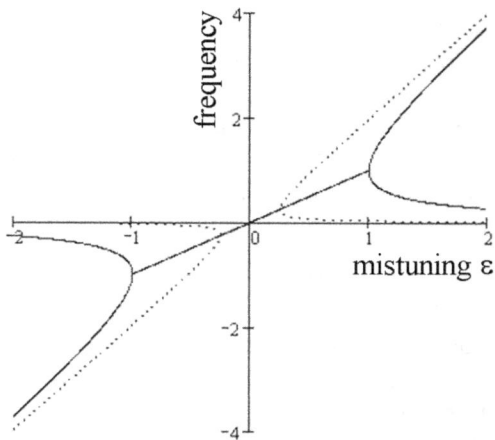

Fig. 18.1. Eigenfrequencies for a hypothetical purely resistive bridge, in two

Fig. 18.2

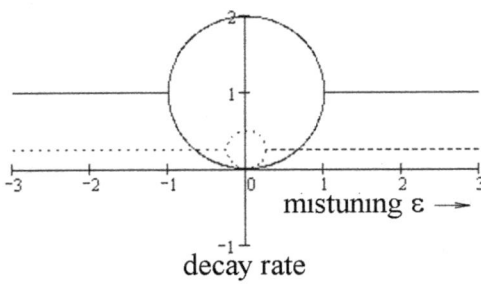

Fig. 18.2. Decay rates for the same hypothetical purely resistive bridge, in two Cartesian co-ordinates. $\xi = 0$ and $\eta = 1$. $H_1 = H_2 = 0$. $B_h = 0.25\eta$. Solid = 'vertical' modes; Dotted = 'horizontal' modes.

The model used to generate general motion decay curves of the partial (as a superposition of the system's four normal modes at one partial number n), takes into account mistuning as the only factor that introduces the antisymmetric modes. In the physical case "imperfection" in the hammer strike, including strike misalignment, especially when there are grooves worn in the hammer face, will contribute to antisymmetric mode excitement.

We can use the mode decay rates themselves as a suitable index for the initial amplitudes of the modes, for a given mistuning, in the absence of localised horizontal perturbations. Thus at zero mistuning the antisymmetric decay rates are zero, so the antisymmetric amplitudes are zero. At larger mistunings both the decay rates and amplitudes tend towards unity for the vertical, or 0.25 for the horizontal. The presence of a local horizontal perturbation H_1 or H_2 will introduce a horizontal antisymmetric component.

The idea of different "antennae" for the radiation of sound (suggested by Weinreich's work, already discussed) leads to the conjecture that the motion in each of the transverse "antennae" co-ordinates contains possible components from both vertical and horizontal normal modes of the system, since the antennae axes may be inclined at an angle α to the system's normal mode axes. There are then numerous possible configurations for the system. The variables include:

- The alignment of the normal mode axes relative to the strike axis, at the boundary of each string;
- The degree of anisotropy in boundary admittance for each individual string;
- The 'mistuning' between the strings (measured in the vertical direction);
- The alignment of the "antennae" axes relative to the strike axis;
- The relative decay rates of the normal modes.

When the reactive part of the boundary admittance is isotropic for both strings, the system is considerably simplified. There are then no Cartesian individual string mode co-ordinate axes. The effect of having "antennae" co-ordinates rotated relative to the actual hammer strike axis is then the same as having hammer irregularity or strike path error in the action. It merely alters the initial amplitudes of the mode motions in the antennae directions. With regard to what the tuner hears, the response of the system is qualitatively the same as that of the one plane model.

However, when the reactive part of the boundary admittance is *anisotropic* for one or both strings, the situation is different. The important features in the behaviour of the system with various degrees of mistuning, then depends on whether one, or both strings are false.

Degrees of falseness over the compass

When the boundary of a string is only slightly anisotropic then that string will exhibit a correspondingly slow beat rate when treated as an individual single string (corresponding to wedging two of the three strings of a trichord in the actual piano situation). Given that the reactive frequency perturbation caused by the boundary is small, the anisotropy as a ratio between vertical and horizontal perturbations, must be correspondingly large, for a significant frequency difference to occur between vertical and horizontal modes. This is not a condition we would necessarily expect the bridge-soundboard system as a whole to exhibit, but will be possible as a result of localised conditions at the boundary of the string, notably at the point of string side-drafting against the side draft pin.

If we consider falseness as a "failing" to maintain isotropic reactive boundary admittance at the bridge boundary, then we can make some general observations. The physical construction of the boundary at the bridge is the same from the mid compass to the top treble, and the string tension is designed to be constant as far as possible, so the statistical likelihood of this "failing" occurring, and its physical severity, we would expect to remain about the same throughout this part of the compass. However, the false beat rate is the difference between the vertical and horizontal frequency perturbations, multiplied by the unperturbed frequency.

Thus we would expect that for equal scaling tension, and more or less unchanging physical local construction at the bridge boundary of each string, the range of possible false beat frequencies for any partial number, will increase exponentially as we rise up the compass of the instrument. An increase with position in the treble direction is indeed what we find in practice. A string at the top treble, say a top F81 (F7) with a false beat rate in the order of 6 per second in its fundamental, we could say has a certain fixed degree of physical "failing" at the boundary, that we can call F. For the same physical failing F at the boundary of the string F33 (F3) in the mid compass, the resultant false beat rate would be in the order of 1 per second, or less.

This is of course only one factor contributing to the higher incidence of falseness in the higher compass. A tendency to reactive anisotropy in the boundary condition may also depend on other factors such as the ratio of string diameter to side draft pin diameter, downbearing angle, partial frequency, and the relationship between many other factors affecting the physical setup.

One string false – "hiding" falseness

The localised physical factors at the boundary of an individual string, such as the side draft arrangement, may contribute a range of effects from nothing, through to a fast false beat rate. What is important is the false beat rate in relation to the overall partial decay rate. Most false beat rates of one beat every few seconds will be audible, but close scrutiny of the single string soundscape can often reveal still slower beats. Where a false beat rate exists in one string, the behaviour of the system as a function of mistuning exhibits some interesting features that could be utilised in tuning practice. For a realistic string configuration with equal vertical resistive and reactive bridge admittance, and horizontal components that have the same reactive value, but a quarter of the vertical resistive value, the eigenfrequencies and their decay rates again appear in pairs resembling the curves for the equivalent Weinreich model in one plane:

Fig.18.3

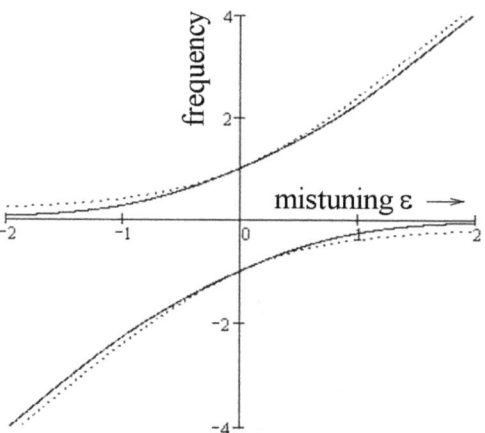

Fig. 18.3. Eigenfrequencies of the system when the bridge is equally resistive and reactive in the 'vertical', is isotropic in its reactive part, but has a 'horizontal' resistive component a quarter that of the 'vertical'.

Fig. 18.4

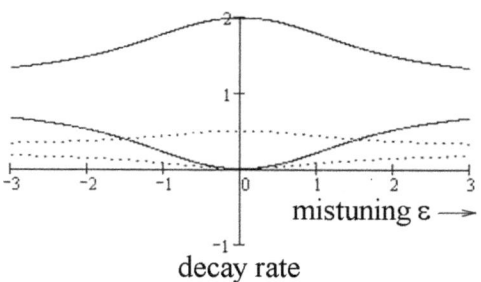

Fig. 18.4. The decay rates for the same conditions as above. Solid = 'vertical' decay rates; Dotted = 'horizontal' decay rates.

The introduction of anisotropy in one string has the effect of shifting the curves along the mistuning axis, relative to each other. The following is an example of the effect of introducing a false beat rate of around 1 Hz in string 1:

Fig. 18.4

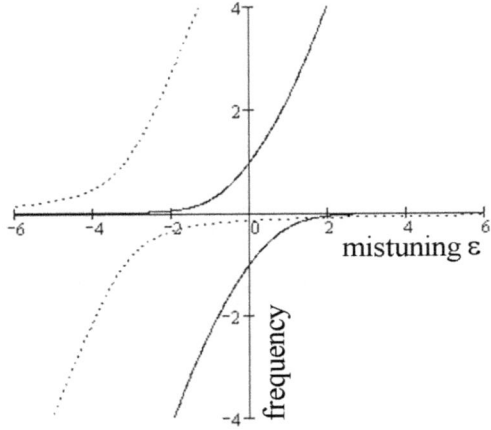

Fig. 18.4. Eigenfrequencies when anisotropy is introduced into the boundary condition of string 1. Solid = 'vertical' modes; Dotted = 'horizontal' modes. The value of the mistuning at which eigenfrequencies are at minimum difference, is now different for each pair.

Fig. 18.5

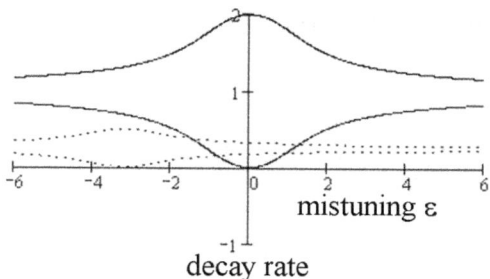

decay rate

Fig. 18.5. Decay rates of the modes under the same conditions as above. Solid = 'vertical' modes; Dotted = 'horizontal' modes.

The mistuning necessary for a minimum difference between the frequencies of the vertical modes, is now different to that required for the horizontal modes. If we were aiming to reduce beat rates to a minimum, there is no one mistuning position at which we could achieve the same result we could have done, if both boundaries had been isotropic in their reactive components, and there had been no falseness. The shape of the decay curve depends on both the real eigenfrequencies and their decay rates, and it is the decay curve that best illustrates what happens in practical tuning terms, at various mistunings.

The false string on its own exhibits a regular beat pattern:

Fig.18.6

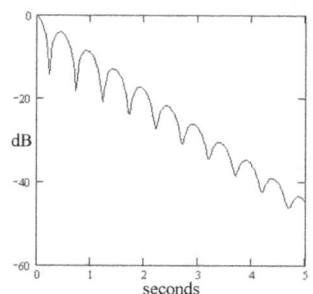

Fig. 18.6. Regular beat pattern exhibited by the theoretical single false partial of string 1.

The following figures illustrate some theoretical behaviour of the coupled unison system when a second string whose reactive boundary is isotropic, is added to the same false string, with increasing degrees of mistuning between the two strings.

Fig. 18.7

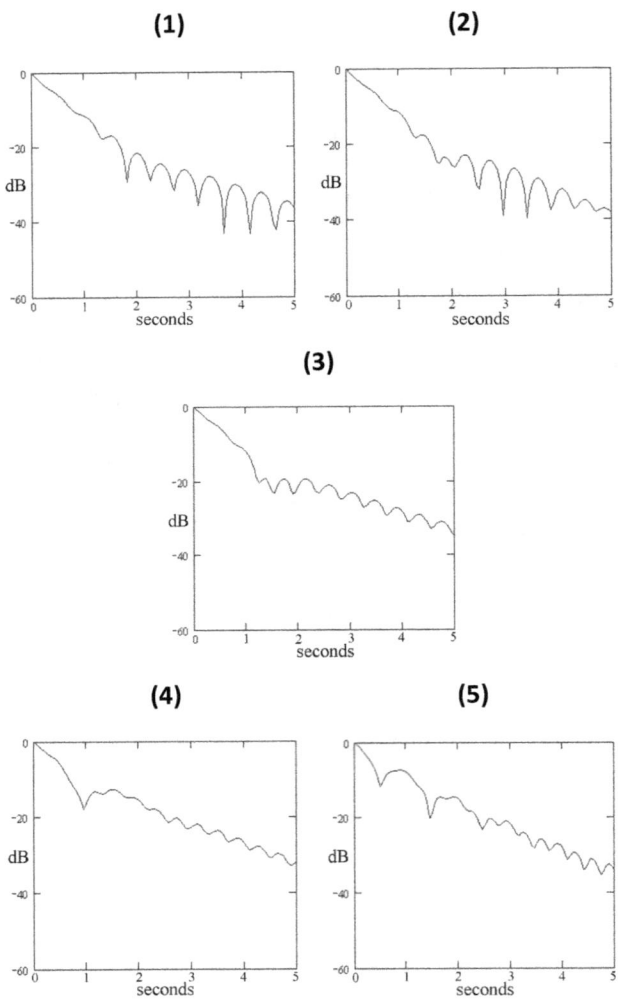

Fig. 18.7. Theoretical decay curves for an example unison with one false string. The mistuning between the 'vertical' individual string modes is: (1) 0 Hz; (2) 0.1 Hz; (3) 0.2 Hz; (4) 0.5 Hz; (5) 1 Hz. The false beat behaviour is probably best 'hidden' in this example when the mistuning is 0.5 Hz (Fig. 8.4). There is then a "single null" type overall decay pattern similar to that found

for 'pure' (non false) strings, and the false beat amplitude is reduced to a minimum.

In this example the false beat rate is about 2 Hz. Dual overall decay rates are clearly present for the mistunings less than 1 Hz. The false beat rate has its presence suppressed at the beginnings of all the decays, and its amplitude throughout the decay curve is minimised at Fig. 8(4), when the mistuning is 0.5 Hz. At this position the overall decay shape relates to the 'single beat' pattern we encountered above in the Weinreich one plane model with a resistive and reactive bridge - the single beat 'null' occurs at about 1 second. At a mistuning of 1 Hz, Fig. 8(5), a superposition of two beat rates is evident, the slower beat decaying in amplitude as the faster beat grows.

If we reduce the angle of inclination of the individual string mode axes to the strike axis the beat amplitude in the isolated false string is reduced. The possible "hiding" or reduction of the false beat in the unison is correspondingly improved:

Fig.18.8

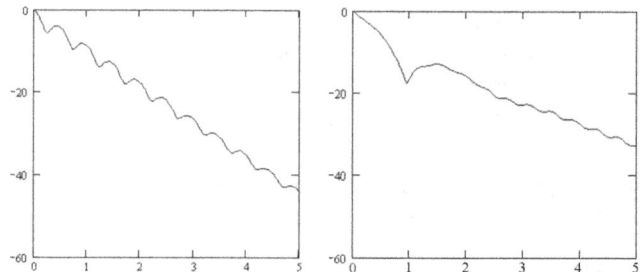

Fig. 18.8. Theoretical decay curves for the isolated false string, and the unison with a mistuning of 0.5 Hz, with a reduced angle of inclination of the individual string mode axes to the strike axis.

The decay curve of the unison at 0.5 Hz mistuning is now very close to the "single null" form, and a considerable improvement on the false string itself, in isolation. The fast beat remains very small in amplitude, even late in the decay.

This relatively simple model illustrates the principle of "hiding" falseness in the coupled unison system, by showing the kind of mechanisms that can allow this, over and above simply attenuating beat amplitudes through superpositioning. Whilst this example is given for only one set of parameters, many other, more complicated beat patterns are possible with other configurations. It illustrates the basic principle that where we are dealing with a false partial in one string of the two string unison, it is often possible with the right degree of mistuning, to minimise the amplitude of beat patterns that cannot be entirely "tuned out" of the unison partial, or to delay their appearance until later in the decay curve, when they may be less prominent relative to other partials.

In some cases where the single string falseness is less severe (perhaps due to suitable angles between the axes and sufficiently slow false beat rate in relation to decay time), it may be possible to position the mistuning so that beating actually vanishes (to the ear at least) in the coupled unison. To achieve this, we would in this model also expect a relatively small inclination of the false string's normal mode axes to the strike axis, and a relatively small reactive anisotropy at its boundary.

A string may be "false" in more than one of its partials, so the task of tuning with falseness may be one of dealing with several instances simultaneously. If both strings are false, or all three in a trichord, the decay curves are irregular and complex, with no one tuning condition obviously ideal. Nevertheless there is usually some position of mistuning at which the result is preferable. The situation is one of having to achieve the "best" result with regard to the "whole picture". As with the unison without noticeable falseness, the "whole picture" includes many partials in the overall spectrum, and a considerable art is required to judge this 'best result' through sensitive trial and error.

Experienced tuners will already be familiar with this technique of "hiding" falseness, which is possible, as we have seen, only because we are in general not dealing in the piano unison with heterodyne beat patterns, but with the behaviour of a coupled system.

Two strings false – the perception of the unison

Because traditional tuning theory deals only in terms of specific set beat rates or no beats, falseness in strings is usually regarded by tuners as a kind of fault in pianos that it would be better not to have. This is also because false beat rates cannot be manipulated with anything like the same degree of control that we can exercise over the beats between "pure" partials, that have exponential individual decays. This makes the handling of falseness a more difficult affair, especially where both (or all three) strings of a unison are false.

Contrary to this, however, falseness cannot always be regarded simply as a "fault". Falseness is a matter of degree, and it is contextual. Even if the string itself is perfectly uniform in its physical properties, a perfect reactive isotropy at the boundary is probably difficult to achieve in practice. What is important, is what a false beat rate will be in relation to the overall decay time.

If the partial has decayed before a 'false' beat takes effect, the partial is effectively 'pure'. The more slight or gradual any fluctuation is, the less likely we are to hear it. Nevertheless, close aural examination of the partials of a single string, one by one, often reveals 'slow' fluctuations in partials, which may be immediately apparent, or may only become apparent over periods in the order 5 or 10 seconds. These slower rises and falls are not qualitatively much different in their effect on the spectrum from the gradual changes of relative amplitude between partials that constantly takes place as the string decays, and due to different partials having different decay rates.

Falseness is not simply and exclusively "a beat", like a heterodyne beat. A false beat, as we have seen, may have a beat amplitude different to the overall amplitude of the false partial itself. The false beat amplitude may also decay at a rate different to the overall decay rate of the false partial. These amplitudes and decay rates are relative to others in the spectrum, and the way that the whole is perceived as a tone in the psychology of aural perception, is extremely complex and largely obscure. It is therefore possible, in some cases, for falseness to be present in unisons without too much detriment to the tone and clarity of the instrument. It is not the case that falseness *per se* should necessarily be equated in every instance with tonal detriment.

The treble of a Blüthner grand, for example, may be very false in 'traditional' tuner's terms, and yet still be greatly superior in tone and quality to many lesser instruments that may well exhibit less falseness. The unstruck aliquot strings may be particularly false, given that they are supported on a separate, smaller bridge, and in sections have the string boundary passing through the top of what amounts to a tall, thin, metal tower, whose 'horizontal' reactive properties are bound to differ substantially from the 'vertical'. On the other hand, on an inferior instrument, a lesser degree of falseness, in terms of its spread of incidence over the compass, the false beat rates, and the beat amplitudes, may be highly detrimental, and will be equated with a poor tone and quality.

Musicians do not necessarily hear beats – but they do hear a "good tone" or a "poor tone". When the tone approaches "bad", due to falseness, the musician may well not distinguish between a failing in the piano and a failing in the tuning, because both manifest as beating.

All things considered, the task in any instance where falseness is present in a unison, is to reduce the degree by which fluctuations in the decay curve disturbs the overall tone of the unison. The tuning of unisons in which two or more strings are false, is thus an extension of the approach for one false string. We are adjusting the behaviour of a coupled system, and aim to set the tuning for as little fluctuation in the decay curve as possible, which may involve a degree of 'mistuning' between the individual strings.

Falseness generalised

In the previous chapters we said that any variable in the mode equations for a single string, that causes two close mode frequencies to occur for one partial number, results in falseness. In the spectrum the general motion of any such two modes is perceived as one false partial. It is still the wave velocity on the string and the boundary conditions that determine the basic structure of normal mode frequency distribution in the spectrum, which approximates to harmonic. It is therefore not surprising that the system should be sensitive to boundary conditions, and when in two lateral dimensions, particularly sensitive to lateral anisotropy in the boundary conditions.

Other factors such as twisting and kinking of the string can exist but are perhaps not so obvious. Whatever the system configuration, it will be comprised of superposed normal modes that individually must take a general form like $Ae^{i\omega t}$, and cannot beat. For a false partial to occur, i.e. a partial with modulating amplitude, there must be at least two normal modes of the system, slightly differing in frequency (any modulation mathematically resolves to more than one frequency). A kink somewhere in the middle of the string can cause the eigenfunction of the standing wave to be non sinusoidal, but this in itself will not necessarily create the conditions for falseness.

It can only cause this if it results in different wave velocities in different planes, resulting in different frequencies at the boundary system. It is theoretically possible to have a very large kink that does not have this effect. Similar arguments are true of twisting - the result then depends on the ability of the string throughout its speaking length to maintain physical and internal stress uniformity under torsion.

A kink at the boundary would be a different matter, however, and might more easily cause falseness if it affects the wave reflection process. Another factor not so often discussed, is longitudinal modes in the string, which, conjecturally, may also possibly have an effect on the lateral modes at certain frequencies, causing two modes to occur at slightly different frequencies. Whatever the actual physical scenario, there must be a superposition of normal modes with close frequencies, in order to produce a false beat pattern. The allowed ranges of possible real frequency and decay rate differences, are subject to the same general limits no matter what the physical cause of falseness, so the generic kinds of decay curves that result, tend to recur. The mathematical model for false beat phenomena, whatever its configuration, will thus tend to yield the same generic decay patterns.

We saw earlier that there are generic curves that occur in the case of two strings in one plane, or for two strings in two planes, when the strings are not "false". These included regular beat patterns of varying beat amplitude, with various beat amplitude decay rates relative to overall decay rates. Notably, there are the 'single beat' and 'two beat' patterns. In the case of the 'false' unison, we will tend to encounter the same decay shapes with additional beat patterns superposed, but may also find irregular overall beat rates that are really the superposition of more than one beat rate.

Chapter 19 - The Trichord

The one plane, two string Weinreich model gives an insight into the behaviour of the bichord, and the introduction of a second plane allows "false" partial behaviour to be taken into account. The trichord can initially best approached in the same simplified way, primarily as a "one plane" model, but with the second plane necessary when we include "false string" phenomena. Following the same principles as before, the matrix for the one plane trichord is a 3 X 3 :

$$\Omega = \begin{pmatrix} 2\varepsilon_1 + \zeta & \zeta & \zeta \\ \zeta & \zeta & \zeta \\ \zeta & \zeta & 2\varepsilon_2 + \zeta \end{pmatrix}$$

Here, ε_1 and ε_2 are the mistunings between strings 1 and 2, and 2 and 3, respectively. The system has three distinct normal modes at zero mistuning. The first mode is symmetric like the two string system, but with *three* strings moving coherently in phase, so that for the given partial the force on the bridge at any time is three times what it would be for one string, and the partial decay rate is three times that for a single string. As before, $\zeta = \xi + i\eta$, where η is the partial decay rate for the single string, so this mode's decay rate is 3η.

The second distinct mode has two strings moving coherently in phase with each other, but 180 degrees out of phase with the third string. The individual amplitude of the third string is greater than that of the other pair, so that net force in the bridge is nil and the decay rate is zero. It is irrelevant which strings constitute the 'in phase' pair, and which one is the 'out of phase' string - the complex frequency of the mode will be unaffected by the choice, and all parts of the system will still move with the same frequency. The three different possibilities are degenerate mode versions.

The third distinct mode has two strings moving antisymmetrically, 180 degrees out of phase with each other, while the third string does not move. Because this is a normal mode, every part of the system is "moving" with the same frequency – the stationary string is not absented from the system. Whilst the net force on the string remains zero, the presence of the string exerts a continuous effect on the system. We can think of the modes as similar to those for a system of three pendulums coupled by two springs. In one of the pendulum system's three normal modes, the two outer weights move out of phase with each other whilst the weight in the middle remains stationary. In the three string system, however, the coupling is resistive also, and *which* string remains stationary is again irrelevant for the mode.

As with the two string system, theoretically, only the symmetric mode will ensue if the mistuning is zero and the hammer strike is "perfect". When we start to introduce mistuning we start to superpose normal modes. There are now *two* mistunings to consider, the mistuning between strings 1 and 2, and the mistunings between strings 2 and 3. The mistuning between strings 1 and 3 is merely a function of the first two. Because of the proximity of the strings the system is symmetrical with respect to the mistunings.

For any partial the trichord tuning condition consists at any time of three "string pair" relationships, defined by the two mistunings. Each string "pair condition" can be called *"unison"* for zero mistuning in the pair, or *"mistuned"* for non-zero mistuning. Whenever the condition is *unison + mistuning* the trichord is in a "special" condition, and we can show the behaviour of the eigenfrequencies for this condition, as the sole mistuning is altered. The eigenfunctions' behaviour as a function of mistuning is again most clearly seen for the hypothetically extreme case of a purely resistive bridge:

Fig. 19.1

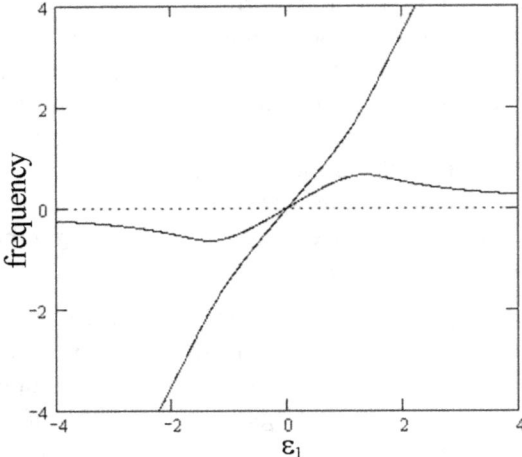

Fig. 19.1. (A) The eigenfrequencies of the trichord as a function of ε_1 when mistuning $\varepsilon_2 = 0$, for the hypothetical purely resistive bridge. The dotted line is eigenfrequency 3.

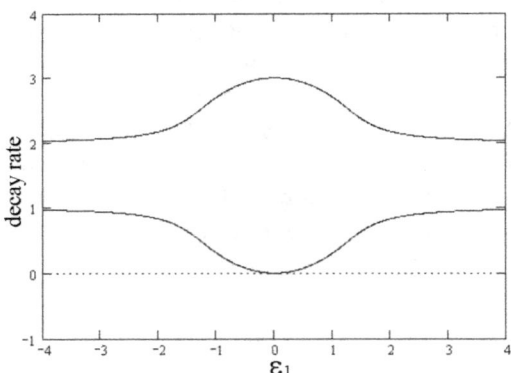

(B) The decay rates of the eigenfrequencies of the trichord as a function of ε_1 when mistuning $\varepsilon_2 = 0$. The dotted line is the decay rate of eigenfrequency 3.

In both the real and imaginary functions, the dotted line corresponds to an eigenfrequency that is unaffected by the mistuning ε_1, and is thus determined purely by the other 'mistuning' ε_2, whose value remains zero.

This is the third mode described above, in which two strings are in antisymmetric motion and the displacement of the third string remains zero. The individual frequency of the third string can be changed without changing the frequency of the system mode, which is determined solely by the two strings in motion.

Unlike the two string unison the mode frequencies do not coalesce inside a "critical tuning" range, and are only equal at exactly zero mistuning. However, there is now a *wider* range, between which the difference in the eigenfrequencies is reduced by about half, from what it would be with uncoupled behaviour.

The consequences of coupling as far as tuning is concerned are best illustrated through the decay curves. As before, in the computer generation of general motion curves, mistuning is treated as the only cause of initial antisymmetric motions, and the decay rates are used as a factor determining antenna amplitudes of the modes. Thus a mode with zero decay rate must also have zero initial amplitude, and a mode with a decay rate of 2η will have twice the initial amplitude of motion whose decay rate is η.

The following theoretical curves are for partials with a decay time of 15 seconds, for various combinations of mistunings between the two pairs of strings, and for a bridge admittance with equal resistive and reactive components:

Fig. 19.2

One string pair at zero mistuning- various mistunings in the other pair

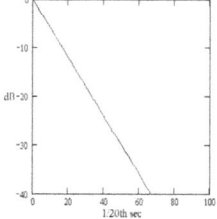
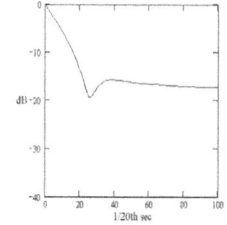
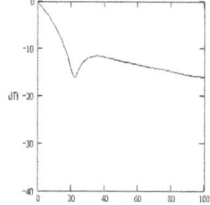

Mistunings zero.

The decay is exponential, and the decay rate is 3 times the single string rate

Mistunings 0.2 Hz and 0 Hz.

Single null. Very slowly decaying 'aftersound'.

Mistunings 0.4 Hz and 0 Hz.

Single null. 'Aftersound' beginning to decay faster.

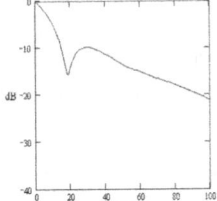 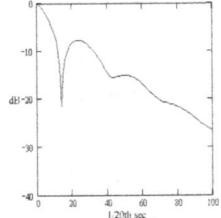 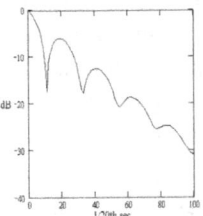

Mistunings 0.6 Hz and 0 Hz.	Mistunings 0.8 Hz and 0 Hz.	Mistunings 1 Hz and 0 Hz.
Single null. 'Aftersound' decays still faster.	Single null plus much reduced secondary beat.	Regular "heterodyne like" beat, approximately 1 per second.

One string pair 0.1 Hz mistuning

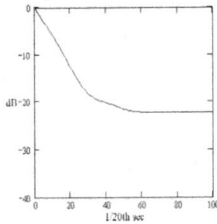 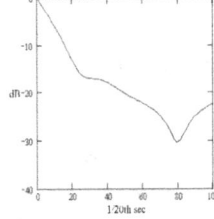 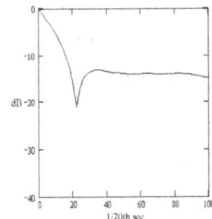

Mistunings 0 Hz and 0.1 Hz.	Mistunings 0.2 Hz and 0.1 Hz.	Mistunings 0.4 Hz and 0.1 Hz.
'Dual decay' pattern – no beat and distinct 'aftersound'.	'Dual decay' pattern and distinct beat rising after 4 seconds.	Single beat and distinct 'aftersound'.

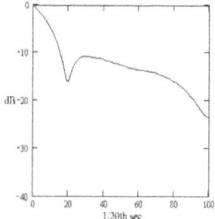 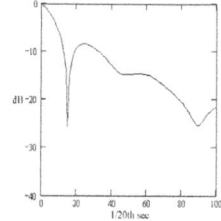 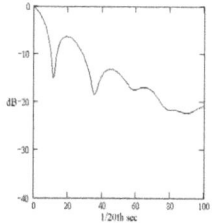

Mistunings 0.6 Hz and 0.1 Hz.

Single beat, steady 'aftersound', and another beat beginning 3 or 4 seconds later.

Mistunings 0.8 Hz and 0.1 Hz.

Pronounced single beat, followed by shallow beat, and another more pronounced beat.

Mistunings 1 Hz and 0.1 Hz.

Regular beat developing into irregular pattern.

One string pair 0.3 Hz

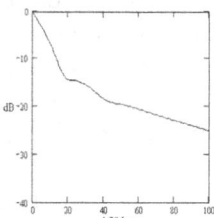 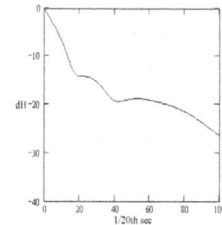 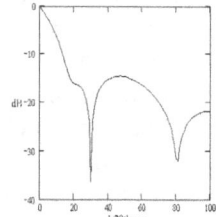

Mistunings 0.3 Hz and 0.3 Hz.

Slight single beat with distinct 'aftersound'.

Mistunings 0.4 Hz and 0.3 Hz.

The 'aftersound' now decays like a slow beat.

Mistunings 0.6 Hz and 0.3 Hz.

Irregular fall into a very pronounced beat, followed by another.

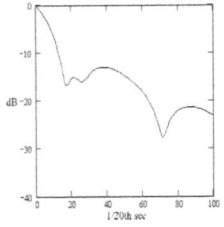 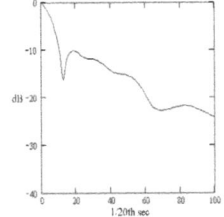 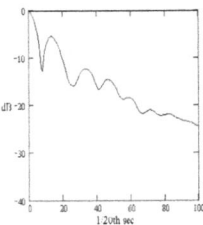

Mistunings 0.8 Hz and 0.3 Hz.

Irregular beats.

Mistunings 1 Hz and 0.3 Hz.

Distinct first beat followed by irregular beats.

Mistunings 1.5 Hz and 0.3 Hz.

Irregular beat pattern.

In the absence of coupled behaviour, we would expect regular beat rates equal to the mistunings, so that for unison groups with both mistunings non zero, there would be two superposed regular beat patterns. The fact that the system is coupled yields quite different results. When there is no mistuning between one pair of strings, the generic decay patterns are similar to what we might expect for a two string unison (a bichord). However, the relationship to mistuning is different. A beat in the trichord is definitely replaced with a "single null" when the mistuning has been reduced to 0.6 Hz, whereas in a bichord with the same decay time we would need a smaller mistuning around 0.3 Hz to achieve a similar result. In the trichord, we can now have a mistuning of as much as 0.8 Hz (for the 15 second decay time) before the generic "single null" pattern starts to transform into the "regular beat" pattern, whereas this happens in the bichord at around 0.4 – 0.5 Hz.

With a slight mistuning of 0.1 Hz replacing the "perfect unison" string pair in the trichord, the generic patterns are very similar to the "single beat" and "dual decay", but with some additional irregularity. We can still have up to around 0.8 Hz mistuning between the strings of the remaining pair, without replacing the "single null" pattern with a pronounced regular beat pattern.

When the "perfect unison" is replaced with a mistuning of 0.3 Hz, the decay is generally irregular, except when the other mistuning is also close to 0.3 Hz, in which case two strings of the trichord are still effectively close to "perfect" unison.

In fine unison tuning the partial decay curves within the unison's spectrum do not in general fall into the two groups *no beat* (*exponential*) and *regular beat*, but rather, are more likely to be the *single null* or slowly changing *irregular* patterns. Within these, dual overall decay rates may also occur.

What we also see here is an important principle regarding the implementation of unison trichords. The use of *three* strings to make a unison group, might make the statistical chances of a given mistuning (between any two strings) occurring in one unison group, greater than for a *two* strings unison group. It might seem at first sight, therefore, that in using trichords, fine tuning stability would have to be compromised for the sake of the advantages in power and sustain the three strings offer. However, in terms of the partial decay curve shape having as little fluctuation as possible, the trichord is remarkably tolerant of mistuning in one string pair provided the other pair's mistuning remains very small. For this particular decay time (15 seconds), a mistuning of say, 0.7 Hz, can be tolerated without breaking away from the plain "dual decay" or "single null" decay patterns, provided one pair of strings remains within about 0.1 Hz mistuning. Even well outside this range (0.3 Hz) the resultant decay curve for any partial is not necessarily "worse" than we would expect from a bichord.

The fact that the three strings and bridge constitute a coupled system, means that the decay of the closely tuned trichord, can be different from the sum of the three individual unison pairs that comprise it. In tuning terms, this kind of behaviour suggests coupled behaviour is an advantage that allows "better" tuning results than might otherwise be expected. The decay curve for any partial, for a given condition of small mistuning, tends to be less severe in terms of amplitude of fluctuation, than it would be if we were dealing with uncoupled strings and heterodyne interference beats.

In a sense, we can think of a trichord as "absorbing" or "hiding" some of the tuning discrepancy that occurs some time after fine tuning. This is an advantage for the musician. For the piano tuner, however, there is a more significant consequence of trichord behaviour. This is that the trichord with all three strings sounding, behaves differently to any of its component string *pairs*, when they are tuned as bichords, using the tuner's wedge.

Chaotic trichords?

The extended dynamical matrix model for the full trichord with two transverse degrees of freedom (vibrations in two planes) is a 12 X 12 matrix. This is not only much more complicated, but in computer analysis of its behaviour it appears to be capable of bifurcation behaviour in the frequencies and decay rates, when a *reactive* bridge component is present. This has not been fully investigated, and perhaps could be due to computation artefacts, but if the results are genuine aspects of the behaviour of the matrix, the trichord in two planes could be especially interesting. This would not be because it is neatly solvable. Rather, bifurcation behaviour is often indicative of chaotic, complex dynamics. In other words, the full trichord may be a complex, chaotic system. This would make the precise behaviour of a given trichord difficult or impossible to predict in advance.

Chapter 20 - Further comments on false partials

Some false partials not only beat, but appear to beat with varying, or cyclically varying frequency. The behaviour of ETDs measuring partial frequencies, often suggests this also. Perceived apparent pitch change alone, is not of course a reliable indicator of an actual frequency change in one partial. Nevertheless, cyclical frequency change of a single partial is not precluded in the theory.

Frequency variation in false partials

Two scalar simple harmonic frequencies can be summed to produce one simple harmonic frequency at the average of the two components, which is beat modulated with a beat frequency equal to the difference between the component frequencies. This is the well known principle of scalar heterodyne beating. In the case of vector motions for normal modes this reduction of two frequencies to one, cannot necessarily be carried out. If the audible partial is mapped from two vector normal mode motions through a vector not aligned with a normal mode, then the resultant partial frequency can vary cyclically between the limits set by the two normal mode frequencies themselves.

To illustrate the principle in a more straightforward visual way, let us look at a simplified schematic representation of the transverse motion at the bridge in one plane, presented graphically. The following Lissajous figures show the developing path of a point representing the transverse position of the bridge boundary of the string, when component frequencies in the y and z mode axis directions differ. For visual clarity the frequencies are now set to just 1 Hz and 1.1 Hz, which reduces the number of loops, and similarly the initial amplitudes of the modes in each co-ordinate are set equal. The precise pattern formed by any locus will of course depend on the frequency difference between the modes, the relative initial amplitudes and displacements, and decays. Here we have set the decay as negligible in one beat cycle, and the other parameters are arranged so that the locus pattern is symmetrical – half way through the beat cycle the path retraces itself exactly back to the starting position, ready for a new cycle.

Fig. 20.1

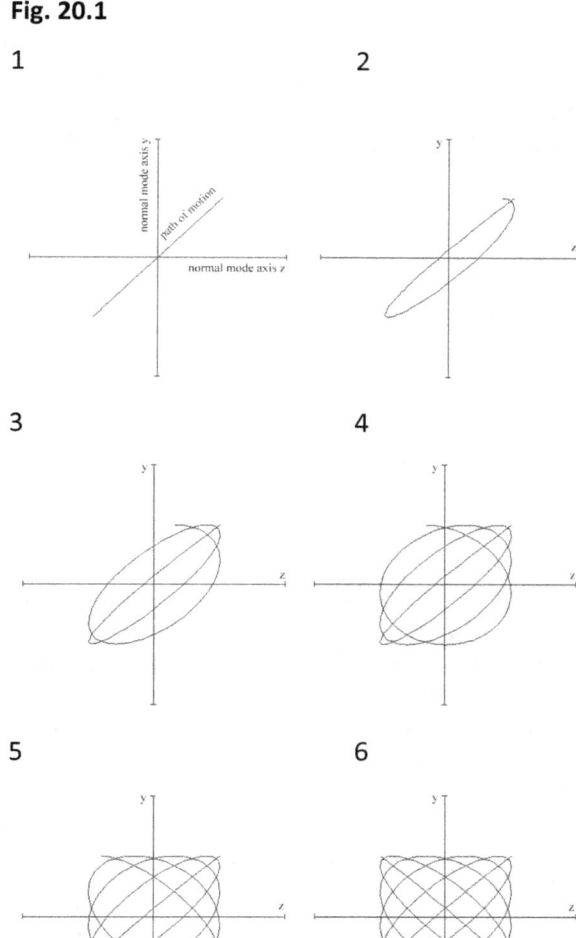

Fig. 20.1. Path of a point representing the moving bridge boundary of the string, when motion begins with equal displacements in the y and z axes. 1: The frequencies in the y and z axes are the same, so the motions in y and z remain in phase, and the locus is a straight line at 45 degrees to the mode axes. 2: The frequencies are 1 Hz in y and 1.1 Hz in z – the motions in each axis start to become out of phase. This is the path covered at 1 second. 3: The path at 2 seconds. 4: Path at three seconds. 5: Path at four seconds. 6: Path at five seconds. The pattern is now complete – the point subsequently retraces the existing locus in the reverse direction for the next five seconds.

Fig. (20.2) shows the path at $t = 5$ seconds. The path started at the top right corner, with equal displacements in z and y, and finishes at the top left corner of the pattern. The point would subsequently retrace the same locus backwards to the starting point. Now let us consider measuring motion on axes at a full 45 degrees inclination (for clarity) to the mode axes. We can refer to these measurement axes as "antennae" axes:

Fig. 20.2

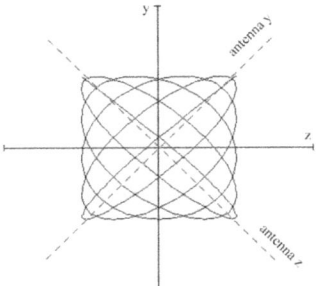

Fig. 20.2. The motion begins with large amplitude in the y antenna axis, and small in the z antenna axis. The amplitude in the y antenna axis subsequently reduces as the amplitude in the z antenna axis increases. Finally, at 5 seconds, the initial situation is fully reversed – the amplitude in the z axis is the large amplitude, while the amplitude in the y axis is small. The amplitude in the mode axes themselves, remains unaffected. In this illustration it is not only unaffected, but is also constant, because there is no decay. After 5 seconds the path is retraced in reverse, so the amplitude in z decreases as the amplitude in y increases, until at 10 seconds the initial conditions are recreated.

Motion in antenna y will thus consist of a frequency with a beat modulated amplitude, and so will motion in antenna z, but the beat patterns will be 180 degrees out of phase with each other. The frequency of motion in the y mode axis here is 0.1 Hz slower then the frequency in the z mode axis. When an antenna axis is aligned somewhere between the mode axes, the motion in the antenna axis direction will approximate to a simple harmonic frequency somewhere between the two extremes 1 Hz and 1.1 Hz in the mode axis directions. In the case of a string *in situ* the beating in one antenna can thus occur in a "signal" with a different frequency to the other "signal" in which beating occurs. If the resultant audible partial were mapped from a superposition of two such axes, the result would be a beat pattern whose alternate beats are *at two different pitches*, depending on the parameters.

Falseness and boundary conditions

There are well over two hundred strings in a piano, and many different manufacturers, models, ages and conditions of pianos, so the precise conditions, and hence the precise behaviour of the piano string *in situ*, cannot be ascertained from a few observations alone. Even the most accurate measurements of the behaviour of one or two individual strings, will not give a very complete picture of the behavioural possibilities of piano strings in general.

Tuners will already know from experience how conditions can vary significantly from string to string, section to section, and piano to piano. To glean sufficient information to put together a generalised picture of piano string behaviour empirically, would require an examination of perhaps a whole section of the compass, across a substantial "cross-section" of instruments. What we can say is that piano tuners know from experience that the incidence and severity of falseness generally increases with diminishing piano quality, and with increasing piano age, and is more noticeably widespread in the higher compass (there are straightforward reasons for this).

Nevertheless, falseness as the generic class of single string behaviour that we have been discussing, is never entirely absent from the compass of even the best instrument. It may seem to be however, because what tuners consider to be falseness is usually restricted to only more noticeable incidences where the severity in terms of beat frequency and beat amplitude, is sufficient to constitute an obstacle to "ideal" tuning results.

In the piano we know the overall structural design of the whole system that carries wave energy away from the strings, but its precise localised behaviour relative to the individual string may be more difficult to define. The bridge, soundboard, entire set of strings, iron frame and casework, are not physically separated components of the instrument. They form one acoustical system whose "centre of co-ordinates" can sensibly be taken to be the massive iron frame. The proportionally very large mass of the iron frame makes it an inert acoustical "ground" relative to which all other acoustical motion can be measured. Relative to the iron frame, transverse wave motion across the soundboard is not the only motion that needs to be considered. As we have said, there is evidence that the sound received by the ears from the piano is radiated from different directional 'antennae' in the piano, which means that the main surface area of the soundboard is not necessarily the only significant radiator. The entire piano is an acoustical system, but nevertheless there is a "subsystem" - a part of the overall system that comes under the control of the tuner, and this is the string *in situ*. It can be considered as one system in which the strings are connected via the bridge, which itself is a representative of the rest of the system to which it is attached.

In looking at the behaviour of a single string we must therefore take into account the variations in the conditions at the wooden bridge boundary of the strings. This depends both on the main structure of the entire strings-bridge-soundboard system and on localised factors that may pertain only to the individual string. At the boundary point of an individual piano string there may be complicating factors that contribute to 'boundary conditions'. The bridge end of the speaking length of the string terminates in an arrangement where the string physically continues but is drafted sideways past the first side draft pin. Although it is intended to be laterally rigid, this pin may or may not be perfectly rigidly fixed in the bridge surface, and even if it is in terms of construction, it may not be *acoustically* rigid.

The small dimensions of the pin coupled with the high tension in the string it drafts sideways, leads to a significant pressure in the wood around the pin hole. The pin is itself an elastic component, and the wood at the surface of the bridge into which it is inserted must also be considered for its localised elastic properties in the immediate vicinity of the pin.

It is conceivable that a less than perfectly rigid fixing could even lead to discontinuous degrees of freedom at the boundary. Where there is an actual fault such as loss of downbearing, the string may not fully contact the bridge surface as well as the pin, and may pass the pin above the point it enters the wood. The pin then effectively acts as a cantilevered support of the string, and its own elastic properties become a major cause of anisotropic boundary conditions. This can lead to severe and falseness. Even if there are no 'faults', the pin and string constitute a tensioned system in both horizontal and vertical directions, with a possible frictional component in the horizontal direction offered by the bridge surface.

The precise point at which boundary conditions can be said to begin can be somewhat obscured by the physical arrangement. The continuation of the string in the part affixed to the bridge is itself part of the boundary arrangement. Where the speaking length joins this, in the vicinity of the side draft pin, there can be a small but not necessarily insignificant distortion in the speaking length of the string, as a small curvature to the string on within the speaking length of the string. It must be remembered that the piano string is not flexible, but is a stiff wire with significant thickness.

A slight distortion in the string at the limit of the string's speaking length, may arguably feature as part of what can be considered boundary conditions. Also, at the speaking length edge of the bridge, the side draft pin is necessarily set back from the edge of the bridge so that a cut out is necessary to bring the vertical boundary point offered by the wood of the bridge close to the horizontal point offered by the tip of the side draft pin inserted in the wood. The point offered by the edge of the wood is, from structural necessity, not precisely the same as the point offered by the pin.

Apart from localised anomalies at the boundary generally, the bridge itself is a solid mass attached both to the soundboard and to another boundary with the backstringing. Despite the relatively large mass and physical dimensions, the boundary necessarily moves in order to transmit energy from the strings to the soundboard, and thus transmits also to the backstringing. This fact is utilised in duplex scaling.

The soundboard is not a flat plate but a surface arched in two dimensions to support downbearing force from the strings, so that together with the strings it is a component of one tensioned system. There will also be a series of torsional forces on the bridge originating from the side drafting of the strings on its surface. A rough estimate for the total downbearing for a system with an average downbearing angle of θ and an overall tension T will be $T\sin\theta$. On a large grand piano with say, 200 KN total tension, this gives almost 3.5 KN total downbearing force per degree downbearing, so clearly the system as whole is in a state of considerable stress, and corresponding elastic deformation.

Chapter 21 - Inharmonicity

"Traditional" tuning theory assumes harmonic partials whose frequencies are integer multiples of the fundamental frequency. Real piano strings are a *dispersive* wave medium, meaning that partial wave velocity is dependent upon wavelength. The allowed wavelengths are still fixed by the string boundary conditions, and as a result the partial frequencies *approximate* to harmonic frequencies. Although approximately harmonic, the deviation from true harmonic frequencies is nevertheless great enough to be significant in practice. Because the partial frequencies are not *harmonic*, the partials are said to be *inharmonic*.

Inharmonicity in pianos is caused principally by string stiffness, a consequence of the physical properties of the metal from which the strings are made. It can be seen from the handling of piano wire that it is far from flexible, and that it has strong 'spring like' properties more like those of a metal bar than a flexible string. The effect of this stiffness in piano strings is that shorter wavelengths have faster wave velocity, and any given wavelength has a higher frequency than it would had the string been perfectly flexible. The partial frequencies are raised from the harmonic positions by a number of cents proportional to the square of the partial number. The constant of proportionality is the *coefficient of inharmonicity*, *b*, for a particular string, given by:[87]

$$b \approx \frac{1731\pi^2 Q d^2}{128 v_0^2 l^4 \rho}$$

1

where Q = Young's modulus for the string material

---

[87] Derived from Young, RW, 'Inharmonicity of Plain Wire Piano Strings', *JASA*, Vol 24, No 3, 267 – 273, May, 1952, p. 268.

d = string diameter

l = speaking length

ρ = string density

$v_0 = c/2l$, the *ideal* fundamental frequency, in which

$c = \sqrt{T/\rho S}$ for tension T and cross sectional area S.

The deviation δ, of any partial from its corresponding harmonic frequency, in cents, is then

$$\delta = bn^2$$

2

Alternatively, the frequency v_n in Hz, of a partial with partial number n is given by :[88]

$$v_n = nv_0\sqrt{1 + \frac{n^2\pi^3 Qd^4}{64l^2 T}}$$

3

Each string has its own dimensions, and thus its own coefficient of inharmonicity, so the coefficients will vary across the piano compass as the physical parameters of the strings vary. We can see from equation 1 that the coefficient of inharmonicity of a string is inversely proportional to the *fourth power* of the string's *speaking length*, so in a piano the scaling factors for the speaking lengths can have a marked effect on inharmonicity. The speaking length is the most critical design parameter that can increase or decrease a string's coefficient of inharmonicity, when other parameters are constant.

Copper wound bass strings can be considerably inharmonic in the higher part of their spectrum, and because of their relative low frequency fundamental, this part of the spectrum falls within audible range, or even within the ear's most sensitive *critical band*. The greater inharmonicity of bass strings is not however due mostly to stiffness added by the copper windings – the increase in stiffness due to the windings is relatively small.

---

[88] *Ibid.* There are some other formulations that differ, depending on method of derivation and approximations.

Rather, the windings provide the string with a much larger mass per unit length, allowing the strings to be much shorter for a given fundamental frequency. It is the decreased length, for the given frequency, that primarily causes greater inharmonicity in the bass strings, as a design feature of the instrument. Inharmonicity leads to tuning problems concerning the matching of spectra between strings with different physical parameters, and also problems concerning the psychological perception of the *pitch* of individual bass strings, which is dependent on the spectrum's partial structure.

As far as unisons are concerned, all the physical parameters of the strings except tension should normally be the same for a unison group, so the coefficient of inharmonicity for all the strings of a given unison group would normally be the same. Inharmonicity may become significant in unison tuning if the strings do not have the same physical parameters. This would normally be the case if they are significantly different gauges, and/or if a mixture of old and new strings is present. Even if the strings are made from the same material, the Young's modulus may vary between new and old metal.

General formula for inharmonic beat rates

We first consider the spectrum of an interval as a superposition of two independent spectra, one for each note that forms the interval. This is the 'traditional' approach. A general 'pure', or *Justly Intoned* interval superposed from two *harmonic* spectra has associated with it a harmonic ratio R_U/R_L, for a 'beatless' condition in which the partials at each meeting in the superposed spectrum coincide exactly. R_U and R_L will be the lowest partial numbers in any harmonic spectrum that are separated by this particular musical interval. Thus, for the interval of a perfect fifth, the lowest partials in any harmonic spectrum that are separated by this interval, are partials with partial numbers 2 and 3, so the numbers 2 and 3 make up the harmonic ratio for a perfect fifth. By convention, we will designate R_U as the larger of the two numbers that make up the harmonic ratio, and R_L as the smaller. Thus the harmonic ratio R_U/R_L for any interval is always greater than unity. The ratio R_U/R_L will also be the *frequency ratio* of the fundamentals for this interval, where the spectra are harmonic. The first meeting of partials in the superposed spectrum of the interval will be a meeting of the R_Uth partial of the lower note with the R_L th partial of the upper. Upper meetings occur at the nR_U th partial of the lower note with the nR_L th partial of the higher note, for any integer n.

A *tempered* interval superposed from two *inharmonic* spectra also has *associated with it* the same harmonic ratio as it would if it had not been tempered or inharmonic. The R_U and R_L still determine which partial numbers will meet in the superposed spectrum, but the partials will no longer exactly coincide at any meeting. Also, the ratio R_U/R_L will no longer be the frequency ratio of the fundamentals.

At any meeting of partials in the spectrum of a tempered interval, the beat frequency will be the difference in the partial frequencies at the meeting. Thus, when the upper note of the interval has a partial frequency of P_U at the first meeting, and the lower note has a partial frequency of P_L at the same meeting, the beat frequency B at the meeting will be:

$$B = P_U - P_L$$

4

If the lower note's partial frequency P_L is greater than P_U, then B will be negative, indicating that the interval is narrow, according to its beat rate at this meeting. This means raising the frequency of the upper note, or lowering the lower note, will *reduce* the beat rate.

For any frequency ratio defining the octave, we can designate its twelfth root as the equally tempered semitone size, which we can call E. We can define any interval in terms of the number N of the equally tempered semitones it contains. Thus for the correct size interval between fundamentals we can write the fundamental frequency F_U of the upper note in terms of the fundamental frequency F_L of the lower note as:

$$F_U = F_L E^N$$

5

With no inharmonicity the partial frequencies at the first meeting are:

$$P_U = R_L F_L E^N$$

and

$$P_L = R_U F_L$$

6

Inharmonicity will raise the partial frequencies by an amount that depends on the coefficient of inharmonicity *b* for the particular note producing the partial, and the partial number *n*. In general, we can write the effect of inharmonicity on any note as a function $\delta(W_S, n)$, where W_S is the set of physical parameters determining inharmonicity for the note at position *S* in the scale. We multiply the harmonic frequency by the value of this function to get the inharmonic frequency for any parameters W_S and Harmonic Number *n*. Thus with inharmonicity included we have:

$$P_U = R_L F_L E^N \delta(W_U, R_L)$$

and

$$P_L = R_U F_L \delta(W_L, R_U)$$

7

where W_U and W_L are the physical parameters determining inharmonicity for the upper and lower notes respectively.

The beat rate $P_U - P_L$ at the first meeting is now:

$$B = R_L F_L E^N \delta(W_U, R_L) - R_U F_L \delta(W_L, R_U)$$
$$= F_L [R_L E^N \delta(W_U, R_L) - R_U \delta(W_L, R_U)]$$

8

The equally tempered semitone ratio *E* can now also be written as:

$$E = 2^{1/12} [\delta(W_L, 2)]^{1/12}$$
$$= [2\delta(W_L, 2)]^{1/12}$$

9

So we finally obtain the beat rate formula:

$$B = F_L \left(R_L \{2\delta(W_L, 2)\}^{N/12} \delta(W_U, R_L) - R_U \delta(W_L, R_U) \right)$$

10

The way in which the inharmonicity function δ will vary from note to note depends on the scaling factor used for determining string speaking lengths. It also depends on the string diameters, but of course for equal tension the diameters are themselves determined by the scaling factor. We can use the formula above to generate beat rates where the longest speaking length in the scale is given, the diameter of this string is given, and the shortest speaking length in the piano is given. If say, the longest scale speaking length of F33 (F3) is L_L, and the shortest of C88 is S, we can determine the scaling factor Δ for the stringing over this 55 note range of the compass as:

$$\Delta = \left(\frac{L_L}{S}\right)^{\frac{12}{55}}$$

11

where $1/\Delta$ is the length change factor over a rising octave and $1/\Delta^{1/12}$ will be the length change factor over a rising semitone. This length change multiplier for a semitone we will designate as Δ_0. We can also use formula 10 to generate beat rates where there is no inharmonicity, in which case δ is unity.

We are interested in complete *sets* of beat rates throughout the scale tuning region, for fifths, fourths, thirds and sixths. We can maintain complete generality by considering not just these four types of interval, but all twelve interval sizes within the octave, from semitone to octave.

First, we need to define a 2 X 13 array containing the harmonic ratios for all these intervals. We designate this array $\mathbf{R}$, and define it:

$$\mathbf{R} = \begin{pmatrix} 1 & 1 \\ 16 & 15 \\ 9 & 8 \\ 6 & 5 \\ 5 & 4 \\ 4 & 3 \\ 45 & 32 \\ 3 & 2 \\ 8 & 5 \\ 5 & 3 \\ 7 & 4 \\ 15 & 8 \\ 2 & 1 \end{pmatrix}$$

12

such that $\mathbf{R}_{i1}/\mathbf{R}_{i2}$ is the harmonic ratio of interval i. Thus, for example, $i = 1$ represents the unison, $i = 8$ represents the perfect fifth, and $i = 13$ represents the octave. For any interval i, there will be 13 beat rates, one for each position in the scale octave at which the interval can occur (including the position an octave above the lowest position). The beat rate for an interval i in the jth position in the scale will thus be B_{ij}. We can thus generate a table of all the beat rates of all the intervals in the scale as one array $\mathbf{B}$.

As an octave is the largest interval for which we will find beat rates, the total part of the compass over which we are calculating beat rates for an octave of scale notes is *two* octaves. The inharmonicity over this range of notes will be different for each note, so the ratio for a 'beatless' octave will also vary depending on which octave in this two octave range we are dealing with. We *could* take the ratio between the bottom note and its 4th partial to define a double octave, and then take the 24th root to define an equally tempered semitone size, but as a matter of convention we will keep our original definition, and define the equally tempered semitone size as the 12th root of the interval from the lowest note to its 2nd partial. Thus the semitone size E will be a constant. This approach will give perfectly reliable results for intervals actually contained within the octave length of the actual scale region.

The speaking lengths of the strings are a discrete function of the string positions j, where the longest string (at the bottom of the scale region) has $j = 1$, the semitone above $j = 2$, etc. The array of speaking lengths is then given by:

$$L_j = \frac{L_L}{(\Delta_0)^{j-1}}$$

13

The frequency of F33 (F3) as the lowest note at A440 Hz concert pitch is:

$$v_0 = \frac{220}{2^{1/3}}$$

14

We designate the string diameter for F33 (F3) as D_L.

It is sufficient to calculate the rest of the diameters for ideal fundamental frequencies (as though there is no inharmonicity) as there is in practice a limited range of diameters available. The array of ideal fundamental frequencies will be:

$$v_j = v_0 2^{\left(j-1/12\right)}$$

15

For steel density ρ the tension T on one string of F33 (F3) will be:

$$T = 4L_1^2 v_0^2 \pi (D_L/2)^2 \rho$$

16

We attempt to maintain as close as possible to equal tension F throughout the range of strings. This is normal design practice. Thus we can calculate the array of ideal diameters, as if any diameter were possible in practice:

$$DI_j = \sqrt{\frac{F}{\pi \rho L_j^2 v_j^2}}$$

17

Standard wire gauges in this region increase by 0.025mm per available gauge, so to convert ideal diameters to actual available diameters, we need to use a function:

$$\Lambda(x, y)$$

that rounds a number x to y decimal places. Then for the 0.025mm increments the array for real diameters is:

$$D_j = \Lambda\left(\frac{DI_j}{0.025 X 10^{-3}}, 0\right) 0.025 X 10^{-3}$$

18

The array of frequencies of inharmonic partials for all notes at positions j, with partial numbers n, will be:

$$f_{jn} = nf_{j0}\sqrt{1 + \frac{n^2 \pi^3 Q D_j^4}{64 L_j^2 F}}$$

19

This is the array of *harmonic* frequencies nf_{j0}, for each note at position j, each one multiplied by the multiplier

$$\sqrt{1 + \frac{n^2 \pi^3 Q D_j^4}{64 L_j^2 F}}$$

This multiplier is the function $\delta(W_S, n)$, the multiplier in B that alters the otherwise harmonic frequencies of the partials. For all notes $j = 1...25$ and the harmonic numbers $n = 1...45$ the array of these multipliers is:

$$I_{jn} = \sqrt{1 + \frac{n^2 \pi^3 Q D_j^4}{64 L_j^2 F}}$$

20

Thus the whole array **B** for intervals $i = 1...13$ and note positions where $j = p = 1...13$ will be given by:

$$B_{ip} = v_p \left(R_{i2} (2I_{p2})^{((i-1)/12)} I_{(p+i-1)(R_{i2})} - R_{i1} I_{(p)(R_{i1})} \right)$$

21

We only need to set the value $Q = 0$ to represent a perfectly flexible string to obtain from **B** a table of 'traditional' beat rates for ideal strings, for all intervals (Fig 21.1).

Fig 21.1

$\mathbf{B} =$

	1	2	3	4	5	6	7	8	9	10	11	12	13
1	0	0	0	0	0	0	0	0	0	0	0	0	0
2	-18.9	-20	-21.2	-22.4	-23.8	-25.2	-26.7	-28.3	-30	-31.7	-33.6	-35.6	-37.7
3	-3.5	-3.8	-4	-4.2	-4.5	-4.7	-5	-5.3	-5.6	-6	-6.3	-6.7	-7.1
4	-9.4	-10	-10.6	-11.2	-11.9	-12.6	-13.3	-14.1	-15	-15.8	-16.8	-17.8	-18.8
5	6.9	7.3	7.8	8.2	8.7	9.2	9.8	10.4	11	11.7	12.3	13.1	13.9
6	0.8	0.8	0.9	0.9	1	1.1	1.1	1.2	1.3	1.3	1.4	1.5	1.6
7	44.5	47.1	49.9	52.9	56.1	59.4	62.9	66.7	70.6	74.8	79.3	84	89
8	-0.6	-0.6	-0.7	-0.7	-0.7	-0.8	-0.8	-0.9	-0.9	-1	-1.1	-1.1	-1.2
9	-11	-11.7	-12.3	-13.1	-13.9	-14.7	-15.6	-16.5	-17.5	-18.5	-19.6	-20.8	-22
10	7.9	8.4	8.9	9.4	10	10.6	11.2	11.9	12.6	13.3	14.1	15	15.8
11	22.2	23.5	24.9	26.4	28	29.6	31.4	33.3	35.3	37.4	39.6	41.9	44.4
12	17.8	18.9	20	21.2	22.4	23.8	25.2	26.7	28.3	30	31.7	33.6	35.6
13	0	0	0	0	0	0	0	0	0	0	0	0	0

The array **B** to 1 decimal place. These are the beat rates for the entire scale from F33 (F3) to F45 (F4) with no inharmonicity (by setting Young's modulus $Q = 0$). Row numbers are i values and column numbers j values. Thus row 1 contains beat rates for unisons, row 2 for semitones, row 5 for major thirds, row 8 for fifths, *etc*. Column 1 contains rates for intervals whose lower note is F33 (F3), column 2 for intervals whose lower notes are F#34 *etc*. Negative values indicate narrow intervals, positive ones, wide.

We can now produce theoretical beat rates for a given top string length, and a given lowest scale string length and diameter for F33 (F3).

This model assumes a constant scaling factor and plain steel strings throughout, which is a somewhat simplified scaling, but will illustrate a general trend. Many smaller instruments may have covered strings within the scale region.

For simplicity we will assume a top string C88 (C8) speaking length of 0.05m, and we take the diameter of the F33 (F3) wire to be 0.001m. Actual pianos would not necessarily adhere to these conditions – they would require slightly different computation specific to the individual instrument design, but the model here is sufficient to illustrate the general principles.

We begin with an F33 (F3) speaking length of 2m to represent a concert grand piano. A table of the main beat rates with inharmonicity, extracted from the array **B**, is given in Fig (21.1).

Fig 21.2

	Maj 3rds	4ths	5ths	Maj 6ths	Min 3rds	Min 6ths
F	7	0.8	-0.6	8	-9.4	-10.8
F#	7.4	0.9	-0.6	8.5	-9.9	-11.4
G	7.8	0.9	-0.6	9	-10.5	-12
G#	8.3	1	-0.7	9.6	-11.2	-12.7
A	8.8	1	-0.7	10.2	-11.8	-13.3
A#	9.3	1.1	-0.7	10.8	-12.5	-14
B	9.9	1.2	-0.8	11.5	-13.2	-14.6
C	10.5	1.3	-0.8	12.2	-14	-15.5
C#	11.2	1.4	-0.8	13.1	-14.8	-16.2
D	11.9	1.5	-0.9	13.9	-15.5	-16.6
D#	12.7	1.6	-0.9	14.9	-16.5	-17.6
E	13.5	1.7	-0.9	15.9	-17.3	-17.8
F	14.2	1.9	-0.9	17.1	-18.5	-18.9

Theoretical beat rates with inharmonicity, to one decimal place, when the F33 (F3) speaking length is 2m. These are close to the 'traditional' rates. Negative values are narrow intervals, according to the beat rates, *i.e.*, raising the fundamental frequency of the interval's upper note will initially reduce the beat rate.

As we might expect, the beat rates are close to their "traditional" values. As in practice, a good indication of the quality of the scale is the progression of the beat rates in the major thirds. These we can represent in the form of a bar graph, Fig 21.3.

Fig 21.3

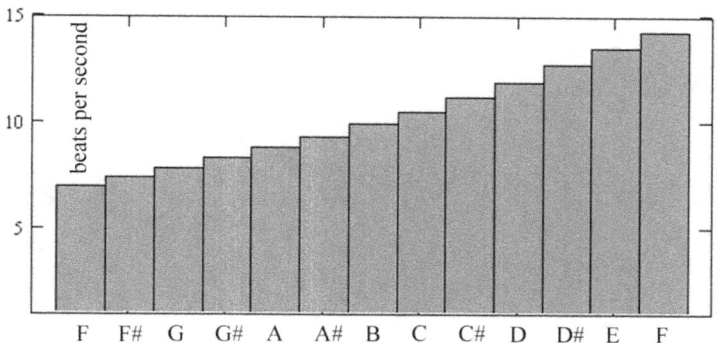

Beat rates for the major thirds in the scale for an F33 (F3) speaking length of 2m. The lower note of each third appears on the horizontal axis. The beat rates progress steadily.

When we decrease the speaking length of F33 (F3) to 1m, keeping the other parameters the same, the beat rates in the main intervals are as shown in Fig 21.4. In practical terms the rates for the fourths and fifths are not significantly affected. The change to the thirds can be seen in the bar representation Fig 21.5.

Fig 21.4

	Maj 3rds	4ths	5ths	Maj 6ths	Min 3rds	Min 6ths
F	6.9	0.8	-0.6	7.9	-9.5	-11.7
F#	7.3	0.8	-0.6	8.4	-10.3	-12.5
G	7.7	0.9	-0.6	8.9	-11	-13.4
G#	8.1	0.9	-0.6	9.2	-11.7	-14.3
A	8.9	1.1	-0.6	10	-12	-15.2
A#	9.4	1.2	-0.7	10.6	-12.7	-16.2
B	10	1.3	-0.7	11.2	-13.4	-17.4
C	10.6	1.1	-0.8	11.8	-14.3	-18.6
C#	10.8	1.2	-0.8	12.6	-15.1	-20
D	11.4	1.3	-0.8	13.3	-17	-21.5
D#	12	1.3	-0.9	14.1	-18.2	-23.1
E	12.7	1.4	-0.9	14.2	-19.4	-24.9
F	14.3	1.9	-0.8	15.8	-19.1	-26.5

Beat rates when the F33 (F3) speaking length is reduced to 1m. There is little change, and there is still progression.

Fig 21.5

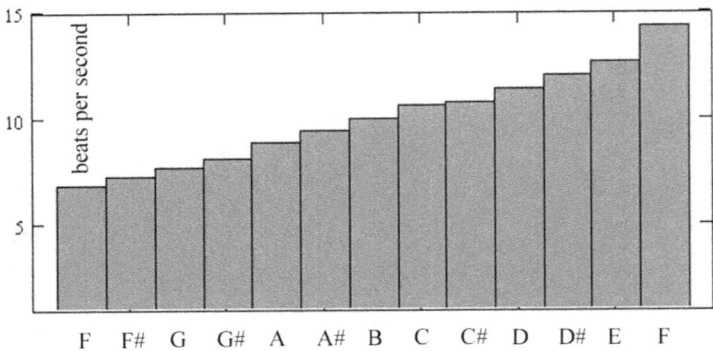

The beat rates in the thirds for an F33 (F3) speaking length of 1m, all other parameters unaltered. There is 'progression' but now we can see the progression is slightly uneven.

We begin to encounter greater changes once the F33 (F3) speaking length is reduced to less than 1m. The following figures show how the beat rates behave as the length is reduced in 0.1m stages to 0.5m. In practice, design for string lengths in these ranges would require larger diameters for the F33 (F3) to achieve a workable tension. Maintaining a 0.001m diameter is sufficient for the time being to illustrate in a general way how beat rates can be affected in theory.

Fig 21.6

	Maj 3rds	4ths	5ths	Maj 6ths	Min 3rds	Min 6ths
F	6.8	0.7	-0.5	7.7	-9.9	-13.1
F#	7.2	0.8	-0.6	8.1	-10.5	-14.1
G	7.6	0.8	-0.7	8.5	-11.2	-15.2
G#	8.4	0.9	-0.6	9.3	-11.4	-15.1
A	8.5	0.9	-0.7	9.5	-12.1	-16.2
A#	9	1	-0.7	10	-13.5	-19
B	9.5	1	-0.8	10.6	-14.5	-20.6
C	10	1.1	-0.9	11.1	-15.5	-22.4
C#	11.2	1.1	-0.8	12.3	-15.3	-21.6
D	11.2	1.2	-0.8	12.3	-16.2	-23.4
D#	11.8	1.2	-0.9	13	-18.8	-28.5
E	12.4	1.3	-1.1	13.6	-20.1	-31.2
F	13	1.3	-1.2	14.2	-21.6	-34.3

Beat rates for an F33 (F3) speaking length of 0.9m.

Fig 21.7

	Maj 3rds	4ths	5ths	Maj 6ths	Min 3rds	Min 6ths
F	6.7	0.7	-0.6	7.1	-10.2	-14.7
F#	7	0.5	-0.6	7.5	-10.9	-17.2
G	7.4	0.8	-0.7	8.2	-10.9	-17
G#	7.8	0.8	-0.7	8.6	-12.4	-18.5
A	8.3	0.6	-0.8	8.6	-13.3	-20.2
A#	8.1	0.5	-0.8	9	-14.2	-24.3
B	9.2	0.9	-0.9	10	-15.1	-23.5
C	9.7	0.9	-0.9	10.5	-16.2	-25.8
C#	10.2	0.5	-1	10.2	-17.4	-28.4
D	10.8	1	-0.8	11.6	-16.6	-30.2
D#	11.3	1	-1.1	12.1	-19.8	-33.3
E	11.9	1	-1.2	12.6	-21.4	-36.8
F	12.4	0.2	-1.3	11.7	-23	-40.7

Beat rates for an F33 (F3) speaking length of 0.8m. The rates for the major thirds are now significantly slower than their 'traditional' rates.

Fig 21.8

	Maj 3rds	4ths	5ths	Maj 6ths	Min 3rds	Min 6ths
F	5.9	0.3	-0.7	6.4	-10.9	-19.7
F#	6.1	0.3	-0.8	6.2	-12.8	-21.7
G	7.1	0.3	-0.7	7	-12.5	-23.1
G#	6.7	0.2	-0.9	7.2	-13.4	-25.6
A	7.9	0.6	-0.8	7.6	-14.2	-24.5
A#	8.2	0.2	-0.9	7.8	-15.3	-30.3
B	7.5	0.1	-1.1	8	-16.5	-33.7
C	9.1	0.6	-1	8.4	-17.5	-31.9
C#	9.5	-0.1	-1.1	8.5	-18.9	-39.9
D	8.3	-0.2	-1.5	8.6	-20.4	-44.6
D#	10.5	0.6	-1.2	9.1	-21.7	-41.7
E	10.9	-0.4	-1.3	9.1	-23.5	-53
F	9	-0.6	-1.9	9	-25.4	-59.5

Beat rates for an F33 (F3) speaking length of 0.7m. The major third beat rates are slower still, and now the higher fourths are *narrow* according to their beat rates – raising the fundamental frequency of the upper note will *slow* the beat rate initially.

Fig 21.9

	Maj 3rds	4ths	5ths	Maj 6ths	Min 3rds	Min 6ths
F	4.8	-0.2	-0.9	4.1	-12.5	-30.8
F#	6	-0.2	-1	4.3	-13.2	-29.7
G	5.1	-0.4	-1.1	4.1	-14.3	-36.7
G#	5.1	-1.1	-1.4	3	-17.8	-41.3
A	5.4	-0.6	-1.2	4.1	-16.5	-43.8
A#	5.4	-1.5	-1.6	3.8	-20.7	-49.3
B	5.7	-0.8	-1.5	4	-18.9	-46.9
C	5.6	-2	-1.6	3.5	-24.1	-58.8
C#	6	-1.1	-1.7	3.7	-21.8	-55.7
D	5.8	-1.4	-1.9	3.1	-28	-70.1
D#	8.9	-1.5	-2	3.3	-25.2	-66.1
E	5.9	-1.9	-2.3	2.4	-27.5	-83.6
F	9.6	-2	-2.4	5.1	-29	-78.4

Beat rates for an F33 (F3) speaking length of 0.6m. The major thirds are now very slow compared to their 'traditional' rates. All the fourths are narrow according to their beat rate, and the fifths are mostly at around two beats per second.

Fig 21.10

	Maj 3rds	4ths	5ths	Maj 6ths	Min 3rds	Min 6ths
F	2.1	-2.4	-1.7	-1.9	-19.4	-58.5
F#	2.2	-2.6	-1.8	-2	-17.2	-56.5
G	-0.3	-3	-2	-3.1	-22.7	-69.8
G#	2	-3.2	-2.1	-3.2	-24	-67.2
A	-1.1	-3.8	-2.4	-4.6	-26.5	-83.3
A#	1.6	-4	-2.6	-4.7	-28	-79.9
B	1.7	-2.8	-2.2	-2.8	-24.3	-83.9
C	1.1	-4.9	-3	-6.6	-32.8	-94.8
C#	1.3	-3.5	-2.6	-4.3	-28.1	-99.4
D	0.5	-6.1	-3.6	-9	-38.3	-112.4
D#	0.6	-4.3	-3.1	-6.2	-32.5	-117.5
E	-0.4	-7.4	-4.3	-11.8	-44.7	-133
F	-0.3	-5.3	-3.7	-8.5	-37.6	-138.7

Beat rates for an F33 (F3) speaking length of 0.5m. At this length the traditional conceptions about equal tempering have completely broken down. In terms of beat rates, some of the major thirds are *narrow*, and all the fourths, and major sixths, are *narrow*, with relatively high beat rates.

Fig 21.11

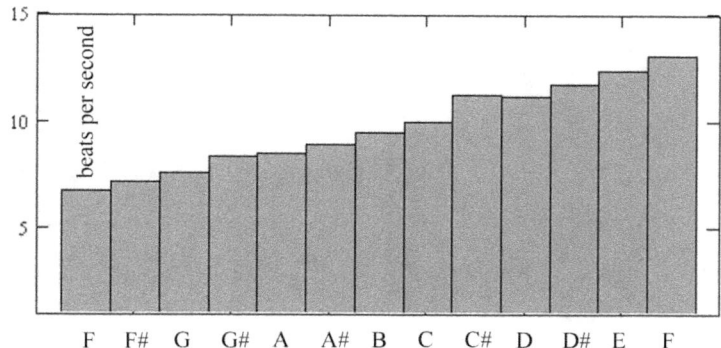

Beat rates for thirds when F33 (F3) speaking length is 0.9m. Progression begins to break down, and the rate for D-F# is slower than C#-F.

Fig 21.12

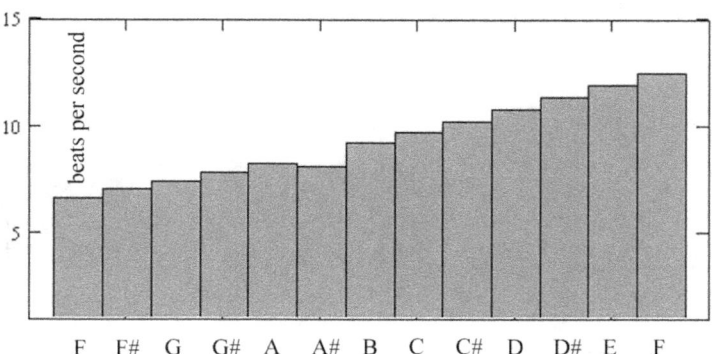

Beat rates for thirds when F33 (F3) speaking length is 0.8m. The rates fail to 'progress' at the third A#-D

Fig 21.13

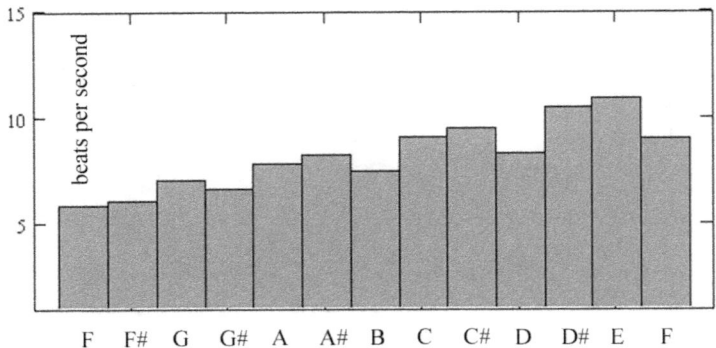

Beat rates for thirds when F33 (F3) speaking length is 0.7m. 'Progression' of beat rates now fails, although there is a regular pattern and an upward trend.

Fig 21.14

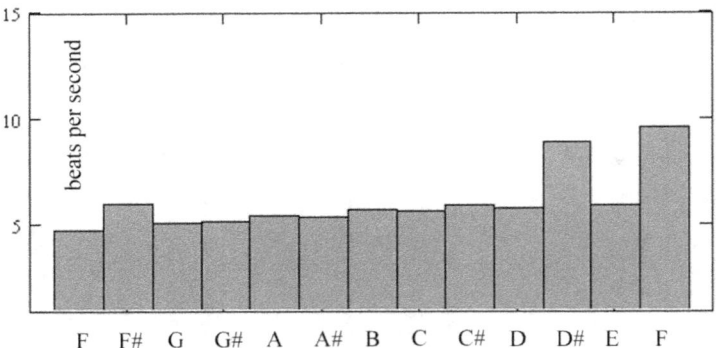

Beat rates for thirds when F33 (F3) speaking length is 0.6m. The rates are now very slow, but with irregular fast rates at F# to A#, D# to F# and top F to A.

Fig 21.15

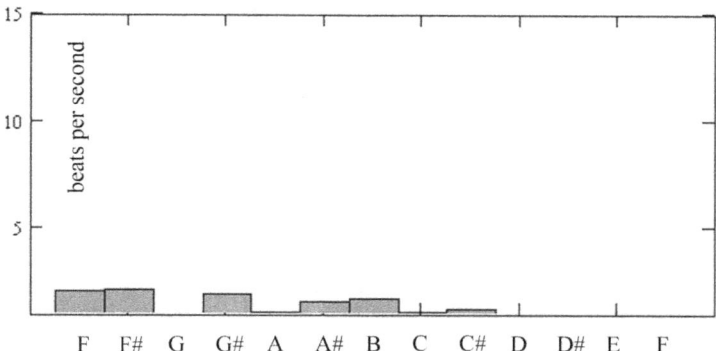

When the speaking length of F33 (F3) is 0.5m the major thirds in practical tuning terms are almost pure (beatless) and some are narrow. The major thirds rates bear no real resemblance to 'traditional' rates.

What we see here is an inverse correlation between the scale string lengths, and the deviation of the scale's beat rate characteristics from the "traditional" conception. String speaking lengths under 1m for the F33 (F3) cause the scale to exhibit definite beat rate deviations from the "traditional" predictions. For lengths under around 0.8m the notions of beat rate progression in the major thirds, and the "rules" about which intervals behave as narrow or wide, begin to break down. The precise details shown here for any given length only pertain to the specific parameters on which this particular model is built. If we use different rules for scaling the string lengths and different diameters we will get different results.

Using a constant scaling factor from F33 (F3) to C88, the system is also critically dependent on the length of the top string C88. We are not dealing with a continuous functional relationship between these string lengths and the beat rates, because the string lengths also determine the diameters, which change in discrete jumps from one gauge to the next, several adjacent notes having strings of the same diameter. This in itself is in part responsible for the failure of beat rates to progress.

It is worthwhile showing the beat rate dependence on the top string length. Let us consider an F33 (F3) speaking length of 0.75m and an F33 (F3) diameter of 1.25mm. With top string C88 speaking length of 0.04m, 0.05m and 0.06 m respectively, the beat rates are:

Fig 21.16

	Maj 3rds	4ths	5ths	Maj 6ths	Min 3rds	Min 6ths
F	6.7	0.8	-0.5	7.4	-10.7	-16.5
F#	7	0.8	-0.5	7.8	-11.4	-18.1
G	7.4	0.5	-0.5	8.2	-12.3	-19.9
G#	7.9	0.9	-0.4	8.6	-11.7	-18.4
A	8.3	0.9	-0.4	9	-14.1	-23.6
A#	8.7	1	-0.6	9.5	-15.2	-26.1
B	9.2	1	-0.6	9.9	-16.4	-29
C	11.2	1.1	-0.4	11.8	-14.9	-25.9
C#	10.2	1.1	-0.3	10.8	-15.9	-28.4
D	10.7	1.2	-0.3	11.3	-20.5	-38.8
D#	11.2	1.2	-0.7	11.7	-22.2	-43.4
E	11.7	1.3	-0.7	12.2	-24.2	-48.6
F	15.6	1.3	-0.1	15.7	-20.5	-41.3

Beat rates for F33 (F3) diameter 1.25mm, F33 (F3) speaking length 0.75m and C88 speaking length 0.04m.

Fig 21.17

	Maj 3rds	4ths	5ths	Maj 6ths	Min 3rds	Min 6ths
F	6	0.4	-0.7	6.1	-10.9	-19.6
F#	6.2	0.3	-0.7	6.3	-12.7	-23.6
G	7.2	0.3	-0.8	7.1	-12.5	-23.3
G#	6.8	0.3	-0.8	6.7	-13.4	-25.9
A	8	0.3	-0.7	7.7	-14.3	-27.8
A#	7.4	0.2	-1	7.1	-15.4	-30.9
B	7.6	0.2	-1	7.2	-18.5	-38.1
C	9.2	0.1	-1.1	8.5	-17.7	-37.1
C#	8.2	0	-1.2	7.5	-19.1	-41.5
D	10.2	-0	-1	9	-20.4	-44.6
D#	8.8	-0.2	-1.4	9.1	-22.1	-50
E	8.9	-0.3	-1.5	7.3	-27.6	-56.1
F	11.7	-0.4	-1.2	9.4	-25.5	-60.4

Beat rates for F33 (F3) diameter 1.25mm, F33 (F3) speaking length 0.75m and C88 speaking length 0.05m.

Fig 21.18

	Maj 3rds	4ths	5ths	Maj 6ths	Min 3rds	Min 6ths
F	5.8	0	-0.8	5.5	-12	-23.6
F#	6.1	0.3	-0.9	5.8	-11.8	-23.8
G	6.4	-0.1	-0.9	5.8	-13.8	-28.2
G#	6.7	-0.1	-1	6.1	-13.5	-28.3
A	6.2	-0.2	-1.1	6.1	-15.9	-33.8
A#	7.3	-0.2	-1.2	6.4	-16.9	-33.7
B	6.6	-0.4	-1.3	6.4	-18.4	-40.5
C	7.9	-0.4	-1.4	6.6	-19.6	-40.2
C#	8.3	0	-1.3	7.8	-18.8	-43
D	8.6	-0.7	-1.6	6.7	-22.6	-48.1
D#	9	-0.1	-1.5	8.1	-21.6	-51.4
E	9.2	-1.1	-1.9	6.7	-26.3	-57.6
F	9.7	-0.4	-1.7	8.3	-24.9	-61.5

Beat rates for F33 (F3) diameter 1.25mm, F33 (F3) speaking length 0.75m and C88 speaking length 0.06m.

Fig 21.19

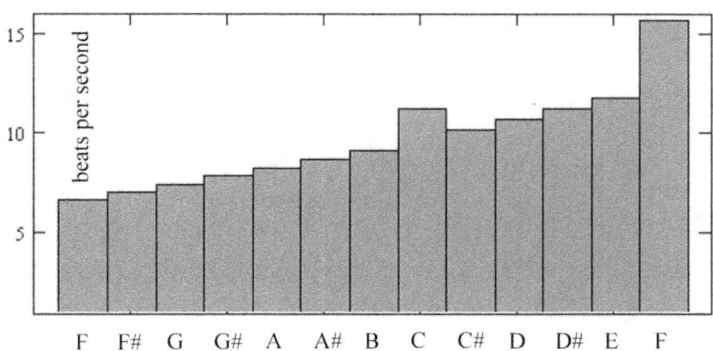

Major thirds beat rates for F33 (F3) diameter 1.25mm, F33 (F3) speaking length 0.75m and C88 speaking length 0.04m.

Fig 21.20

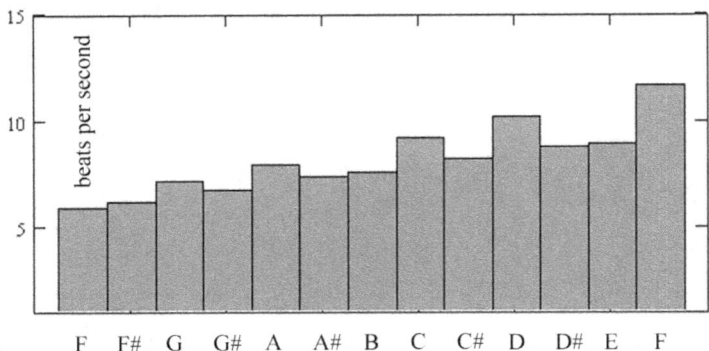

Major thirds beat rates for F33 (F3) diameter 1.25mm, F33 (F3) speaking length 0.75m and C88 speaking length 0.05m.

Fig 21.21

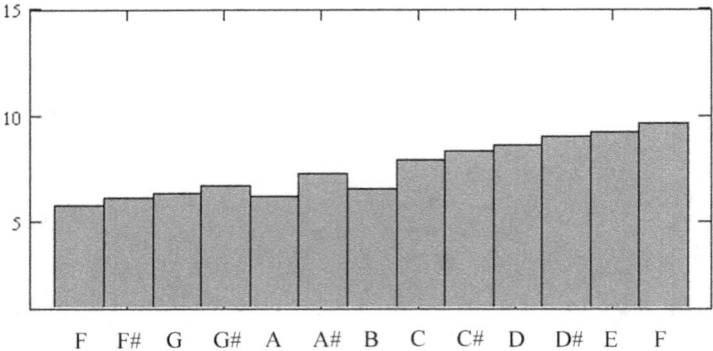

Major thirds beat rates for F33 (F3) diameter 1.25mm, F33 (F3) speaking length 0.75m and C88 speaking length 0.06m.

In general, these results show that from theoretical prediction, there will be in tuning scales with shorter string lengths, a possible result of:

1) Generally slower than "traditional" beat rates in most of the thirds;
2) Faster than "traditional" beat rates in the fifths;
3) Possibly faster than "traditional" beat rates in the fourths;
4) The possibility of fourths that are *narrow*, according to the beat rate

This is assuming the scale is "frequency correct" in its fundamentals, according to "traditional" theory. From the results above we see that the effect of small string lengths *is very critically dependent* upon scaling factor, and the length of the instrument's top string. This means that some very small pianos may have tuning characteristics reasonably close to "normal", whilst others of about the same size will deviate so greatly that the predications of "traditional" theory for the beat rates, are altogether lost.

Tuning strategy

It is important to recognise at once that these results do not mean that one is necessarily advised to adopt slower thirds and faster fifths as a tuning strategy. However, in general, the solutions to scales in pianos where there are rapid changes of inharmonicity over the scale, due to the reverse curvature at the end of the long bridge, or due to the break occurring in the scale, will involve slower thirds at the bottom of the scale. The solutions often also include faster sounding fifths, but the situation is more complicated because the fifths may be wide at one adjustable partial whilst narrow at the other.

The primarily aim is to produce, as much as possible, *evenly progressing beat rates* in the thirds and sixths in the scale, and not to have the fifths beating unpleasantly fast, in fact, to have the fifths beating as slowly as possible without compromising any other fifth, fourth, or the progression of the thirds. How to address high rates of inharmonicity change in a piano's scaling, is dealt with in the chapters on scale tuning and small piano syndrome.

It is equally important to recognise that in many pianos a high rate of change of inharmonicity across the scale, occurs at the end of the long bridge, before the bass break, where the curve of the long bridge is compromised in order to keep it from reaching the edge of the soundboard. On smaller instruments where inharmonicity is high, this typically occurs at the bottom end of the scale area, or even close to the middle of the scale area. The general model here does not deal with these features.

In general, these features will have a major effect on tuning, such that if reasonable progression of beat rates in the fast intervals is to be achieved without severely compromising any fifths or fourths, then one cannot expect to tune the scale simply by attempting to apply the beat rates of "traditional" theory. The tuning strategy must be much more dynamic than this, the task very much approached as a puzzle solving exercise in which the tuning behaviour of the piano is discovered and worked around, or exploited, as tuning proceeds.

Octaves and inharmonicity

To what *extent* are octaves affected by inharmonicity? Each note whose fundamental is tuned to the second partial of the note an octave below, will naturally form a "stretched" octave with the lower note. The amount of "stretch", *i.e.* the amount by which the fundamental frequency of the upper note exceeds twice that of the lower, will be determined by the inharmonicity in the lower note, *i.e.*, the amount by which the lower note's second partial is more than twice the frequency of its fundamental. Consider a simple scenario of tuning F45 (F4) as a 'beatless' octave to F33 (F3), and then F57 (F5) as a 'beatless' octave to F45 (F4), and so on. We would then have a sequence of F33, F45, F57 and F69 (F3, F4, F5 & F6) tuned in "beatless" octaves. The "stretching" from F33 to F69 (F3 to F6) as a triple octave interval will be the sum of the "stretching" in each of the single octaves. What happens if we now test the tuning of F69 (F6) relative to F33 (F3)?

We will be testing the fundamental of F69 (F6) against the eighth partial of F33 (F3). Only if the sum of the individual octave "stretchings" equals the natural "stretching" in the spectrum of F33 (F3), from its fundamental to its eighth partial, can the interval F33 to F69 (F3 to F6) be expected to be "beatless". This is unlikely to occur as a matter of course. The inharmonicity in the strings changes with their lengths and gauges, so the amount of "stretching" in the individual octaves will differ, from octave to octave. It will generally increase, the higher up the compass we travel from the mid-range, as the strings become shorter. The amount of "stretching" between any one partial of F33 (F3) and another partial an octave above, will also generally increase as the partial numbers become higher, but it is unlikely to match the "stretching" in the spectrum of F33 (F3). The match or mismatch will depend upon the size and scaling of the instrument.

We can construct another mathematical model to demonstrate this. We can use the same arrays as we did above for string diameters and speaking lengths over the part of the compass from F33 to C88 (F3 to C8). Thus we have:

$$\text{String diameters} = D_j$$

$$\text{Speaking lengths} = L_j$$

Where $j = 1$ is the parameter for F33, $j = 2$ for F#34, and so on.

For an F33 harmonic fundamental frequency of f_0, and utilising Eqn. 3 for the frequency of an inharmonic partial, we have for the fundamental frequencies of F33, F45, F67 and F69 respectively, each tuned as an octave from the F immediately below :

$$V_{F33} = f_0 \sqrt{1 + \frac{\pi^3 Q D_1^4}{64 L_1^2 T}}$$

$$V_{F45} = 2 f_0 \sqrt{1 + \frac{2^2 \pi^3 Q D_1^4}{64 L_1^2 T}}$$

$$V_{F67} = 2 V_{F45} \sqrt{1 + \frac{2^2 \pi^3 Q D_{13}^4}{64 L_{13} T}}$$

$$V_{F69} = 2 V_{F45} \sqrt{1 + \frac{2^2 \pi^3 Q D_{25}}{64 L_{25} T}}$$

22

The 8th partial of F33 will be:

$$P_{F33} = 8 v_0 \sqrt{1 + \frac{8^2 \pi^3 Q D_1}{64 L_1 T}}$$

23

The beat rate between F33 and F69 will be:

$$V_{F69} - P_{F33}$$

24

which will give negative values when the interval is narrow.

For C88 (C8) speaking length of 0.05m and an F33 (F3) diameter of 0.001m, we can generate from these formulae the beat rate between F33 (F3) and F69 (F6), for various F33 (F3) speaking lengths, representing different sizes of instrument. The following table illustrates the resulting beat rates.

F33 speaking length	Beat rate
2m	1.3 Hz
1m	-0.7 Hz
0.9m	-2.8 Hz
0.8m	-6.1 Hz
0.7m	-13.8 Hz
0.6m	-30.6 Hz

For a hypothetical instrument of concert size with a 2m F33 (F3) speaking length, and a top string speaking length of 0.05m, the beat rate is *positive*, indicating the 3 octave interval F33 (F3) to F69 (F6) is beating *wide*, as a result of the beatless single octaves within it. This is of course a *theoretical* result for only one scaling factor. For a 1m F33 (F3) speaking length, the interval has become narrow, but still only by a very small amount. At 0.8m speaking length it is narrow by over 6 beats per second, which, if the 8th partial of F33 (F3) is loud and clear, would probably demand reduction by stretching the individual octaves. As the speaking length of F33 (F3) is shortened further, the beat rate rises steeply until at a length of 0.6m it is over 30 Hz, which would probably be impossible to hide even by stretching the intervening octaves to a point at which they were unacceptably wide.

The practical counterpart of this general theoretical trend, is found in the actual tuning differences between instruments of different sizes. The need to stretch octaves just for the sake of matching the tuning with the inharmonicity, as much as possible, depends very much on the individual instrument's scaling characteristics, rather than just on the size of the instrument. Even so, the general trend occurs in practice. Smaller instruments tend to require more stretching than larger ones, and differences in instrument size become more critical from this point of view, between smaller instruments. Theoretically, at least, the largest of instruments would not necessarily be the least problematical.

Scale tuning sequences

The following are summaries of the example scale tuning sequences in the book. The beat rates from "traditional" theory are given.

RNC scale sequence (Ch. 5)

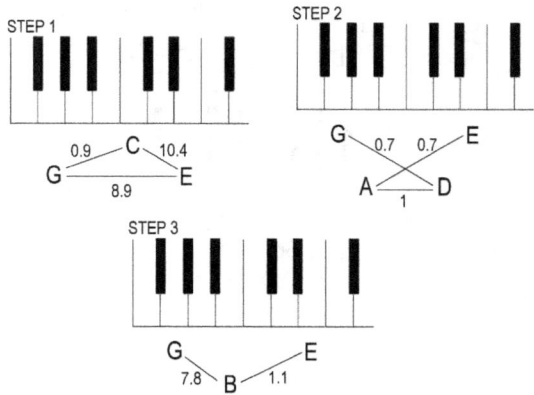

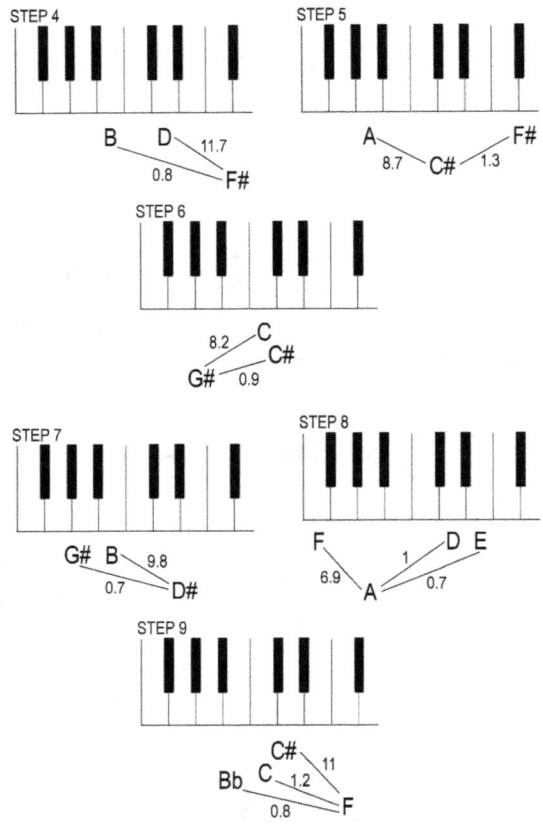

Scale sequence - example number 1 (Ch. 9)

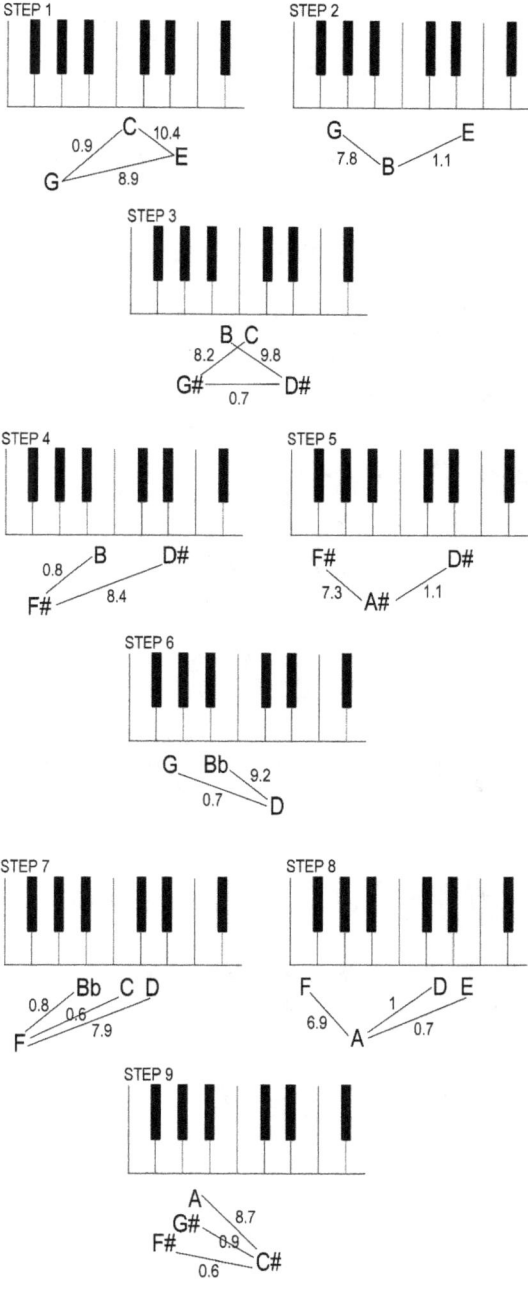

Scale sequence - Example number 2 (Ch. 9)

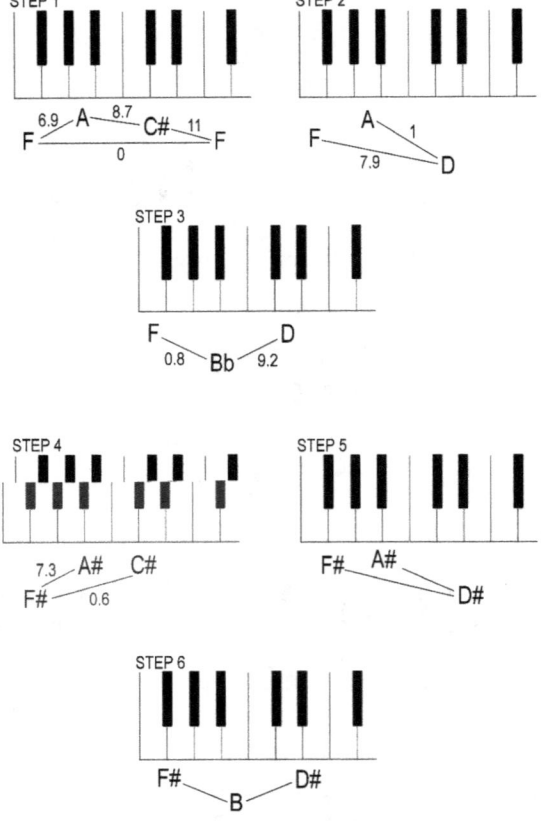

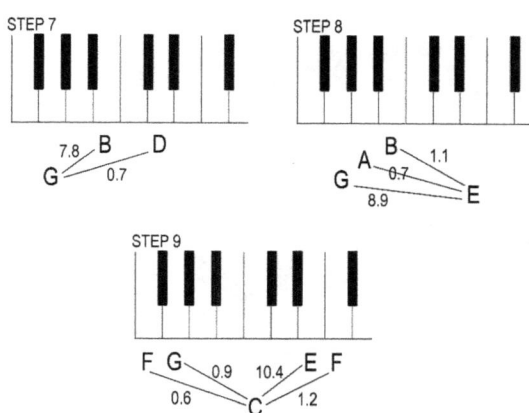

Glossary of some key concepts

A440 "pitch standard" - Refers to the note A49 (A4), as having a frequency of 440 Hz.

Absolute pitch – The ability to easily recognise and name note pitches without the need of a known reference note. The word "absolute" designates the opposite of "relative", *relative pitch* being the ability to identify notes from their relative intonation with other known notes. Absolute pitch is often ambiguously called "perfect pitch". Absolute pitch is often confused with pitch acuity, which is the ability to distinguish small differences of pitch. Pitch acuity is not confined to those with absolute pitch sense.

Acoustical consonance – The tuning condition of an interval in which beating in the adjustable partials is reduced (roughly speaking) to a "minimum". Optimum acoustical consonance is not generally an exact condition, but is "clear" in practical terms.

Adjustable partial – a partial that can contain beating whose beat rate is adjustable in fine tuning.

Agraffe – A brass "stud" terminating the speaking length of the string on some pianos, containing holes through which the strings pass. The agraffe design on upright pianos includes a bridge.

Aliquot strings – Unstruck strings that resonate because they are tuned to unison with, or an harmonic interval above, their corresponding struck strings. The aliquot string is a fourth string beside and above the trichord, with its own tuning pin (wrest pin). Many Blüthner grand pianos have aliquot strings.

Amplitude – A measure of magnitude of a vibration, wave or sound.

Antinode – A point on the vibrating string where maximum transverse motion contributing to a specific partial, occurs.

Aperiodic – An aperiodic wave or motion does not repeat its form or motion exactly in time.

Art (of piano tuning) - The relationship between art and science is of course much debated, and it turns out to be especially important in the subject of piano tuning. Because the words *art* and *science* are used in different ways in different contexts, I define here how I am using the words in the book. Before I do this, however, it is important to point out that there are those who would assert that *everything*, including things like *fine art* and all its meaning for those who love it, beauty, and even the Divine, can be reduced to, explained by, understood in terms of, and ultimately perhaps even reproduced by, *science*. Such belief is known as *scientism*, and it has to be appreciated that in order for a distinction between science and art to be worth emphasising, *scientism* has to be at least temporarily suspended.

So here then, is the essential difference between an *art* and a *science* as I use the words in the book. An *art* is a skilled, creative activity, *producing* results that are at least in part dependent on the individual, subjective and/or personal influence of the practitioner. A *science*, on the other hand, is an activity that *finds* results that are independent of the individual, subjective and/or personal influence of its practitioners. (Where science *produces* results, it becomes *technology*). In these terms, piano tuning is an art, even though it does *use* the findings of science in its techniques.

Attenuation – The reduction of partial amplitude, fluctuation amplitude, or beat amplitude.

B

Backstringing – The length of string from the soundboard bridge to the hitch pin.

Bass bridge – The wooden bridge fixed to the soundboard, terminating the speaking length of the bass strings.

Beats / beating – A regular pattern of repeating fluctuation in the loudness of a partial.

Beatless – Refers to the tuning of an interval when it is tuned to optimum acoustical consonance.

Beat rate – The number of beats per second in a beating partial.

Beat amplitude (Beat depth) – The amplitude of a beat pattern appearing in a partial, as distinct from the amplitude of the partial itself.

Bichord – A course of two strings tuned to unison, constituting one note.

Bridge – The structure that terminates the ends of the speaking lengths of the strings. May be designed to reflect wave motion back along the string, or to partially transmit motion to the soundboard.

Bridge coupling – Strings on the piano may share the same soundboard and may share the same bridge. The strings, bridge and soundboard, therefore must often be considered together as one *coupled oscillator system*, the strings being the primary oscillators, and the bridge-soundboard being the coupling oscillator. Coupled oscillator systems are generally capable of behaviour not found in systems where the oscillators are not coupled. "Traditional" tuning theory treats the oscillators as uncoupled.

Bridge pin – Each string is drafted sideways past two angled pins inserted into the surface of the soundboard bridge, as it passes over the bridge. These ensure a firm fixing of the string to the bridge. Insecurity of the bridge pin can lead to *falseness*.

Buck – see *Crown*.

C

C532.3 "pitch standard" - Refers to the note middle C40 (C4), whose second partial is at the frequency 523.3 Hz.

Capo d'Astro – An inverted bridge, part of the iron frame, applied to the treble stringing in many grand pianos.

Cent – A micro-interval equivalent to $1/100^{th}$ of an equally tempered semitone. Scientifically, a ratio of approximately 1.0005777895:1. It is often used to give an indication of comparative interval sizes or perceived pitch separations. It is not, however, a scientific unit of pitch. Pitch itself is a subjective sensation, rather than a scientific quantity.

Check intervals – Tuning intervals used to "check" the tuning of other intervals.

Chip up – A "traditional" term referring to the process of tuning the strings on a *strung back* by plucking them.

Circle of Pythagoras – The mathematical concept of *circulating* perfect fifths each with a ratio 3:2 is elegantly represented by using a diagrammatic device, an actual circle, known as a *temperament circle*. Whilst there is no primary evidence for the device's use by the original Pythagoreans, a circle divided into 12 arcs, 11 of which represent perfect fifths with ratios 3:2, has become known as the Pythagorean Circle.

Comma – see *Pythagorean comma* or *Syntonic comma*. In applied temperament theory or tempering practice, the difference between the *Pythagorean comma* and the *Syntonic comma* (about 2 cents) is ignored.

Compound octave – A multiple octave. In tuning, typically a double or triple octave.

Complex network – A network is *an interconnected system of things*. In piano tuning theory the "things" are musical intervals. The network is a concept of great importance in contemporary thought and science, the principles of networks being ubiquitous, and applicable to extremely diverse fields. Many networks have multiple connections between the connected "things", and are *complex* in their behaviour. The network of musical intervals on the piano is complex in the way it affects the tone properties of the connected intervals.

Covered strings – Strings that are copper wound to increase their mass per unit length. They are used in the bass, and sometimes tenor sections of the stringing.

Crown – Refers to the two-dimensional curve or arching of the soundboard against the *downbearing* of the strings. The crown can change with changes of temperature and humidity. Also known as *buck*.

Coupling – see *Bridge coupling*.

D

Damping – The process of stopping the vibrations of strings.

Diesis – The micro-interval, around 41 cents, between three harmonic major thirds (with ratios 5:4) and an octave (ratio 2:1).

Decay – The reduction of energy, amplitude or loudness over time, as energy from a partial or tone is lost.

Decay tone – The part of a piano tone after the initial *transient*.

Downbearing – Where the strings pass across a bridge, they pass through an angle that causes a component of the tension to create a force acting perpendicular to the bridge. This is the *downbearing angle*, and the force is the *downbearing force*.

Duplex scaling – The utilisation of the backstringing length in many grand pianos as an additional wave carrying length of string, designed to enhance the tone characteristics of the piano by increasing the decay time of certain partials. The backstringing lengths are fixed by additional duplex bridges on the iron frame so that they form aliquots of their corresponding speaking lengths. The backstringing would otherwise be muted and/or not arranged in aliquot lengths. The controlled scaling of the piano is thus in two parts, hence *duplex* scaling.

E

Equal temperament – the tuning system adopted on pianos in which all the intervals of any species are the same musical "size".

ETD – Abbreviation for *Electronic Tuning Device*.

F

Falseness – see *False beats*.

False beats – Beats originating in the partials of single strings, that may also be inherited by intervals of more than one string. An important cause of false beating is insecurity of the bridge pin, and other faults at the soundboard bridge termination of the speaking length. False beats have beat *rates* that are essentially *unadjustable*, that is, their rate cannot be significantly altered within the normal fine tuning range of string tension changes. Inherited false beats may, however, be *attenuated* through the fine tuning.

Fast intervals – Intervals with beat rates in the central compass typically around 6 – 12 Hz. Specifically refers to the thirds and sixths.

FFT – Fast Fourier Transform. The digitally produced *Fourier transform*, showing the *frequency components* (frequency ingredients) of a tone.

Fourier transform – A mathematical function; the *Fourier transform* of a wave function in the *time domain* representation (appearing as a wave form with time on the horizontal axis) becomes the *spectrum* of *Fourier components* in the *frequency domain* representation (appearing as a set of peaks with frequency on the horizontal axis). The *harmonics* or *partials* of "traditional" piano tuning theory are the Fourier components of the wave function assumed by the theory.

Fourier component – See *Fourier transform*.

Frequency – The number of cycles per second, or vibrations per second, in an oscillation.

Fundamental – Refers to the partial of a piano tone with the lowest frequency.

G

Generic decay patterns – The patterns found in the fluctuations of adjustable partials in actual piano tone do not all resemble the regular "beat" patterns of "traditional" theory. They do, however, classify into a number of recognisable *generic* pattern types, including the "regular beat" pattern.

Grand piano – A piano in which the soundboard is horizontal, and that has a spine and bent side.

H

Harmonic – Refers to a *partial* of a periodic wave or vibration.

Harmonic series – Refers to the frequency arrangement of *harmonics*, i.e. partials of a periodic tone, in which the frequency of partial n is n times the fundamental frequency. Also refers to the arrangement of musical pitches associated with this. To find the pitches of the partials of a note, which are (approximately) in an harmonic series, start with the note itself, whose pitch is that of the *fundamental*, or 1st partial of that note's approximate harmonic series. Partial number 2 is then at the pitch an octave above. Partial 3 is a perfect fifth above partial 2. Partial 4 is a perfect fourth above that. The intervals between the first six partials, from partial 1 to partial 6, are *octave, perfect fifth, perfect fourth, major third, minor third*.

Harmonic interval – a musical interval between two (piano) tones whose fundamental frequencies are in the *harmonic ratio* for that interval. Alternatively, may be an interval tuned as *pure* or *beatless*.

Harmonic ratio – Each musical interval has an harmonic ratio associated with it. This is the integer ratio approximately appearing between string or pipe lengths, when the strings or pipes produce that musical interval between them, their other parameters being equal. The common harmonic ratios in tuning are: *minor third* - 6/5; *major third* - 5/4; *perfect fourth* - 4/3; *perfect fifth* - 3/2; *minor sixth* – 8/5; *major sixth* – 5/3; *octave* – 2/1; *major tenth* – 5/2; *double octave* – 4/1; *major seventeenth* – 5/1; *triple octave* – 8/1.

Hitch pin – The pin inserted into the iron frame terminating the string at the back (in a grand piano) or the bottom (in an upright piano).

Hysteresis – Refers to the "lag" between changes of tension in the top stringing and changes in the speaking length, due to friction at the top bridge, agraffe, or *capo d'astro*.

I

Interval – In piano tuning the use of the word *interval* generally implies a musical interval with both notes sounding simultaneously. When sounded one after the other, an interval is characterised by the relative pitch intonation between the two notes. When the notes are played together, the *tone* of the interval as an entity in its own right also comes into existence. This tone can have properties and qualities that are not found in either of two notes when sounding separately.

The number of semitone steps in the main tuning intervals are:

Minor third – 3

Major third – 4

Perfect fourth – 5

Perfect fifth – 7

Minor sixth – 8

Major sixth - 9

Octave – 12

Major tenth – 16

Major seventeenth - 28

Intonation – The pitch relationship of one note relative to another. In the case of "absolute pitch" sense, it may also be used to describe the perceived pitch of a note relative to its expected pitch. In both cases, pitch is a "sensation" that is, a physiological-psychological response to the incoming sound information. It is not a property of the sound itself that can be absolutely measured. See *pitch*.

Inharmonicity – Normal mode frequency dispersion due to string stiffness. The partial frequencies are raised above the values they would have if they had fallen in an harmonic series. The higher the partial number, the more they are raised.

Inheritance – False beats originating in the behaviour of a single string can be inherited by the partials of intervals containing that string.

J

Just (tuning or interval) - Refers in practice to the tuning of an interval to *optimum acoustical consonance*. Equivalent terms also used: "pure", and "beatless". In temperament theory it refers to an interval whose tuning is defined by an *harmonic ratio*.

Just intonation – A term often used to describe the tuning system of a scale so tuned that the interval of each note with the root note is justly intoned, or tuned *just*.

K

Kinetic friction – Friction that occurs when two surfaces are sliding against one another.

L

Latency – Refers to the tendency for the tension in the speaking length to change after tuning, even when the pin appears to be properly *set*. It could occur with changes of the bridge or soundboard due to changing overall tension.

Lever – The tuning lever (sometimes tuning crank) is the primary tool used for turning the wrest (tuning) pin. Professional levers have interchangeable tuning heads, for fitting onto different gauge pins.

Long bridge – The wooden bridge fixed to the soundboard, that terminates the speaking length of the treble strings, at the hitch pin end.

M

Matrix – A rectangular array of mathematical quantities or expressions that is treated as a single entity and manipulated according to special rules. The mathematical properties of matrices often encapsulate the generic behaviour possibilities of systems they represent, and can be powerful tools for exploring the behaviour. Weinreich used matrix representation to model the behaviour of unison strings coupled by the bridge and soundboard.

Meeting – Refers to two partials of separate strings (or notes) that have close (or the same) frequencies. These are important because they create *adjustable* partials.

Mistuning – The frequency difference between two partials, each from a single string sounding alone. The mistuning is not necessarily the same as the difference between the two frequency components (*eigenfrequencies*) generated by the system comprised of the two strings coupled by the bridge and soundboard.

Mode – see Normal mode.

Mode coupling – see *Bridge coupling*.

Muting felt – A strip of felt sometimes inserted between the trichords when tuning the scale. it leaves the middle string of each trichord free to vibrate, so the scale can be tuned on single strings.

N

Network – See *complex network*.

Node – A point on the vibrating string where no transverse motion, or minimum transverse motion, contributing to a specific partial, occurs.

Normal mode – A motion of a vibrating body in which all parts are vibrating with simple harmonic motion at the same frequency. Normal modes *superpose* to produce complex motion.

Non adjustable partial – A partial that does not contain beating whose beat rate is (significantly or usefully) adjustable in fine tuning.

O

Oblique strung – Stringing in which the strings are angled to improve the length of the bass strings, but are not *overstrung*.

Octave stretching – The tuning of octaves so that the fundamental frequency of the upper note is higher than the frequency of the second partial of the

lower note.

Octave-tenth – See *seventeenth*.

Overstrung (overstringing) – Stringing design in which the bass strings on the bass bridge are on a different plane distance from the soundboard, and cross over the treble strings. The design enables longer bass strings. Sometimes called *cross stringing*. *Double overstrung* pianos (typically miniature grands) have two sets on two different planes.

P

Papp's wedge – A sprung double wedge for muting strings in upright piano tuning.

Partial – A single, discrete, audible ingredient of piano tone. Not necessarily a single, discrete *Fourier component*.

Perfect pitch – see *Absolute pitch*.

Periodic – A periodic wave or vibration repeats its form or motion exactly in time.

Psycho-acoustics – The study of the perception of sound.

Pure – Refers to the tuning of an interval when it is tuned to optimum *acoustical consonance*.

Pure tone – A fundamental ingredient of sound, whose wave form is sinusoidal.

Pitch – A subjective aural "sensation" affected by the frequencies present in the sound. It is commonly described in terms of "high" or "low", correlating to higher and lower frequencies. The sense and perception of pitch is affected by physiological and psychological variables. Outside science it is quite commonly (but erroneously) assumed, through psychological ascription, to be an objective property of the sound phenomenon itself.

Pitch acuity – The aural sensitivity to, and ability to perceive and recognise small pitch differences.

Pitch ambiguity – Refers to situations in which the acoustical parameters that affect pitch sensation create an ambiguous pitch. In such instances the psychological factors in pitch sensation become more dominant. *Shepard's scale* is a well known demonstration of pitch ambiguity specially created to illustrate the effect through just one strategy. Pitch ambiguity is most prevalent in pianos in the lower bass, particularly on smaller instruments and poorer quality instruments.

Pitch flattening – Refers to the tendency for the upper treble of the piano compass to sound rather flat in its intonation with the bass and central compass, even when the individual octave tuning sounds good.

Pitch raise – A term used by piano tuners referring to the raising of string frequencies by more than a particular amount. Generally this may be greater than 4 Hz, but often the term is used to refer to a larger rise corresponding to a musical pitch change of at least a quarter of a semitone.

Pitch sharpening - Refers to the tendency for lower bass of the piano compass to sound rather sharp in its intonation with the central compass and treble, even when the individual octave tuning sounds good.

Principle of superposition – Refers to the principle of superpositioning simple waves (or vibrational *modes*) to create complex waves (or vibrational modes). Mathematically, a complex wave will analyse into a set of simple sinusoids. *Superpositioning* (adding together) the simple sinusoids will produce the complex wave.

Progression – Refers to the steady increase in the beat rates of intervals, in particular in the thirds and sixths, with rising position in the compass. Alternatively refers to the steady decrease with falling position.

Pythagorean Circle – see *Circle of Pythagoras*

Pythagorean comma (*ditonic comma*) – The micro-interval (about 24 cents) between 7 octaves (whose ratios are 2:1) and 12 harmonic perfect fifths (whose ratios are 3:2)

R

Recipe – see *Tone recipe*.

Recoil – The torsional springing back of the tuning pin after turning force is released.

Relative pitch – The pitch of one musical tone relative to another; The ability to perceive and judge one pitch relative to another; The ability in musicianship, to name one or more notes, when another reference note is given. This involves aurally recognising intervals, and some knowledge of music theory.

Return string – A single length of wire that serves for two strings, by passing round the hitch pin, rather than terminating in an eye.

S

Scale – The chromatic scale of 13 notes or more in the central compass, usually tuned first, whose intervals are equally tempered.

Schisma - The micro-interval between the *syntonic comma* and the *Pythagorean comma*.

Science – see *Art*.

Setting the pin – The art of placing the rotational position of the pin and the tensions in the various portions of the string, so that the tuning is stable.

Seventeenth – Refers to the major seventeenth, an interval of two octaves plus a major third. Also called the major *octave-tenth*.

Shared string – A single *return string* that is included in two different trichords. The two trichords are usually, but not always, adjacent.

Side draft pin – see *Bridge pin*.

Simple harmonic motion – Motion in which the acceleration is inversely proportional to the displacement from the point of equilibrium (and acting towards it).

Slow intervals – Intervals with beat rates in the central compass typically around 0.5 – 1 Hz, but in any case less than around 3 Hz. Refers specifically to perfect fourths and perfect fifths.

Soundboard – The large area of (generally spruce) in the piano, with special construction, on which the strings act through the wooden bridges. It functions as an amplifier.

Soundscape – The tone of a note or interval is a complex recipe of constantly changing ingredients, and when aural acuity is refined to begin to perceive this, the tone may be called the soundscape of the note or interval. In particular, it has some qualities that may be likened to a live seascape.

Speaking length – The length of string terminated at one end by the top bridge, agraffe, or *capo d'astro* bar, and at the other end by the soundboard bridge.

Species – Refers to a musical interval of a given number of semitones. All intervals of a given *species* in equal temperament tuning, will have, as much as possible, the same "size", or pitch interval or relative pitch intonation between its notes. However, in different positions in the compass, different intervals of the same species will have different tone properties, especially *beat rates*.

Spectrum – Refers to the set of partials (or harmonics) that are the ingredients of a tone recipe (more accurately, in piano tone, the *decay tone recipe*). It specifically refers to the partials considered as a set of separate partials, rather than to the overall perceived tone quality or recipe. The counterpart of the spectrum is the *soundscape*.

Spring - The torsional springing back of the tuning pin, top stringing and tuning lever, as turning force on the tuning lever is removed.

Stability – Refers to the stability of tuning, *i.e.* its lack of tendency to fall (or rise) out of tune.

Standing waves – The wave motion produced on the string (and in other media) when travelling waves moving in opposite directions coincide. Standing waves create nodes and antinodes, and are *normal modes* of motion.

Static friction – The friction between two surfaces when they are not in relative motion.

Straight strung – Stringing in which the strings are roughly parallel over the whole compass.

Strung back – The composite component of the piano consisting of the soundboard, iron frame, grid (if present), and strings.

Syntonic comma – The micro-interval (about 22 cents) between four harmonic perfect fifths (ratio 3:2), and one octave (ratio 2:1) plus an harmonic major third (ratio 5:4). Also known as the *comma of Didymus*.

T

Temperament – In practice, a tuning system in which some or all of the intervals are deliberately tuned away from *acoustical consonance*. In temperament theory, a controlled system of arranging the network of musical intervals that preserves the ratio 2:1 for the octave.

Temperament theory – The theory concerned with how musical intervals defined by ratios, "fit together". It is essentially an idealised, arithmetic science or theory of relationships between the integer ratios associated with musical intervals, and the ratio relationships generated as a result of their combinations in further ratios. It is closely related to what was once called the *science of music* or *harmonics*, which in turn was part of a long tradition of beliefs originating from pre-Socratic ideas about cosmology and cosmogony, including what became known as the *harmony of the spheres*.

Temperament circle – A circular visual device with notes marked around the circumference, representing a musical *temperament*.

Tenth – Refers to the major tenth, an interval of one octave plus a major third.

Tone – The perceived sonic quality or *timbre* of a string, note or interval. *Tone quality* is used in the book to denote the subjective sensation of tone, whilst *tone property* refers to the measurable, acoustical properties that affect perceived tone. Both tone quality and certain tone properties can be aurally perceived.

Tone property - see *tone*.

Tone quality – see *tone*.

Tone recipe – A complex tone consists of a mixture or *superposition* of simpler "ingredients", the simplest of which are which are "pure tones". The simplest discrete audible ingredient of a *tone recipe* may be a *partial*, which may contain more than one pure tone. See also *Principle of superposition*.

Top bridge – The raised bridge cast as part of the iron frame terminating the speaking length of the strings at the wrest pin (tuning pin) end.

Top stringing – The length of string from the top bridge, agraffe or capo d'astro, to the wrest pin (tuning pin).

"Traditional" piano tuning theory – This is not really a tradition in the usual cultural sense, but bears similarities in that information and ideas have been passed down generations both by aural transmission, and to a certain extent through written reiteration of ideas without the critical re-evaluation. It has its roots in 19th century acoustical theory. There are a good number of sources for the theory still extant, by far the most prominent being Helmholtz's *On the sensations of tone*. Earlier propagation of the theory was largely through Braid White's *Piano tuning and allied arts*, first published in 1917. Since then the theory has been propagated through numerous texts that have not always credited this influential source.

The main thrust of the theory is in conceptualising piano tuning as the tuning of specific beat rates in specific intervals, these beat rates typically being quoted to several decimal places. Tuning is thereby represented largely as the activity of simply trying to achieve these rates with the greatest possible precision. The theory itself is valid as an elementary and idealised model, but has often been given spurious status as a reliable explanation of the art of piano tuning.

The shortcomings of the theory in relating to the work of a master piano tuner are twofold. The first is that it fails to represent with sufficient fidelity, the detailed and complex nature of real piano string and tone behaviour encountered in tuning practice. In place of this, it substitutes a supposed precision of fixed beat rate values. The second is that it implies the most important musical qualities in the piano, which are tone and intonation, can be reduced, and represented as a simple set of numbers.

Transient – The initial noise at the beginning of a piano tone, largely the sound of the hammer striking the string.

Travel – The (perceived) rotational distance through which the lever turns before a change to the soundscape begins to be effected.

Trichord – A group of three strings tuned to unison, constituting one note.

Tuning – The art of adjusting the tensions on the strings for optimum tone and intonation, and in accordance with the principle of equal temperament.

Tuning circle – see *temperament circle*

Tuning lever – see *lever*.

Tuning head – see *lever*.

U

Upright piano – Any piano in which the soundboard is vertical.

Unison – Often refers to the physical group of two or three strings (bichord or trichord) constituting one note.

Una corda – The left pedal on most grand pianos and on some uprights. It shifts the action sideways (usually, but not always, from left to right) so that the hammers strike two of the three strings in the trichords, and one of the two strings in the bichords.

W

Waterfall – A "3-D" graph showing the frequency components of a sound as a function of time. The digitally generated form is also sometimes called a *time FFT*.

Wedge – Tool used for muting one or two strings. See *Papp's wedge*.

Weinreich model – A seminal mathematical representation of the two-stringed unison introduced in the 1970s by Weinreich. It differs from the "traditional" model by taking into account *bridge coupling* between the strings.

Wolf – When a perfect fourth or perfect fifth is tuned so that it is wider or narrower than its acoustically consonant tuning by a comma or more, it is generally known as a wolf interval, wolf, or specifically a wolf fifth, or wolf fourth. The origin of the term "wolf", though it implies a "howling", is uncertain.

Select bibliography

Askenfelt, A, (Ed), *Five lectures on the acoustics of the piano*, Stockholm, 1990

Backus, J, *The acoustical foundations of music*, NY, 1969, 1977

Barbour, J Murray, *Tuning and temperament*, NY, 1972

Bear, MF; Connors, BW; Paradiso, MA, *Neuroscience*, Baltimore; Philadelphia, 2001

Benade, AH, *Fundamentals of musical acoustics*, NY, 1976

Booth, George W, *Pianos, piano tuners and their problems*, London, 1996

Campbell M, and Greated C, *The musician's guide to acoustics*, London, 1987

Capleton, B, 'False beats in coupled piano string unisons', *Journal of the Acoustical Society of America*, 115, 2, 2004, pp. 885-892.

Capleton, B, 'Piano tuning techniques and the tuning characteristics of pianos considered as an effect of inharmonicity and mode coupling', *Journal of the Institute of Musical Instrument Technology*, 4, 3, 1991, pp. 91-126.

Capleton, B, 'Piano unison tuning', www.amarilli.co.uk/academic, 2006-.

Cartling, B, 'Beating frequency and amplitude modulation of the piano tone due to coupling of tones', *Journal of the Acoustical Society of America*, 117, 4, April 2005, pp. 2259-2267.

Chase Hundley, T, Benioff, H, and Martin, DW, 'Factors contributing to the multiple rate of piano tone decay', *Journal of the Acoustical Society of America*, 64,5, 1973, pp. 1303-1309.

Cree Fischer, J, *Piano tuning / regulating and repairing*, (1907), reprinted as *Piano tuning. A simple and accurate method for amateurs* (Dover, NY, 1975)

Fenner, K, *Causes of variable tuning characteristics of pianos*, Europiano Publications, 3, European Union of Piano Craft Guilds, Das Musikinstrument, Frankfurt, 1977

Funke, O, *Theorie und Praxis des Klavierstimmens*. (*Piano tuning in theory and practice*). Tr. Wisbey, RW; Wehlau, CH, Frankfurt, 1961

Gafurio, Franchino, *Practica musicae*, Milan, 1496

Gelfand, SA (Ed), *Hearing: an introduction to psychological and physiological acoustics*, NY, 2004

Hart, HC, Fuller, MW, and Lusby, WS, 'A precision study of piano touch and tone', *Journal of the Acoustical Society of America*, VI, 1934, p. 80

Haynes, B, *A history of performing pitch: the story of "A"*, Oxford, 2002

Helmholtz, H, *On the sensations of tone*, Tr. Ellis, NY, 1954

Howard, DM; Angus, J, *Acoustics and psychoacoustics*, Oxford, 2001

Howe, AH, *Scientific piano tuning and servicing*, NY, 1941

Kent, EL, 'Influence of Irregular Patterns in the Inharmonicity of Piano-Tone Partials upon Tuning Practice', *Journal of the Acoustical Society of America*, 35, 11, Nov 1963 p. 1909

Kirk, RE, 'Tuning Preferences for Piano Unison Groups', *Journal of the Acoustical Society of America*, 31, 1959, pp. 1644–1648

Krumhansi, CL, *Cognitive foundations of musical pitch*, NY, Oxford, 1990

Lindley, Mark, *Lutes, viols & temperaments*, CUP, 1984

Lloyd, LS and Boyle, H, *Intervals scales and temperaments*, London, 1963, 1978

Martin, DW, 'Decay rates of piano tones', *Journal of the Acoustical Society of America*, 19, 4, 1947, pp. 535-541

Mersenne, Marin, *Harmonie Universelle*, Paris, 1636-7, Tr Roger E Chapman, 1957

Mori, T and Bork, I, 'The influence of inharmonicity on piano tuning', *Journal of the Acoustical Society of America*, 105, 2, Sep 1999, p. 1181

Morse, PM, *Vibration and sound*, NY, 1948

Morse, PM, and Ingard, KU, *Theoretical acoustics*, NY, 1968

Moyer, AE, *Musica scientia: musical scholarship in the Italian renaissance*, NY, 1992

Pfeiffer, W, *The piano hammer*, Frankfurt, 1978

Plack, CJ, *The sense of hearing*, Mahwah NJ; London, 2005

Plack, CJ; Oxenham, AJ; Fay, RR, *Pitch: neural coding and perception*, NY, London, 2005

Poole, LMG, 'Further thoughts on piano tuning', *American Journal of Physics*, 47, 6, 1979, p. 564.

Railsback, OL, 'Scale temperament as applied to piano tuning', *Journal of the Acoustical Society of America*, 9, 3, Jan 1938, p. 274

Raman, CV, Ed. Ramaseshan, S, *Scientific papers of CV Raman*, Volume II, Acoustics, Bangalore, 1988

Rasch, RA; Heetvelt, V, 'String inharmonicity and piano tuning', *Journal of the Acoustical Society of America*, 76, 1 (Supp) 1984, p. S22.

Reblitz, AA, *Piano servicing tuning & rebuilding*, NY, 1976

Riemann, Hugo, *Geschichte der musiktheorie*, Berlin, 1898

Sanderson, AE, 'The Mathematics of Piano Tuning', *Journal of the Acoustical Society of America*, 85, 1, (Supp), 1989, p. S65

Schuck, OH, and Young, R W, 'Observations on the vibrations of piano strings', *Journal of the Acoustical Society of America*, 15, 1, 1943, pp. 1–11

Smith, H, *The art of tuning the piano*, London, c. 1900

Spillane, D, *The piano. Scientific and practical instructions relating to tuning, regulating, and toning*, NY, 1893

Steblin, RA *History of key characteristics in the eighteenth and early nineteenth centuries*, Epping, 1983.

Tanaka, H, Mizutani, K, and Nagai, K, 'Experimental analysis of two-dimensional vibration of a piano string measured with an optical device', *Journal of the Acoustical Society of America*, 105 (2), 1999, p.1181

Weiss and Taruskin, *Music in the western world*, NY, 1984

Wolf and W. Sette, S K, 'Some application of modern acoustic apparatus', *Journal of the Acoustical Society of America*, 6, 1935, pp. 160–168

Wolfenden, S, *A Treatise on the Art of Pianoforte Construction*, London, 1916

Weinreich, G, 'Coupled piano strings', *Journal of the Acoustical Society of America*, 62, 6, 1977, pp. 1474-1484

White, W Braid, *Piano tuning and allied arts*, Boston Mass., 1917, 1972

White, W Braid, *Theory and practice of pianoforte building*, NY, 1909

White, W Braid, *Theory and practice of piano construction. With a detailed practical method for tuning* NY, 1975 (reprinted from above)

Young, R W, 'Inharmonicity of plain wire piano strings', *Journal of the Acoustical Society of America*, 24, 3, 1952, pp. 267-273

Zwicker, E; Fasti, H, *Psychoacoustics, facts, and models*, Berlin, NY, c. 1990

www.ingramcontent.com/pod-product-compliance
Lightning Source LLC
Chambersburg PA
CBHW070753300426
44111CB00014B/2393